Advances in

BOTANICAL RESEARCH

Incorporating Advances in Plant Pathology

VOLUME 45

RAPESEED BREEDING

Advances in

BOTANICAL RESEARCH

Incorporating Advances in Plant Pathology

Advances in

BOTANICAL RESEARCH

Incorporating Advances in Plant Pathology

RAPESEED BREEDING

Edited by

SURINDER KUMAR GUPTA

SK University of Agricultural Sciences & Technology, Chatha, Jammu, India

Series Editors

M. DELSENY

Laboratoire Génome et Développement des Plantes, CNRS IRD UP, Université de Perpignan, France

J.-C. KADER

Laboratoire Physiologie Cellulaire et Moléculaire des Plantes, CNRS, Université Paris 6, France

VOLUME 45

2007

ELSEVIER

AMSTERDAM • BOSTON • HEIDELBERG • LONDON
NEW YORK • OXFORD • PARIS • SAN DIEGO
SAN FRANCISCO • SINGAPORE • SYDNEY • TOKYO

Academic Press is an imprint of Elsevier

Academic Press is an imprint of Elsevier
525 B Street, Suite 1900, San Diego, California 92101-4495, USA
84 Theobald's Road, London WC1X 8RR, UK

This book is printed on acid-free paper. ♾

For information on all Academic Press publications
visit our Web site at www.books.elsevier.com

ISBN-13: 978-0-12-374098-4
ISBN-10: 0-12-374098-3

PRINTED IN THE UNITED STATES OF AMERICA
07 08 09 10 9 8 7 6 5 4 3 2 1

CONTENTS

History, Origin, and Evolution

S. K. GUPTA AND ADITYA PRATAP

Breeding Methods

B. RAI, S. K. GUPTA AND ADITYA PRATAP

The Chronicles of Oil and Meal Quality Improvement in Oilseed Rape

ABHA AGNIHOTRI, DEEPAK PREM AND KADAMBARI GUPTA

Development and Practical Use of DNA Markers

KATARZYNA MIKOLAJCZYK

Self-Incompatibility

RYO FUJIMOTO AND TAKESHI NISHIO

Fingerprinting of Oilseed Rape Cultivars

VLADISLAV ČURN AND JANA ŽALUDOVÁ

Haploid and Doubled Haploid Technology

L. XU, U. NAJEEB, G. X. TANG, H. H. GU, G. Q. ZHANG, Y. HE AND W. J. ZHOU

Breeding for Apetalous Rape: Inheritance and Yield Physiology

LIXI JIANG

Breeding Herbicide-Tolerant Oilseed Rape Cultivars

PETER B. E. MCVETTY AND CARLA D. ZELMER

Breeding for Blackleg Resistance: The Biology and Epidemiology

W. G. DILANTHA FERNANDO, YU CHEN AND KAVEH GHANBARNIA

Development of Alloplasmic Rape

MICHAL STARZYCKI, ELIGIA STARZYCKI AND JAN PSZCZOLA

Honeybees and Rapeseed: A Pollinator–Plant Interaction

D. P. ABROL

Genetic Variation and Metabolism of Glucosinolates

NATALIA BELLOSTAS, ANNE DORTHE SØRENSEN, JENS CHRISTIAN SØRENSEN AND HILMER SØRENSEN

Mutagenesis: Generation and Evaluation of Induced Mutations

SANJAY J. JAMBHULKAR

Rapeseed Biotechnology

VINITHA CARDOZA AND C. NEAL STEWART, JR.

Oilseed Rape: Co-existence and Gene Flow from Wild Species

RIKKE BAGGER JØRGENSEN

Evaluation, Maintenance, and Conservation of Germplasm

RANBIR SINGH AND S. K. SHARMA

Oil Technology

BERTRAND MATTHÄUS

CONTRIBUTORS TO VOLUME 45

D. P. ABROL *Division of Entomology, Sher-e-Kashmir, University of Agricultural Sciences and Technology, Faculty of Agriculture, Udheywalla, Jammu 180002, Jammu and Kashmir, India*

ABHA AGNIHOTRI *Plant Biotechnology, TERI, Habitat Place, Lodhi Road, New Delhi, India*

RIKKE BAGGER JØRGENSEN *Risø National Laboratory, Biosystems Department, Denmark*

NATALIA BELLOSTAS *Department of Natural Sciences, Biochemistry and Natural Product Chemistry, Faculty of Life Sciences, Copenhagen University, Thorvaldsensvej 40, DK-1871, Frederiksberg C, Denmark*

VINITHA CARDOZA *Department of Plant Sciences, University of Tennessee, Knoxville, Tennessee 37996*

YU CHEN *Department of Plant Science, University of Manitoba, Winnipeg, Canada*

JENS CHRISTIAN SØRENSEN *Department of Natural Sciences, Biochemistry and Natural Product Chemistry, Faculty of Life Sciences, Copenhagen University, Thorvaldsensvej 40, DK-1871, Frederiksberg C, Denmark*

VLADISLAV ČURN *Faculty of Agriculture, University of South Bohemia, Czech Republic*

W. G. DILANTHA FERNANDO *Department of Plant Science, University of Manitoba, Winnipeg, Canada*

ANNE DORTHE SØRENSEN *Department of Human Nutrition, Group for Paediatric and International Nutrition, Faculty of Life Sciences, Copenhagen University, Rolighedsvej 30, DK-1958, Frederiksberg C, Denmark*

RYO FUJIMOTO *Graduate School of Agricultural Science, Tohoku University, Japan*

KAVEH GHANBARNIA *Department of Plant Science, University of Manitoba, Winnipeg, Canada*

H. H. GU *Institute of Crop Science, College of Agriculture and Biotechnology, Zhejiang University, China*

KADAMBARI GUPTA *Plant Biotechnology, TERI, Habitat Place, Lodhi Road, New Delhi, India*

S. K. GUPTA *Division of Plant Breeding and Genetics, SK University of Agricultural Sciences and Technology, Chatha, Jammu, India*

Y. HE *Institute of Crop Science, College of Agriculture and Biotechnology, Zhejiang University, China*

SANJAY J. JAMBHULKAR *Nuclear Agriculture and Biotechnology Division, Bhabha Atomic Research Centre, Mumbai, India*

LIXI JIANG *Institute of Crop Science, College of Agriculture and Biotechnology, Zhejiang University, 268 Kaixuan Road, Hangzhou 310029, China*

BERTRAND MATTHÄUS *Federal Research Center for Nutrition and Food, Institute for Lipid Research, Piusallee 68/76, 48147 Münster, Germany*

PETER B. E. MCVETTY *Department of Plant Science, University of Manitoba, Winnipeg, Canada*

KATARZYNA MIKOLAJCZYK *Plant Breeding and Acclimatization Institute, Research Division in Poznan, Strzeszynska 36, 60-479 Poznan, Poland*

U. NAJEEB *Institute of Crop Science, College of Agriculture and Biotechnology, Zhejiang University, China*

TAKESHI NISHIO *Graduate School of Agricultural Science, Tohoku University, Japan*

ADITYA PRATAP *Division of Plant Breeding and Genetics, SK University of Agricultural Sciences and Technology, Chatha, Jammu, India*

DEEPAK PREM *Plant Biotechnology, TERI, Habitat Place, Lodhi Road, New Delhi, India*

JAN PSZCZOLA *Plant Breeding and Acclimatization Institute, Poland*

B. RAI *School of Agriculture, Banaras Hindu University, Varanasi, India*

S. K. SHARMA *National Bureau of Plant Genetic Resources, New Delhi, India*

RANBIR SINGH *National Bureau of Plant Genetic Resources, New Delhi, India*

HILMER SØRENSEN *Department of Natural Sciences, Biochemistry and Natural Product Chemistry, Faculty of Life Sciences, Copenhagen University, Thorvaldsensvej 40, DK-1871, Frederiksberg C, Denmark*

ELIGIA STARZYCKI *Plant Breeding and Acclimatization Institute, Poland*

MICHAL STARZYCKI *Plant Breeding and Acclimatization Institute, Poland*

C. NEAL STEWART, JR. *Department of Plant Sciences, University of Tennessee, Knoxville, Tennessee 37996*

G. X. TANG *Institute of Crop Science, College of Agriculture and Biotechnology, Zhejiang University, China*

L. XU *Institute of Crop Science, College of Agriculture and Biotechnology, Zhejiang University, China*

JANA ŽALUDOVÁ *Faculty of Agriculture, University of South Bohemia, Czech Republic*

CARLA D. ZELMER *Department of Plant Science, University of Manitoba, Winnipeg, Canada*

G. Q. ZHANG *Institute of Crop Science, College of Agriculture and Biotechnology, Zhejiang University, China*

W. J. ZHOU *Institute of Crop Science, College of Agriculture and Biotechnology, Zhejiang University, China*

PREFACE

The changes in the genetic makeup of any crop play a major contributory role toward increasing the production. The breeding technologies/methodologies aim at manipulating the genetic makeup of a given variety so as to maximize the productivity under a given set of environmental conditions. Despite the fact that the recent advances made in rapeseed, the need and the opportunities to increase its production, oil, and protein yield are as great today as they have ever been. There is no single publication available that deals exclusively with rapeseed breeding, production, and utilization. Realizing the importance of this crop in Europe and rest of the countries of the world, there is an urgent need to update the knowledge of the recent technologies developed so far in enhancing the rapeseed production at global level. This volume assumes special significance in the context that it deals with such aspects as breeding, human and animal nutrition, honeybee–rapeseed plant interaction, and oil refinement technologies, and so on, and is therefore relevant and well timed.

This volume broadly covers 18 chapters, which have been well prepared by leading scientists of the world with long experience and incisive knowledge of the subject. It also contains the topics of interest not only to breeders but to nutritionist, biotechnologist, and industrialist as well. The well-chosen topics deal exhaustively with past accomplishments and future possibilities in rapeseed breeding.

I am highly indebted to Prof. Nagendra Sharma, vice-chancellor, Sher-e-Kashmir University of Agricultural Sciences & Technology (SKUAST), Jammu, India for encouraging me to carry out rapeseed breeding research with all the required facilities and Prof. H. R. Kalia, former vice-chancellor, Himachal Pradesh Krishi Vishvavidyalaya, Palampur, HP, India for his valuable and critical suggestions during the preparation of this volume. I am also grateful to all the contributors of different chapters of this volume for their ready response. Help rendered by Prof. Inomata and Prof. Takeshi Nishio, Tohoku University, Japan; Prof. Weijun Zhou, deputy director, Institute of Crop Science, China; Dr. B. E. Peter McVetty, University of Manitoba, Canada; Dr. R. W. Daniels, the Scotts Company, France; Prof. S. K. Sharma, director, National Bureau of Plant Genetic Resources and Dr. Abha Agnihotri, TERI, New Delhi, India; and Prof. B. Rai, former dean, Banaras Hindu University, India in the form of relevant literature and technical inputs is thankfully acknowledged. I am highly thankful to Dr. Cesar Gomez Campo, Universidad Politecnica, Madrid, Spain for providing several species of *Brassica* and wild allies for carrying out rapeseed research in this University. Christine

Minihane, acquisions editor, and Patricia Gonzalez, developmental editor, Elsevier deserve special thanks for bringing this volume into life. Dr. Aditya Pratap, assistant professor, Division of Plant Breeding Genetics, SKUAST, Jammu, India for outstanding and formidable volume of correspondence and Sh. Madan Mohan Gupta, father-in-law of the Editor for proof reading are thankfully acknowledged and my wife for her tolerance of long stretch of months end committed to this Volume 45.

S. K. Gupta
Editor

CONTENTS OF VOLUMES 34–44

Contents of Volume 34

Contents of Volume 35

Contents of Volume 36

Contents of Volume 37

Contents of Volume 38

Contents of Volume 39

Contents of Volume 40

Contents of Volume 41

Contents of Volume 42

Contents of Volume 43

Contents of Volume 44

History, Origin, and Evolution

S. K. GUPTA AND ADITYA PRATAP

Division of Plant Breeding and Genetics,
SK University of Agricultural Sciences and Technology, Chatha,
Jammu, India

ABSTRACT

The oilseed Brassicas are the world's third most important source of vegetable oils and their production has witnessed a steady upward movement during the past 25 years. During this period, the production share of European countries has also increased considerably, particularly after the introduction of double-low (low erucic acid, low glucosinolate) cultivars. Besides improvement in the nutritional profile of the *Brassica* oil and its meal, the conventional breeding as well as modern biotechnological tools have led to the improvement of various agronomically important quantitative and qualitative characters. The nuclear restriction fragment length polymorphism technology has greatly aided in determining the degree of genetic variability among various Brassicas as well in studying their evolution pattern. The oldest references regarding origin and cultivation of rapeseed come from Asia, though the evolution of this crop took place in many countries throughout the globe, including China, India, Switzerland, Germany, Australia, Denmark, the Netherlands,

Advances in Botanical Research, Vol. 45
Incorporating Advances in Plant Pathology

0065-2296/07 $35.00
DOI: 10.1016/S0065-2296(07)45001-7

Rome, and Europe. However, most of the work on development of modern and high-quality rapeseed known as canola was done in Canada. Lack of consistency in names, inclusion of too many forms in one species, and the entirely different forms of present day Brassicas from their ancestors make this genus a complex member of Cruciferae and poses several taxonomic and classification problems. Still many attempts have been made to establish the origin of various *Brassica* species and their interrelationships through cytogenetic, chemotaxonomic, and molecular studies. The present chapter focuses on such historical and evolutionary developments in rapeseed.

I. INTRODUCTION

Oilseed rapes are the world's third most important source of vegetable oils after palm and soybean (Beckman, 2005). The rapeseed production has witnessed a steady upward movement during the past 25 years and presently, it contributes about 14% of the global vegetable oils. More recently, the introduction of low erucic acid varieties enhanced its value as edible oil, particularly among the health conscious consumers and varieties with low glucosinolates increased the value of its defatted meal for use as a feed for livestock. The development of double-low varieties (canola) (Downey and Rakow, 1987) has made rapeseed one of the major plant oil sources at the global level, and now there is a constant tendency to increase its share in production of oilseeds (Bartkowiak-Broda *et al.*, 2005).

Rapeseed of commercial interest is grown in the cooler areas of the world covering mainly North America, northern part of Europe, Canada, China, and India (Fig. 1). This is obtained from the species of *Brassica*, members of Cruciferae, and within the genus are some 160 species, mainly annual and biannual herbs (Weiss, 1983). Oleiferous rapes are generally derived from two *Brassica* species, *Brassica napus* L. and *B. campestris* L. syn. *B. rapa* L. To distinguish between them, *B. rapa* is known as turnip rape and *B. napus* as Swede rape with winter types existing in both of them. *B. campestris* is also referred to as toria, sarson, summer turnip rape, Polish rape, and so on. Similarly, different names are also given to *B. napus* such as Argentine rape, Swede rape, and colza. All the rapeseed contributing cultivated *Brassica* species are highly polymorphic including oilseed crops, root crops, and vegetables such as Chinese cabbage, broccoli, and Brussels sprouts. However, our discussion in this and the subsequent chapters shall concentrate mainly on the oilseed rapes.

II. AREA AND PRODUCTION

The increase in global rapeseed production over the years has been catalyzed mainly by the ever-increasing population pressure and diversification of uses of *Brassica* products. The increase in demand has been met by conventional

Fig. 1. World rapeseed distribution. Area in dark represents region of origin and area in light represents region of cultivation (from: http://www2.mpiz-koeln.mpg.de/pr/garten/schau/BrassicarapaLvarsilvestri/Brassica rapa...).

and modern plant-breeding approaches, including those associated with seed quality improvement and modification of fatty acid profiles according to the domestic and industrial use. As a result, there has been a paradigm shift in the country-wise production patterns of this crop in these years (Table I). Asia alone produced about 70% of the world's rapeseed–mustard production during 1948–1952. However, there has been a remarkable increase in the production share of European countries over the periods, and Canada and Europe together produced more than half of the global rapeseed (48,545 thousand tons) during 2005.

The last 25 years also witnessed a tremendous increase in the area under rapeseed cultivation in major producing countries, specifically after the introduction of double-low varieties (Table II). In Canada and entire Europe, *B. napus* and *B. campestris* are the major species grown with both spring-sown and winter-sown types existing in both (Bunting, 1984). However, in countries like India and China, the production is also shared by other species, viz., *B. oleracea* and *B. juncea*. Although used widely as edible oil in Asia, rapeseed oil has gradually become important domestic and industrial oil in the western nations as a result of breeding for improved oil and meal quality and better processing techniques. Today China, India, Canada, and Europe are the top producers, although there is a good potential for these crops to be successfully grown in Australia, United States, and South America, where both the area and production have sharply increased over the years.

TABLE I

Production of Rapeseed (in Thousand Tons) over Selected Years in Major Producing Countries During 1990–2005

S.No.	Country	1990	1992	1994	1996	1998	2000	2001	2002	2003	2004	2005
1	Australia	98	178	264	623	1690	1775	1756	871	1703	1496	1405
2	Bangladesh	217	243	239	246	254	249	238	233	218	211	191
3	Canada	3266	3872	7233	5062	7643	7205	5017	4407	6771	7728	8447
4	China	6958	7653	7492	9201	8300	11,380	11,331	11,252	11,420	13,182	13,050
5	Czech Republic	380	430	452	520	680	844	973	710	388	935	769
6	Denmark	793	406	372	251	359	293	212	218	354	468	342
7	France	1976	1795	1772	2902	3734	3477	2878	3320	3361	3993	4534
8	Germany	2088	2617	2896	1970	3388	3586	4160	3849	3638	5276	5052
9	Hungary	106	44	53	138	73	179	205	208	108	291	283
10	India	4125	5862	5328	5999	4703	5788	4187	5083	3918	6500	6800
11	Pakistan	233	220	197	254	333	297	262	244	353	401	347
12	Poland	1206	758	756	449	1099	958	1064	953	793	1,633	1434
13	Sweden	422	283	214	132	123	121	106	159	129	227	198
14	United Kingdom	1258	1213	1243	1415	1567	1157	1157	1468	1771	1609	1902
15	United States	54	72	219	219	709	909	908	697	686	608	717
16	World[a]	23,325	25,231	30,170	31,428	35,762	37,388	36,018	32,897	39,427	46,139	48,545

[a]World Oil: Rapeseed Statistics and Graphics (http://worldfood.apionet.or.jp/graph/num.cgi).

Source: FAOSTAT (http://faostat.fao.org/site/340/DesktopDefault.aspx?PageID=340).

TABLE II

Area Under Rapeseed (in Thousand Hectares) over Selected Years in Major Countries

S.No.	Country	1990	1992	1994	1996	1998	2000	2001	2002	2003	2004	2005
1	Australia	73	107	356	408	1247	1459	1332	1298	1211	1351	962
2	Bangladesh	338	340	336	336	344	329	318	303	297	279	242
3	Canada	2529	3045	5766	3451	5428	4859	3785	3426	4689	4938	5154
4	China	5503	5976	5783	6734	6527	7494	7095	7143	7221	7330	7220
5	Czech Republic	137	168	190	226	264	324	343	312	251	259	267
6	Denmark	271	170	170	106	117	99	79	84	107	122	112
7	France	680	665	671	875	1145	1186	1083	1036	1082	1125	1231
8	Germany	722	1001	1058	854	1007	1078	1138	1296	1268	1283	1344
9	Hungary	60	35	28	93	52	116	110	129	71	105	123
10	India	4967	6553	6289	6546	7041	6027	4477	5073	4419	7150	7300
11	Pakistan	307	286	268	320	339	327	295	263	353	386	360
12	Poland	500	417	370	283	466	437	443	439	426	538	544
13	Sweden	168	137	128	65	55	48	45	67	59	84	82
14	United Kingdom	390	421	491	415	534	402	451	432	542	558	593
15	United States	31	53	140	142	437	608	590	520	433	338	451
16	World[a]	17,587	19,634	22,596	21,862	25,398	24,733	23,288	21,716	25,465	26,781	27,349

[a]World Oil: Rapeseed Statistics and Graphics (http://worldfood.apionet.or.jp/graph/num.cgi).

Source: FAOSTAT (http://faostat.fao.org/site/340/DesktopDefault.aspx?PageID340).

III. HISTORY

The crop Brassicas have been very important as food crops in the form of vegetables, oilseeds, feed and fodder, green manure, and condiments and have played a great role in the human history by contributing a good share of food in one form or another. Two species, *B. juncea* and *B. campestris* having a range of morphotypes, are the crops of antiquity in India where much before the Christian era, they were used for many purposes including oil for cooking and frying, spice for seasoning food articles, vegetables, and for religious ceremonies (Mehra, 1966). Since time immemorial, the *Brassica* crops have been a part and parcel of human agriculture system, and at present also they occupy a predominant place in the world's agrarian economy.

The name rape is derived from the Latin word "rapum," which means turnip. Though exact time and place of origin of rape is yet uncertain, a host of oldest references of this crop with regards to its cultivation comes from India, China, and Japan. Early history suggests that rapeseed has been cultivated for several thousand years with its origins in Asia. Sanskrit writings of 2000–1500 BC directly refer to oleiferous *B. napus* forms (sarson types) and mustard. Seeds of *B. juncea* have been found in the archaeological sites in India dating back to ca 2300 BC (Prakash, 1980; Weiss, 1983). The Chinese word for rapeseed was first recorded ca 2500 years ago and the oldest archaeological discoveries may date back as far as to ca 5000 BC (Yan, 1990). The original use of rapeseed was intended to use its oil for burning lamps, and to some extent, it was also used as edible oil by poor people. In later centuries, it gradually became an important cooking and lighting oil.

The Greek, Roman, and Chinese writings of 500–200 BC refer to rapiferous forms of *B. rapa* and also describe their medicinal values (Downey and Robellen, 1989). Seeds of *B. juncea* have been excavated from Chanhudaro, a site of Indus Valley civilization that existed in the plains of Punjab along the river of Indus ca 2300–1750 (Piggot, 1950). Species from the genus *Brassica* were cultivated in ancient Rome and also in Gallia (Fussel, 1955), and seeds of these species had also been found in the old German graves and Swiss constructions from the Bronze Age (Neuweiller, 1905; Schiemann, 1932; Witmack, 1904). In Dodoneus's "Herbalist" (1578), a mention has been made regarding the growing of *B. rapa* var. *rapifera* in 1470 as a winter crop. In his "Herball," Gerarde (1597) had very clearly differentiated between turnips (*B. rapa*) and navews (*B. napus*). Rape has been recorded as an oilseed crop in Europe at least since the Middle Ages, but it is still uncertain which species was cultivated (Appelquist and Ohlson, 1972).

Domestication of rapeseed in Europe appears to have started in the early Middle Ages, although the true turnip was probably introduced by Romans.

Since many other oil-yielding plants, particularly olive tree, were available in southern Europe, *B. rapa* initially spread mainly as turnip rape crop within Europe. However, in more prosperous countries like the Netherlands, the farmers used almost all *Brassica* seeds to produce vegetable oils. Oil was extracted from "raepsaet, koolaet, and mostaert saet" according to a Dutch reference of fourteenth century, which means "the seeds of *B. rapa*, *B. oleracea*, and *B. nigra* (or *Sinapis alba*)" (Reiner *et al.*, 1995). As *B. rapa* was most intensively grown at that time, it can be concluded that this crop was the major source of producing large quantities of vegetable oils. Seeds of *B. rapa* were first recorded in Europe in 1620 by the Swiss botanist Casper Banhin. However, Boswell (1949) was of the view that these existed much earlier than this. As per some anonymous authors, rapeseed was grown in Europe as early as in the thirteenth century.

In the Netherlands, the commercial plantings of rapeseed were recorded in the early sixteenth century. It had limited industrial use at that time until the development of steam power, when it was discovered that rape oil was an excellent lubricant for steam engines.

B. rapa was the dominant species in the western Canada in the early 1970s. It is comparatively a recent introduction in Canada and the United States and is found as an occasional weed or volunteer in the cultivated fields (Muenscher, 1980; Munz, 1968). In late 1980s, large acreages of *B. rapa* and *B. napus* were grown in the Prairie Provinces and these crops gradually started getting established. However, the production area sown to *B. rapa* decreased to about 15–20% in 1990s (The Biology of *Brassica rapa*, 1999). In Austria, the annual wild-type *B. rapa* is found as a weed in rye and potato crops situated in relatively cool and high areas with an altitude of about 1000 m (Holzner, 1981). In the 1970s, the information on its distribution had been very uncertain due to incross and possibility of its escape from culture (Reiner *et al.*, 1995).

Canola is a modern, high-quality form of rapeseed. It originated in Canada through genetic modification of rapeseed by conventional plant breeding and emerged in the 1970s as a viable oilseed, equipped with the appropriate genetics to transform the oil and meal from unacceptable to highly desired products for both human as well as livestock consumption (Shahidi, 1990). Since then, rapeseed gained a remarkable reputation in the oilseed market throughout the globe. Today, the fatty acid profile of canola is considered as the most desirable of all vegetable oil profiles by nutritionists (Stringam *et al.*, 2003).

Although superior edible oils had been developed by 1971, the presence of high amount of glucosinolates in the meal still remained a major concern in the expansion of market of the vegetable oil derived from rapeseed. Prior to

1960, the erucic acid (a long-chain fatty acid) content of rapeseed oil was not of particular interest while evaluating the oil use for edible purposes. However, during the 1960s, reports started pouring in, which indicated that there could be negative health effects, particularly heart abnormalities, which are associated with the consumption of erucic acid in rapeseed oil. The concern was further aggravated by the establishment of the European Economic Community (EEC) in 1960 with France, West Germany, Italy, the Netherlands, Belgium, and Luxemburg as the founder members, which stressed on the use of varieties with erucic acid content less than 5%. As a result, the traditional rapeseed oil started being considered as unsafe for human health. This led to the concentration of rapeseed-breeding efforts toward the development of such varieties in late 1960s and early 1970s. The application of gas–liquid chromatography (Craig and Murphy, 1959) led to the identification of low erucic acid plants in *B. napus* and *B. campestris* with the first low erucic acid plants in them identified in 1968 and the first *B. campestris* variety in 1971. In 1977, the cultivation of such varieties was made mandatory.

This represented a major improvement in the quality of rapeseed oil. The Canadian plant breeders were quick enough in the identification of plants with low eicosenoic acid and erucic acid content (Downey and Harvey, 1963; Stefansson *et al.*, 1961). While oil quality in rapeseed was being improved, they were also simultaneously working toward the improvement of quality of the defatted protein meal by-product (Krzymanski and Downey, 1969). The rapeseed meal is an excellent source of protein with a favorable balance of amino acids. However, its use was limited by its high glucosinolate content, which is a constituent of most of the plants of Brassicas. Traditional rapeseed varieties contained high levels of glucosinolates in the meal, which when fed to livestock in sufficient quantities led to the problems related with nutrition, digestion, and thyroid. The development of fast and accurate chemical methods led to the identification of plants of the *B. napus* cultivar Bronowski from Poland, which were essentially free of the harmful glucosinolates normally found in rapeseed. The low glucosinolate genes were then incorporated in the well-adapted and high-yielding cultivars of *B. napus* and subsequently transferred to *B. campestris*.

In 1974, Dr. Baldur Stefansson from the University of Manitoba successfully developed the first "double-low" variety with reduced levels of both erucic acid and glucosinolates (www.canola-council.org). This led to the evolution of a greatly improved crop, which met specific quality requirements of an oilseed fit for human as well as livestock consumption. As a result of these improvements, the FDA gave GRAS (generally recognized as safe) status to rapeseed oil in 1985 for use in the US food products.

The word "canola" was coined and trademarked for such type of rapeseed products, low in both glucosinolates and erucic acid to distinguish them from traditional rapeseed. The name canola was initially registered by the Western Canadian Oilseed Crushers' Association for reference to oil, meal, protein extractions, seed, and seed hulls from or of varieties with 5% or less erucic acid in the oil and 3 mg/g of glucosinolates (www.canola-council.org). Later, the control of the term was transferred to the Rapeseed Association of Canada in 1980, which subsequently changed its name to Canola Council of Canada (2006). The new target of achieving the ideal glucosinolate level at 15 μmol is underway. Keeping the above facts in view, it may be conveniently inferred that all canola is rapeseed but all rapeseed is not canola.

IV. ORIGIN AND EVOLUTION

The *Brassica* genus is a very complex member of the Cruciferae family, and as such it contains many cultivated plants and wild species. It therefore poses several taxonomic and classification problems. Also, there is a lack of consistency in the names of different oil-yielding Brassicas throughout the globe, which aggravates the problem further. The scientific nomenclature is highly confusing, which makes it difficult for many to decide as to what particular scientific name should be used for a particular plant. Bailey (1922) listed the following reasons responsible for the chaotic nomenclature of Brassicas (Kalia and Gupta, 1997):

1. Inclusion of too many forms in one species, thereby weakening the definitions
2. Attempts to identify oriental cultivated Brassicas with species known in Europe
3. Mixing of seeds

Furthermore, during the process of domestication, man has modified the entire plant and the present day Brassicas are entirely different from their ancestors. Also the occurrence of similar plant forms in more than one *Brassica* species resulted in considerable confusion and misclassification by early botanists (Downey and Robellen, 1989).

The cytogenetic relationships between the rapeseed species as well as their closest allies were first explained systematically by U (1935) about 70 years ago (Fig. 2). These relationships show that *B. campestris* ($2n = 20$, AA) of rapeseed, and *B. nigra* ($2n = 16$, BB) and *B. oleracea* ($2n = 18$, CC) are the primary species and *B. napus* ($2n = 28$, AACC) and *B. carinata* ($2n = 34$,

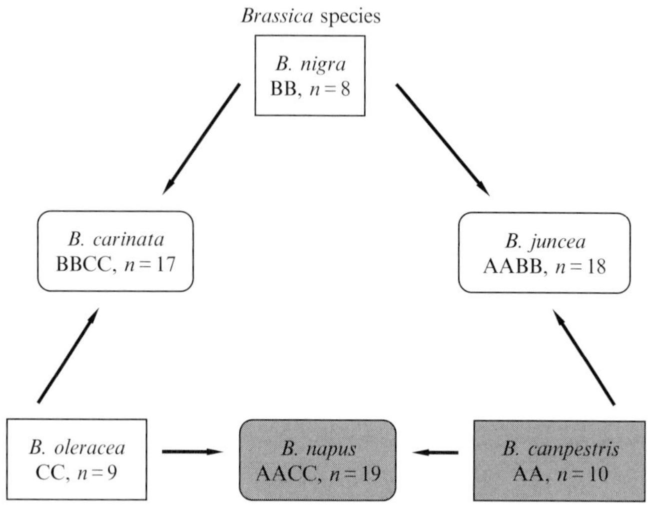

Fig. 2. Evolution of cultivated *Brassica* species and its relatives (U, 1935).

BBCC) of rapeseed and *B. juncea* ($2n = 36$, AABB) are the amphidiploids resulting from paired crossings between the primary species. Morinaga (1928, 1929a,b, 1934a,b) discussed that crop Brassicas include six cotydemes—three elementary ones with 16, 18, and 20 chromosomes as diploid and three with higher chromosome numbers of 34, 36, and 38 as tetraploid, the latter having evolved through interspecific hybridization in nature between any two of the elementary taxa (Table III).

Morinaga and his associates carried extensive cytogenetic studies in oilseed Brassicas and clarified the relationships between them (Prakash and Hinata, 1980). According to the hypothesis of Morinaga (1934a,b), the three species with the higher chromosome number, *B. napus* L., *B. juncea* L. Czern. and Coss., and *B. carinata* A. Braun, are amphidiploids combining in pairs the chromosome sets of the low chromosome number species *B. nigra*, *B. oleracea*, and *B. rapa*. U (1935) verified the hypothesis with successful resynthesis of *B. napus*. Resynthesis of *B. juncea* and *B. carinata* was accomplished by Frandsen (1943, 1947). Robellen (1960) suggested that the low chromosome number species might have developed from the ancestral species, which could have even lower chromosome numbers. Also the chromosome analysis of the monogenomic species revealed that only six chromosomes were distinctly different, the remaining being homologous with one or another of the basic set of six.

Hooker and Thompson (1861) placed two-valved, erect fruited form of yellow-seeded sarson (*S. glauca* Roxb.) and toria (*S. dichotoma*) in one group

TABLE III
Genus Brassica and Their Ecotypes

Species	Subspecies	2*n* Chromosome number	Common name	Use
B. nigra Koch	–	16	Black mustard	Condiment
B. oleracea	acephala	18	Kale	Vegetable/fodder
	aboglabra	18	Chinese kale	Vegetable
	botrytis	18	Cauliflower	Vegetable
	capitata	18	Cabbage	Vegetable/fodder
	gemmifera	18	Brussels sprouts	Vegetable
	gongylodes	18	Khol rabi	Vegetable
	italica	18	Broccoli	Vegetable
B. campestris				
	chinensis	20	Pak-choi	Vegetable
	japonica	20		Vegetable
	narinosa	20		Vegetable
	oleifera	20	Turnip rape	Oilseed
	pekinensis	20	Chinese cabbage	Vegetable
	rapa	20	Turnip	Vegetable/fodder
B. napus				
	oleifera	38	Rape	Oilseed
	rapifera	38	Rutabaga	Vegetable
B. juncea				
	rugosa	36	Chinese mustard	Vegetable
	oleifera	36	Indian mustard	Oilseed
B. carinata				
		34	Ethiopian mustard	Vegetable/oilseed

Source: Kalia and Gupta (1997).

and named it as *B. campestris*, which was later referred to as *B. campestris* subsp. *napus* by Hooker and others. They considered the three- to four-valved, nodding fruited form as one species (*B. trilocularis*) and four-valved, erect fruited form as another (*B. quadrivalvis*) (Kalia and Gupta, 1997). Duthie and Fuller (1882) considered the two-valved, erect fruited form of yellow sarson as *B. campestris* subsp. *napus* var. *glauca*. They also considered brown sarson and toria as the varieties of *B. napus*. The same nomenclature was also adopted by Watt (1889). He however suggested that brown-seeded sarson and toria should be treated as subsp. *campestris* and *napus*, respectively, since they have hispid and glabrous leaves.

Prain (1898) combined the forms of yellow seeds sarson having two-valved nodding fruits, three- to four-valved nodding fruits, and four–valved erect fruits with brown-seeded sarson into one group, which he called as *campestris* var. sarson. In this monograph, he considered toria as *B. campestris* subsp.

napus var. *oleifera* but adopted the term dichotoma on account of its priority. He also considered toria as *B. napus* var. dichotoma because of the marked differences in the time of growth and also mode of cultivation.

Olsson (1954) suggested that all the 20 chromosome forms of leafy, oleiferous, and rapiferous *Brassiccas* should be grouped into one species, *B. campestris*. This was in support of the views of Howard (1940) that the name *B. campestris* should be reserved for the forms with $2n = 20$ and *napus* for the forms with $2n = 38$. He proposed that the names sarson and toria should be *B. campestris* L. var.sarson and *B. campestris* L. var.toria, respectively. Singh (1958) considered yellow and brown sarson as varieties, whereas Prakash (1973) considered them as the forms of subsp. *oleifera*. Toxopens *et al.* (1984) suggested a classification and nomenclature of *B. campestris* at intraspecific level in which he proposed that the name of *B. campestris* should be changed to *B. rapa*. At present, the following nomenclature of sarson and toria is being followed (Kalia and Gupta, 1997):

- *B. campestris* var. yellow sarson for yellow sarson
- *B. campestris* var. brown sarson for brown sarson
- *B. campestris* var. toria for toria

A. *B. CAMPESTRIS*

The name *B. campestris* was assigned to an annual weed growing in "non-loamy fields of Europe" by Linnaeus (1753) in "Species Planatarum." It was described as a plant with rough, stiff hairs when young, and just like *B. rapa* by DeCandolle *et al.* (1824). However, when it was realized that *B. campestris* and the turnip rape *B. rapa* have to be classified as same species, a confusion was created in nomenclature and the wild type was often subordinated under *B. rapa* (Reiner *et al.*, 1995).

There are well-defined groups of *B. rapa* based on their morphological characteristics: (1) The oleiferous *rapa*, of which a specific form is known as canola; (2) the leafy type *B. rapa*, including the Chinensis group (Pak-choi, celery mustard), the pekinensis group (Chinese cabbage), and the perviridis group (tendergreen); and (3) the rapiferous type, consisting the rapifera group (turnip, rapini) and the ruvo group (turnip, broccoli, Italian turnip) (Hortus Third, 1976; Prakash and Hinata, 1980). Wild *B. rapa* subsp. *oleifera* (including var. *sylvestris*) is considered as the species from which var. *rapa* L. (cultivated turnip) and var. *sylvestris* (Lam.) Briggs (turnip rape) originated. Historically, *B. campestris* appears to be the most widely distributed oilseed crop among the Brassicas. At least 2000 years ago, it was distributed from the Atlantic islands in the west to the eastern shore of China and Korea and from

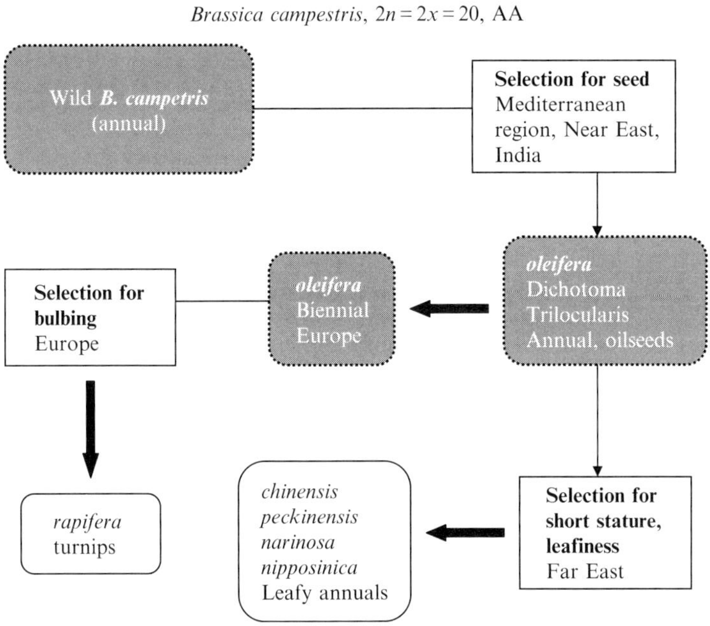

Fig. 3. Evolution of *B. campestris* and its relatives (modified from McNaughton, 1979).

northern Norway south to the Sahara and northern India with a primary center near the Himalayan region (Hedge, 1976). Presently, it is native throughout Europe, Russia, Central Asia, and Near East (Prakash and Hinata, 1980). However, its secondary centers of origin are located in the European-Mediterranean area and in Asia. *B. campestris* as a crop is believed to have evolved from wild *B. campestris* strains, which are found to occur from western Europe to China (Fig. 3). Protein studies of different forms of *campestris* by Denford (1975) suggested that two different forms of *B. campestris* exist, that is, eastern (leafy forms of China, Chinese cabbage, Pak-choi, and so on) and western (Indian forms, European turnip, and turnip rape). This indicated that the original progenitor of the Indian and European forms was the same and that the Indian brown sarson evolved in the northwest of the Indian subcontinent from the original stock as suggested by the Russian workers (Sinskaia, 1928; Vavilov and Bukinich, 1929), who regarded India as one of the independent centers of origin. However, there is some doubt as to whether the Asian and Near Eastern types arose from the independent center of origin in Afghanistan and then moved eastward as *B. rapa* became domesticated. Prakash and Hinata (1980) also suggested that oleiferous *B. campestris* subspecies developed in two places giving rise to two different races, one European and other Asian. There is a lot of evidence that European oilseed type *B. rapa* must be very close to the turnip type *B. rapa* genetically because it was produced out of it only some 100 years ago. On the other hand, in China, "Lintao Caizi" very well known to the world as *B. chinensis* (leafy type *B. rapa*, $n = 10$) is used as an oilseed crop. This can be interpreted as parallel to the evolution of the oilseed type out of turnip type *B. rapa* in Europe (Sun *et al.*, 1991). However, the Asian oleiferous types, known as brown and yellow sarson and toria, differ from the European turnip rape in plant morphology and also tend to vary in the patterns of fatty acid compositions and glucosinolates (Downey and Robellen, 1989). The protein and other chemotaxonomic analyses also support the separated European and Asian centers of origin, although the Indian forms of *B. campestris* are conspecific with the European turnip rape as well as domesticated and wild turnips (Denford, 1975; Vaughan *et al.*, 1976). The old German books from the sixteenth century also support the view of origin of the oilseed type *Brassica* crops in Europe. There has been a mention of difficulties in harvesting the oilseed crops by the chronist Konrad Heresbach (Schroder-Lembke, 1976).

Direct comparison of restriction fragment length polymorphism (RFLP) patterns and phylogenetic analysis on 17 cultivated and 5 wild populations of *B. rapa* were conducted by Song *et al.* (1988). On the basis of phylogenetic analysis, they were able to designate the 17 cultivated forms into two distinct groups as European group and East Asia group. All the five wild populations examined were clustered with the European group (Song *et al.*, 1990), suggesting that the European group might be more ancient than the East Asia group. The phylogenetic groupings seem to correspond with the respective geographic distribution of the cultivated and wild forms of *B. rapa*.

Out of different ecotypes of *B. campestris*, brown sarson appears to be the oldest (Singh, 1958). The earliest reference with regards to brown sarson has been found in the Chandogya Upanishada (III, 14.3, ca 1000 BC), wherein the word "sarshap" has been used in reference to its oil (Prakash, 1961). Alam (1945) inferred that the Indian brown sarson and toria closely resemble one or the other groups of *B. campestris* forms of Afghanistan, and Persian synonyms are available for these crops. However, since no remains are there so far of brown sarson seeds from any place, it is difficult to determine the exact place and time of origin of brown sarson or its introduction in northwest of India, from where it migrated eastward and differentiated into other ecotypes (Hinata and Prakash, 1984). The yellow-seeded *B. campestris* is grown only in a limited area in eastern parts of India only. However, it is not found anywhere in wild state, though its probable progenitor could be brown sarson (Singh, 1958).

Typically *B. campestris* is found in costal lowlands, high montane areas (the slopes of high valleys or mountain ranges), and alpine and high sierras. In Canada, where it is comparatively a recent introduction, it is found on disturbed land, typically in crops, fields and gardens, along roadsides, and on wastelands (Warwick and Francis, 1994).

B. *B. NAPUS*

B. napus is an amphidiploid and is comparatively of recent origin (Olsson, 1960). It is uncertain to maintain if *B. napus* is found wild or not, since wild forms of this crop are difficult to find (Hinata and Prakash, 1984). However, if wild *napus* exists, it must be a European-Mediterranean species that originated in the area of overlap between *B. oleracea* and *B. campestris* (Fig. 4; McNaughton, 1976). Though the origins of *B. napus* are ambiguous, these are proposed to involve natural interspecific hybridization between the two diploid species, *B. oleracea* ($n = 9$) and *B. rapa* ($n = 10$) (U, 1935). Song and Osborn (1992) on the basis of their analysis of chloroplast and mitochondrial DNA suggested that *B. montana* ($n = 9$) might be closely related to the prototype that gave rise to both the cytoplasms, that is, of *B. rapa* and *B. oleracea*. They also suggested that *B. napus* has multiple origins and the most cultivated forms of *B. napus* were derived from a cross in which a closely related ancestral species of *B. rapa* and *B. montana* was the maternal donor. The parental origins of *B. napus* were also investigated using six microsatellite markers located in the chloroplast genome by Allender *et al.* (2005). They also suggested that either of *B. rapa* or *B. montana* or a common ancestor could have been the maternal parent of *B. napus*. Though they suggested that *B. oleracea* was not the parental parent of most of *B. napus* accessions, a small number of accessions shared *B. oleracea* haplotype. Similarly, the phylogenetic analysis based on nuclear RFLP data also suggests that *B. napus* has multiple origins (Song *et al.*, 1993). The various cytoplasm types found in *B. napus* accessions correspond to the progenitor diploid species which provide a strong evidence for the multiple origins of this crop (Song *et al.*, 1997).

In *B. napus* as well as *B. campestris*, a range of morphological forms are found, both having annual and biennial types. Keeping this in view, Olsson (1960) suggested that *B. napus* could have arisen several times by spontaneous hybridization of different forms of *B. campestris* and *B. oleracea*. The majority of the cultivated *B. napus* accessions appear to have arisen by an

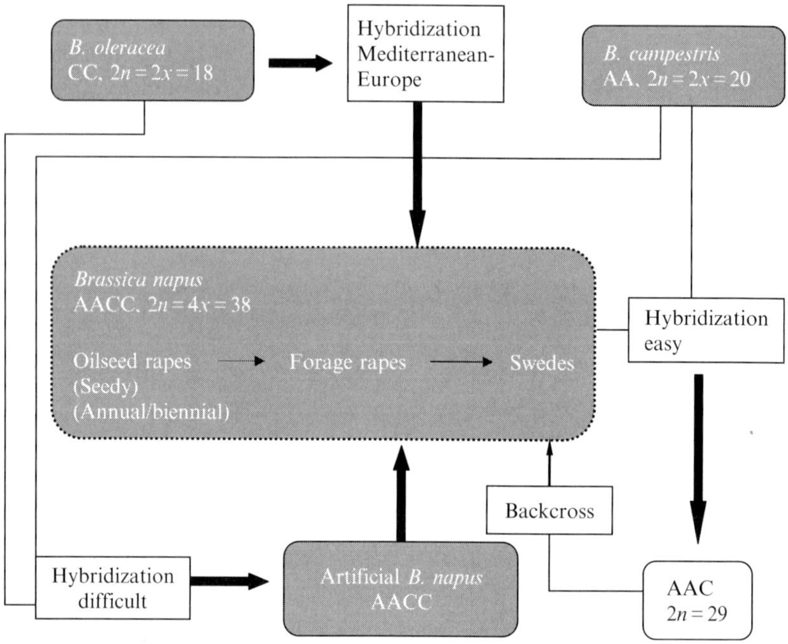

Fig. 4. Evolution of *B. napus* and its relatives (modified from McNaughton, 1979).

interspecific cross in which a wild 9 or 10 chromosome species having the *B. montana* cytoplasm type or the M/N type was the female donor (Song *et al.*, 1997). Thus, the Swede rapes could have originated in the medieval gardens where turnips and kales grew side by side.

The spontaneous formation of *B. napus* is likely to have been an extremely rare event, not only due to postfertilization barriers but also due to differences in floral morphology leading to discrimination by insect pollinators thereby reducing the chances of the interspecific hybridization.

Mizushima and Tsunoda (1967) inferred that *B. napus* was found in the coast of northern Europe because *B. oleracea* extended its territory up to northern Europe from the Irano Turanean regions with its high adaptability to low temperatures. Sinskaia (1928) and Schiemann (1932) were also of the view that it might have originated in the Mediterranean region or in the western or northern Europe. In Europe, production of oleiferous *B. napus* might have started during the Middle Ages. In Asia, it was introduced during the nineteenth century. The Chinese and Japanese germplasm was developed by crossing European *B. napus* cultivars with indigenous *B. rapa* cultivars (Shiga, 1970).

V. CONCLUSION

Oilseed rapes have a range of morphotypes and accordingly vary in their origin, cultivation, use, and history. The evolution of rapeseed has witnessed a shift in their morphophysiological traits from their original forms to present day canola, which is highly suitable for human as well as animal consumption. Extensive cytogenetic and crossability studies aided by the use of molecular and isozyme markers have helped the plant breeders in studying their evolution. However, more precise strategies are required in future to evolve more promising and canola type cultivars, guided by the demands of the market and industry.

REFERENCES

Alam, Z. (1945). Nomenclature of oleiferous Brassicas cultivated in Punjab. *Indian Journal of Agricultural Sciences* **15**, 173–181.

Allender, C., Evered, C., Lynn, J. and Graham, K. (2005). Tracing the origins of *Brassica* napus using chloroplast microsatellites. *In* "Plant and Animal Genomes XIII Conference, January 15–19, 2005" p. 411. Town & Country Convention Center, San Diego, CA.

Appelquist, L. A. and Ohlson, R. (1972). "Rapeseed: Cultivation, Composition, Processing and Utilization." Elsevier, Amsterdam, The Netherlands, London and New York.

Bailey, L. H. (1922). The cultivated Brassicas I. *Gentes Herbarum* **1**, 53–108.

Bartkowiak-Broda, I., Mikolajczyk, K., Spasibionek, S. and Cegielska-Taras, T. (2005). Genetic and breeding research timing At increasing the value of rapeseed oil as a source of renewable energy. *In* "Alternative plants for sustainable agriculture" (S. Jezowski, K. M. Wojciechowicz and E. Zenkteler, eds.), pp. 129–139. Institute of Plant Genetics PAS, Poland.

Beckman, C. (2005). Vegetable oils: Competition in a changing market. Bi-weekly Bulletin. *Agriculture and Agri-Food Canada* **18**(11), Available at http://www.agr.gc.ca/mad-dam/e/bulletine/v18e/v18n11_e.htm.

Boswell, V. R. (1949). Our vegetable travelers. *National Geographic Magazine* **96**, 145–217.

Bunting, E. S. (1984). Oilseed rape in perspective with particular reference to crop expansion and distribution in England, 1973–(1983). *In* "Aspects of Applied Biology (6), Agronomy, Physiology, Plant Breeding and Crop Protection of Oilseed Rape" (D. H. Scarsbrick and R. W. Daniels, eds.), pp. 11–21. Collins Publishing, London.

Canola Council of Canada (2006). (Available at: http://www.canola-council.org/portal.html).

Craig, B. M. and Murphy, N. L. (1959). Quantitative fatty acid analysis of vegetable oil by Gas Liquid Chromatography. *Journal of the American Oil Chemists Society* **36**, 549–552.

DeCandolle, A. P. (1824). Translated into German by Berg, C.W.1824. Die verschiedenen Arten. Unterarten und Spielarten des Kohls und der Rettige, Welche in Europa gebauet warden, Leipzig.

Denford, K. E. (1975). Isoenzyme studies in members of the genus *Brassica*. *Botaniska notiser. Lund* **128**, 455–462.
Dodoneus, R. (1578). A nievve Herball. Antwerp. Translated by H. Lyte, London.
Downey, R. K. and Harvey, B. L. (1963). Methods for breeding for oilseed quality in rape. *Canadian Journal of Plant Science* **43**, 271–275.
Downey, R. K. and Rakow, G. (1987). Rapeseed and mustard. *In* "Principles of Cultivar Development, Vol. 2, Crop Species" (W. R. Fehr, ed.), pp. 437–486. Macmillan Publishing Company, New York.
Downey, R. K. and Robellen, G. (1989). Brassica species. *In* "Oil Crops of the world" (G. Robellen, R. K. Downey and A. Ashri, eds.), pp. 339–362. McGraw Hill Publishing Company, New York.
Duthie, J. F. and Fuller, J. B. (1882). "Field and Garden Crops of NWP and Oudh," pp. 28–29. Department of Agriculture and Commerce, Roorkee.
Frandsen, K. J. (1943). The experimental formation of *B. juncea*. *Dansk Bot. Archiv* **11**, 1–17.
Frandsen, K. J. (1947). The experimental formation of Brassica napus L. va. Oleifera DC. and *Brassica carinata* Braun. *Dansk Bot. Arkiv* **12**, 1–16.
Fussel, G. E. (1955). History of cole (*Brassica* sp). *Nature* **176**, 48–51.
Hedge, I. C. (1976). A systematic and geographical survey of the world cruciferae. *In* "The Biology and Chemistry of Cruciferae" (J. G. Vaughan, A. J. McLeod and B. M. G. Jones, eds.), pp. 1–45. Academic Press, New York.
Hinata, K. and Prakash, S. (1984). Ethnobotany and evolutionary origin of Indian oleiferous *Brassicae*. *Indian Journal of Genetics* **44**, 102–112.
Holzner, W. (1981). "Acker-Unkräuter-Bestimmung, Verbreitung, Biologie und Ökologie." Leopold Stocker Verlag, Graz, Stuttgart.
Hooker, J. D. and Thompson, T. (1861). Praceursores and floram indicum cruciferae. *Journal of the Linnean Society* **5**, 128–181.
Hortus Third, B. L. (1976). "A Concise Dictionary of Plants Cultivated in the United States and Canada." L. H. Bailey Hortorium p. 1290. MacMillan Publishing Co. Inc., New York.
Howard, H. W. (1940). Nomenclature of *Brassica* species. *Current Science* **9**, 494–495.
Kalia, H. R. and Gupta, S. K. (1997). Importance, nomenclature and origin. *In* "Recent Advances in Oilseed Brassicas" (H. R. Kalia and S. K. Gupta, eds.), pp. 1–11. Kalyani Publishers, New Delhi, India.
Krzymanski, J. and Downey, R. K. (1969). Inheritance of fatty acid composition in winter forms of rapeseed (*Brassica napus*). *Canadian Journal of Plant Science* **49**, 313–319.
Linnaeus, C. (1753). Species Planatarum. Holmiae (Stockholm) (Reprint London, 1957).
McNaughton, I. H. (1976). Turnip and relatives. *In* "Evolution of Crop Plants" (N. W. Simmonds, ed.), pp. 45–48. Longman, London and New York.
McNaughton, I. H. (1979). Swedes and rapes. *In* "Evolution of Crop Plants, (Reprint 1979)" (N. W. Simmonds, ed.), pp. 53–56. Longman, London and New York.
Mehra, K. L. (1966). History and ethnobotany of mustard in India. *Advancing Frontiers of Plant Sciences* **19**, 51–59.
Mizushima, U. and Tsunoda, S. (1967). A plant exploration in *Brassica* and allied genera. *Tohoku Journal of Agricultural Research* **17**, 249–277.
Morinaga, T. (1928). Preliminary note on interspecific hybridization in *Brassica*. *Proceedings of Imperial Academy Tokyo* **4**, 620–622.
Morinaga, T. (1929a). Interspecific hybridization in *Brassica* I. The cytology of F_1 hybrids of *B. nepella* and various other species with 10 chromosomes. *Cytologia* **1**, 16–27.

Morinaga, T. (1929b). Interspecific hybridization in *Brassica* II. The cytology of F_1 hybrids *B. cerna* and various other species with 10 chromosomes. *Japanese Journal of Botany* **4**, 277–280.

Morinaga, T. (1934a). Interspecific hybridization in *Brassica* VI. The cytology of F_1 hybrids of *B. juncea* and. *B. nigra. Cytologia* **6**, 62–67.

Morinaga, T. (1934b). On the chromosome number of *Brassica juncea* and *Brassica napus* on the hybrid between these two and on offspring of the hybrid. *Japanese Journal of Genetics* **9**, 161–163.

Munz, P. A. (1968). "A California Flora," 1681 p. University of California Press, Berkeley and Los Angeles.

Muenscher, W. C. (1980). "Weeds," 2nd Edn., 586 p. Cornell University Press, Ithaca and London.

Neuweiller (1905). Die prahistorishe Pflanzenreste Mitteleuropas, Zurich.

Olsson, G. (1954). Crosses within the campestris group of the genus *Brassica. Hereditas* **40**, 398–418.

Olsson, G. (1960). Species crosses within the genus Brassica II. Artificial *Brassica napus* L. *Hereditas* **46**, 351–386.

Piggot, S. (1950). "Prehistoric India to 1000 BC." Penguin Books.

Prain, D. (1898). The mustard cultivated in Bengal. *Agricultural Ledger* **5**, 1–80.

Prakash, S. (1961). Oilseeds. *In* "Foods and Drinks in Ancient India," pp. 265–266. Munshi Ram Manohar Lal, New Delhi.

Prakash, S. (1973). Artificial *Brassica juncea* Coss. *Genetica* **44**, 249–263.

Prakash, S. (1980). Cruciferous oilseeds in India. *In* "*Brassica* Crops and Wild Allies. Biology and Breeding" (S. Tsunoda, K. Hinata and C. Gomez Campo, eds.), pp. 151–163. Japan Scientific Society Press, Tokyo.

Prakash, S. and Hinata, K. (1980). Taxonomy, cytogenetics and origin of crop *Brassica*, a review. *Opera Botanica* **55**, 11–57.

Reiner, H., Holzner, W. and Ebermann, R. (1995). The development of turnip type and oilseed type Brassica rapa crops from the wild type in Europe-An overview of the botanical, historical and linguistic facts: Rapeseed Today and Tomorrow, Ninth International Rapeseed Congress, Cambridge, UK, July 4–7, 1995, Vol. 4, pp. 1066–1069.

Robellen, G. (1960). Beitrage zur analyse des *Brassica*-Genoms. *Chromosoma* **11**, 205–228.

Schiemann, E. (1932). Entstehung der kulturpflan zen Handlab. *Vererbwis Lfg* **15**.

Schroder-Lembke, G. (1976). Die Entwicklung des Raps-und Rubsenanbaus in der deutschen Landwirtschaft. Z. fur Agrargeshichte und Agrarsoziologie **24**, 145–160.

Shahidi, F. (1990). Rapeseed and canola: Global production and distribution. *In* "Canola and Rapeseed: Production, Chemistry, Nutrition and Processing Technology" (F. Shahidi, ed.), p. 13. Van Norstrand Reinhold, New York.

Shiga, T. (1970). Rape breeding by interspecific crossing between *Brassica napus* and *Brassica campestris* in Japan. *Japan Agricultural Research Quarterly* **5**, 5–10.

Singh, D. (1958). "Rape and Mustard." The Indian Central Oilseeds Committee, Bombay.

Sinskaia, E. N. (1928). The oleiferous plants and roots of the family Cruciferae. *Bulletin of Applied Botany, Genetics, and Plant Breeding* **10**, 1–648.

Song, K. M. and Osborn, T. C. (1992). Polyphylatic origins of *B. napus*: New evidence based on organelle and nuclear RFLP analysis. *Genome* **35**, 992–1001.

Song, K. M., Osborn, T. C. and Williams, P. H. (1988). *Brassica* taxonomy based on nuclear restriction length polymorphisms (RFLPs) 2. Preliminary analysis of subspecies with *B. rapa* (syn. *campestris*) and. *B. oleracea. Theoretical and Applied Genetics* **76**, 593–600.

Song, K. M., Osborn, T. C. and Williams, P. H. (1990). *Brassica* taxonomy based on nuclear restriction length polymorphisms (RFLPs) 3. Genome relationships in *Brassica* and related genera and the origin of *B. oleracea* and *B. rapa*. *Theoretical and Applied Genetics* **79**, 497–506.

Song, K. M., Osborn, T. C. and Williams, P. H. (1997). Taxonomy based on nuclear RFLP analysis. *In* "Recent Advances in Oilseed Brassicas" (H. R. Kalia and S. K. Gupta, eds.), pp. 12–24. Kalyani Publishers, New Delhi, India.

Song, K. M., Tang, K. L. and Osborn, T. C. (1993). Development of synthetic *Brassica* amphidiploids by reciprocal hybridization and comparison to natural amphidiploids. *Theoretical and Applied Genetics* **86**, 811–821.

Stefansson, R. R., Hougen, F. W. and Downey, R. K. (1961). Note on the isolation of rape plants with seed oil free from erucic acid. *Canadian Journal of Plant Science* **41**, 218–219.

Stringam, G. R., Ripley, V. L., Love, H. K. and Mitchell, A. (2003). Transgenic herbicide tolerant canola. The Canadian experience. *Crop Science* **43**, 1590–1593.

Sun, W. C., Pan, Q. Y., An, X. H. and Yang, Y. P. (1991). *Brassica* and *Brassica* related oilseed crops in Gansu, China. *In* "Proceedings GCIRC, Eighth International Rapeseed Congress" (D. I. McGregor, ed.), Vol. 4, pp. 1130–1135. Saskatoon.

The Biology of *Brassica rapa* (1999). Regulatory Directive Govt. of Canada, p. 20 (available at:http://maltawildplants.com?CRUC/Docs/BRSRA/Brassicarapa.pdf).

Toxopens, H., Oost, E. H. and Reuling, G. (1984). Current aspects of the taxonomy of cultivated *Brassica* species. The use of *B. rapa* L. versus *B. campestris* L. and a proposal for a new intraspecific classification of B. rapa L. *Cruciferae Newsletter* **9**, 55–58.

U, N. (1935). Genome analysis in *Brassica* with special reference to the experimental formation of *Brassica napus* and peculiar mode of fertilization. *Japanese Journal of Botany* **7**, 389–452.

Vaughan, J. G., Phelan, J. R. and Denford, K. E. (1976). Seed studies in the Cruciferae. *In* "The Biology and Chemistry of the Cruciferae" (J. G. Vaughan, A. J. McLeod and B. M. G. Jones, eds.), pp. 119–144. Academic Press, New York.

Vavilov, N. I. and Bukinich, D. D. (1929). Agriculture in Afghanistan.*The Bulletin of Applied Botany of Genetics and Plant Breeding* **33**, 378–382.

Warwick, S. I. and Francis, A. (1994). Guide to the Wild Germplasm of *Brassica* and Allied Crops, Part V: Life History and Geographical Data for Wild Species in the Tribe Brassiceae (Cruciferae). Technical Bulletin 1994 Centre for Land and Biological Resources Research, Agriculture and Agri-Food Canada.

Watt, U. (1889). Brassica. *In* "Dictionary of Economic Products of India," Vol. 1, pp. 520–534. Government of India Central Printing Office, Calcutta.

Weiss, E. A. (1983). Rapeseed. *In* "Oilseed Crops" (E. A. Weiss, ed.), pp. 161–215. Longman, London and New York.

Witmack, L. (1904). Über die in Pompej gefundenen Pflanzenreste. *Englers Bot. Jahrb. Bd* **33**.

Yan, Z. (1990). Overview of rapeseed production and research in China. *In* "Proceedings of International Canola Conference, April 1990," pp. 29–35. Potash and Phosphate Institute, Atlanta, GA.

Breeding Methods

B. RAI,* S. K. GUPTA[†] AND ADITYA PRATAP[†]

*School of Agriculture, Banaras Hindu University, Varanasi, India
[†]Division of Plant Breeding and Genetics,
SK University of Agricultural Sciences and Technology, Chatha,
Jammu, India

Advances in Botanical Research, Vol. 45
Incorporating Advances in Plant Pathology

0065-2296/07 $35.00
DOI: 10.1016/S0065-2296(07)45002-9

ABSTRACT

Oleiferous rapes are interesting breeding material since they have a complete range of breeding systems ranging from complete range of cross-pollination to self-pollination. A strong intergenomic interaction, also accounting for the evolution of mating systems in *Brassica* genus, affects the mode of pollination and therefore accounts for the choice of breeding methods in a particular rapeseed species. Seed and oil yield and early maturity have been the primary objectives of breeding in the Indian subcontinent, while quality received greater attention in the European countries, and accordingly the breeding methods also differed. However, more recently, major breeding efforts have been shifted toward the development of "double-zero" or "double-low" cultivars throughout the globe, and the desirable genes within as well as outside the genus are being tapped for the purpose, aided by conventional and biotechnological breeding methods. For this, both interspecific and intergeneric hybridizations have a great potential for creating new variability. Induced mutagenesis has also been fruitful to create variability for earliness, compact plant type, and yellow seed color. Conventional breeding methods have received due attention by rapeseed breeders throughout the globe and a large number of varieties have been bred utilizing these methods. Lately, artificial synthesis of amphidiploids and development of herbicide tolerant cultivars has also gained momentum. The present chapter focuses on various breeding methods and issues in rapeseed related to breeding for various biotic and abiotic stresses and herbicide tolerance, male-sterility and fertility restoration, development of hybrids, synthetics and composites, quality aspects, and creation of new variability through hybridization and mutagenesis.

I. INTRODUCTION

Oilseed Brassicas include number of crop species which have an amalgam of breeding systems raging from complete cross-pollination to a high level of self-pollination. Therefore, they are quite interesting material from the breeding point of view. The different crop species of this group of crop, *Brassica campestris* var. toria, lotni brown sarson, Banarasi rai (*B. nigra*), *taramira* (*Eruca sativa*), and so on, are highly cross-pollinated (because of the presence of self-incompatibility, presence of bright yellow opened petals, entomophily, high sucrose content ranging from 40 to 60% in their nectaries to attract honeybees and the extrorse anther condition which turns away from the stigmatic surface at the time of dehiscence), whereas *B. juncea*, gobhi sarson (*B. napus*), karan rai (*B. carinata*), tora brown sarson (*B. campestris*), and so on are predominantly self-pollinated (because of the absence of self-incompatibility), light pale yellow petal color, low sucrose content (5–11) in the nectaries and introrse anther condition. However, even in the self-pollinated group, due to stray pollen contamination and visit by honeybees, bumble bees, and so on, the extent of outcrossing varies from 14 to 30% (Rai and Singh, 1976; Rakow and Woods, 1987; Rambhajan *et al.*, 1991; Singh, 1958). The self-incompatibility is of homomorphic sporophytic type (Bateman, 1955) and is genetically controlled by a series of "S" alleles. The presence of same allele in the pollen and stigma will inhibit the germination of the pollen grains or will prevent the pollen tube from penetrating the stigmatic surface of the style and effecting fertilization.

The evolution of mating system in genus *Brassica* is also very interesting. There is strong intergenomic interaction affecting the mode of pollination. The three primary, monogenomic species (*B. campestris* with AA, *B. nigra* with BB, and *B. oleracea* with CC genomes) are highly cross-pollinated while their amphidiploid products (*B. juncea*, AABB; *B. napus*, AACC; and *B. carinata*, BBCC) are predominantly self-pollinated. The commercially cultivated species *B. campestris*, however, contains both self-compatible (yellow sarson, tora brown sarson) and self-incompatible forms (toria, lotni brown sarson). In this crop species, lotni brown sarson appears to be the logical progenitor of its different cultivated forms. The evolution in this crop species has followed two independent pathways. On one hand, toria type has evolved as an escape from the onslaught of the biotic and abiotic stresses but retained its self-incompatibility gene complex. The early maturity (75–100 days) makes it a better material to survive the stresses imposed by frost injury, aphid infestation, and the threat from *Alternaria* leaf blight disease in comparison to its parental form lotni brown sarson, which usually takes 125–140 days for crop maturity and suffers heavily on account of these stresses.

There is very good morphological similarity, chromosomal homology, and cross-compatibility between toria and lotni brown sarson. The only visible difference between them is their relative number of days taken to crop maturity. On the other hand, tora brown sarson has evolved from lotni types, this is primarily because of the cultivator's preferences for the bold seeds, uniform types, and tall growing plants, which are considered very suitable for the mixed or intercropping systems being followed by the farmers. However, in the long process of human selection for uniformity, the self-incompatibility gene complex has been lost. Later, as a result of macromutation(s) in tora brown sarson, the yellow sarson types have evolved and have been retained by the farmers for better seed and quality values. However, in India these types are being replaced by the *B. juncea* types because of their better yield performance, stability of production, and comparatively better tolerance to various biotic and abiotic stresses.

Under European and Canadian conditions, both winter and summer (spring planted) forms of *B. campestris* and *B. napus* are being grown. But in *B. juncea*, only the spring form has evolved. Winter types of Gobhi-sarson (*B. napus*) are largely grown under north European, Chinese, and Canadian conditions. But because of the short crop-growing period and comparatively better winter hardiness, spring types of *B. campestris* are usually preferred and are largely grown in Sweden, Finland, and some parts of Canada and northwest China. In Indian subcontinent, the spring types of *B. juncea* and *B. campestris* cultivars are largely grown. Serious attempts are now being made to introduce *B. napus* for cultivation in northwest India but on the whole, *B. juncea* is the dominant commercial species and covers nearly 75% of the total cropped area under oilseed Brassicas.

It is, thus, clear that toria, lotni brown sarson, *taramira*, *B. nigra*, and *B. tournefortii* are highly cross-pollinated crops, and maintain very high degree of heterozygosity. Panmixis generation after generation in nature eventually frustrates the efforts of enforced inbreeding or the fixation of genotypes. Hence, in such outbreeding population, breeding superior performing cultivars with high yield would obviously require adoption of a breeding procedure which maintains the balanced heterozygosity for the optimum plant productivity. This could be accomplished through selection (mass-selection, recurrent selection, disruptive selection, and so on), breeding of synthetic and composite varieties, and ultimately by developing superior performing hybrids. On the other hand, for breeding purpose, the predominantly self-pollinated crops, such as yellow sarson, mustard (*B. juncea*), gobhi sarson (*B. napus*), *B. carinata*, and so on, should be treated as often cross-pollinated crops. The appropriate breeding procedures for this group of crops could be the pureline selection with progeny testing, pedigree selection, mass pedigree selection, backcrossing, and mutation breeding approach as are discussed in this chapter.

II. BREEDING OBJECTIVES FOR VARIETAL DEVELOPMENT

In the Indian subcontinent, genetic improvement of seed yield is the prime breeding objective, while in the western world, breeding for quality receives greater attention. In the Asian countries, centuries of rapeseed and mustard cultivation have led to the development of local land races of *B. juncea* and *B. campestris*, and these now form the basic raw material for the breeders. In these crops, high number of siliquae/plant and more number of seeds/siliquae have been observed to be important yield attributes associated with its higher yield expression and could form suitable criteria to breed for high seed yield (Anand *et al.*, 1975; Nagalakshmi, 1992; Ramanujam and Rai, 1963; Shabana *et al.*, 1990). Dry matter accumulation at rosette stage and leaf area index (LAI) have also been observed to be positively associated with seed yield (Olsson, 1990).

Early maturing varieties (80–90 days) are usually required in the Indian subcontinent for fitting in the relay-, multiple-, and intercropping systems. These are suitable for escaping frost injury and for growing in the drought-prone or dryland areas with scanty rainfall. Development of high yielding, early maturing varieties is also a major breeding objective in central China and in western Canada where frost free days in growing season are usually less than 100 days. The early maturing varieties complete their life cycle during this period and escape the frost injury.

All over the world, breeding for resistance to diseases and insect pests has become as important breeding objective. In the Indian subcontinent, *Alternaria* leaf blight, white rust, downy, and powdery mildews are the major diseases, while in the western countries, blackleg (*Leptosphaeria maculans* Desm.) is important in Canada and Australia. Some other diseases which could cause considerable economic losses to these crops are clubroot (*Plasmodiophora brassicae*), root rot (*Rhizoctonia solani* Kuhn), stem root, and so on. In some areas, *Sclerotinia sclerotiorum* (Lib.) could pose an equal or even greater threat to cultivation of Brassicas than the blackleg disease. Races of white rust (*Albugo candida*) that could attack *B. campestris* (Race 7) and *B. juncea* (Race 2) have been identified (Pidskalny and Rimmer, 1985).

European and Canadian *B. napus* cultivars are resistant to all known races of white rust, but many Chinese varieties are susceptible to Race 7 (Fan *et al.*, 1983a,b). *B. juncea* varieties possess comparatively better field tolerance to leaf blight caused by *Alternaria brassicae* than that of the *B. campestris* selection (Rai *et al.*, 1976). *B. carinata* selections have also been observed to show comparatively better tolerance to leaf blight than other *B. campestris* or *B. juncea* selections (Bansal *et al.*, 1990).

In the Indian subcontinent, mustard aphid (*Lipaphis erysimi* Kalt.), mustard sawfly (*Athalia proxima*), and leaf miner (*Bagrada cruciferarum*) are the important insect pests that cause considerable economic losses. *B. juncea* selections are reported to possess better tolerance to mustard aphid than *B. campestris* selection (Rai and Sehgal, 1975; Rai *et al.*, 1987).

In *B. campestris*, two potential sources of dwarfing genes have been reported, and it has been suggested that they could be utilized meaningfully in developing semidwarf cultivars of toria and sarson for cultivation under high population densities for obtaining high seed yield (Rai and Kumar, 1978; Rai and Singh, 1993; Tyagi *et al.*, 1983).

The comparatively better salt tolerance of *B. juncea* than of *B. campestris* has made it a better choice for its cultivation under the salt-affected soils of northwestern Indian states. At Indian Agricultural Research Institute, New Delhi, a number of promising somaclones have been spotted which could give better performance than the local varieties for cultivation under such conditions.

The relative ability of the spring rapeseed cultivars to withstand the onslaught of frost at flowering time is considered important in northwest India and Sweden. There is a variable reaction of the Indian and Swedish cultivars for the frost tolerance (Aberg, 1984). Several biotechnology groups are now working to transfer genes for tolerance to glyphosate, chlorosulfuron, and other herbicides into the agronomic background of the various oilseed *Brassica* varieties.

In Europe and Canada, breeding for oil and cake better suited to human nutrition, and livestock feeding has received higher research priority than anywhere in the Asian countries. While a high erucic acid rape oil is liked by industry, zero or low glucosinolate (00) oil is usually required for the human consumption. The rapeseed oil with zero erucic acid content is more or less parallel to groundnut or sesamum oil in its fatty acid composition. Consumers in east Indian states usually prefer mustard oil with pungency for frying fish or preparing pickles, while those in the west Indian states (Haryana, Punjab, Rajasthan, and so on) prefer oil with low pungency (Rai, 1976). Now the development of "00" or canola quality varieties is being taken up in China, India, and Pakistan.

A. GENETIC RESOURCES

The success of any crop breeding program normally depends on the extent of favorable genes available in the genetic stocks handled by the breeders. At international level, IBPGR collects, maintains, and handles the genetic diversity of a number of agrihorticultural crops. In India, National Bureau of Plant Genetic Resources (NBPGR) conserves about 19,600 accessions of different oilseed crops including 4584 of the oilseed Brassicas and its wild allies. These are now being conserved under long-term storage in gene banks for its possible use in future (Singh and Rana, 1994). During 1993–1994, total of 132 accessions of oilseed Brassicas and its wild allies species were introduced and total 365 accessions were grown and characterized with 22 descriptor attributes at NBPGR, New Delhi. At present, 4584 entries of oilseed Brassicas are being maintained under All-India Rapeseed-Mustard Research Project, state Agriculture Universities, and institutes for its possible usages at various oilseed Brassicas breeding research centers in India and abroad. Under field glass house conditions, these genetic stocks are being maintained by either sib-mating or selfing. Under field conditions, an isolation distance of 400 m is required to be maintained. In the insect proof cages, glass house chambers, sib-mating of the culture is affected by

TABLE I
Some Sources of Donors for Important Agronomic Attributes in Rapeseed

Agronomic attributes	*B. campestris*/*B. napus*
Earliness	TS 29, Bhawani, TL 15
Tolerance to	*B. napus*: GSL 1, Gullivar
Aphid	*B. hirta*: PI 751516
Leaf miner	*B. campestris*: Chamba 1,2
Mustard sawfly	PT 303
Tolerance to	
Alternaria leaf blight	*B. napus*: Karat
White rust	Newton, Blaze, Metapolka
Downy mildew	GSB 7027
Resistance to frost injury	Span. Torch, Candle, FR 80, DBS 1
Sources of zero erucic acid	*B. napus*: Tower, Altax, Global, Wester
	B. campestris: Torch, Tobin, Span, Candle
High oil content	BSH-1, TK 101, PT 5, PZT 25,48
Yellow seed coat	*B. campestris*: Tobin, Bele, Torch

introducing honeybee. Pure stocks of the self-incompatible inbred lines have to be maintained by bud-pollination and by selfing under muslin-cloth bags in self-compatible lines. Over years, a number of genetic stocks have been identified for desirable agronomical attributes like earliness, tolerance or resistance to diseases and insect pests, shattering, frost tolerance, and so on (Table I). Some of these genetic stocks are now being utilized in crossing programs in India in intervarietal and interspecific crosses to create new genetic variability, and some are being utilized as base population for selection work. The exotic cultivars have so far not been used for direct commercial cultivation in India because of their late maturity and low yield.

B. SOURCES OF CREATING NEW GENETIC VARIABILITY

In rapeseed, hybridization is accomplished by emasculating the flower buds that are due to open the following day. Next day, stigmata of emasculated buds are dusted with the freshly dehisced pollen from the stamen of selected plant. Under storage conditions, pollen viability has been observed to last up to 35 days (Chiang, 1974). In oilseed Brassicas, because of the cross-pollinating nature of the primary species, enough variability is available, but for searching new desired genes or gene complexes for resistance to diseases, insect pests, male-sterility, fertility restoration, and so on, it requires to resort to purposeful intervarietal or distant hybridizations. Intraspecific crosses (i.e., in case of *B. campestris*; crosses between tora, lotni brown sarson, tora brown sarson, and yellow sarson cultivars) are much successful and the success rate for such crosses, if carefully made, is greater than 90% and a single emasculated and pollinated bud may yield 10–20 crossed seeds per siliquae. However, the success rate of interspecific hybridization depends much on the genetic relationship, genomic constitution of parental species used, and also on the direction of cross.

In general, the interspecific hybridization is more successful, if an amphidiploid species (*B. juncea*, *B. napus*, or *B. carinata*) is used as the female parent, which has one genome in common with the pollen parent. Hybrids between monogenomic primary species are rather more difficult to be obtained with success rate of 0.002 and 0.03 hybrids per pollinated flower (Downey *et al.*, 1980; Mahapatra and Bajaj, 1987; Quazi, 1988).

The basic understanding of crossability relationship among the oilseed *Brassica* species is important to the breeders of these crops because there are good possibilities of transferring agronomically important attributes like diseases and insect pest resistance, cytoplasmic male sterility (CMS), fertility restoration, desirable quality attributes, and so on. Rao (1990) observed that out of six possible combinations between *B. juncea*, *B. napus*, and *B. carinata* including reciprocals, the *B. juncea* × *B. napus* cross was easier to be made. *B. juncea* × *B. napus* hybrid plants were observed to be more vigorous than their reciprocal. Good success was also obtained in *B. napus* × *B. carinata*, but the hybridization between *B. carinata* × *B. juncea* in either direction was rather difficult, primarily because of their nonsynchrony of flowering rather than any of its crossability barriers. In elaborate fraction I protein analysis and restriction pattern of chloroplast DNA studies, *B. nigra* and *B. campestris* have been identified as the female parental genomes in *B. carinata* and *B. juncea*, respectively (Uchimiya and Wildman, 1978), and *B. oleracea* as the female parent of *B. napus* (Ichikawa and Hirai, 1983; Prakash and Chopra, 1991; Raut and Prakash, 1985).

It has now been possible to transfer blackleg resistant genes from *B. juncea* to *B. napus* because of possible recombination between A and C genomes in *B. juncea* crosses and A and B genomes in *B. carinata* crosses (Sacristan and Gerdemann, 1986). A line completely resistant to blackleg disease caused by *L. maculans* was selected from the F_3 progenies of cross *B. juncea* × *B. napus* (Roy, 1984). Resistance to *P. brassicae* has also been transferred from *B. napus* to *B. oleracea* (Chiang and Crete, 1983). Same resistance has also been transferred to *B. napus* from *B. campestris* by subsequent backcrossing with *B. napus*. When a white rust resistant line of *B. carinata* was crossed with that of *B. juncea*, the F_1 was observed to be resistant to white rust with some additional resistance to *A. brassicae*. This has shown the possibility of transferring white rust resistance from *B. carinata* to *B. juncea* (Singh and Singh, 1987, 1988). The Swedish rapeseed cultivar 821 has been developed from the cross *B. napus* × *B. chinensis* (He *et al.*, 1987). The triazine resistance has been transferred from *B. napus* to *B. oleracea* (Ayotte *et al.*, 1986, 1987). CMS has been transferred from *B. juncea* to *B. napus* through interspecific hybridization followed by four generations of blackcrossing (Mathias, 1985).

The CMS was transferred from radish to *B. oleracea* (Bannerot *et al.*, 1974; McCollum, 1988). The fertility restorer genes for Polima type CMS system of *B. napus* have been found in the *B. juncea* var. Zem (Fan Tai, 1985). Genes for earliness have also been introgressed from *B. juncea* and *B. carinata* to *B. napus* varieties. The genes for high linoleic acid have been transferred from *B. juncea* to *B. napus* through selection in F_2 generation (Roy and Tarr, 1985, 1986). There are good possibilities of incorporating shattering resistance from *B. juncea* and *B. carinata* to *B. napus* cultivars (Prakash and Chopra, 1988; Rao, 1990). Wide hybridization has been reported with some degree of success in the crosses of *B. spinescens* ($2n = 16$) × *B. campestris* ($2n = 20$) and for the production of *B. napus* × *Raphanobrassica* hybrids (Agnihotri *et al.*, 1990a,b,c) by embryo rescue and ovary culture techniques. Protoplast fusion has helped in obtaining somatic hybrids of *B. oleracea* with *Moricandia arvensis* which possess intermediates C3–C4 photosynthesis carbon metabolism (Toriyama *et al.*, 1987).

From above-mentioned examples, it is clear that both interspecific and intergeneric hybridizations have much potential for creating new variability for rapeseed improvement. The fact is that the available natural variability of oilseed *Brassica* Landraces/germplasm has not yet been fully tapped and exploited with a few exceptions. If we search systematically, the needed characteristics could be found within the species of interest from close relatives. Few examples include spotting of the needed early maturing selections of *B. napus* in India and Canada, vary widely in seed size, oil content in *B. juncea* in India, resistance to white rust, and blackleg in *B. napus* and *B. campestris* in France and Australia. But wherever usable variability is not available in the working germplasm, the induced mutagenesis could as well be explored and utilized.

C. INDUCED MUTAGENESIS FOR CREATING NEW VARIABILITY

Induced mutagenesis is a useful tool for creating new variability hitherto not available, and a number of studies utilizing ionizing radiations (X-rays and cobalt 60) and chemicals such as ethylemethane sulfonate (EMS) have been used. Usually 60- to 80-Kr doses of γ-rays are quite effective. Induced mutagenesis has been used to obtain mutant lines with 3% linolenic acid in *B. napus* (Rakow, 1973; Robbelen and Nitsch, 1974), for spotting seed color mutant in mustard (Verma and Rai, 1980a), and for tolerance to leaf spot disease (Verma and Rai, 1980b). It has also been very helpful in developing a number of rapeseed varieties in Sweden. Induced mutagenesis has also been used to create new variability for earliness, compact plant type, and yellow seed color in mustard at Bhaba Atomic Research Centre, Trombay, India. Two high yielding lines of mustard (TM2 and T4) have emanated from this program.

III. BREEDING METHODS

A. SELECTION IN GENETIC IMPROVEMENT FOR YIELD IN CROSS- AND SELF-POLLINATED OILSEED BRASSICAS

Selection procedures in *B. campestris* var. toria vary from mass selection to recurrent selection and in predominantly self-pollinated ones, the desirable plants are visually selected from genetically variable broad-based population, viz., landraces, segregating populations, germplasm complexes, gene pools, and so on, and bulked. The bulked seed is grown cycle after cycle. One cycle of mass selection in toria has given a yield improvement of 8.2% (Chaubey, 1979). In India, the toria commercial varieties Type 9, ITSA, Type 36, PT 30, BK 23, M 27, TS-29; in lotni brown sarson, BS-2 BS 70, BSH-1; and in *taramira*, Type 27 have all emanated as a result of mass selection either from the local landraces or from the genetically variable germplasm entries. Disruptive selection has been utilized to develop Suphala cultivar from the local landrace, Gurgaon brown sarson (GBS).

Segregating populations or the progenies from the crosses could also make good base population for initiating recurrent selection program. In this method, the desirable individual open-pollinated plants (around 3000) are harvested and threshed separately. A part of their seed is saved and a part of it is planted in a progeny row, evaluated visually, and superior rows are selected, tagged, and harvested separately. After harvesting and threshing, the seeds are analyzed for their 1000 seed weight, oil content, glucosinolates, protein content, and so on. Thereafter, equal quantities of the reserved seeds from the selected plant are composited. This way, the first cycle of recurrent selection is completed and this composited seed is grown again in field in isolation where intercrossing takes place among the plants within the composited populations. The second cycle of recurrent selection starts with the harvesting of the single plants (around 1000) from this population. A bulk seed sample is harvested from the remaining plants of the population for use in replicated yield trial to determine response to selection in each recurrent cycle for character under improvement, viz., oil content, seed yield, or tolerance to a disease. Recurrent cycle selection is continued till reasonable level of improvement is achieved.

Pureline selection usually followed in *B. napus*, *B campestris* var. yellow and brown sarson involves the isolation of superior performing lines, from a genetically broad-based population based on their progeny performance.

B. HYBRIDIZATION AND PEDIGREE SELECTION

Hybridization aims at recombining a number of desirable attributes from different parents in a single commercial variety. In oilseed rapes, it may be effectively utilized for concentrating favorable genes for high yield, high oil content, disease and insect pest tolerance, and so on. When the desirable gene is available from unadapted or wild population, backcrossing would be the right procedure, but if the favorable gene is available from an adapted or cultivated material's background, then pedigree method of selection would be the most appropriate procedure. The Indian mustard variety "Prakash" was developed by following pedigree selection in the cross "RL 18 × T9" and "Pusa Bold" was evolved from the progenies of cross "Varuna × BIC 1780." In this method, 5–10 F_1 plants are grown to obtain F_2 seed and 1000–3000 F_2 plants are grown and harvested individually, from which F_3 progenies are raised. In F_4 generation, selection is practiced. The variation among F_4 families is good indication for effectiveness of further selection. This method has been utilized to develop a low erucic acid, high yielding, and winter hardy *B. napus* variety from a cross between the high erucic acid winter *B. napus* var. "Rapol" and the low erucic acid spring *B. napus* var. "Oro."

In *B. napus*, backcross technique could effectively be utilized in transferring zero erucic acid and zero glucosinolate content or the cytoplasmic male-sterility attribute in the agronomic background of the commercial cultivars. This approach has been used to transfer the low glucosinolate content of *B. napus* var. "Bronowski" into a number of commercial cultivars of gobhi sarson.

The development of spring *B. napus* var. "Westar" is good example of the combination of both backcrossing and pedigree selection. Once around in 1985, this variety was very popular and occupied more than 90% of the total western Canadian rapeseed production acreage.

In India, the first national *B. campestris* var. toria, PT 303, is a product of the intercrossing between B54 × DS 17 M (Rai, 1987). In Haryana, hybridization pedigree selection has been used to develop promising selection RH 7846 and RH 8113, which apart from giving very high yield are also resistant to mustard aphid and *Alternaria* leaf blight disease.

C. DEVELOPMENT OF COMMERCIAL HYBRIDS

Development of commercial hybrids is the ultimate goal of breeding self-pollinated crops to exploit the nonadditive gene action present in yield heterosis. The basic requirement for developing commercial hybrids in crops like rapeseed is the availability of proven experimental hybrids (preferably with more than 20% standard yield heterosis), stable performing male sterile (A), maintainer (B), and fertility restoring (R) lines, good synchrony of flowering in seed and pollen parent, and adequate seed-setting on male-sterile seed parent through natural cross-pollination. High level of heterosis for seed yield in both spring and winter forms of *B. napus*, that is, quantitatively, 40% heterosis for yield has been reported in summer rape and 60–70% for winter form (Grant and Beversdorf, 1985; Lefort-Buson and Datte, 1982; Sernyk and Stefansson, 1983). In India, 10–72% parent yield heterosis has been reported in gobhi sarson, 20–107% in *B. campestris* (Das and Rai, 1972; Doloi, 1977; Labana *et al.*, 1975; Srivastava and Rai, 1993; Yadava *et al.*, 1974), and a range of 14–30% natural outcrossing is usually observed in these crops. So, this is sufficient to justify efforts to develop CMS lines and search for usable fertility restorer lines for producing the hybrids.

In oilseed Brassicas, a number of CMS sources, viz., *B. carinata* CMS, *B. juncea* CMS, *B. oxyrrhina* CMS, *B. tournefortii* CMS, *Raphanus*-based Ogura CMS, *B. napus*-based *Polima* CMS, *Settiana* CMS, *Siifolia* CMS, and so on, are now well known and some of them are being worked with rather intensively. Out of these CMS sources, fertility restoration has been identified in *Raphanus*-based Ogura CMS and Polima CMS in the western countries, and it has been detected in the CMS-based crosses in *B. tournefortii*, *B. juncea* CMS, Polima CMS, and S*iifolia* CMS in India. During 1992–1993 and 1993–1994, a total of 784 and 3292 CMS F_1 crosses were made, out of which, 54 complete and 54 partial fertility restoring crosses have been detected. The fertility restoration in *tournefortii* CMS has been observed to be female genotype specific and photoperiod dependent. Complete fertility under Ludhiana (India) conditions is retained till mid-February in short-day conditions but fertility reversal starts after that under long-day conditions.

In Ogura type *Raphanus*-based CMS, the European variety could restore fertility. Bannerot *et al.* (1977) introduced the nuclear genes of *B. napus* into *Raphanus* cytoplasm to develop *B. napus* CMS source, which is stable under a wide range of environments but restorer genes have not yet been identified. The Polima CMS is most promising CMS in *B. napus*. Restorer genes are now available and fertility restoration in the hybrids is satisfactory. The first commercial hybrids of gobhi sarson based on Polima CMS were tried in Canadian trials in 1986 under north Indian conditions. The level of pollen fertility in two gobhi sarson CMS hybrids, that is hybrid 10 and hybrid 2, has been observed to be more than 90% (Singh, 1986). The Punjab Agricultural University, Ludhiana in India recommended to release first CMS-based gobhi sarson hybrid "PGSH-51" for its cultivation in Punjab in India. This hybrid has given 21.2% better yield over the check variety GSL-I. The actual seed yield of this hybrid obtained in farmer's field was 19.1 q/ha as compared to 16.4 q/ha obtained from the check, GSL-I (Banga, 1994). Under experimental conditions at Ludhiana, the seed setting on the CMS lines of *Siifolia* has been observed comparatively better than that observed on B. carinata CMS or on *B. tournefortii*-based CMS lines. The average yield depression observed on growing the advanced generation seeds of 25 CMS-based hybrids of *B. juncea* at Banaras Hindu University, Varanasi was 12% with a range of 5.2–24.6%. In the experiment rapeseed hybrids, the average yield depression was 14.7% and the range varied from 3.4 to 25.6%.

After the successful release of gobhi sarson hybrid in India, much stress is now laid on the development of commercial hybrids in mustard (*B. juncea*). Initial efforts of testing four experimental hybrids in mustard during 1993–1994 have not met with much of success as the two experimental hybrids (PHR-16 and FMH-3) could give only a marginal increase in yield (4.5–6.5%) advantage over the national check, Kranti. In China also, a number of hybrids have been developed such as Zhong You Zha 93–1, yielding about 10% higher than the double-zero cultivars. Two new hybrids, viz., Hyola 42 (early maturing) and Hyola 51 (late maturing) were developed in Australia (Anonymous, 1999).

D. COMMERCIAL HYBRIDS IN *B. CAMPESTRIS*

High level of exploitable yield heterosis has been reported in *B. campestris* hybrids (Das and Rai, 1972; Hutchenson *et al.*, 1981). In this species, CMS system has been developed by backcrossing *B. campestris* cv. "Yukina" into the *Diplotaxis muralis* cytoplasm (Hinata and Konno, 1979). The *B. campestris* "yukina" CMS was stable and restorer genes have been identified for this CMS source but the genes for maintenance of CMS will have to be transferred into the background of the adapted commercial cultivars of *B. campestris* before the hybrids could be put to test in *B. campestris*. In china, where adequate labor is available, gobhi sarson hybrids have been produced utilizing this type of male sterility (Lee and Zhang, 1983).

E. SELF-INCOMPATIBILITY AND HYBRID SEED PRODUCTION

On the basis of sporophytic type of self-incompatibility, a theoretical model of triple cross [(A × B) × C]×[(D × E) × F] technique has been suggested by Thompson (1964) to exploit heterosis and produce commercial hybrid in these crops. However, practically it has not been put to commercial use. The difficulties of producing commercial quantities of selfed seeds of self-incompatible parental lines of the hybrid and also difficulties in detecting the breakdown of self-incompatibility in the production plots during flowering duration make it rather economically unviable and unprofitable. Self-incompatibility is a good outbreeding mechanism in nature, but unfortunately, due to very high self-incompatibility, heterozygosity, and as a result of high inbreeding depression, it frustrates the efforts to produce and maintain the homozygous lines which could produce the hybrid cultivars. It is a difficult task to maintain the inbred lines through continued selfing, primarily because of big loss of vigor of the inbred population to grow and produce seeds.

F. SYNTHETIC AND COMPOSITE VARIETIES

Where production of hybrids is not an economically viable preposition and there are considerable difficulties in producing the hybrid seed, utilization of synthetic and composites could fit in and serve the purpose. In the Indian subcontinent, development of composite varieties is being viewed as a possible way out for stepping up the average yield level of oilseed Brassicas. As these crops are largely grown under the rainfed conditions, these are subjected to all sorts of biotic and abiotic stresses (Rai, 1979). However, in Canada efforts to develop synthetics in spring *B. napus* were not very successful for commercial utilization.

The composite breeding program in toria or other cross-pollinated *Brassica* species usually involves the production of number of intervarietal hybrids or by making their blends. This is followed by evaluation of inbreeding depression in seed yield from F_2 and later generations and evaluation of the performance of the experimental composites against the ruling checks (Rai, 1982).

The evaluation of genetic diversity in relation to expression of heterosis in *B. campestris* has shown that intercrossing between toria and sarson varieties would be better than crossing among the toria and toria or sarson varieties. However, under Indian conditions, if one has to choose the parental population based on their geographical distribution, then crossing between East Indian toria cultures with that of the exotic (e.g., Canadian) spring *B. campestris* cultivars would be better and more productive than crossing among the Indian toria varieties themselves (Agrawal, 1976). In India, two commercial composites, viz., PT 303 from Pantnagar and Toria Ludhiana composite (TLC-1) from Ludhiana have been released for commercial cultivation. TLC-1 has been developed by compositing the seeds of ten toria populations by following simple recurrent selection in its advanced generation.

Development of synthetic varieties requires the development of inbred lines, their testing for general combining ability, making their all possible cross combinations, predicting the F_2 performance constituting a number of experimental synthetics, testing their yield levels in yield trials over locations, and finally releasing those which excel the standard checks. In India, one of the experimental synthetic populations "Syn 65" has been released for cultivation in the name of cv. "Sangam." In lotni brown sarson (*B. campestris*), Pusa Kalyani has been developed and released as cultivar from the IARI research station Kanpur (UP) utilizing this breeding approach.

G. ARTIFICIAL SYNTHESIS OF AMPHIDIPLOIDS FOR COMMERCIAL CULTIVATION

The commonest choice for direct interspecific hybridization is to double the chromosome number in the sterile hybrids and to establish fertile amphidiploids. This provides stability and could help in the preservation of gene complexes of both the component species by enhancing the preferential pairing of the homologous chromosomes. Artificial synthesis of some of the commercially cultivated amphidiploid species, viz., *B. juncea*, *B. napus*, and *B. carinata* has been reported long back (Frandsen, 1943; Ramanujam and Srinivasachar, 1943; U, 1935). Artificial synthesis of *B. juncea* and *B. carinata* is comparatively easier than *B. napus*. Synthesis of *B. napus* has added to the usable variability for use in India. Undoubtedly, the derivatives of synthetic *B. juncea*, *B. napus*, and *B. carinata* could provide additional material for widening the range of genetic variability. It is now possible to use some of these resynthesized digenomic strains in interspecific hybridization via backcross to obtain useful genotypes for commercial production programs. At IARI, New Delhi, Raut and Kaul (1982), Raut and Prakash (1985), and Prakash and Raut (1983) have synthesized early maturing and productive amphidiploids of *B. napus* by crossing early indigenous strain of *B. campestris* var. lotni brown sarson with that of *B. oleracea* var. botrytis. The selections obtained are now being field tested for their comparative yield performance with commercial check varieties.

The efforts to produce and improve tetroploids of toria variety T22 (Rajan, 1955; Sikka and Rajan, 1957) and various yellow sarson strains for genetic improvement of seed and oil yield has not met with much of success, though the induced tetraploidy has been of some use in developing high-yielding fodder varieties of Brassicas in west European countries.

H. DEVELOPMENT OF HERBICIDE-TOLERANT CULTIVARS

Herbicides provide an inexpensive and effective means of control of weeds in crop *Brassica*. Development of herbicide resistant cultivar in *Brassica* was started in 1960. The tolerance was cytoplasmically controlled and effective against S-triazine family of herbicide.

Identification of triazine-tolerant biotype of bird's rape mustard led to the development of triazine-tolerant *B. napus* oilseed cultivars through introgression of the tolerant weed biotype cytoplasm in oilseed rape (Fig. 1). Through an interspecific cross and backcross program, the tolerant cytoplasm of *B. campestris* was combined with the nucleus of *B. napus* to produce the first triazine-tolerant cultivar, OAC Trinton (Beversdorf *et al.*, 1980).

Triazine-tolerant *B. napus* cultivars are very useful and indeed essential in fields, where highly competitive weeds such as wild mustard (*S. arvensis* L.), stickweed (*Thlaspi arvense* L.), and quack grass (*Agropyron repens* L.) are found, suggesting that the growth rate and yield of triazine-tolerant cultivars will always be significant than that of recurrent parent (Downey and Rimmer, 1993) (Table II).

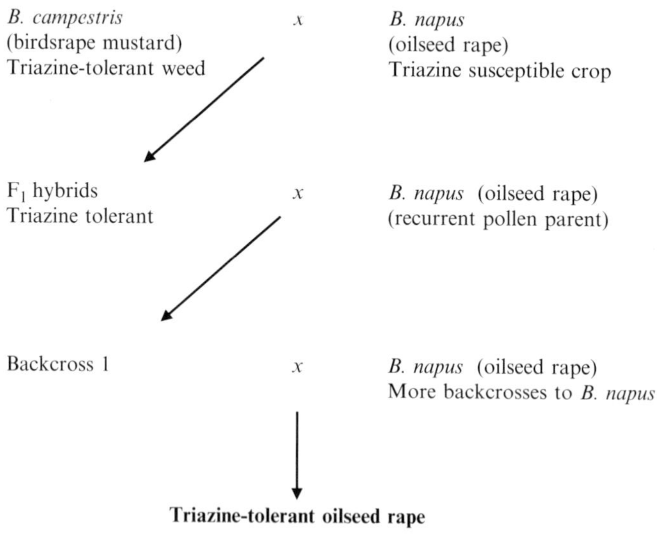

Fig. 1. Schematic diagram of transfer of the triazine-tolerant cytoplasm from a weed biotype to oilseed rape by backcrossing (Beversdorf and Kott, 1987).

TABLE II

Comparative Agronomic Performance of Triazine-Tolerant and Susceptible Cytoplasms in Identical Nuclear Backgrounds (Reciprocal Single-Cross Hybrids) of Oilseed Rape[a]

Genotypes	Yield (t/ha)	Flowering (dap)[b]	Maturity (dap)	Lodging (1–5)[c]	Height (cm)
OAC Triton	2.3d[c]	51ab	94a	3.3bc	112a
Topas	2.6bcd	53bc	96a	1.0a	126bc
Hanna	2.8bc	55cd	96a	1.4ab	127bc
Triton × Topas[d]	2.2d	56cd	95a	1.0a	124b
Topas × Triton[d]	3.0ab	55cd	95a	1.0a	131bc
Triton × Hanna[d]	2.4cd	56cd	95a	2.1bc	114a
Hanna × Triton[d]	3.3a	54bcd	96a	1.5ab	137c

[a]Numbers within columns followed by the same letter are not different at the $p < 0.05$ level according to Duncan's multiple range test.
[b]Days after planting.
[c]Linear scale from erect (1) to prostrate (5).
[d]Female parent (cytoplasmic donor) is designated first.
Source: Beversdorf and Kott (1987).

Rapeseed varieties that are tolerant to herbicides, such as imidazoline, glufosinate, and glyphosate, are now available commercially in the United States and Canada (Neal Stewart, personal communication). Other examples of herbicide resistance include sulfonylurea resistance in *B. napus* (Blackshaw *et al.*, 1994), and bromoxynil resistance in *B. napus* (Zhong *et al.*, 1997).

I. DEVELOPMENT OF *ALTERNARIA* BLIGHT AND APHID-RESISTANT CULTIVARS

Breeding for resistance to diseases and insect pests has now become an important breeding objective. In the Indian subcontinent, *Alternaria* leaf blight, white rust, and downy and powdery mildews are the major diseases in rapeseed, while in the Western countries, blackleg (*L. maculans* dasm.) is important in Canada and Australia. *Alternaria* blight, also known as pod spot is an important disease and is caused by *A. brassicae* (Berk.) Sacc. Yield losses may range up to 70% varying from location to location and year to year. No resistance has been reported against this disease in oilseed Brassicas. However, some of the interspecific crosses have been attempted between *B. napus* and *B. alba* with an objective to transfer *Alternaria* resistance from *B. alba* to *B. napus* (Brim *et al.*, 1987; Chevre *et al.*, 1991; Dueck and Degenhardt, 1975; Rai, 1976).

Tewari (1991) have shown that more distantly related Crucifereae species may be very resistant to black spot. Some other diseases, which could cause considerable economic losses to these crops, are clubroot (*P. brassicae*), root rot (*R. solani* Kuhn), stem rot, and so on. In some areas, *S. sclerotiorum* (Lib.) could pose an equal or even greater threat to the cultivation of Brassicas than blackleg disease. Races of white rust (*A. candida*) that could attack *B. campestris* (Race 7) and *B. juncea* (Race 2) have been identified (Pidskalny and Rimmer, 1985).

Europian and Canadian *B. napus* cultivars are resistant to all known races of white rust but many Chinese varieties are susceptible to Race 7 (Fan *et al.*, 1983a,b). *B. juncea* varieties possess comparatively better field tolerance to *A. brassiceae*-caused leaf blight than that of the *B. campestris* selections (Rai, 1976). *B. carinata* selections have also been observed to show comparatively better tolerance than other *B. campestris* or *B. juncea* selections (Bansal *et al.*, 1990).

In the Indian subcontinent, mustard aphid (*L. erysimi* Kalt.), mustard sawfly (*Sathalia proxima*), and leaf minor (*B. cruciferarum*) are the important insect pests that affect and cause economic losses. *B. juncea* selections possess better tolerance to mustard aphid than *B. campestris* selections (Rai and Sehgal, 1975; Rai *et al.*, 1987).

J. BREEDING FOR DESIRABLE OIL AND MEAL QUALITY

The most valuable product of the oilseed Brassicas is the oil which has cash value of two and three times that of the high-protein oil meal obtained after the extraction of the oil from seed. Usually, there is a negative genetic association (−0.80) between oil and the protein content. Yellow seed color is usually preferred in oilseed Brassicas because it has significantly thinner seed coat and less seed fiber than brown seed coat. So, in market, yellow seeded varieties are preferred. In *Brassica* breeding program, yellow seed coat is considered a desirable attribute; however, in the marketing circles, the fatty acid composition of the oil usually determines its value for edible or industrial usages. In India, the oil obtained from the commercial cultivars now being grown are usually characterized by high amount of erucic acid, while in western Europe, Canada, and Australia, the cultivars now being grown are either "zero erucic acid ones" or contain very low (less than 2%) erucic acid in their oil. The glucosinolate in these varieties has been retained in nature because it provides protection to the plants against diseases and insect pests. But the feeding value of rapeseed cake becomes very limited due to its presence in the seed meal and is considered undesirable from the livestock feeding point of view. In high concentrations, in nonruminants like swine and poultry, it hydrolyzes to form thiocynates, isothiocynates, or nitriles and can adversely affect iodine uptake by the thyroid gland and can reduce their weight gains (Fenwick *et al.*, 1983).

A high amount of erucic acid is, however, desirable for certain industrial usages. In Sweden, a high erucic white mustard (*Sinapis alba*) cv. "mustang" has been released (Persson, 1986). The erucic acid content of this variety is 54% as compared to 40% of the old Swedish var. "Trico." The available variability of erucic acid in the rapeseed oil has been observed to vary between 1% and 60% (Krzymanski and Downey, 1969). From human consumption point of view, usually zero or low erucic acid oil is better and is required to be commercially produced. So, when for the first time, the nutrition experts questioned the validity of the nutritional value of erucic acid in the rapeseed oil for human food, its genetic elimination from the fatty acid profile of the rapeseed oil was recognized and it formed an important breeding objective in the western world. Since then, some systematic and meaningful work began in this direction. The development of gas liquid chromatography (GLC) (Craig and Murphy, 1959) and this technique which used only half of the cotyledon to test the erucic acid content provided the quick means for screening vary large populations necessary to identify genetically changed plants with low or zero erucic acid level. With this available technique, the selected plants with half cotyledon intact were grown and were carried forward and rape plants with seeds free from erucic acid were isolated (Downey and Harvey, 1963; Stefansson *et al.*, 1961). Utilizing this technique, a number of cultivars containing zero or low erucic acid content have been bred and released for commercial cultivation in Canada, Sweden, France, and West-Germany, and the farmers in these countries are taking advantage of this technology. To provide better oil and to meet the export needs in India, due emphasis is now being laid to develop canola quality rapeseed varieties. In the major Asian mustard-producing countries like China, India, and Pakistan, breeding for 00 quality has now received an important research priority. Some useful work is now under way and it is hoped that it could be possible to develop such cultivars for commercial production in these Asian countries in coming years.

Major fatty acid in rapeseed follows a biosynthetic pathway similar to other oilseeds. The rapeseed oil is distinct from the groundnut and safflower oil in the sense that from the oleic stage, it has also attached branch to synthesize ecosenoic and erucic acid. In fatty acid evolutionary process, it is usually considered to be a primitive type of oil biosynthesis because it still retains and maintains the ability to synthesize the long-chain fatty acids (20:1, 22:1) and has very high concentration of these unsaturated fatty acids. In zero erucic acid varieties, with the loss of chain of ecosenoic and erucic acids either due to mutation or selection, the oleic and linoleic acid contents get increased. In *B. campestris*, an increase from 26.6 to 54.8% for oleic and 17.5–31.1% for linoleic acid was observed. After the elimination of erucic acid, the fatty acid profile of rapeseed oil becomes more or less parallel to that of either groundnut or the sesamum oil and is considered to be better oil from human nutrition point of view. In the genetic improvement of the fatty acid profile for human consumption, efforts have also been made to reduce the percentage of linolenic acid from 8 to 10% to less than 3%. In Canada, utilizing these mutants with low linolenic acid content, it has been possible to develop cultivars with less than 3% linolenic acid combined with linoleic acid contents of more than 22%. Development of near zero lines of *B. napus* has also been attempted (Roy and Tarr, 1986).

As indicated before, for making rapeseed-mustard meal better suited and amenable for the poultry and swine feed stocks, it is necessary that the glucosinolate content is either reduced or altogether eliminated from the meal through appropriate breeding techniques. With the development of high performance quantitative GLC technique (Brazezinski *et al.*, 1986; McGregor *et al.*, 1983; Sang and Truscott, 1984; Spinks *et al.*, 1984), now it is not only possible to obtain the profiles of glucosinolates but also feasible to measure their absolute quantitative levels (McGregor and Downey, 1975). It has also been observed that the genes controlling glucosinolate content in rapeseed are either pleiotropic or in linkage with the seed filling stage and have a positive correlation with 1000 seed weight (Oliveri and Parrini, 1986). It has also been observed that genotypes with high seed yield could be lost, if very strict selection is followed for the quality characters in the early segregating generations. So, it would be better to keep population heterozygous for quality characters and select plants for these characters in advanced generations (Yongming Zhou and Houlilin, 1991).

The first zero glucosinolate and zero erucic acid spring *B. napus* cultivar was released for commercial cultivation in Canada in 1974, and since then a number of canola quality varieties have been bred and released for their cultivation. Some of early generation varieties were Tower, Regent, Andor, Westar in *B. napus*, and Candle and Torch in *B. campestris*. The canola cultivars of *B. campestris* and *B. napus* must contain less than 2% erucic acid and less than 30 μmol equivalent of the butenyl isothiocyanate per g of moisture free, oil free meal. To accomplish this limit for commercial production, breeding trial targets are aimed at 0.5% erucic acid in the oil and less than 10 μmol/g of meal in the breeding lines/genotypes.

At present, the breeding efforts in development of canola quality double-low *B. napus* cultivars by improving the oil composition (developing high oleic, low linolenic, low saturated fat cultivars) and enhancing vitamin levels are underway in many countries of the world including Germany (Luhs *et al.*, 2003), Canada (Raney *et al.*, 2003a,b), United States (Corbett and Sernyk, 2003),

Australia (Gororo *et al.*, 2003), France (Carre *et al.*, 2003), and Poland (Spasibionek *et al.*, 2003).

IV. CONCLUSION

The current trends in the rapeseed breeding research indicate that to maintain the tempo of progress in quality and yield improvement work in these crops, much expanded research efforts would be needed to solve the emerging and challenging problems ahead. Now, there would be far greater need for the collection, computerization, and creation of new usable genetic variability, greater application of cytoplasmic male-sterility techniques, chromosomal mapping studies, and assimilation of many new and novel ideas for tackling new problems. It would also be desirable to make broad-based gene pools of different cross-pollinated species of oilseed Brassicas and maintain at least one gene pool having all the available collection of the group. Such a population is likely to have more natural recombination hitherto not available in nature and provide an opportunity to break some of the existing undesirable genetic linkage, and provide good base population for future recurrent selection programs.

Exploitation of heterosis in oilseed Brassicas would require more intensive and concerted efforts for effective utilization of cytoplasmic male sterility. In years to come, meaningful basic work would also be needed on the stability of sterility in CMS lines, understanding of the mechanism of fertility-restoration, extent of cytoplasmic-penality that would normally be expected on using CMS lines from a very distant wild types/species of genera and on perfectization of hybrid seed production techniques. In the Indian subcontinent, some of recently introduced materials of the disease-resistant or canola quality lines from outside may not be high yielders per se under the local condition, but they could possibly make good parental lines for production of hybrids and if that possibility exists, it should be explored. The production of doubled-haploids lines, either by microspore or anther culture techniques, is now possible to rapidly produce homozygous inbred lines from the promising *B. napus* and *B. campestris* genotypes as well. Such inbreds could be produced and used to develop more productive hybrid cultivars. The inputs available from biotechnology may more purposefully be utilized in solving the pressing problems of male sterility, fertility restoration, crossability, and oil and seed meal quality.

An exciting and challenging area of rapeseed breeding research would be to develop cultivars with built-in genetic resistance to devastating insect pests like mustard aphids and *Alternaria* leaf blight disease in the Indian subcontinent and for the important diseases like blackleg (*L. maculans* Desm.), clubroot (*P. brassicae*), *S. sclerotiorum*-caused damage, and white rust diseases in some of the western countries where these diseases threaten production.

Development of fertilizer responsive, nonlodging, compact plant types with high population densities would be more rewarding breeding preposition in the years to come. Incorporation of dwarfing genes and development of semidwarf varieties of *B. campestris* could pay rich dividends. Presently, in the Indian subcontinent, much emphasis is being laid on the genetic improvement of yield but the future will see much expanded genetical and breeding investigations to improve quality characteristics of the commercial cultivars to meet the export needs.

The search for genes governing thermo-and photoinsensitivity as also better photosynthetic activity would receive far greater attention than what is at present. The incorporation of such genes in rice and wheat has proved useful in expanding the areas of their production. So, why not in rapeseed? The cultivars with these genes could be grown over a wide range of crop-growing conditions and this would help in increasing the overall production of these important oilseed crops.

REFERENCES

Aberg, E. (1984). Results from the scientific cooperation between Indian and Swedish institution regarding the use of cruciferous oilcrops. Sveriges, Lantbruka suneversitet Swedish Uni. Agric. Sci. Dep. Plant Husb. Rep. 137 Uppsala.

Agnihotri, A., Lakshmikumaran, M., Shivannam, K. R. and Jagannathan, V. (1990a). Embryo rescue of interspecific hybrids of *Brassica spinescens* × *B. campestris* and DNA analysis. Current plant science and biotechnology agriculture. *Progress in Plant Cellular and Molecular Biology* 1990, 270–274.

Agnihotri, A., Gupta, V., Lakhmikumaran, M., Shivanna, K. R., Prakash, S. and Jagannathan, V. (1990b). Production of *Eruca-Brassica* hybrids by embryo rescue. *Plant Breeding* **104**, 281–289.

Agnihotri, A., Shivanna, K. R., Raina, S. N., Lakshmikumaran, M., Prakash, S. and Jagannathan, V. (1990c). Production of *Brassica napus* × Raphanobrassica hybrids by embryo rescue. *Plant Breeding* **105**, 292–299.

Agrawal, B. L. (1976). "Studies on Genetic Diversity and Heterosis in Rapeseed (*Brassica Campestris*)." Unpublished Ph. D. Thesis, GBPUAT, Pantnagar, India.

Anand, I. J., Singh, J. N. and Khanna, P. P. (1975). Inter-relationship and diversity in yellow sarson (*Brassica campestris* var. *Sarson Prain*). *Indian Journal of Agricultural Science* **45**, 252–258.

Anonymous, P. P. (1999). "Breeding and Quality Analysis of Rapeseed," pp. 1–36. Agtrans Research, Peter Chudleigh Toowong, Brisbane, China.

Ayotte, R., Harney, P. M. and Machado, V. S. (1986). Transfer of triazine resistance from *Brassica napus* to *B. oleracea*. *Cruciferae Newsletter* **11**, 95–96.

Ayotte, R., Harney, P. M. and Machado, V. S. (1987). Transfer of triazine resistance from *Brassica napus* to *B. oleracea*. I. Production of F_1 hybrids through embryo rescue. *Euphytica* **36**, 615–624.

Banga, S. S. (1994). "Personal communication." Department of Plant Breeding, Punjab Agricultural University, Ludhiana 12 Aug. 1994.

Bannerot, H., Boulidar, L., Canderon, Y. and Tompe, J. (1974). Transfer of cytoplasmic male sterility from *Raphanus sativus* to *B. oleracea*. Proc. Eucarpia meeting on Cruciferae. *Crop Science* **25**, 52–54.

Bannerot, H., Boulidar, L. and Chupeau, Y. (1977). Unexpected difficulties met with the radish cytoplasm in *B. oleracea*. *Cruciferae Newsletter* **2**, 16.

Bansal, U. K., Sequin Swartz, G., Rakow, G. F. W. and Petrie, G. A. (1990). Reaction of *Brassica* species to infestation by *Alternaria brassicae*. *Canadian. Journal of Plant Science* **70**, 1159–1162.

Bateman, A. J. (1955). Self incompatibility system in angiosperms III. *Cruciferae Heredity* **9**, 53–68.

Beversdorf, W. D. and Kott, L. S. (1987). Development of triazine resistance in crops by classical plant breeding. *Weed Science* **35**, 9–11.

Beversdorf, W. D., Weiss-Lerman, J., Erickson, L. R. and Souza-Machado, V. (1980). Transfer of cytoplasmically inherited triazine resistance from birdrape to cultivated oilseed rape (*Brassica Campestris* L. and *B. napus* L.). *Canadian Journal of Genetics and Cytology* **22**, 167–172.

Blackshaw, R. E., Kanashiro, D., Moloney, M. M. and Crosby, W. L. (1994). Growth, yield and quality of canola expressing resistance to acetolactate synthase inhibiting herbicides. *Canadian Journal of Plant Science* **74**, 745–751.

Brazezinski, W., Mendelewski, P. and Munse, B. G. (1986). Comparative study on determination of glucosinolates in rapeseed. *Cruciferae Newsletter* **11**, 128–129.

Brim, H., Plessi, J. and Renard, M. (1987). Resistance of some crucifers to *Alternaria brassicae* (Berk.) Saacc. *In* "Proceedings of the 7th Rapeseed Conference," pp. 1222–1227. Paris.

Carre, P., Dartenuc, C., Evrard, J., Judde, A., Labalette, F., Raoux, R. and Renard, M. (2003). Frying stability of rapeseed oils with modified fatty acid composition. *In* "Proceedings of 11th International Rapeseed Congress," Vol. 2, pp. 540–543. The Royal Veterinary and Agricultural University, Copenhagen, Denmark, 6–10th July.

Chaubey, C. N. (1979). Mass selection in toria. *Indian Journal of Genetics* **39**, 194–201.

Chevre, A. M., Eber, F., Brun, H., Plessis, J., Primard, C. and Renard, M. (1991). Cytogenetic studies of *Brassica napus-Sinapsis alba* hybrids from ovary culture and protoplast fusion. Attempts to introduce *Alternaria* resistance into rapeseed. *Proceedings of International Rapeseed Conference* **8**, 346–351.

Chiang, M. S. (1974). Cabbage pollen germination and longevity. *Euphytica* **23**, 579–584.

Chiang, M. S. and Crete, R. (1983). Transfer of resistance to Race 2 of *Plasmodiophora brassicae* from *B. napus* to cabbage (*B. oleracea* var. *capitata*) the inheritance of resistance. *Euphytica* **32**, 479–483.

Corbett, P. and Sernyk, L. (2003). Global opportunites for naturally stable canola/rapeseed oils. *In* "Proceedings of 11th International Rapeseed Congress," Vol. 2, pp. 524–527. The Royal Veterinary and Agricultural University, Copenhagen, Denmark, 6–10th July.

Craig, B. M. and Murphy, N. L. (1959). Quantitative fatty acid analysis of vegetable oil by gas liquid chromatography. *Journal of the American Oil Chemists Society* **36**, 549–552.

Das, B. and Rai, B. (1972). Heterosis in intervarietal crosses of toria. *Indian Journal of Genetics* **32**, 197–202.

Doloi, P. C. (1977). "Levels of self-incompatibility, heterosis and inbreeding depression in *Brassica Campestris*." Unpublished Ph.D. Thesis, Govind Ballabh Pant University of Agriculture and Technology, Pantnagar, Nainital, India.

Downey, R. K. and Harvey, B. L. (1963). Methods of breeding for oil quality in rape. *Canadian Journal of Plant Science* **43**, 271–275.

Downey, R. K. and Rimmer, S. R. (1993). Agronomic improvement in oilseed Brassicas. *Advances in Agronomy* **50**, 1–66.

Downey, R. K., Klassen, A. J. and Stringam, G. R. (1980). Rapeseed and mustard. *In* "Hybridization of Crop Plants" (W. R. Fehr and H. H. Hadley, eds.), pp. 495–509. American Society of Agronomy Inc., Madison, WI.

Dueck, J. and Degenhardt, K. (1975). Effect of leaf age and inoculum concentratio-non reaction of oilseed Brassica species to *Alternaria brassicae*. *Proceedings of the American Phytopathological Society* **2**, 168 (Abstract).

Fan, Z. and Tai, W. (1985). A cytogenetic studies of monosomics in *Brassica napus*. *Canadian Journal of Genetics and Cytology* **27**, 683–688.

Fan, Z., Rimmer, S. R. and Stefansson, B. R. (1983a). Inheritance of resistance to *Albugo candida* in rapid cycling population of *Brassica Campestris*. *Phytopathology* **77**, 527–532.

Fan, Z., Rimmer, S. R. and Stefansson, B. R. (1983b). Inheritance of resistance to *Albugo candida* in rape. *Canadian Journal of Genetics and Cytology* **25**, 420–424.

Fenwick, G. R., Heaney, R. K. and Mullim, W. J. (1983). Glucosinolates and their breakdown products in food and food plants. *CRC Critical Reviews in Food Science and Nutrition* **18**, 123–201.

Frandsen, K. J. (1943). The experimental formation of *Brassica juncea* (Czern and Coss). *Dansk Botanisk Arkiv* **11**, 1–17.

Gororo, N., Salisbury, P., Rebetzke, G., Burton, W. and Bell, C. (2003). Genotypic variation for saturated fatty acid content of victorian canola. *In* "Proceedings of 11th International Rapeseed Congress," Vol. 1, pp. 215–217. The Royal Veterinary and Agricultural University, Copenhagen, Denmark, 6–10th July.

Grant, I. and Beversdorf, W. D. (1985). Heterosis and combining ability estimates in spring planted rape (*Brassica napus*). *Canadian Journal of Genetics and Cytology* **27**, 472–478.

He, Y. H., Yang, R. F. and Luo, S. Q. (1987). Selection of Swede rape cultivar 821 with high potential and multiple resistance: Study of characteristic structures. *Oil Crops China* **2**, 11–15.

Hinata, K. and Konno, N. (1979). Studies on male sterile strain having the *Brassica campestris* nucleus and the *Diplotaxis muralis* cytoplasm. I. On the breeding procedure and some characteristics of the male sterile strain. *Japanese Journal of Breeding* **29**, 305–311.

Hutchenson, D. S., Downey, R. K. and Campbell, S. J. (1981). Performance of a naturally occurring subspecuies hybrid in *B. Campestris* var. *oleifera*. *Canadian Journal of Plant Science* **61**, 895–900.

Ichikawa, H. and Hirai, A. (1983). Search for a female parent in the genesis of *Brassica napus* by chloroplast DNA restriction pattern. *Japanese Journal of Genetics* **58**, 419–424.

Krzymanski, J. and Downey, R. K. (1969). Inheritance of fatty acid composition in winter forms of *Brassica napus*. *Canadian Journal of Plant Science* **49**, 313–319.

Labana, K. S., Badwal, S. S. and Chaurasia, B. D. (1975). Heterosis and combining ability in *B. juncea*. *Crop Improvement* **2**, 46–51.

Lee, S. J. and Zhang, Y. (1983). The utilization of genetic male sterility in *Brassica* male sterility in Shanghai, China. *In* "Proceedings of 6th International Rapeseed Congress," pp. 360–364. Paris, France.

Lefort-Buson, M. and Datte, Y. (1982). Genetic study of some agronomic characters in winter oilseed rape (*Brassica napus*). *I. Heterosis. Agronomie* **2**, 315–322.

Luhs, W., Weier, D., Marwede, V., Frauen, M., Lekband, G., Becker, H. C., Frentzen, M. and Friedt, W. (2003). Breeding of oilseed rape (*Brassica napus* L.) for modified tocopherol composition- synergy of conventional and modern approaches *In* "Proceedings of 11th International Rapessed Congress, Copenhagen, Denmark 1, 194–197. The Royal Veterinary and Agricultural University," Copenhagen, Denmark, 6–10th July.

Mahapatra, D. and Bajaj, Y. P. S. (1987). Interspecific hybridization in *B. juncea* × *B. hirta* using embryo rescue. *Euphytica* **36**, 321–326.

Mathias, R. (1985). Transfer of mustard (*Brassica juncea* L.) into rapeseed (*Brassica Napus* L.). *Plant Breeding* **95**, 371–374.

McCollum, G. D. (1988). CMS (ESG-508) and CMS (FSG-512) cytoplasmic male sterile cabagge germplasm with raddish cytoplasm. *Horticultural Science* **23**, 227–228.

McGregor, D. I. and Downey, R. K. (1975). A rapid and simple assay for identifying low glucosinolate in rapeseed. *Canadian Journal of Plant Science* **55**, 191–196.

McGregor, D. I., Mullim, W. J. and Fenwick, G. R. (1983). Review of analysis of glucosinolate analytical methodology for determining glucosinolate composition and content. *Journal Association of Official Analytical Chemists* **66**, 825–849.

Nagalakshmi, T. V. (1992). "Analysis of genetic divergence, combining ability and heterosis in Indian mustard (*B. juncea*)." Ph. D. Thesis, BHU, Varanasi, India.

Oliveri, A. M. and Parrini, P. (1986). Relationship between glucosinolate content and yield component in rapeseed. *Cruciferae Newsletter* **11**, 126–127.

Olsson, G. (1990). Rape yield. Production components. *Svensk Frotidning* **59**, 168–169.

Persson, C. (1986). High erucic acid from white mustard (*Sinapis alba*) for technical use. *Cruciferae Newsletter* **11**, 134.

Pidskalny, R. S. and Rimmer, S. R. (1985). Virulence of *Albugo candida* from turnip rape (*Brassica campestris*) and mustard (*B. juncea*) on various crucifers. *Canadian Journal of Plant Pathology* **7**, 283–286.

Prakash, S. and Chopra, V. L. (1988). Introgression of resistance to shattering in *Brassica napus* from *Brassica juncea* through non-homologous recombination. *Plant Breeding* **101**, 167–168.

Prakash, S. and Chopra, V. L. (1991). Cytogenetics of crop brassicas and their allies. *In* "Chromosomal Engineering in Plants: Genetics, Breeding and Evolution, Part B," pp. 19–61. Elsevier, Amsterdam.

Prakash, S. and Raut, R. N. (1983). Artificial synthesis of Brassica napus and its prospects as an oilseed crop in India. *Indian Journal of Genetics* **42**, 282–290.

Quazi, M. H. (1988). Interspecific hybrids between *B. napus* and *B. oleracea* developed by embryo culture. *Theoretical and Applied Genetics* **75**, 309–318.

Rai, B. (1976). Considerations in the genetic improvement of oil quality in rapeseed. *Oilseed Journal* **6**, 13–15.

Rai, B. (1979). "Heterosis Breeding," pp. 1–183. Agro-Biological Publications, Azad Nagar, New Delhi, India.

Rai, B. (1982). Breeding strategy for developing high yielding varieties of toria (*Brassica campestris* var. *toria*). *In* "Research and Development Strategies for Oilseeds Production in India," pp. 131–135. ICAR, New Delhi.

Rai, B. (1987). PT 303 the first national variety of toria. *Indian Farming* **37**, 16–15.

Rai, B. and Kumar, A. (1978). Rapeseed and mustard production programme. *Indian Farming* **28**, 27–30.

Rai, B. and Sehgal, V. K. (1975). Field resistance of Brassica germplasm to mustard aphids (Lipaphis Erysimi kalt). *Science and Culture* **41**, 444–445.

Rai, B. and Singh, A. (1976). Commercial seed production in rapeseed. *Indian Farming* **26**, 15–17.

Rai, B. and Singh, D. (1993). A note on the potential sources of dwarfing genes in Indian rapeseed (*Brassica campestris*). *Indian Journal of Genetics* **53**, 153–156.

Rai, B., Kolte, S. J. and Tiwari, A. N. (1976). Evaluation of oleiferous *Brassica* germplasm for resistance to *Alternaria* leaf blight. *Indian Journal of Mycology and Pathology* **6**, 76–77.

Rai, L., Rai, B., Sanaghmitra, M. and Rao, B. P. (1987). Control of aphids in mustard and safflower crops. *Indian Farming* **37**, 16–17.

Rajan, S. S. (1955). The effectiveness of mass-pedigree system of selection improvement of seed setting in autotetroploids toria. *Indian Journal of Genetics* **15**, 47–49.

Rakow, G. (1973). Selektion auf linol und Linolen sauregehalt in rapssamen nach mutagener behind lung. Z. Pflanzenzuchtg. **69**, 62–82.

Rakow, G. and Woods, D. L. (1987). Outcrossing in rape and mustard under Saskatchewan prairie conditions. *Canadian Journal of Plant Science* **67**, 147–151.

Ramanujam, S. and Rai, B. (1963). Analysis of yield components in yellow sarson. *Indian Journal of Genetics* **23**, 312–319.

Ramanujam, S. and Srinivasachar, D. (1943). Cytogenetic investigations in genus *Brassica* and the artificial synthesis of *Brassica juncea*. *Indian Journal of Genetics* **3**, 73–88.

Rambhajan, Chauhan, Y. S. and Kumar, K. (1991). Natural cross-pollination in Indian mustard. *Cruciferae Newsletter* **14/15**, 24–25.

Raney, J. P., Olson, T. V., Rakow, G. and Ripley, V. L. (2003a). *Brassica juncea* with a canola fatty acid composition from an interspecific cross with *Brassica napus*. *In* "Proceedings of 11th International Rapeseed Congress," pp. 281–283. The Royal Veterinary and Agricultural University, Copenhagen, Denmark, 6-10th July.

Raney, J. P., Olson, T. V., Rakow, G. and Ripley, V. L. (2003b). Selection of near zero aliphatic glucosinolate *Brassica juncea* from an interspecific cross with *B. napus*. *In* "Proceedings of 11th International Rapeseed Congress," pp. 284–286. The Royal Veterinary and Agricultural University, Copenhagen, Denmark, 6–10th July.

Rao, M. V. B. (1990). "Widening variability in cultivated digenomic *Brassica* through interspecific hybridization." Ph. D. Thesis, IARI, New Delhi, India.

Raut, R. N. and Kaul, T. (1982). Synthesis of new genotypes of *Brassica napus* suitable for cultivation in India. *Current Science* **51**, 838–839.

Raut, R. N. and Prakash, S. (1985). Synthetic brassicas: New oilseeds for greater production. *In* "Genetic Manipulations for Crop Improvement," p. 326. Oxford and IBH, New Delhi.

Robbelen, G. and Nitsch, A. (1974). Genetische und physiologische undersuchungen and polyen-fettsaure-mutanten von Raps. I auslese und beschreibung neuer mutanten. *Z Pflansenzuchtg* **75**, 93–105.

Roy, N. N. (1984). Interspefific transfer of *Brassica juncea* type high blackleg resistance to *Brassica napus*. *Euphytica* **33**, 295–303.

Roy, N. N. and Tarr, A. W. (1985). IXLIN- an interspecific source for high linoleic acid content in rapeseed. *Plant Breeding* **95**, 201–209.

Roy, N. N. and Tarr, A. W. (1986). Development of new zero linolenic acid (18:3) lines of rapeseed (*Brassica napus* L). *Z Pflanzenzuchtg* **96**, 218–233.

Sacristan, M. D. and Gerdemann, M. (1986). Different behavior of *Brassica juncea* and *B. carinata* as source of interspecific transfer to *B. napus*. *Plant Breeding* **97**, 304–314.

Sang, J. P. and Truscott, R. J. W. (1984). Liquid chromatographic determination of glucosinolates in rapeseed as desulphoglucosinolates. *Journal Association of Official Analytical Chemists* **67**, 829–833.

Sernyk, J. L. and Stefansson, B. R. (1983). Heterosis in summer rape (*B. napus*). *Canadian Journal of Plant Science* **63**, 407–413.

Shabana, R., Shrief, S. A., Ibrahim, A. F. and Gisler, G. (1990). Correlation and path analysis for new released double zero spring rapeseed cultivars grown under competitive systems. *Journal of Agronomica and Crop Science* **165**, 138–143.

Sikka, S. M. and Rajan, S. S. (1957). Breeding better Brassicas. *Indian Oilseeds Journal* **I**, 73–81.

Singh, D. (1958). "Rape and Mustard, Indian Central Oilseeds," p. 105. Committee, Hyderabad, India.

Singh, D. and Singh, H. (1988). Inheritance of white rust resistance in interspecific crosses of *B. jumcea* × *B. carinata*. *Crops Research* **1**, 189–193.

Singh, H. and Singh, D. (1987). A note on the transfer of resistance to white rust from Ethiopian mustard to Indian mustard. *Cruciferae Newsletter* **12**, 95.

Singh, J. (1986). "Genetic analysis of yield and its component characters in *Brassica napus*." Ph. D. Thesis, Meerut University, Meerut (UP), India.

Singh, R. and Rana, R. S. (1994). Genetic resources programme on oilseed Brassica: Introduction and evaluation at NBPGR during 193–94. Paper Presented at the First All India Rapeseed-Mustard Research Workers Group Meeting Gwalior 19–22, Aug., 1994.

Spasibionek, S., Krzymanski, J. and Bartkowiak-Broda, I. (2003). Mutants of *Brassica napus* with changed fatty acid composition. *In* "Proceedings of 11th International Rapeseed Congress," Vol. 1, pp. 221–224. The Royal Veterinary and Agricultural University, Copenhagen, Denmark, 6–10th July.

Spinks, E. A., Sones, K. and Fenwick, G. R. (1984). The quantitative analysis of glucosinolates in cruciferous vegetables, oilseeds and forage crops using high performance liquid chromatography. *Fette Seifen Anstrichm* **86**, 228–231.

Srivastava, K. and Rai, B. (1993). Expression of heterosis for yield and its attributes in rapeseed. *Indian Journal of Agriculture Sciences* **63**, 243–245.

Stefansson, R. R., Hougen, F. W. and Downey, R. K. (1961). Note on the isolation of rape plants with seed oil free from erucic acid. *Canadian Journal of Plant Science* **41**, 218–219.

Tewari, J. P. (1991). Resistance to *Alternaria brassicae* in crucifers. 10 BC/WPRS Bull. **14**, 154–161.

Thompson, K. F. (1964). Triple cross hybrid kale. *Euphytica* **13**, 173–177.

Toriyama, K., Hinata, K. and Kameye, T. (1987). Production of somatic hybrid plants Brassico-moricandia through protoplast fusion between *Moricandia arvensis* and *Brassica oleracea*. *Plant Science* **48**, 123–128.

Tyagi, D. V. S., Rai, B. and Verma, R. B. (1983). A note on the bunchy dwarf mutant in toria. *Indian Journal of Genetics* **43**, 374–377.

U, N. (1935). Genome analysis in *Brassica* with special reference to the experimental formation of *B. napus* and particular mode of fertilization. *Japanese Journal of Botany* **7**, 389–452.

Uchimiya, H. and Wildman, S. G. (1978). Evaluation of fraction I protein in relation to origin of amphidiploid *Brassica* species and other members of cruciferae. *The Journal of Heredity* **69**, 299–303.

Verma, V. D. and Rai, B. (1980a). Note on induced mutagenesis for spotting out usable sources of resistance to *Alternaria* leaf spot in Indian mustard. *Indian Journal of Agriculture Sciences* **50**, 278–280.

Verma, V. D. and Rai, B. (1980b). Mutation in seed coat colour in Indian mustard. *Indian Journal of Agriculture Sciences* **50**, 545–548.

Yadava, T. P., Singh, H., Gupta, V. P. and Rana, R. K. (1974). Heterosis and combining ability in raya for yield and its components. *Indian Journal of Genetics* **34A**, 648–695.

Yongming, Z. and Houli, L. (1991). Selective strategies on the breeding for quality in *Brassica napus*. *Cruciferae Newsletter* **14/15**, 58–59.

Zhong, R., Zhu, F., Liu, Y. L., Li, S. G., Kang, L. Y. and Luo, P. (1997). Oilseed rape transformation and the establishment of a bromoxynil-resistant transgenic oilseed rape. *Acta Botanica Sinica* **39**, 22–27.

The Chronicles of Oil and Meal Quality Improvement in Oilseed Rape

ABHA AGNIHOTRI, DEEPAK PREM AND KADAMBARI GUPTA

Plant Biotechnology, TERI, Habitat Place, Lodhi Road, New Delhi, India

Advances in Botanical Research, Vol. 45
Incorporating Advances in Plant Pathology

0065-2296/07 $35.00
DOI: 10.1016/S0065-2296(07)45003-0

ABSTRACT

Almost all plant parts of rapeseed are used in a range of human livelihood activities: seeds for edible oil, the leaf as vegetable and animal fodder, the oilseed cake as high-value animal feed and the dried stalk as domestic fuel. The nutritional quality of the two economically important products, that is oil and cake, is of prime importance due to its direct and indirect impact on human health. *Brassica* oil especially that of oilseed rape (*Brassica napus*) is nutritionally superior to most of the other edible oils due to the lowest amounts of harmful saturated fatty acids (SFAs) and a good proportion of mono- and polyunsaturated fatty acids (FAs). In addition, it is also a source of the two essential FAs, linoleic and linolenic, that are not present in some of the other edible oils. The meal is a rich source of good-quality proteins as well; however, the value of the conventional oil and meal from varieties being grown in India or other Asian countries gets restricted at the global level due to the presence of higher amounts of a long carbon chain unsaturated FA (erucic acid) in the seed oil and sulphur-containing compounds (glucosinolates) in the meal, both of which are undesired. This chapter presents a comprehensive overview of nutritional implications of oil and meal quality of rapeseed, inheritance of the chief biochemical determinants, the analytical techniques for estimation of nutritional quality parameters and the breeding efforts towards attaining desired nutritional quality in globally prominent rapeseed species with specific relevance to the Indian perspective.

I. INTRODUCTION

Oils or fats act as a vehicle for some of the important vitamins and also play a significant role in metabolic functions. Therefore, these are an integral part of our diet, providing most concentrated form of energy. At the molecular level they are composed of triglycerides that contain one glycerol molecule linked by covalent bonds to three fatty acid (FA) molecules. The physical and chemical properties of oil are directly dependent on the composition of its FAs that make up the triglycerides and the occurrence of double bonds between the carbon molecules that make up the FAs. On the basis of occurrence of the double bond, FAs can be classified as saturated FAs (SFA), monounsaturated (MUFA) and polyunsaturated FAs (PUFA). The SFA (such as palmitic acid, C16:0) increases the levels of low-density lipoprotein (LDL) in the blood that has a significant role in cholesterol deposition, and are thus undesired for human nutrition (Gurr, 1992). The MUFA (oleic acid, C18:1) being thermostable provide a longer shelf life and are preferred for cooking and deep frying (Prabhu, 2000). It also reduces cholesterol and is thus beneficial for health (Bonanone *et al.*, 1992). The rapeseed-mustard oil also provides the two essential PUFAs, linoleic and linolenic (C18:2 and C18:3, respectively), that need to be supplemented in the diet (Newton, 1998) and are not present in some of the other edible oils such as sunflower and groundnut (Prakash *et al.*, 2000). Very high amounts of PUFAs make the oil amenable to oxidation, thereby reducing its shelf life; however, it is hypocholestermic and needs to be supplied in the diet at low levels. Therefore, the oils containing high amounts of MUFA, moderate amounts of PUFA and low amounts of SFA are considered good for human consumption and thus with regard to this rapeseed-mustard oil is one of the healthiest vegetable oils.

The rapeseed-mustard meal, used as a valuable animal feed, contains about 40% proteins with a well-balanced aminogram (Miller *et al.*, 1962). However, the feeding value of traditional rapeseed-mustard meal has been limited due of the presence of sulphur-containing compounds called glucosinolates (Fenwick *et al.*, 1983). The glucosinolates as such do not cause much harm but their breakdown products adversely affect iodine uptake by the thyroid gland in non-ruminant animals such as swine and poultry, and reduce palatability and body weight gain (Bille *et al.*, 1983; Fenwick *et al.*, 1983). Therefore, an oilcake with a minimum amount of or free of glucosinolates is highly desirable for animal feed.

The 'Canola' (commonly known as double-low or 'OO') is a trade name designating rapeseed having less than 2% erucic acid in the seed oil and less than 30-μM glucosinolate/g of the de-oiled cake. It has about 6% SFAs, 65% oleic, 20% linoleic and 9% linolenic acid, and is considered as having the ideal FA composition of edible oils that is preferred internationally for human consumption (Downey, 1990). While the changeover to canola quality rapeseed cultivars in Europe, Canada and Australia took place in early eighties, such changeover in the major rapeseed-mustard-growing Asian countries, India and China, is at nascent stage. The biochemical composition of presently cultivated traditional rapeseed-mustard varieties in India, containing high erucic acid (up to 50%) in the seed oil and high glucosinolate (up to 250 μM) in the meal, does not match the internationally accepted standards (Agnihotri and Kaushik, 2002), thus necessitating the need to work towards achieving this goal. Some of the desired quality parameters include high oil content and manipulation of FAs composition in the seed oil for specific purposes; crude fibre, protein content and amino acids composition in meal. Other minor constituents of importance comprise phytic acid, sinapine and phenolic compounds that are in general included in breeding programme as a part of monitoring and characterization. The detailed characteristics of rapeseed-mustard oil and meal have been reviewed by Agnihotri and Kumar (2004), some of the important quality parameters are discussed below.

A. OIL CONTENT

The quality of seed, to a large extent, is dependent on the oil content that directly affects the economic value of the crop. The moisture content is also an important parameter as it has a bearing on the seed storage (Bandel *et al.*, 1991) but it has not been investigated as a breeding objective and has mainly been used for documentation of seed quality characteristics (Anonymous, 2001, 2002, 2003, 2004).

The oil content for *Brassica* oilseeds ranges from 35 to 44% (Downey and Rimmer, 1993). Most of the cultivated varieties of Indian rapeseed-mustard have oil content ranging from 39 to 42% and a simple switch-over to yellow-seeded cultivars can bring about 2% increase in oil content (Banga, 1996). It is possible to develop cultivars with increased oil content; however, it results at the expense of reduction in either carbohydrate or proteins accumulation. The energy expense for increased oil accumulation is greater if the oil content is enhanced by a decrease in the carbohydrates as compared to the proteins (Mitra and Bhatia, 1979). Bhatia and Mitra (1992) have proposed that an increase of 5% in oil content, as a result of carbohydrate reduction in the seed, enhances the photosynthate requirement by 4.6%, while a similar increase in oil as a result of reduced protein accumulation results in 1.8% increase in photosynthate requirement. Although the genetic basis for such association is not clearly established, a similar negative correlation has been shown to exist between seed oil and protein or carbohydrate content in *Brassica napus* (Grami *et al.*, 1977).

B. FA PROFILE

Most of the vegetable oils contain oleic and linoleic acid as the predominant unsaturated FAs; however, the traditional rapeseed-mustard oil is an exception (Fig. 1). In addition to trace amounts of SFAs, palmitic and stearic, the MUFA oleic (9.6–17.5%) and two PUFAs linoleic (11.2–20.2%) and linolenic acid (8.1–20%), it also contains 44–51% of erucic acid (Banga, 1996; Kaushik and Agnihotri, 2000). The FA profiles of the commonly used edible oils is summarised in Table I.

Gopalan *et al.* (1974) have reported the occurrence of cardiac lipidosis with accumulation of erucic acid in mammalian system. High erucic acid content has also been reported to cause impaired myocardial conductance and increased blood cholesterol (Renarid and McGregor, 1976). Sauer and Kramer (1983) and Kramer *et al.* (1998) have reported high erucic acid *B. napus* oil to be less metabolisable. Therefore, high erucic acid-containing oil is nutritionally undesired and efforts have been directed towards development of *Brassica* cultivars having oil free of or with low levels of erucic acid along with high levels of oleic, moderate amounts of linoleic and low levels of linolenic acids (Downey and Rimmer, 1993).

Erucic acid content has been reported to be negatively correlated to linoleic and oleic acid in oilseed Brassicas (Ahuja *et al.*, 1984; Craig, 1961; Downey and Craig, 1964; Jonsson, 1973; Singh *et al.*, 1996; Stefansson, 1983; Stefansson *et al.*, 1961). Therefore, reduction in erucic acid content enhances

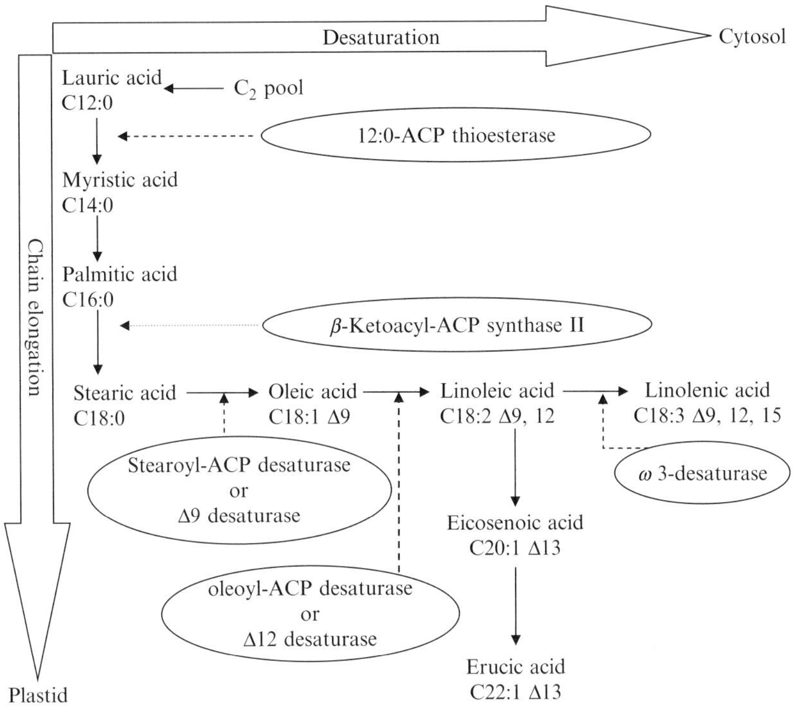

Fig. 1. Fatty acid biosynthesis pathway in rapeseed-mustard. Source: Jonsson (1977), Murphy (1999).

the proportion of oleic and linoleic acids, but the extent of increase is species specific. Reduction in erucic acid content leads to nearly 60% oleic acid accumulation in *B. napus* (Downey, 1990); however, for a similar reduction an increase of only up to 45% was recorded in *B. juncea* (Agnihotri and Kaushik, 1998; Banga *et al.*, 1988). In addition to this, both linoleic and linolenic acids are positively correlated since both of these are produced by a common biosynthetic desaturation pathway (Diepenbrock, 1983). Recent research confirms that each step in the FA biosynthetic pathway is under individual enzymatic control (Smith *et al.*, 2000). Thus alternative approaches of mutagenesis or RNAi induced silencing shows promise to induce low level of linolenic with high linoleic acid or low levels of both.

C. PROTEIN CONTENT

It has been demonstrated that 35% of all proteins consumed by humans is derived from animals, and 70% of this is provided by ruminant animals (Minson, 1997). The quality of the proteins is in turn influenced by the nature of protein sources available to animals (Avery, 1998), and is measured in terms of the provision of essential amino acids for body protein synthesis

TABLE I

Variability for Fatty Acid Profile in Oilseed Brassicas Compared to Other Common Vegetable Oils

Species	Fatty acid composition (%)[a]							References
	Palmitic	Stearic	Oleic	Linoleic	Linolenic	Eicosenoic	Erucic	
B. napus	3.4 ± 0.93 (2–4.9)	1.4 ± 0.41 (0.8–1.6)	48.3 ± 17.9 (9.9–70)	19.8 ± 4.9 (13.5–32.7)	11.8 ± 4.5 (3.3–22.7)	4.6 ± 3 (1.2–10)	16.6 ± 21.1 (0–53.6)	Appelqvist, 1969; Banga, 1996; Downey, 1983; Kaushik and Agnihotri, 2000; Scarth *et al.*, 1988
B. juncea	4.4 ± 2.6 (2.1–6.5)	1.06 ± 0.47 (0.6–2)	33 ± 15.3 (8–49)	26.5 ± 9.6 (11.2–40.3)	12.8 ± 4.9 (2.2–24.2)	6.2 ± 3.34 (1.2–9.2)	18.4 ± 21.8 (0–53.1)	Appelqvist, 1970; Banga, 1996; Downey, 1983; Downey and Rimmer, 1993; Kaushik and Agnihotri, 2000
B. rapa	2.5 ± 0.7 (2–3.8)	1 ± 0.1 (0.9–1.2)	29 ± 18.9 (12.9–58.6)	17 ± 4.7 (12–24)	9.1 ± 0.7 (8.2–10.3)	7.2 ± 4.13 (1–9.6)	32.3 ± 22 (0.3–55.5)	Appelqvist, 1969; Downey, 1983; Kaushik and Agnihotri, 2000
B. carinata	4.3 ± 1.3 (3.2–6.6)	1.4 ± 0.8 (0.9–2.8)	11.8 ± 4.1 (8.8–19)	18.4 ± 2.2 (16.2–18)	10.9 ± 2.1 (8.8–13.9)	6.8 ± 0.5 (6.1–7.5)	39.5 ± 11.8 (20.6–53.2)	Barro *et al.*, 2001; Downey and Rimmer, 1993
Soybean	15.3	9.2	23.6	48.2	8.7	0	0	Hymowitz *et al.*, 1972
Sunflower	5.8	5.2	16	71.5	0.2	0.1	0	Earle *et al.*, 1968
Groundnut	0	3.7	64.3	17.2	0	1.2	0	Worthington and Hammons, 1971
Maize	11.5	2.2	26.6	58.7	0.8	0	0	Beadles *et al.*, 1965
Safflower	7.6	2	10.8	79.6	0	0	0	Knowles, 1968
Cotton	23.6	2.5	17.9	54.2	0	0	0	Anderson and Worthington, 1971

[a]FA composition data based on traditional cultivars and advanced breeding lines presented as means ± standard deviation, values in parenthesis represent the range.

(Chadd *et al.*, 2002). The defatted *Brassica* meal, containing about 40% protein with a well-balanced aminogram, is an excellent source of proteins valued for animal nutrition (Miller *et al.*, 1962). It is particularly rich in lysine and methionine, which are essential amino acids not found in cereal grains. The lysine content of rapeseed-mustard, although tends to be lower than that of soy meal, has a higher proportion of sulphur amino acids (Chadd *et al.*, 2002). Although some variation in the protein content of rapeseed-mustard can be due to cultivar, soil type and environmental factors (Bell, 1995), it is a reasonably concentrated protein source providing 430–450 g/kg crude protein. A comparison of rapeseed-mustard oil meal amino acid profile with some other oilseed crops is given in Table II.

The protein content of the *Brassica* oil meal is economically important; however, as elaborated under Section I.A, the bioenergetic constrains limit the development of genotypes with both high oil as well as high protein content. Hence, the emphasis on breeding for improved oil content and FA profile modifications has resulted in a trade-off decrease in seed protein content in *B. napus* genotypes; therefore, there is an urgent need to focus on meal protein content as a breeding objective (Malabat *et al.*, 2003). Moreover, due to the high costs and low precision of the Kjeldahl nitrogen-based protein estimation method, scientists were inclined to breed for higher oil content while maintaining the *status quo* for protein levels (Downey and Rakow, 1987). With the availability of automated and accurate elemental nitrogen analysers and alternate near-infrared spectroscopy (NIRS)-based methods, greater emphasis can now be placed on using a selection pressure for high protein content without reduction in oil content (Downey and Rimmer, 1993; Leckband *et al.*, 2003).

D. GLUCOSINOLATE CONTENT

The use of oilseed Brassicas meal as an inexpensive protein-rich food and feed gets restricted by the presence of sulphur-containing compounds, called glucosinolates. Glucosinolates are a class of secondary metabolites, characteristically found in the vegetative tissues and the seeds of cruciferous plants (Fenwick *et al.*, 1983). At present, over 100 different kinds of glucosinolates have been reported, and are found in 15 botanical families of dicotyledonous plants; *Akaniaceae*, *Bataceae*, *Brassicaceae*, *Bretschneideraceae*, *Capparaceae*, *Caricaceae*, *Euphorbiaceae*, *Gyrostemonaceae*, *Limnanthaceae*, *Moringaceae*, *Pentadiplantdraceae*, *Resedaceae*, *Salvodoraceae*, *Tropaeolaceae* and *Tovariaceae*.

Glucosinolates are found in all parts of the plant and up to 15 different kinds of glucosinolates have been found in the same plant (Zukalova and Vasak, 2002). Their concentrations differ according to the tissue type,

TABLE II

Comparative Amino Acid Composition of Meal of Oilseed Brassicas Versus Some Other Oil Meals

Component(s)	Oleiferous Brassicas	Soybean	Cottonseed	Peanut	Sunflower	Coconut
Crude protein[a]	35.6	47.5	41.4	49.1	42.2	21.9
Arginine[b]	6.2	7.3	11.0	10.4	6.9	10.9
Histidine	2.7	2.7	2.8	2.2	2.2	1.8
Isoleucine	4.0	4.5	3.1	3.6	3.4	3.4
Leucine	7.2	7.7	6.0	5.8	5.5	6.2
Lysine	5.8	6.4	4.2	3.4	2.8	2.6
Methionine	2.1	1.4	1.6	1.1	1.9	1.6
Methionine + cystine	4.6	3.0	3.3	2.5	3.5	2.9
Phenylalanine	4.0	5.0	5.3	4.8	3.9	3.8
Phenylalanine + tyrosine	7.2	8.9	8.3	8.5	6.4	6.5
Threonine	4.5	3.9	3.3	2.6	3.2	3.1
Tryptophan	1.3	1.4	1.2	1.0	1.0	0.9
Valine	5.1	4.8	4.3	4.0	4.1	4.9

[a]Crude protein value expressed as a percentage of oil meal; source: Bajjalieh (2002).
[b]Amino acids are expressed as a percentage of crude protein; source: Bajjalieh (2002).

physiological age and health of the plant. Generally, the glucosinolate content is highest in the seeds (up to 10% of the dry weight), and the levels in the leaf, stem and root are ~10 times lower.

The predominance of different types of glucosinolates in oilseed Brassicas is species specific. The major species-specific variations have been recorded for allyl (sinigrin), 3-butenyl (gluconapin) and 2-hydroxy-3-butenyl (progoitrin) glucosinolates with sinigrin being present in *B. juncea, B. nigra* and *B. carinata*; gluconapin in *B. juncea*; pentenyl and gluconapin in *B. rapa* and progoitrin in *B. napus* (Agnihotri and Kumar, 2004). In addition to this, the relative proportion of different glucosinolates is also dependent on the species, for example various kinds of glucosinolates present in *B. juncea*, in decreasing order of their abundance are gluconapin, sinigrin, progoitrin, napoleiferin and glucobrassicanapin (Agnihotri and Kumar, 2004).

The general structure of glucosinolates is shown in Fig. 2. It consists of a thioglucosidic link to the carbon of a sulphonated oxime. The R group (side chain) and the sulphate group have anti-stereo-chemical configuration. The R group is derived from amino acids and is highly variable in properties, from lipophilic to marked hydrophilic (Ettlinger and Lundeen, 1956). It can be aliphatic (alkyl, alkenyl, hydroxyalkenyl, ω-methylthioalkyl), aromatic (benzyl, substituted benzyl) or heterocyclic (indolyl). The sulphate group imparts strong acidic properties and thus the glucosinolates occur in nature as anions counterbalanced by a cation. The cation is usually potassium, being one of the most abundant cations in plant tissues. The sulphate group and the thioglucose moiety impart non-volatile and hydrophilic properties to all glucosinolates.

The precise localisation of glucosinolates is not known, but experimental evidence suggests that they are most probably present in vacuoles (Luthy and Matile, 1984; Matile, 1980). Myrosinase, a glycoprotein enzyme responsible for hydrolysis of glucosinolates, is stored in a tonoplast-like membrane-bound organelle called the idioblast (Thangstad *et al.*, 1991). Thus, the two components of the system are separated until autolysis or tissue damage brings them into contact. On mechanical injury, myrosinase catalysed hydrolysis of glucosinolates occurs to form thiocyanates, isothiocyanates and/or nitriles (Fig. 3).

During seed development at crop maturity, glucosinolates are actively transported to the seed embryo against a concentration gradient (Fieldsend and Milford, 1994; Zhao *et al.*, 1993a,b). Therefore, although no primary physiological role has been embryo as yet, they probably contribute sulphur for essential amino acid synthesis in young seedlings. The breakdown products of glucosinolates are involved in the protective mechanism of the plant system, since these act either as toxins or repellent against pests and diseases (Mitten, 1992; Wallsgrove *et al.*, 1999). Although glucosinolates as such do not cause

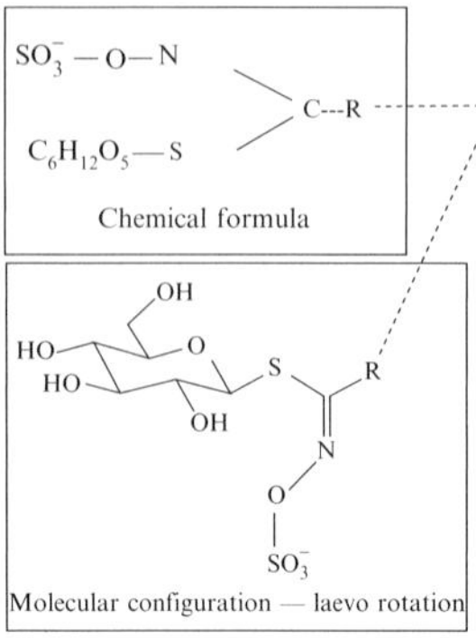

Side chain (R group)	Systematic name	Trivial name	Type
$CH_2 = CH - (CH_2)_2$	3- Butenyl - GSL	Gluconapin	Aliphatic
$CH_2 = CH - CH_2$	2- Propenyl (allyl) - GSL	Sinigrin	Aliphatic
$CH_2 = CH - CHOH - CH_2$	2-Hydoxy-3- butenyl - GSL	Progoitrin	Hydroxy-Aliphatic
$CH_2 = CH - CH_2 - CHOH - CH_2$	2- Hydroxy- 4 pentenyl - GSL	Napoleiferin	Hydroxy-Aliphatic
$CH_2 = CH - (CH_2)_3$	4- Pentenyl - GSL	Glucobrassicanapin	Aliphatic

Fig. 2. The general structure of glucosinolates and the composition of side chains for glucosinolates present in *B. juncea*. The natural forms of glucosinolates exhibit laevo rotation in solution and have a large number of homologs and β-hydroxylated analogs Sources: http://boneslab.chembio.ntnu.no/paal/glucosin.htm, Zukalova and Vasak (2002); Ettlinger and Lundeen (1956).

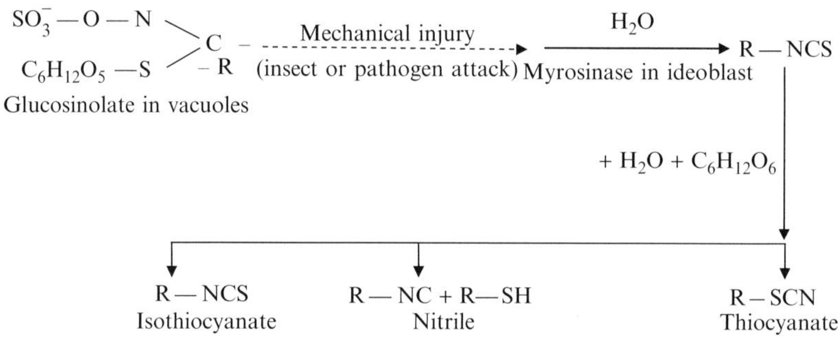

Fig. 3. The reactions involved in breakdown of glucosinolates. Source: Zukalova and Vasak (2002), Kaushik and Agnihotri (1999).

harm as a part of mammalian diet (McMillan *et al.*, 1986; Vermorel *et al.*, 1986) but their breakdown products, thiocyanates, isothiocyanates and/or nitriles, are undesirable in animal feeds. They adversely affect iodine uptake by thyroid gland in non-ruminant animals such as swine and poultry, and reduce palatability and feed efficiency in terms of development and weight gain (Bille *et al.*, 1983; Fenwick *et al.*, 1983). Therefore, the presence of high amounts of glucosinolates severely limits the use of the traditional *Brassica*-defatted meal.

To avoid the glucosinolate hydrolysis products to accumulate in *Brassica* meal, the myrosinase enzyme is heat inactivated as one of the first steps in oil extraction process. However, in most developing countries where oil expellers are used for oil extraction, heat treatment of seed before extraction is usually not done. Therefore, even the extracted edible oil has relatively large proportion of glucosinolates and their breakdown products, which impart the characteristic pungency in the oil. In view of these facts, a prime breeding objective for oilseed brassica breeders has been to develop varieties having low glucosinolate content.

E. SEED COAT COLOUR

Seed coat colour is of importance since it determines several quality parameters. In oilseed Brassicas, it varies from yellow to black with intermediate shades. The variability in seed coat colour is due to deposition of polyphenols in the palisade layer. The palisade cells are parenchyma of the testa in *Brassica* species (Leung *et al.*, 1979). Yellow seed coats exhibit the lowest amount of polyphenols while the brown-seeded varieties contain maximum polyphenols. In addition, yellow-seeded varieties generally have lower chlorophyll content and are thus preferred over dark seed coat by the oil industry. This is because high chlorophyll content in the seeds causes discoloration of the oil, which needs to be removed by technical process (Jonsson, 1977). Besides this, yellow-seeded varieties possess higher oil content, lower fibre and higher protein content as compared to the dark-seeded varieties (Jonsson and Bengstsson, 1990; Stringam *et al.*, 1974). Higher protein and lower crude fibre content are desired for better digestibility and absorbance. Further, the yellow hull obtained from yellow-seeded varieties closely matches the color of other feedstuff used in poultry and livestock thus allowing modification of feed formulations without changing its appearance (Downey and Bell, 1990). Yellow-seeded *B. rapa* and *B. napus* cultivars contain up to 2.5% more oil content (Daun and DeClercq, 1988), 3–5% lower fibre and 2.6–5% higher protein content in comparison to brown-seeded cultivars (Liu *et al.*, 1991; Shirzadegan and Robbelen, 1985; Stringam *et al.*, 1974). Therefore, owing to inherent advantages of yellow-seeded varieties over dark-seeded varieties, emphasis is now being laid at developing genotypes with yellow seed coat colour.

II. GENETIC CONTROL OF SOME BIOCHEMICAL CONSTITUENTS

A. OIL CONTENT

The accumulation of photosynthates, such as oil, in the seed is a complex physiological process involving several genes. Combining ability studies (line × tester) have indicated both additive and non-additive gene effects that govern oil content in *B. juncea* (Gupta *et al.*, 1985; Kumar *et al.*, 1982).

In contrast to this, Singh and Yashpal (1991) have emphasised on non-additive gene effects responsible for oil content using diallel cross analysis. Therefore, selection may not be the best approach for enhanced oil content and utilising hybrid vigour or mutation for generation of novel genotypes may yield better results for generation of high oil-bearing genotypes, since in several countries the seed is purchased on the basis of its oil content (Dhillon *et al.*, 1992).

B. ERUCIC ACID CONTENT

The synthesis of FA at cellular level occurs in the plastids or in the endoplasmic reticulum. The major pathways involved in FAs or lipid metabolism in plants have been reviewed by Murphy (1999) and for rapeseed-mustard this is depicted in Fig. 1. The genes involved in the plant FA synthesis are prime targets for FA profile modifications for developing designer oil-bearing crops.

The FA synthesis genes code for three classes of enzymes; the plastid localised acyl–acyl carrier protein (ACP) responsible for adding carbon to the FA chain utilising acetyl-CoA as substrate, the endoplasmic reticulum localised desaturases responsible for production of double bonds, and the cytosol localised thioesterases responsible for chain termination (Broun *et al.*, 1998). Apart from these, the genes coding for enzymes responsible for formation of the triacylglycerol molecules through esterification of glycerol and FA molecules known as the Kennedy pathway (Cao and Huang, 1987) have also become interesting targets for FA accumulation changes (Voelkar *et al.*, 1992).

Stefansson *et al.* (1961) and Downey (1964) identified genotypes with a genetic block in the biosynthesis of eicosenoic and erucic acid in summer rape (*B. napus*) and summer turnip rape (*B. rapa*), respectively. Since then, the genetic control of erucic acid has been studied in good detail in different *Brassica* species and is summarised in Table III (Prem, 2006). It is suggested that erucic acid content in *B. napus* is governed by two genes with additive effect (Chen and Beversdorf, 1990a,b; Downey and Harvey, 1963; Kirk and Hurlstone, 1983; Luhs and Friedt, 1995; Siebel and Pauls, 1989). It is also reported to be controlled by the embryo genotype (Harvey and Downey, 1964; Anand and Downey, 1981; Jonsson, 1977). This has led to the development of the half seed technique for evaluation of FA profile in segregating populations and has been applied to several oilseed Brassicas (Downey and Rimmer, 1993) including *B. juncea* (Agnihotri, 1999). In *B. carinata*, two genes with additive effect and no dominance have been reported (Fernandez *et al.*, 1988). Inheritance studies in *B. juncea* are contradictory since two

TABLE III
Inheritance of Erucic Acid Content in Oilseed Brassicas

Species	Number of genes	Inheritance	References
B. juncea	Two genes	Additive	Kirk and Hurlstone, 1983; Potts and Males, 1999
		Additive and dominance	Monpara and Jaisani, 2000
B. napus	Two genes (codominant)	Additive	Chen and Beversdorf, 1990a,b; Chen and Heneen, 1989; Downey and Harvey, 1963; Harvey and Downey, 1964; Jonsson, 1977; Kirk and Hurlstone, 1983; Krzymanski and Downey, 1969; Luhs and Friedt, 1995; Siebel and Pauls, 1989; Stefansson and Hougen, 1964
	Two genes	Additive	Li and Qui, 1987
	One gene	Partial dominance to over dominance	Moller *et al.*, 1985
	Five genes	Additive	Anand and Downey, 1981
B. rapa	One major gene	Additive	Dorell and Downey, 1964
B. carinata	Two genes	Additive	Fernandez-Escobar *et al.*, 1988

genes showing dominance and acting in additive manner have been proposed by Kirk and Hurlstone (1983), in contrast to one pair of genes (Shpota and Podkolzina, 1986), two gene pairs with additive effect and partial dominance (Liu and Liu, 1989) and partial dominance (Chauhan *et al.*, 2002). Therefore, on the basis of these studies, low erucic acid biosynthesis in *B. juncea*, similar to that in *B. napus*, is known to be under the control of two recessive genes with additive effect.

C. GLUCOSINOLATE CONTENT

The enzymes involved in glucosinolate biosynthetic pathway are yet to be understood completely (Glendening and Poulton, 1990). However, the biosynthesis of glucosinolates in *Brassica* follows the general pathway presented in Fig. 4 (Halkier and Du, 1997). The genetics of glucosinolate content is complicated primarily because the glucosinolate production pathway involves several steps and it does not involve stable intermediate compounds. The work on the genetics for glucosinolate content for various oilseed Brassicas has been summarised in Table IV (Prem, 2006).

For the first time in 1967, Polish *B. napus* cv. 'Bronowski' was discovered containing glucosinolate content of about 12-μM/g oil-free meal. It also contained low erucic acid in the seed oil (7–10%). The discovery of 'Bronowski'

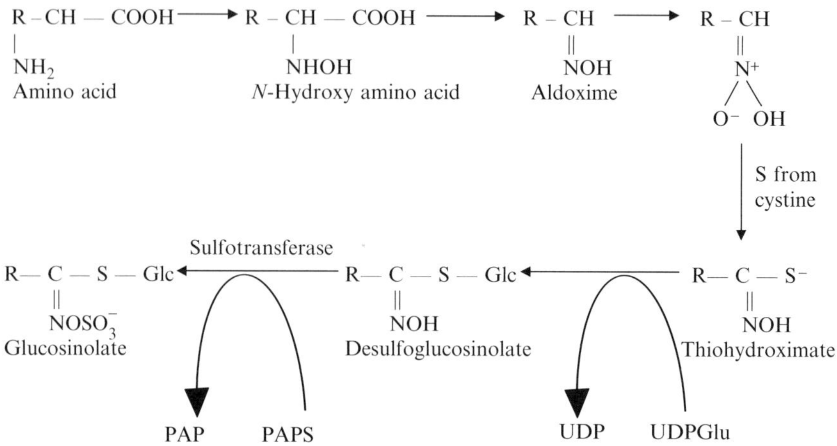

Fig. 4. General pathway for biosynthesis of glucosinolates in crop Brassicas. Glucose is transferred in its active form by uridine-biphosphate-glucose (UDPGlu) and the sulphate group is then transferred by phospho-adenosine-phosphosulphate (PAPS). Sources: Halkier and Du (1997), Glendening and Poulton (1990).

from Poland revolutionised the work in the direction of glucosinolate inheritance in oilseed brassicas (Finlayson *et al.*, 1973) and has provided the source of low glucosinolate genes for practically all cultivated oilseed *Brassica* species. Kondra and Stefansson (1970) have proposed that in *B. napus* the maternal genotype rather than the embryo genotype controls the seed glucosinolate content, and the low glucosinolate levels are controlled by as many as 11 recessive alleles that do not show independent segregation. The formation of individual glucosinolates is thought to occur through the break at the end of the biosynthetic pathway (Lichter *et al.*, 1988). Further, Lein (1970) have determined an additional influence of cytoplasm on glucosinolate synthesis. Glucosinolates also show tissue level variation within the same plant (Inglis *et al.*, 1992; Milford *et al.*, 1989; Schilling and Friedt, 1992) and the leaf glucosinolate quantity and profile is weakly correlated to the seed glucosinolate level in small seedlings (Glen *et al.*, 1990). This suggests that the glucosinolate content in the leaves and the seed may be under different genetic control (Mitten, 1992). There have been some reports indicating that the genes for glucosinolate contents in vegetative tissue are pleiotropic and/or linked with the grain-filling stage (Uppstrom, 1983). The glucosinolate profile at the seedling stage may serve as a tentative tool to predict glucosinolate profile of seeds, but its authenticity is doubtful since the differentiation processes that the plant undergoes after the seedling stage are not clearly understood, and these may produce unpredictable and drastic changes (Razin and Cedar, 1991).

TABLE IV
Genetics of Glucosinolate Content in Brassica Species

Species and glucosonolate type/content	Gene action	References
B. juncea		
High glucosinolate	Nuclear genes with additive effect	Love *et al.*, 1990
Low glucosinolate	At least three partial recessive genes, maternally controlled	Ahuja and Banga, 1992
Low glucosinolate	Six to seven recessive genes	Sodhi *et al.*, 2002
B. rapa		
Gluconapin and progoitrin	Alleles with partial dominance	Kondra, 1967
Absence of gluconapin and progoitrin	Single recessive gene for each compound and 2 genes segregating independently	
B. nigra		
Sinigrin	Modulating role of cytoplasm	Hemingway *et al.*, 1961
B. napus		
High glucosinolate	Three to five genes	Kondra and Stefansson, 1970; Lein, 1970
	Additive and non-additive effect	Gupta and Labana, 1989
	Three major genes with partial dominance with maternal influence	Mou and Liu, 1988, 1990; Zhou and Liu, 1987
	Two genes with partial dominance with maternal influence	
	Additive, dominance and additive × additive interaction, predominance of non-additive gene effect	Gupta *et al.*, 1993, 1989
Progoitrin	Additive and dominance effect and significant epistatic interactions	Rahman and Poulsen, 1995
Total glucosinolate content		
Absence of sinigrin	Two to three dominant genes	Gland *et al.*, 1981
Total glucosinolate		
Higher gluconapin, glucobrassicanapin, progoitrin	Three, four to five, four loci, respectively, with dominance or partial dominance	Kondra and Thomas, 1975
Aliphatic glucosinolate	Six unlinked genes	Magrath *et al.*, 1993
	Maternal control	Lein, 1970; Magrath and Mithen, 1993
	Additive and non-additive gene	Krzymanski *et al.*, 1995
Total glucosinolate	High heritability (87–95%)	Rucker and Robbelen, 1994
Low glucosinolate	Four to five recessive genes acting in additive manner	

D. YELLOW SEED COAT

As with the rest of the agronomic traits, breeding for yellow-seeded varieties is complicated due to multiple gene inheritance, maternal effects and environmental factors (Shirzadegan, 1986; Van Deynze and Pauls, 1994). The seed coat colour in *B. rapa* is controlled by one (Teutonico and Osborn, 1994) to multiple genes (Schwetka, 1982), while in *B. napus* a three-gene model with maternal genotype controlling seed coat colour has been proposed (Shirzadegan, 1986; Van Deynze and Pauls, 1994). The seed coat colour in *B. juncea* is controlled by two independent dominant genes (Anand *et al.*, 1985; Vera and Woods, 1982; Vera *et al.*, 1979), whereas in *B. carinata* it is under monogenic dominance (Getinet and Rakow, 1997). It has been elucidated that brown seed coat colour is dominant over yellow seed coat colour. A dominant repressor gene is present in yellow seed coat plants that inhibits expression of seed coat pigment synthesis genes and is absent in brown-seeded plants (Getinet and Rakow, 1997). The *B. juncea* varieties grown for commercial cultivation in India are brown seeded while about 26 years ago, the first yellow-seeded *B. rapa* var. Candle was released for commercial cultivation in Canada (Downey, 1990). Attempts have been made to develop artificially synthesised yellow-seeded varieties of *B. rapa, B. napus* and *B. juncea* (Abraham and Bhatia, 1986; Jonsson and Bengstsson, 1990; Rahman *et al.*, 2001) by using the yellow-seeded forms that exist in the natural germplasm of *Brassica* species.

III. TECHNIQUES FOR ESTIMATION OF BIOCHEMICAL COMPOSITION

Most of the desired biochemical and agro-morphological traits are predominantly under polygenic control, mostly recessive genes being inherited independently. Therefore, a large sample population needs to be screened for selecting the desired genotype for which efficient analytical techniques are an essential prerequisite. The analytical techniques being used for breeding double-low traits in oilseed Brassicas have been extensively reviewed by Agnihotri (1999) and Agnihotri and Kumar (2004). The available protocols for biochemical estimation in oilseed Brassicas are summarised in Table V (Prem, 2006).

As evident, till date the nuclear magnetic resonance (NMR)- or solvent extraction- and the gas chromatography (GC)-based estimation of FA methyl esters (FAMEs) have remained the preferred methods for oil content and FA profile estimations, respectively. The reverse phase high performance

TABLE V

Techniques for Estimation of Biochemical Components in Oilseed Brassicas

Parameter/trait and method or basic principal	Advantage/drawback	References
Seed oil content		
Solvent extraction	Accurate but requires large seed sample (2–5 g) and is destructive	Anonymous, 2000
Nuclear magnetic resonance (NMR)	No sample preparation required, rapid, accurate and non-destructive but requires expensive equipment	Tiwari *et al.*, 1974
Near-infrared reflectance spectroscopy (NIRS)	No sample preparation required, rapid, accurate and non-destructive but requires expensive equipment and is species specific	Greenwood *et al.*, 1999; Mika *et al.*, 2003; Velasco *et al.*, 1999
Seed FA profile		
Chromatographic methods		
Erucic acid co-precipitation with SFAs as lead or magnesium salts followed by separation of free FA by fractional crystallisation in ethanol	Time consuming and lack sensitivity	Kaufmann and Fiedler, 1938; Stiepel, 1926
Thin layer chromatography (TLC)	Low sensitivity	Stahl, 1969
Paper chromatography for separation of FA using 95% acetic acid as mobile phase	Quick screening method for selection but suitable only for initial screening	Thies, 1971
Gas chromatography—quantitative estimation of methyl esters of individual FA	Accurate estimation of FA profile however cumbersome sample preparation since it required extraction of oil prior to esterification	Appelqvist, 1968; Conacher and Chadha, 1974; Craig and Murty, 1958, 1959; McGregor, 1974, 1977; Stringam and McGregor, 1980
GC estimation with modified sample preparation		
Methanolic hydrochloric acid used for esterification	Oil extraction required prior to esterification	Downey and Craig, 1964

(*continues*)

TABLE V (*continued*)

Parameter/trait and method or basic principal	Advantage/drawback	References
Sodium methoxide used for esterification	Oil extraction required prior to esterification	Hougen and Bodo, 1973; Stringam and McGregor, 1980
One step transesterification using acetylchloride in methanol–benzene (4:1, v/v)	Reduced reaction time thus rapid and accurate estimation	Kaushik and Agnihotri, 1997; Lepage and Roy, 1986
Non-chromatographic methods		
Determination of erucic acid based on the solubility of oil in absolute ethanol or mixture of methanol and *n*-propanol (1.7:2, v/v)—time required for warm alcoholic solution to turn opaque on cooling related to erucic acid content	Not suitable for breeding purpose due to low sensitivity	McGregor, 1977
NIRS	No sample preparation required, rapid and accurate but species specific and requires large seed sample with high representative variability for standardisation; The equipment used is expensive	Pallot *et al.*, 1999; Reinhardt and Röbbelen, 1991; Velasco *et al.*, 1999, 1995a, 1997a
Meal protein content		
Kjeldahl nitrogen estimation-based protein content evaluation	Low sensitivity, time – labour consuming, involves hazardous reagents and is destructive	Anonymous, 1995
Combustion Nitrogen Analysis (CNA) or DUMAS nitrogen estimation-based protein evaluation	Accurate and rapid but destructive and involves expensive instruments	Simonne *et al.*, 1997
NIRS	No sample preparation required, rapid, accurate and non-destructive but requires expensive equipment and is species specific	Kumar *et al.*, 2003; Velasco and Mollers, 2002

Meal glucosinolate content		
Spectrophotometric methods based on measurement of glucosinolate degradation products or glucosinolate–reagent colour complex—methods for estimation of total glucosinolate		
Steam distillation and titration of volatile isothiocyanates combined with UV spectroscopy of oxazolidinethiones	Low sensitivity and low reproducibility of results	Wetter, 1955, 1957
Gas chromatography of volatile isothiocyanates combined with UV spectroscopy of oxazolidinethiones	Low sensitivity for low glucosinolate content since reactions are best if glucosinolate breakdown products are in large volume	Youngs and Wetter, 1967
UV spectroscopy of thiourea derivatives of the isothiocyanates	Low sensitivity but good reproducibility of results	Appelqvist and Josefsson, 1967; Daxenbichler *et al.*, 1970; Wetter and Youngs, 1976
Glucosinolate–palladate coloured complex-based spectroscopic determination	Efficient, fast and reliable method however, it involves expensive reagents	Kolovrat, 1988; Theis, 1983
Thymol method based on determination of glucosinolate content using thymol reagent	Low repeatability of results and hazardous reagents involved	Brzezinski and Mendelewski, 1984; DeClercq and Daun, 1989; Tholen *et al.*, 1989; Truscott and Shen, 1987
Glucose-UV test and Glukotest: myrosinase catalysed glucosinolate breakdown glucose estimation	Accurate estimation but enzyme system-based glucose estimation requires stringent maintenance experiment conditions for consistence in results	Bjorkman, 1972; Lein, 1970
Test tape method	Rapid and efficient method suitable for commercial screening	McGregor and Downey, 1975; Comer, 1956; Lein, 1970; Van Etten *et al.*, 1974
ELISA-based colorimetric estimation using glucosinolate-specific antibody complex or glucosinolate–sodium tetrachloropalladate complex	Rapid and efficient method for quantitative estimation	Hassan *et al.*, 1988; Kumar *et al.*, 2004

(*continues*)

TABLE V (*continued*)

Parameter/trait and method or basic principal	Advantage/drawback	References
Spectroscopy-based glucosinolate estimation		
NIRS	Rapid, accurate and non-destructive but requires large sample representing the high variability in glucosinolate content, is species specific and requires expensive equipment	Daun and Williams, 1995; Mika *et al.*, 2003
X-ray fluorescence or reflectance	Non-destructive but requires expensive equipment and the risk of handling hazardous radiations	Schnug and Haneklaus, 1988; Tholen *et al.*, 1993
Chromatographic methods for determination of individual glucosinolates		
GLC technique for analysis of hydrolysis products of glucosinolates using myrosinase enzyme or as their trimethyl-silyl-desulpho derivatives	Accurate estimation of individual glucosinolate	Brzezinski *et al.*, 1986; Persson, 1974; Stominski and Campbell, 1987; Thies, 1980; Underhill and Kirkland, 1971
High performance liquid chromatography (HPLC) based on separation of desulphoglucosinolates	Accurate but requires time consuming enzymatic desulphatation step due to which some glucosinolates may escape detection	Bjerg and Sorenson, 1986; McGregor, 1985; Palmer *et al.*, 1987; Sang and Truscott, 1984; Spinks *et al.*, 1984
HPLC-based estimation of intact glucosinolates		
Reverse phase HPLC of intact glucosinolates	Efficient and accurate estimation of individual glucosinolates	Kaushik and Agnihotri, 1999

liquid chromatography (HPLC)-based estimation of individual glucosinolates and ELISA-based colorimetric estimation of total glucosinolates by glucosinolate–tetrachloropalladate complex are the preferred methods for estimation of glucosinolate profile and total glucosinolate content, respectively. However, initial steps of plant-breeding programmes require the screening of a large number of samples, at times from single plants and frequently from less number of seeds, with flexibility for their future use. Thus rapid, cost effective, non-destructive and reliable multi-trait methods are most sought after for analysis of seed oil and meal quality.

NIRS provides such an alternative to the primary methods for estimation of quality components in many agricultural products (Shenk and Westerhaus, 1993). It offers the main advantage of simultaneous evaluation of multiple seed components such as oil, protein and glucosinolate content (Daun and Williams, 1995; Mika *et al.*, 2003) and FA composition (Velasco *et al.*, 1997a) without seed destruction in oilseed brassicas. However, development of species-specific calibrations or multiple species calibrations using a variable species sample set is most desired to address the species-specific variations in NIR absorbance pattern of the whole seed and specific base line shifts in NIR spectrums (Shenk and Westerhaus, 1993).

The seed coat colour also influences the absorbance or reflectance of the NIR. Thus, the calibrations for any seed component should contain both yellow- and brown-seeded genotypes for achieving a reliable spectroscopic evaluation (Van Deynze and Pauls, 1994). In addition, the environmental condition under which the crop is grown has a profound influence on various seed components. Thus, calibration sets composed of seeds obtained from different environments/year of harvest for the same genotype are also essential to take into account the environmental variations (Dardanne, 1996). However, much of work to utilise this versatile tool has been done for specific *Brassica* species, mostly *B. napus*, and information on the development of calibrations for multi-trait in other *Brassica* species is limited (Velasco *et al.*, 1998). Attempts have been made to develop NIRS calibrations for whole seeds of *B. juncea, B. rapa* and *B. napus* to estimate the seed biochemical components, irrespective of seed coat colour or year of harvest (Prem, 2006).

IV. THE PROGRESS TOWARDS NUTRITIONAL QUALITY IMPROVEMENT IN RAPESEED

Considering the reported adverse effects of consuming high erucic acid-containing oil and high glucosinolate-containing meal, the international efforts for quality improvement were initiated during early 1950s. Later the yellow-seeded cultivars were reported to have a thinner seed coat and lower fibre content, thus attributing to high oil content as compared to the brown/dark-seeded ones. Therefore, emphasis is being made on developing varieties having low erucic acid in the seed oil, low glucosinolate in the meal and yellow seed coat colour. Selection from the available gene pool and hybridisation-based gene transfer remains an important principle for rapeseed-mustard improvement towards the desired nutritional quality (Downey and Rimmer, 1993), and major achievements in this direction are reviewed below.

A. DEVELOPMENT OF GENOTYPES WITH OIL COMPOSITION MODIFICATIONS

The breeding efforts resulted in the development of low erucic acid strains of *B. napus* (Stefansson *et al.*, 1961) and *B. rapa* (Downey, 1964) during early 1960s, followed by those in *B. juncea* (Kirk and Oram, 1981) and *B. carinata* (Alonso *et al.*, 1991). Since the development and cultivation of double-low genotypes of *B. napus*, the maintenance of existing low erucic acid genotypes is more or less a routine exercise in several countries, and experimental work towards development and improvement of low erucic germplasm for other species is being pursued at global level (Rakow and Raney, 2003).

The first *B. rapa* low erucic acid cv. 'Span' was released for commercial cultivation in Canada in 1971 (Downey, 1990). The Canadian low erucic-breeding lines of *B. rapa* have been used to develop agronomically suitable cultivars for cultivation in Europe (Downey, 1990). *B. rapa* var. AC Parkland has higher linoleic and linolenic acids and lower oleic acid contents as compared to *B. napus*; however, it has a significantly lower (1–1.5%) SFA content, and has been utilised as a donor source in breeding for low SFA (>5%) *B. napus* through inter-specific crossing (Rakow and Raney, 2003). Among the Indian *B. rapa* germplasm evaluated for FA profile, the erucic acid content ranged between 50% and 59% and no source for low erucic acid trait could be found (Ahuja *et al.*, 1984). Therefore, transfer of low erucic acid trait has been attempted for *B. rapa* in India by inter-specific crossing of *B. napus* low erucic cv. 'Tower' to *B. rapa* cv. 'TL 15' followed by progeny advancement to F_7 generation (Badwal *et al.*, 1991). However, subsequent status of these low erucic acid lines/germplasm is not available.

Naturally occurring low erucic acid sources have not been reported for *B. carinata* as yet. Analysis of naturally occurring germplasm lines (more than 250 accessions), mostly originating from Ethiopia showed erucic acid range from 28 to 41% (Alemayehu *et al.*, 1999; Rakow, 1995). The first reported low erucic acid lines for this species were developed in 1988 by reciprocal inter-specific hybridisation between a German low erucic *B. napus* genotype 'Duplo' and a Spanish *B. carinata* germplasm 'C-101' followed by recurrent backcrossing to *B. carinata* parent (Fernandez-Sirrano and Alonso, 1988; Fernandez *et al.*, 1988). This development resulted in the generation of early maturing low erucic *B. carinata* genotypes suitable for Europe (Alonso *et al.*, 1991). In Canada, low erucic *B. juncea* × high erucic *B. carinata* inter-specific cross has been utilised to generate low erucic germplasm in *B carinata* (Getinet *et al.*, 1994). However, the reduction of erucic acid in *B. carinata* resulted in a simultaneous trade-off increase in PUFA content (>35% linoleic and >20% linolenic acid) and low levels (<35%) of oleic acid (Rakow, 1995). FA profile mutants of *B. carinata* produced from ethyl methanesulphonate treatment were reported by Velasco *et al.* (1995b, 1997b) for high oleic/low linolenic and high linoleic/low linolenic acid types. These mutants have been used in crossing programmes aimed at developing low erucic and high oleic germplasm, and preliminary results of F_2 segregants from these crosses show that such FA recombinants may be stabilised in good agronomic background in the near future (Nabloussi *et al.*, 2003). Although the prospects of using *B. carinata* as a potential arid and semi-arid region oilseed crop have been discussed with selection of early maturing types, the nutritional quality improvement of Indian germplasm has not been addressed as yet (Malik, 1990; Raut, 1996).

Kirk and Oram (1981) for the first time identified zero erucic genotypes of *B. juncea* christened as ZEM-1 and ZEM-2 from cultivars of Indian and Chinese origin that have been used as donors for low erucic acid in development of many cultivars. Apart from reduction in erucic acid, generation of genotypes with variable FA profile is desired for several multi-purpose edible/non-edible uses. The high oleic, low linolenic acid *B. juncea* genotypes have been reported by several scientists (Oram *et al.*, 1999; Potts *et al.*, 1999). The high oleic acid-containing *B. napus* and *B. juncea* with better shelf life have also been obtained (Stoutjesdijk *et al.*, 1999). Linolenic acid is an EFA having cholesterol reducing role (Eskin *et al.*, 1996); however, high linolenic (>2.5%) acid content is undesirable as it decreases the shelf life of the oil. Both mutagenesis and transgenic modifications have been used to reduce linolenic acid content in *B. napus* (DeBonte and Hitz, 1996; Scarth and McVetty, 1999). In addition, successful attempts have been made to produce *B. napus* lines with reduced linolenic and increased linoleic acid contents through chemical/UV mutagenesis (Cegielska-Taras *et al.*, 1999; Rakow, 1973).

The exotic *B. juncea* strains having low erucic acid were not suitable under Indian agroclimatic conditions due to late maturity and poor seed set (Anonymous, 1994). Gupta *et al.* (1994, 1998) identified low erucic acid genetic stock among Indian accessions of *B. juncea*. Several low erucic acid genotypes of *B. juncea* and *B. napus* have been developed in India either through inter-specific hybridisation (Banga *et al.*, 1988; Khalatkar *et al.*, 1991; Malode *et al.*, 1995) or through selection of transgressive segregants as a result of inter-specific/inter-generic hybridisation (Agnihotri and Kaushik, 1998, 1999a; Agnihotri *et al.*, 1995), followed by pedigree method. The *B. juncea* strains ZEM-1 and ZEM-2 have been utilised to transfer low erucic acid content into high erucic acid-containing Indian cv. RLM 619, RH 30 and RL 1359 (Chauhan *et al.*, 2002). *B. juncea* strains with high oleic acid (up to 50%) in comparison to low oleic acid in the cultivated varieties (10–25%) have also been identified (Agnihotri and Kaushik, 2001). Transfer of low erucic acid (Ahuja, 1990) and high oleic acid (Agnihotri *et al.*, 2006; Kaushik and Agnihotri, 2003) contents in *B. juncea* from *B. napus* has also been reported (Raney *et al.*, 2003a). Some of the low erucic/high oleic acid strains of rapeseed-mustard registered at NBPGR, ICAR are TERI(OE) M21, TERI(OE)R03 and TERI(OE)R09. Several low erucic acid genotypes of *B. juncea*, such as TERI(OE)M21, TERI(OE)M9901, TERI(OE)M9902, LEB-15, LES-39, PBCM-8-2, YSRL-9-18-2, CRL-1359-19, PRQ-9701, and of *B. napus*, such as TERI(OE)R03, TERI(OE)R05, TERI(OE)R15, TERI (OE)R983, TERI(OE)R984, have been developed and tested under the All India Coordinated Research Project on Rapeseed-Mustard (AICRPRM), ICAR but they could not compete in yield with the existing high erucic varieties (Agnihotri and Kaushik, 2003a; Anonymous, 2002, 2006; Chauhan *et al.*, 2000). Further, Chauhan *et al.* (2002) studied the genetic inheritance of six generations of *B. juncea* obtained by crossing var. Varuna with low erucic acid Indian strain TERI (OE) M21 (INGR No. 98001) and have selected lines with high oleic and low erucic acid in Indian mustard. Simultaneously, efforts have also been made to develop low erucic acid/high oleic acid-containing *B. napus* suitable for mustard-growing belts, and TERI-Unnat (INGR No. 98001) was identified for release by AICRPRM, ICAR (Anonymous, 2001). Recently, the low erucic acid *B. juncea* strain LES-39 developed by IARI has also been released. However, most of these strains need improvement in their agronomic attributes and efforts are underway to utilise the developed strains and incorporate low erucic acid traits in the background of genetically superior genotypes.

Advances in recombinant DNA technologies have provided an alternative method for developing varieties with modified quality traits such as seed oil or protein composition either through manipulation of existing pathways or by addition of novel biochemical pathways (Murphy, 1996). The use of seed-specific antisense technology has allowed for the selective modulation of key enzyme activities in the developing seed, while keeping the rest of the genetic background of the plant intact (Kuntzon *et al.*, 1992). Using this concept, transgenic lines of *B. napus* var. Westar having high palmitic and stearic acid have been developed (Hitz *et al.*, 1995). Australian scientists have reported development of cosuppression system based on post-transcriptional gene silencing of endogenous desaturase gene that resulted in development of high oleic acid genotypes of rapeseed-mustard (Stoutjesdijk *et al.*, 1999). This RNAi approach has also shown great potential for total endogenous desaturase silencing and using this concept total silencing of the Δ12 desaturase gene in *B. napus* has been achieved, resulting in the production of genotypes accumulating 89% oleic acid in the seed oil (Smith *et al.*, 2000).

The rapeseed-mustard oil normally contains low levels of stearic acid and lauric acid at a concentration of 1–2% and 0.1–0.2%, respectively. High lauric rapeseed can be used as a substitute in detergent markets, leading to replacement of conventional lauric oils derived from coconut or palm kernel, whereas high stearic rapeseed is a useful substitute in margarine markets and replaces conventional hydrogenated vegetable oil. The two most notable achievements in oil modification through transgenic technology to date are the 40% stearic and 40% lauric rapeseed varieties (Laurical) first produced and entered in field trials by Calgene in 1993–1994 (Murphy, 1995). Laurical was the first genetically manipulated rapeseed variety given permission for commercial cultivation in 1995 in United States. The Δ9 stearoyl ACP desaturase gene that normally converts stearic to oleic acid was partially inactivated in rapeseed using antisense technology, resulting in the accumulation of seed oil containing up to 40% stearic acid (Kuntzon *et al.*, 1992). This high stearic variety contains an antisense copy of a *Brassica* stearate desaturase gene, which inhibits the function of the normal rapeseed stearic desaturase gene, resulting in an accumulation of stearic acid, rather than their saturation to oleate. The resulting high stearic oil has many advantages over the normal rapeseed oil for the production of certain solid fats such as margarines.

The development of a *B. rapa* strain with enhanced palmitic and stearic acid has also been reported by Swedish scientists with the aim of developing rapeseed-mustard cultivars suitable for production of margarine (Persson, 1985). The possibility of producing moderately high palmitic (up to 17%) and stearic acid (up to 10%) segregants from inter-specific crosses of *B. juncea* and *B. carinata* has also been reported (Gupta, 2006). Reduction in saturated fat content (<5%) in *B. napus* has been targeted through selection in breeding populations (Gororo *et al.*, 2003), inter-specific crosses with *B. rapa* and *B. alboglabra* and mutagenesis (Raney *et al.*, 2003a) in Australia and Canada, respectively. Reduction of SFA in rapeseed by a conjugated haploid mutagenesis, *in vitro* selection and DH production approach has also been reported. The mutant embryos, produced from isolated microspores subjected to UV treatment (254 nm directly at Petri dish for 90 s), were matured at elevated temperature (32°C), resulting in artificially elevated saturate accumulation and subsequent selection of low SFA mutant embryos (Beaith *et al.*, 2005). These FA variants will be useful donors for speciality oil sector.

In addition to the FA modification, presence of adequate amounts of vitamin A (β-carotene) and tocopherols (TOC) are recently emerging important parameters. The rapeseed having high carotenoid canola oil (HCCO) having β-carotene up to 960 μg/g has been developed (Shewmaker *et al.*, 1999) and is further being utilised in breeding programme to transfer this trait in *B. juncea* through transgenic approach at TERI (Dhawan, 2002; Saxena, 2006). This will be extremely beneficial in combating the vitamin A deficiency that poses a serious health hazard specially in developing nations. TOC are the polynutrients that have bioactivity as vitamin E and are considered important as nutraceuticals, being a key player in human health and nutrition. They are found in eight different but related structures as tocopherols or tocotrienols, both acting as anti-oxidants (Bramley *et al.*, 2000; Kalmal-Edin and Appelqvist, 1996). The tocopherol content has been noted to have an inverse relationship with oil content, thus selections for high oil content have resulted in a corresponding decreased levels of TOC (Marquard, 1990). Efforts have been made to increase the γ-TOC (that reduces the oxidation of PUFA thus providing better storage potential) or α-TOC (that are active fat-soluble vitamin E necessary for reproductive processes). In the German joint project NAPUS 2000, both conventional breeding procedures and genetic engineering techniques have been applied to create novel genetic variants for TOC composition (Leckband *et al.*, 2003). Increased TOC levels will form the basis of value addition in canola rapeseed oil in the human health sector (Luhs *et al.*, 2003).

B. DEVELOPMENT OF LOW GLUCOSINOLATE GENOTYPES

The Polish *B. napus* cv. 'Bronowski' containing glucosinolate content of about 12-μm/g oil-free meal and 7–10% erucic acid in the seed oil was discovered in 1967 and subsequently used globally as a source of low glucosinolate. Till date the 'Bronowski' gene is the only known source for low glucosinolate content and no natural germplasm source for stable low glucosinolate genes has been reported in *B. juncea*. The low glucosinolate *B. rapa* germplasm was also reported a few years after the discovery of 'Bronowski' (Downey *et al.*, 1969); however, the attempts to transfer this low glucosinolate trait into cultivated varieties did not succeed (Robbelen, 1980). A low glucosinolate-breeding line of *B. rapa* subsp. *chinensis* had also been reported from the Slovak republic (Jonsson and Uppstrom, 1986) but further utilisation of this source in development of low glucosinolate *B. rapa* or transfer to any other *Brassica* species has not been reported. Instead, the Bronowski gene block has been transferred to *B. rapa* via inter-specific hybridisation followed by repeated backcrossing to the *B. rapa* parent (Downey, 1990; Rakow and Raney, 2003).

Alemayehu *et al.* (1999) have reported the total glucosinolate range of 29- to 165-μmol/g meal in 236 germplasm collections and 180 inbred lines of *B. carinata*. This diversity reveals that there could be a potent source for low glucosinolate content in this species. However, exploitation of *B. carinata* germplasm for development of low glucosinolate genotypes has not been reported as yet. Efforts are being made to use doubled haploidy in conjugation with chemical mutagenesis for developing low glucosinolate *B. carinata*. The DH technique originally developed for *B. napus* (Coventy *et al.*, 1988) has been used for producing haploid embryos, and the mutagenised haploid embryos were evaluated for glucosinolate content by tes-tape method, and preliminary results showed that several embryos had low glucosinolate content (Kott, 1995).

In *B. juncea* Love *et al.* (1990) identified low glucosinolate mustard 'BJ-1058' in a BC_1F_3 progeny through inter-specific hybridisation between *B. juncea* and *B. rapa* containing the 'Bronowski' gene. The line generated, although late in maturity, showed stable incorporation of low glucosinolate gene and has been utilised globally as a useful donor source for low glucosinolate content. In Australia, considerable success has been achieved in development of low glucosinolate genotypes using mutagenesis, inter-specific hybridisation and tissue culture coupled with pedigree selection (Oram *et al.*, 1999). Three strategies have been mainly used for reduction of glucosinolate content, viz. (1) irradiation of ZEM-1 seeds with γ-rays and recovery of LG 1-3 and LG 4-3 breeding lines with reduced glucosinolate content (Oram and Kirk, 1993); (2) re-synthesis of *B. juncea* using *B. nigra* and low glucosinolate *B. rapa*, the developed synthetic *B. juncea* was crossed to cv. Lethbridge to produce low glucosinolate-breeding line Syn Y (Oram *et al.*, 1997) and (3) a somaclonal variant (SCL 4) with reduced glucosinolate content was developed from tissue culture of Indian accession PI 183117 (Palmer *et al.*, 1996).

In India, efforts have been directed towards developing *B. juncea* lines that have low glucosinolate content along with other agronomic and quality traits (Anonymous, 2006). However, the desired success could not be achieved due to the lack of suitable donors and a complex genetics of total glucosinolate biosynthesis (Vageeshbabu and Chopra, 1997). The transgressive segregants (TERI 5 and TERI 6) having low glucosinolate, selected from inter-specific/inter-generic crosses (Agnihotri *et al.*, 1995) and Canadian accession BJ-1058, were utilised to develop low glucosinolate genotypes in the background of popularly grown *B. juncea* var. Pusa Bold (Agnihotri and Kaushik, 2003a). Some of the promising *B. juncea* genotypes developed/identified for reduced glucosinolate content include K 19-11, K19-12, RC 182, RC 184 and CM 1;

however, the stability and heritability of the low glucosinolate trait is yet to be tested (Chauhan *et al.*, 2000). The low glucosinolate genotypes TERI-LGM 06 and LGM-08 were tested under multi-location trials of AICRPRM, ICAR (Anonymous, 2006). Although these low glucosinolate lines can be utilised as useful donor sources in breeding programmes, they need further improvement for agronomic characteristics.

Efforts have also been made for systematic characterisation of Indian and exotic germplasm lines to search for potential low glucosinolate donor sources. More than 1200 germplasm lines of *Brassica* species and related genera, obtained from National Bureau of Plant Genetic Resources, ICAR, were analysed, comprising 651 *B. juncea*, 443 *B. rapa*, 35 *B. napus*, 10 *B. nigra*, 18 *B. carinata*, 6 *B. chinensis*, 63 *Eruca sativus*, 5 *Sinapis alba*, 3 Raphanus sp., 5 *B. tourniforti*, 1 *Crambe asyssinica* and 18 *B. rapa*. Among these, only a few lines (exotic collections) with low or nearly low glucosinolate content (<30-μmol/g meal) were identified: 20 of *B. napus*, 2 *B. juncea*, 1 *B. carinata*, 1 *C. asyssinica* and 2 *E. sativa* (TERI, 2004; http://www.teriin.org/reports/rep206/rep206.htm). The majority of germplasm lines screened had high glucosinolate content 120–180 μmol/g and all Indian accessions of *B. juncea* and *B. rapa* had high glucosinolate (65–363-μmol/g meal). Hence, there is a need for further characterisation of all available Indian accessions of *Brassica* species for their quality to identify any potential donor source for low glucosinolate gene.

C. DEVELOPMENT OF DOUBLE-LOW GENOTYPES

The 'Canola', commonly known as double-low or 'OO' (Downey, 1990), that designates rapeseed having less than 2% erucic acid in the seed oil and less than 30-μmol glucosinolate per gram oil-free meal, is an internationally accepted trade standard considered ideal for food and feed purposes. In addition to low erucic acid and low glucosinolate, yellow seed coat colour is another desired characteristic that adds to the high oil content, high protein, low crude fibre and low polyphenols. Therefore, during recent years, the concept of 'OO' is being expanded to 'OOO' to include this as one of the major breeding objectives.

Efforts towards development of double-low cultivars led to the release of world's first double-low *B. napus* cv. Tower and the world's first double-low *B. rapa* cv. Candle. Several *B. napus* varieties having canola quality and high oil content such as Cyclone (from Denmark), Shiralee (from Australia) and AC Excel (from Canada) are available, and many *Brassica* species having double-low traits have been reported from several countries (Rakow, 1995). Yellow-seeded *B. napus* lines with canola quality, higher protein and oil content have been developed through inter-specific crossing between yellow-seeded *B. rapa* 'yellow sarson' with canola quality *B. napus*, followed by development of doubled haploids and pedigree breeding (Rahman, 2001). In Russia, inter-specific crossing between *B. rapa* × *B. napus*, *B. juncea* × *B. napus* and physical mutagenesis using 100 Kr γ-irradiation, followed by selection, resulted in recovery of yellow-seeded, low erucic, low glucosinolate (OOO) forms of spring rape (*B. napus*) and turnip rape (*B. rapa*). However, low resistance to major fungal diseases and pests and uneven ripening are the major drawbacks of the recovered winter/spring types of rape as well as turnip rape (Bochkaryova *et al.*, 2003). The breeding efforts in canola quality double-low *B. napus* are now mainly being focussed on value addition through improving the oil composition (developing high oleic, low linolenic, low saturated fat cultivars) or oil enrichment (enhancing vitamin levels) for specialty segments in many countries including Canada (Raney *et al.*, 2003a,b), Australia (Gororo *et al.*, 2003), United States (Corbett and Sernyk, 2003), Germany (Leckband *et al.*, 2003; Luhs *et al.*, 2003), France (Carre *et al.*, 2003) and Poland (Spasibionek *et al.*, 2003).

Following the success of inter-specific introgression of low glucosinolate genes, Love *et al.* (1991) developed 'OO' *B. juncea* strain LDZ. This has been further improved and has shown promising results in field trials (Rakow and Raney, 1995). The glucosinolate content of the lines developed from this pedigree has been further reduced to 5.5-μM/g meal by backcrossing and recurrent selection resulting in the development of breeding line J00-6866 (Rakow and Raney, 2003). Along with this, Raney *et al.* (2003a) have reported simultaneous improvement of oil quality and agronomic characteristics in double-low *B. juncea*. A breeding line developed from the material generated by Love *et al.* (1991), namely J90-4253, was crossed with a low linolenic acid *B. napus* genotype and the inter-specific F_1 was backcrossed five times to J90-4235. From the segregating backcross generation, a stable high oleic (56.2%) near double-low (glucosinolate 53-μM/g meal and zero erucic) breeding line TO00-5987 has been identified for high yielding capability (Raney *et al.*, 2003b). In Australia, development of double-low *B. juncea* germplasm has been reported using the low erucic genotypes as starting material (Burton *et al.*, 2003a). Several double-low *B. juncea* cultivars, having erucic acid-free oil and low glucosinolate (8.3- to 24.2-μM/g meal), have shown promising yield potential in multi-location trials (Burton *et al.*, 2003b).

In India, Khalatkar *et al.* (1991) have reported the introgression of 'OO' characteristics in *B. juncea* genotype Pusa Bold. The strain developed was designated as Heera; however, due to exotic nature, it was found unsuitable for Indian agroclimatic conditions. Using Heera as a donor for low glucosinolate genes, attempts have been made to introgress the 'OO' characteristics in Indian mustard cv. Pusa Bold, but the desired success is yet to be achieved (Malode *et al.*, 1995). Agnihotri and Kaushik (1999a, 2003a), using low erucic acid donors TERI (OE) M21 (INGR No. 98001) and ZEM-1, and the low glucosinolate donor line BJ-1058, have reported successful incorporation of double-low characteristics in Indian *B. juncea* cv. Varuna, through a three-way cross [(Varuna × ZEM-1/TERI (OE) M21) × BJ-1058] followed by backcrossing and selections by pedigree method. Three double-low *B. juncea* genotypes, NUDH-YJ-5 (INGR No. 03034), Heera (INGR No. 03033) and TERI-GZ-05 (INGR No. 04078) having yellow seed coat colour have been registered by ICAR, and are being used for further improvement in agronomic traits (Anonymous, 2004, 2006). The use of DH technology along with chemical mutagenesis has resulted in the development of stable low erucic, moderately low glucosinolate and yellow-seeded DH *B. juncea* mutants (Prem *et al.*, 2006). These mutants are being evaluated for their yield potential and form important germplasm source for genetic improvement in *B. juncea*.

Utilising the early maturing, shattering tolerant, low erucic acid *B. napus* line TERI (OE) R03 (INGR No. 98002), and exotic *B. napus* varieties Cyclone and Shiralee as pollen donors, several early maturing canola quality strains of *B. napus* have been developed and registered at NBPGR, ICAR (Agnihotri *et al.*, 2004) that include TERI(OO)R985, TERI(OO)R986 and TERI(OO)R9903 (Agnihotri and Kaushik, 1999b, 2003b). The promising strains of rapeseed that performed better than the national check variety of *B. napus* (GSL-1) under AICRP R&M, ICAR are TERI(OO)R9903, GSC 865-2 and GSC-3A; however, they could not compete with the best national check of *B. juncea*. The double-low *B. napus* GSC-865 and TERI-Uttam-Jawahar [TERI(OO)R9903] have been released for cultivation in the states of Punjab and Madhya Pradesh, respectively. Apart from these, a double-low *B. napus* hybrid 'Hyola 401' has been released for commercial cultivation and is being marketed by Advanta India Ltd. (http://www.advantageindia.com).

V. CONCLUSION

Development of designer canola oil is in progress at the global level to generate speciality oils. Modifications in the composition of FAs have been achieved through various conventional methods in association with biotechnological techniques such as induced mutation, *in vitro* embryo rescue, doubled haploid and genetic engineering, especially post-transcriptional gene silencing. Some of the notable alterations include reduced erucic acid to <2%, increased oleic acid >60%, reduced linolenic acid <5%, and reduced palmitic acid up to 4%, reduced SFA to <5% in rapeseed-mustard oils. Introduction of genetic engineering techniques has brought an enormous increase in oleic acid from an average 15–20% to 90% in canola. Besides the FA profile, low glucosinolate content in the seed meal lends a potential for broader future utilisation in various food and feed applications. In India, until the late 1980s, the emphasis had mainly been on breeding for increased yields rather than the quality. With the launch of the Technology Mission on Oilseeds during mid-1980, which helped in increased production, equal emphasis is being laid on improvement in nutritional quality. The biggest challenge is to combine the double-low characteristics with good yielding capability, all of which are quantitative traits governed by multiple recessive genes. The genetic engineering method to transfer the genes for low erucic acid and low glucosinolate in the high yielding varieties would have been helpful in achieving this target, but is still in nascent stage. A National Network on Improvement of Quality of Oilseed Brassica was set up by ICAR during 1996–1997, involving five actively cooperating centres, to utilise the half seed method and efficient analytical techniques, to expedite the conventional breeding methods and to transfer the genes for improved quality traits in the background of agronomically acceptable Indian mustard. However, despite the initial success achieved in realising nutritionally superior *B. juncea* genotypes with low erucic acid, low glucosinolate as well as yellow seed coat and triple low traits, further efforts are required towards developing exploitable genetic variability in double-low *B. juncea* with adequate agronomic performance under Indian agroclimatic conditions. Notable success has been achieved by developing canola quality *B. napus* strains with early maturity and shatter tolerance suitable for mustard-growing belts in India. Furthermore, development of novel genetic variants for developing designer oilseed crops with modified quality traits such as seed oil or protein composition, and oil enrichment with enhanced vitamin levels is a step ahead in the direction of value addition in canola rapeseed oil in the human health sector which would poise it to maintain a favourable position in the world market.

ACKNOWLEDGMENT

We gratefully acknowledge the guidance provided by Dr. R. K. Downey, Emeritus Scientist, Agriculture and Agri-Food Canada, Saskatoon, Canada for the chapter.

REFERENCES

Abraham, V. and Bhatia, C. R. (1986). Development of strains with yellow seed coat in Indian mustard (*Brassica juncea* Czern and Coss). *Plant Breeding* **97**, 86–88.

Agnihotri, A. (1999). Screening techniques for evaluation of quality traits in oilseed. *Brassicas Journal of Oilseeds Research* **16**(2), 183–195.

Agnihotri, A. and Kaushik, N. (1998). Transgressive segregation and selection of zero erucic acid strains from intergeneric crosses of *Brassica*. *Indian Journal of Plant Genetic Resources* **11**(2), 251–255.

Agnihotri, A. and Kaushik, N. (1999a). Genetic enhancement for double low characteristics in Indian rapeseed mustard. *In* "Proceedings of the 10th International Rapeseed Congress." Canberra, Australia, 26–29 September 1999.

Agnihotri, A. and Kaushik, N. (1999b). Transfer of double low characteristics in Indian *B. napus*. *Journal of Oilseeds Research* **16**(2), 227–229.

Agnihotri, A. and Kaushik, N. (2001). Introgression of high oleic acid content in Indian mustard. *In* "Proceedings Diamond Jubilee Symposium on Hundred Years of Post-Mendelian Genetics and Plant Breeding." IARI, New Delhi, February 22–23.

Agnihotri, A. and Kaushik, N. (2002). Quality considerations in edible oilseeds: Rapeseed and Mustard. *In* "Oilseeds and Oils: Research and Development Needs" (M. Rai, H. Singh and D. M. Hegde, eds.), pp. 315–326. Indian Society of Oilseeds Research, Hyderabad.

Agnihotri, A. and Kaushik, N. (2003a). Towards nutritional quality improvement in Indian mustard (*Brassica juncea* [L]. Czern and Coss) var. Pusa Bold. *In* "Proceedings of the 11th International Rapeseed Congress" (H. Sorensen, J. C. Sorensen, S. Sorensen, N. B. Muguerza and C. Bjergegaard, eds.), Vol. 2, pp. 501–503. The Royal Veterinary and Agricultural University, Copenhagen, Denmark, 6–10th July.

Agnihotri, A. and Kaushik, N. (2003b). Combining canola quality, early maturity and shattering tolerance in *B. napus* for Indian growing conditions. *In* "Proceedings of the 11th International Rapeseed Congress" (Hilmer Sorensen, Jens Christian Sorensen, Susanne Sorensen, Natalia Bellostas Muguerza and Charlie Bjergegaard, eds.), Vol. 2, pp. 436–439. The Royal Veterinary and Agricultural University, Copenhagen, Denmark, 6–10th July.

Agnihotri, A. and Kumar, S. (2004). Value addition in rapeseed-mustard: Status and prospects. *In* "Rapeseed-Mustard Research in India" (A. Kumar and N. B. Singh, eds.), pp. 212–231. National Research Centre on Rapeseed Mustard. ICAR, Sewar, Bharatpur.

Agnihotri, A., Raney, J. P., Kaushik, N., Singh, N. K. and Downey, R. K. (1995). Selection for better agronomical and nutritional characteristics in Indian rapeseed-mustard. *In* "Proceedings of the Ninth International Rapeseed Congress," Vol. 2, pp. 425–427. Cambridge, UK, 4–7 July.

Agnihotri, A., Kaushik, N., Sarkar, G., Prem, D. and Gupta, K. (2004). Genetic enhancement in rapeseed-mustard for quality traits. *In* "Biotechnology for Food and Nutritional Security" (Vibha Dhawan, ed.), pp. 119–143. TERI, New Delhi.

Agnihotri, A., Sarkar, G., Kaushik, N., Gupta, K. and Prem, D. (2007). Introgression of high oleic acid in Indian Mustard through inter-specific hybridisation. *In* "Proceedings of the 12th GCIRC International Rapeseed Congress," to be held at Wuhan, China, 26–30th March 2007. Vol. 1 (Genetics and Breeding) pp. 188–190. Science Press USA Inc., New Jersey.

Ahuja, K. L. (1990). Quality improvement research on rapeseed mustard. *In* "Proceedings of an Indo-Sweden Symposium," pp. 23–33, 4–6.

Ahuja, K. L. and Banga, S. K. (1992). Oil and meal quality. *In* "Breeding oilseed Brassicas" (K. S. Labana, S. S. Banga and S. K. Banga, eds.), pp. 76–93. Narosa Publishing House, New Delhi.

Ahuja, K. L., Labana, K. S., Raheja, R. K. and Badwal, S. S. (1984). Oil content and fatty acid variation in mutants of *Brassica juncea*. *Journal of Oilseeds Research* **1**, 71–75.

Alemayehu, N., Becker, H. and Gebeyehu, G. (1999). Genetic variability in Ethiopian mustard (*Brassica carinata* Braun) for quality characteristics. *In* "Proceedings of the 10th International Rapeseed Congress." Canberra, Australia in proceedings CD, September 26–29.

Alonso, L. C., Fernandez-Serrano, O. and Fernandez-Escobar, J. (1991). The outset of a new oilseed crop: *Brassica carinata* with low erucic acid. *In* "Proceeding of the Eighth International Rapeseed Conference," pp. 170–176. Saskatoon, Canada.

Anand, I. J. and Downey, R. K. (1981). A study of erucic acid alleles in digenomic rapeseed (*Brassica napus* L) *Canadian. Journal of Plant Science* **61**, 199–203.

Anand, I. J., Reddy, W. R. and Rawat, D. S. (1985). Inheritance of seed colour in mustard. *Indian Journal of Genetics* **45**, 34–37.

Anderson, O. E. and Worthington, R. E. (1971). Boron and manganese effects on protein, oil content and fatty acid composition in cottonseed. *Agronomy Journal* **63**, 566–569.

Anonymous (1994). Annual Progress Report 1993–94. All India Coordinated Research Project on Rapeseed- mustard, ICAR. National Research Centre on Rapeseed- mustard, Bharatpur, Rajasthan, India.

Anonymous (1995). Annual Progress Report 1994–95. All India Coordinated Research Project on Rapeseed- mustard, ICAR. National Research Centre on Rapeseed- mustard, Bharatpur, Rajasthan, India.

Anonymous (2000). AOCS official method Am 2–93. Sampling and analysis of vegetable oil source materials-Determination of oil content in oilseeds *American Oil Chemist Society international*, CD-ROM. http://www.aoac.org/pubs/omacd_rom_revised.htm.

Anonymous (2001). Annual Progress Report 2000–01. All India Coordinated Research Project on Rapeseed- mustard, ICAR. National Research Centre on Rapeseed- mustard, Bharatpur, Rajasthan, India.

Anonymous (2002). Annual Progress Report 2001–02. All India Coordinated Research Project on Rapeseed- mustard, ICAR. National Research Centre on Rapeseed-mustard, Bharatpur, Rajasthan, India.

Anonymous (2003). Annual Progress Report 2002–03. All India Coordinated Research Project on Rapeseed- mustard, ICAR. National Research Centre on Rapeseed- mustard, Bharatpur, Rajasthan, India.

Anonymous (2004). Annual Progress Report 2003–04. All India Coordinated Research Project on Rapeseed- mustard, ICAR. National Research Centre on Rapeseed- mustard, Bharatpur, Rajasthan, India.

Anonymous (2006). Annual Progress Report 2005–06. All India Coordinated Research Project on Rapeseed- mustard, ICAR. National Research Centre on Rapeseed- mustard, Bharatpur, Rajasthan, India.

Appelqvist, L. A. (1968). Rapid methods of lipid extraction and fatty acid methyl ester preparation for seed and leaf tissue with special remarks on preventing the accumulation of lipid contaminants. *Arkiv Kemi* **28**, 551–570.

Appelqvist, L. A. (1969). Lipids in Cruciferae. IV. Fatty acid patterns in single seeds and seed populations of various Cruciferae and in different tissues of *Brassica napus* L. *Heriditas* **61**, 9–44.

Appelqvist, L. A. (1970). Lipids in Cruciferae. VI. The fatty acid composition of seeds of some cultivated *Brassica* species and of *Sinapis alba* L. *Fette Seifen Anstrichm* **72**, 783–792.

Appelqvist, L. A. and Josefsson, E. (1967). Method for quantitative determination of isothiocyanates and oxazolidinethiones in digests of seed meals of rapeseed and turnip rape. *Journal of Science and Food Agriculture* **18**, 510–519.

Avery, D. T. (1998). It's time for the world to get realistic about meat. *Outlook on Agriculture* **27**(1), 35–40.

Badwal, S. S., Ahuja, K. L. and Raheja, R. K. (1991). Selection for double low Indian *Brassica campestris* lines from *Brassica napus* × *Brassica campestris*. *In* "Proceedings of the GCIRC Eighth International Rapeseed Congress." Saskatoon, Canada, 9–11 July.

Bajjalieh, N. (2002). Proteins from oilseeds. *In* "Proceedings Protein Sources for the Animal Feed Industry–Expert Consultation and Workshop." Bangkok 29th April–3rd May 2002.

Bandel, V. A., Mulford, F. R., Ritter, R. L., Kantzes, G. J. and Hellman, J. L. (1991). Fact Sheet- 635: Canola production guidelines University of Maryland and USDA http://www.agnr.umd.edu/MCE/Publications/PDFs/FS635.pdf.

Banga, S. K. (1996). Breeding for oil and meal quality. *In* "Oilseed and Vegetable Brassicas: Indian Perspective" (V. L. Chopra and S. Prakash, eds.), pp. 234–249. Oxford and IBH Publishing Company, New Delhi.

Banga, S. S., Banga, S. K., Gupta, M. L. and Sandhya, G. S. (1988). Synthesis of genotypes having specialized fatty acid composition in Indian mustard (*Brassica juncea*). *Crop Improvement* **25**(1), 21–25.

Barro, F., Fernandez-Escobar, J., Vega de la, M. and Martin, A. (2001). Doubled haploid lines of *Brassica* carinata with modified erucic acid content through mutagenesis by EMS treatment of isolated microspore. *Plant Breeding* **120**(3), 262–264.

Beadle, J. B., Just, D. E., Morgan, R. E. and Reiners, R. A. (1965). Composition of corn oil. *Journal of American Oil Chemist Society* **42**, 90–95.

Bell, J. M. (1995). Meal and by-product utilisation in animal nutrition. *In* "Brassica Oilseeds: Production and Utilisation" (D. Kimber and D. I. McGregor, eds.), pp. 301–337. CABI, Wallingford, UK.

Bhatia, C. R. and Mitra, R. (1992). Bioenergetic limitations to breeding better *Brassicas*. Chapter 5. *In* "Breeding oilseed Brassicas" (K. S. Labana, S. S. Banga and S. K. Banga, eds.), pp. 67–75. Narosa Publishing House, New Delhi.

Bille, N., Eggum, B. O., Jacobsen, I., Olseno, O. and Sorensen, N. (1983). Antinutritional and toxic effects in rats of individual glucosinolates (+) myrosinases added to a standard diet I. Effects on protein utilization and organ weight. *Tierphysiol Tierer nahar Futter-mittelkd* **49**, 195–210.

Bjerg, B. and Sorenson, H. (1986). Quantitative analysis of glucosinolates in oilseed rape based on HPLC of desulfoglucosinolates and HPLC of intact glucosinolates. *In* "Proceedings of CEC Workshop Glucosinolates in Rapeseeds Gembloux," 1–3 October, A Paraitre.

Bjorkman, R. (1972). Preparative isolation and 35S-labeling of glucosinolates from rapeseed (*Brassica napus* L.). *Acta Chemica Scandinavia* **26**, 1111–1116.

Bochkaryova, E. B., Gorlov, S. L., Serdyuk, V. V. and Khalilova, L. A. (2003). Trends and results of rapeseed and turnip rape breeding. *In* "Proceedings of the 11th International Rapeseed Congress," Vol. 2, pp. 445–447. The Royal Veterinary and Agricultural University, Copenhagen, Denmark, 6–10th July.

Bonanone, A., Pagnan, A., Biffani, S., Opportuno, A., Sogato, H., Dorella, M., Maiorino, M. and Potts, D. A. (1992). Effects of dietary monounsaturated and polyunsaturated fatty acids on the susceptibility of plasma LDL to oxidative modification. *Arteriosclerosis and Thrombosis* **12**, 529–533.

Bramley, P. M., Elmadfa, I., Kafatos, A., Kelly, F. J., Manios, Y., Roxborough, H. E., Schuch, W., Sheehy, P. J. A. and Wagner, K. H. (2000). Vitamin E. *Journal of the Science of Food and Agriculture* **80**, 913–938.

Broun, P., Shanklin, J., Whittle, E. and Somerville, C. (1998). Catalytic plasticity of fatty acid modifications enzymes underlying chemical diversity of plant lipids. *Science* **282**, 1315–1317.

Brzezinski, W. and Mendelewski, P. (1984). Determination of total glucosinolate content in rapeseed meal with thymol reagent. *Zeitschrift fuer Pflanzenzuchtung* **93**, 177–183.

Brzezinski, W., Mendelewski, P. and Musse, B. G. (1986). Comparative study on determination of glucosinolates in rapeseed. *Cruiceferrae Newsletter* **11**, 128–129.

Burton, W., Saisbury, P. and Potts, D. (2003a). Inheritance of allyl glucosinolate in the development of canola quality *Brassica juncea* for Australia. *In* "Proceedings of the 11th International Rapeseed Congress," p. 280. The Royal Veterinary and Agricultural University, Copenhagen, Denmark, 6–10th July.

Burton, W., Saisbury, P. and Potts, D. (2003b). The potential of canola quality *Brassica juncea* as an oilseed crop for Australia. *In* "Proceedings of the 11th International Rapeseed Congress," pp. 5–7. The Royal Veterinary and Agricultural University, Copenhagen, Denmark, 6–10th July.

Cao, Y. Z. and Huang, A. H. C. (1987). Acyl coenzyme A preference of diacylglycerol acyltransferase from the maturing seeds of Cuphea, maize, rapeseed and canola. *Plant Physiology* **84**, 762–765.

Carre, P., Dartenuc, C., Evrard, J., Judde, A., Labalette, F., Raoux, R. and Renard, M. (2003). Frying stability of rapeseed oils with modified fatty acid composition. *In* "Proceedings of the 11th International Rapeseed Congress," Vol. 2, pp. 540–543. The Royal Veterinary and Agricultural University, Copenhagen, Denmark, 6–10th July.

Cegielska-Taras, T., Szala, L., Naleczynkska, A., Kolodziej, K. and Ogrodowczyk, M. (1999). Selection for high erucic acid content in winter oilseed rape (*Brassica napus* L.) on microspore-derived embryos. *Journal of Applied Genetics* **40**(4), 305–315.

Chadd, S. A., Davies, W. P. and Koivisto, J. M. (2002). Practical production of protein for food animals. *In* "Proceedings Protein Sources for the Animal Feed Industry–Expert Consultation and Workshop." Bangkok, 29th April–3rd May 2002, available at http://www.fao.org/documents/show_cdr.asp?url_file=/docrep/007/y5019e/y5019e00.htm.

Chauhan, J. S., Tyagi, M. K., Kumar, P. R., Tyagi, P., Singh, M. and Kumar, S. (2000). Breeding for oil and seed meal quality in rapeseed-mustard in India-A review. *Agricultural Reviews* **23**(2), 71–92.

Chauhan, J. S., Tyagi, P. and Tyagi, M. K. (2002). Inheritance of erucic acid content in two crosses of Indian mustard. *SABRAO Journal of Breeding and Genetics* **34**(1), 19–26.

Chen, B. Y. and Heneen, W. K. (1989). Fatty acid composition of resynthesized *Brassica napus* L, *B. campestris* L. and *B. alboglabra* Bailey with special reference to the inheritance of erucic acid content. *Heredity* **63**, 309–314.

Chen, B. Y. and Beversdorf, W. D. (1990a). Fatty acid inheritance in microspore-derived populations of spring rapeseed (*Brassica napus* L.). *Theoretical and Applied Genetics* **80**, 465–469.

Chen, J. L. and Beversdorf, W. D. (1990b). A comparison of traditional and haploid-derived populations of oilseed rape (*Brassica napus* L.) for fatty acid composition of the seed oil. *Euphytica* **51**, 59–65.

Comer, J. P. (1956). Semi-quantitative specific test paper for glucose in urine. *Annals of Chemistry* **28**, 1748–1750.

Conacher, H. B. S. and Chadha, R. K. (1974). Determination of docosenoic acids in fats and oils by gas- liquid chromatography. *Journal of American Oil Chemical Society* **57**, 1161–1164.

Corbett, P. and Sernyk, L. (2003). Global opportunities for naturally stable canola/rapeseed oils. *In* "Proceedings of the 11th International Rapeseed Congress," Vol. 2, pp. 524–527. The Royal Veterinary and Agricultural University, Copenhagen, Denmark, 6–10th July.

Coventy, J., Kott, L. and Beversdorf, W. D. (1988). Manual for microspore culture technique for *Brassica napus Department of Crop Science and Technology Bulletin* OAC publication 0489, University of Guelph, Guelph, Ont., Canada.

Craig, B. M. and Murty, N. L. (1958). The separation of saturated and unsaturated fatty acid esters by gas-liquid chromatography. *Canadian Journal of Chemistry* **36**, 1297–1301.

Craig, B. M. and Murty, N. L. (1959). Quantitative fatty acids analysis of vegetable oils by gas chromatography. *Journal of American Oil Chemist Society* **36**, 549–552.

Craig, B. M. (1961). Varietal and environmental effects on rapeseed. III. Fatty acid composition of 1958 varietal tests. *Canadian Journal of Plant Science* **41**, 204–210.

Dardanne, P. (1996). Stability of NIR spectroscopy equations. *NIR News* **7**, 8–9.

Daun, J. K. and DeClercq, D. R. (1988). Quality of yellow and dark seeds in *Brassica campestris* canola varieties Candle and Tobin. *Journal of American Oil Chemist Society* **65**, 122–126.

Daun, J. K. and Williams, P. C. (1995). Use of NIR spectroscopy to determine quality factors in harvest surveys of canola. *In* "Proceedings of the Ninth International Rapeseed Congress Cambridge," pp. 864–866. UK, 4–7 July 1995.

Daxenbichler, M. E., Spencer, G. F., Kleiman, R., Van Etten, C. H. and Wolff, I. A. (1970). Gas liquid chromatographic determination of products from the progoitrins in crambe and rapeseed meal. *Annals of Biochemistry* **38**, 373–382.

DeBonte, L. R. and Hitz, W. D. (1996). Canola oil having increased oleic acid and decreased linolenic acid content and its manufacture using transgenic plants. In Scarth *et al.* 1999. Designer oil canola—a review of food-grade brassica oils with focus on high oleic, linolenic types. *In* "Proceedings of the Xth International Rapeseed Congress." Canberra, Australia, September 26–29, 1999.

DeClercq, D. R. and Daun, J. K. (1989). Determination of the total glucosinolate content in Canola by reaction with thymol and sulfuric acid. *Journal of American Oil Chemist Society* **66**(6), 788–791.

Dhawan, V. (2002). Biotechnology for developing Countries: An important tool in sustainable development. *In* "Proceedings BIO 2002, An International Biotechnology Convention and Exhibition." Toronto, Canada, June 2002.

Dhillon, S. S., Kumar, P. R. and Gupta, N. (1992). Breeding objectives and constrains. *In* "Breeding Oilseed Brassicas" (K. S. Labana, S. S. Banga and S. K. Banga, eds.), pp. 8–20. Narosa Publishing House, New Delhi.

Diepenbrock, W. (1983). Genotypic differences between the contents of linolenic acid in galactolipids and triacylglycerol from seeds of rape plants. *In* "Proceedings of the Sixth International Rapeseed Conference," pp. 321–362. Paris, France.

Dorell, D. C. and Downey, R. K. (1964). The inheritance of erucic acid content in rapeseed (*Brassica campestris*). *Canadian Journal of Plant Science* **44**, 499–504.

Downey, R. K. (1964). A selection of *Brassica campestris* L containing no erucic acid in its seed oil. *Canadian Journal of Plant Science* **44**, 295.

Downey, R. K. (1983). The origin and description of the *Brassica* oilseed crops. *In* "High and Low Erucic Acid Rapeseed Oils Production, Usage, Chemistry, and Toxicological Evaluation" (J. K. G. Kramer, F. D. Sauer and W. J. Pidgen, eds.), pp. 1–20. Academic Press, New York.

Downey, R. K. (1990). *Brassica* oilseed breeding—achievements and opportunities. *Plant Breeding Abstracts* **60**(10), 1165–1170.

Downey, R. K. and Bell, J. M. (1990). New development in canola research. *In* "Canola and Rapeseed Production, Chemistry, Nutrition and Processing Technology" (F. Shadidi, ed.), pp. 37–46. Van Nostrand Reinhold, New York.

Downey, R. K. and Craig, B. M. (1964). Genetic control of fatty acid biosynthesis in rapeseed (*Brassica napus* L.). *Journal of American Oil Chemist Society* **41**, 475–478.

Downey, R. K. and Harvey, B. L. (1963). Method of breeding for oil quality in rape. *Canadian Journal of Plant Science* **43**, 271–275.

Downey, R. K. and Rakow, G. (1987). Rapeseed and mustard. *In* "Principles of Cultivar Development" (W. R. Fehr, ed.), Vol. 2, pp. 437–486. Macmillan, New York.

Downey, R. K. and Rimmer, S. R. (1993). Agronomic improvements in oilseed. *Brassicas Advances in Agronomy* **50**, 1–66.

Downey, R. K., Craig, B. M. and Young, C. G. (1969). Breeding rapeseed for oil and meal quality. *Journal of American Oil Chemists Society* **46**, 121–123.

Earle, R. R., Van Etten, C. H., Clark, T. F. and Wolff, I. A. (1968). Compositional data on sunflower seed. *Journal of American Oil Chemist Society* **45**, 876–879.

Eskin, N. A. M., Macdonald, B. E., Przybylski, R., Malcomson, L. J., Scarth, R., Mag, T., Ward, K. and Adolphe, D. (1996). Canola oil. Chapter 1 in Bailey's industrial oil and fat products. "Edible Oil and Fat Products: Oils and Oilseeds," 5 Edn, Vol. 2, pp. 1–95. John Wiley & Sons, Inc., New York.

Ettlinger, M. G. and Lundeen, A. J. (1956). The structure of sinigrin and sinalbin: An enzymatic rearrangement. *Journal of American Oil Chemist Society* **78**, 4172–4173.

Fenwick, G. R., Heaney, R. K. and Mullin, W. J. (1983). Glucosinolates and their breakdown products in food and food plants. *Critical Reviews in Food Nutrition* **18**, 123–201.

Fernandez, E. J., Dminguez, J. and Martin, A. (1988). Genetics of the erucic acid content in interspecific hybrids of Ethiopian mustard and rapeseed. *Plant Breeding* **100**(4), 310–315.

Fernandez-Escobar, J., Dominguez, J., Martin, A. and Martinenz, J. M. F. (1988). Genetics of the erucic acid content in interspecific hybrids of Ethiopian mustard (*Brasica carinata* Braun) and rapeseed (*B. napus* L.). *Plant Breeding* **100**, 310–315.

Fernandez-Sirrano, O. and Alonsso, L. C. (1988). Breeding Brassica carinata for low erucic acid content. *Cruciferae Newsletter* **13**, 32–33.

Fieldsend, J. and Milford, G. F. J. (1994). Changes in glucosinolates during crop development in single and double low genotypes of winter oilseed rape (*Brassica napus* L.). I. Production and distribution in vegetative tissues and developing pods during development and potential role in the recycling of sulphur within crop. *Annals of Applied Biology* **124**, 531–542.

Finlayson, A. J., Kryzmanski, J. and Downey, R. K. (1973). Comparison of chemical and agronomic characteristics of two *Brassica napus* L. cultivars, Bronowski and Target. *Journal of American Oil Chemist Society* **50**, 407–410.

Getinet, A. and Rakow, G. (1997). Repression of seed coat pigmentation in Ethiopian mustard. *Canadian Journal of Plant Science* **77**(4), 501–505.

Getinet, A., Rakow, G., Raney, J. P. and Downey, R. K. (1994). Development of zero erucic acid Ethiopian mustard through an interspecific cross with zero erucic acid oriental mustard. *Canadian Journal of Plant Science* **74**, 793–795.

Gland, A., Robbelen, G. and Thies, W. (1981). Variation of alkenyl glucosinolates in seeds of *Brassica* species. *Zeitschrift fuer Pflanzenzuechtung* **87**, 96–110.

Glen, D. M., Jones, H. and Fieldsend, J. K. (1990). Damage to oilseed rape seedlings by the field slug *Deroceras reticulatum* in relation to glucosinolate concentration. *Annals of Applied Biology* **117**, 197–207.

Glendening, T. M. and Poulton, J. E. (1990). Partial purification and characterization of a adenosine 5′–phophosulfate: Desulfoglucosinolate sulfotransferase from Cress (*Lipidium sativum*). *Plant Physiology* **94**, 811–818.

Gopalan, C. D., Krishnamurthi, D., Shenolikar, I. S. and Krishnamachari, K. A. V. R. (1974). Myocardial changes in monkey fed on mustard oil. *Nutrition and Metabolism* **6**, 352–365.

Gororo, N., Salisbury, P., Rebetzke, G., Burton, W. and Bell, C. (2003). Genotypic variation for saturated fatty acid content of Victorian canola. *In* "Proceedings of the 11th International Rapeseed Congress," Vol. 1, pp. 215–217. The Royal Veterinary and Agricultural University, Copenhagen, Denmark, 6–10th July.

Grami, B., Baker, R. J. and Stefansson, B. R. (1977). Genetics of protein and oil content in summer rape. Heritability, number of effective factors and correlations. *Canadian Journal of Plant Science* **57**, 937–943.

Greenwood, C. F., Allen, J. A., Leong, A. S., Pallot, T. N., Golder, T. M. and Golebiowski, T. (1999). An investigation of the stability of NIRS calibration for the analysis of oil content in whole seed canola. *In* "Proceedings of the 10th International Rapeseed Congress." Canberra, Australia in proceedings CD, September 26–29.

Gupta, K. (2006). "Evaluation of *Brassica juncea* (L.) Czern × *B. rapa*/*B. carinata* hybrids and their advanced progenies against two major fungal diseases:

Molecular, morphological and biochemical characterization." Ph.D. thesis TERI university, New Delhi.

Gupta, M. L., Banga, S. K., Banga, S. S., Sandha, G. S., Ahuja, K. L. and Raheja, R. K. (1994). A new genetic stock for low erucic acid in Indian Mustard. *Cruciferae Newsletter* **16**, 104–105.

Gupta, M. L., Ahuja, K. L., Raheja, R. K. and Labana, K. S. (1998). Variation for biochemical quality traits in promising genotypes of Indian mustard. *Journal of Research* **25**(1), 1–5.

Gupta, S. K. and Labana, K. S. (1989). Triple test cross analysis for some physiomorphological traits in oilseed rape (*Brassica napus*). *Indian Journal of Genetics* **49**, 385–387.

Gupta, S. K., Thakral, S. K., Yadva, T. P. and Kumar, P. (1985). Combining ability and genetic architecture of oil content in Indian mustard. *Haryana Agricultural University Journal of Research* **15**, 467–470.

Gupta, S. K., Labana, K. S. and Ahuja, K. L. (1993). Inheritance of glucosinolate and oil content in oilseed rape. *Journal of Oilseed Research* **10**(2), 294–295.

Gurr, M. (1992). Dietary lipid and coronary heart disease. *Journal of Lipid Research* **31**, 195–243.

Halkier, B. A. and Du, L. (1997). The biosynthesis of Glucosinolates. *Trends Plant Science* **11**, 425–431.

Harvey, B. L. and Downey, R. K. (1964). The inheritance of erucic acid content in rapeseed (*Brassica napus*). *Canadian Journal of Plant Science* **44**, 104–111.

Hassan, F., Rothnie, N. E., Yeung, S. P. and Palmer, M. V. (1988). Enzyme linked immunosorbent assays for alkenyl glucosinolates. *Journal of Agriculture and Food Chemistry* **36**, 398–403.

Hemingway, J. S., Schofield, H. J. and Vaughan, J. G. (1961). Volatile mustard oils of *Brassica juncea* seeds. *Nature* **192**, 993.

Hitz, W. D., Mauvis, C. J., Ripp, K. G., Reiter, R. J., DeBonte, L. and Chen, Z. (1995). The use of cloned rapeseed genes for the cytoplasmic fatty acid desaturases and the plastid acyl-ACP thioesterases to alter relative levels of polyunsaturated and saturated fatty acids in rapeseed. *In* "D5-Breeding Oil Quality," pp. 470–478. GCIRC 1995, Cambridge, UK.

Hougen, F. W. and Bodo, V. (1973). Extraction and methanalysis of oil from whole or crushed rapeseed for fatty acids analysis. *Journal of American Oil Chemist Society* **50**, 230–234.

Hymowitz, T., Palmer, R. G. and Hadley, H. H. (1972). Seed weight, protein, oil and fatty acid relationship within the genus. *Glycine Tropical Agriculture* **49**, 245–250.

Inglis, I. R., Wadsworth, J. T., Meyer, A. N. and Feare, C. J. (1992). Vertebrate damage to 'OO' and 'O' varieties of oilseed rape in relation to SMCO and glucosinolate concentration in the leaves. *Crop Protection* **11**, 64–68.

Jonsson, R. (1973). Breeding for low erucic acid content in summer turnip rape (*Brassica campestris* L. Var. annua L.). *Zeitschrift fuer Pflanzenzuchtung* **69**, 1–18.

Jonsson, R. (1977). Erucic acid heredity in rapeseed (*Brassica napus* L.) and *Brassica campestris*. *Hereditas* **86**, 15–170.

Jonsson, R. and Bengstsson, L. (1990). Yellow seeded rape and turnip rape. I. Influence of breeding for yellow seeds upon yield and quality properties. *Sver Utsaedesfoeren Tidskr* **80**, 149–155.

Jonsson, R. and Uppstrom, B. (1986). Quality breeding in rapeseed. *In* "Svalof 1886–1986 Research and Results in Plant Breeding" (G. Olsson, ed.), pp. 173–184. LTS Forlag, Stockholm, Sweden.

Kalmal-Edin, A. and Appelqvist, L. A. (1996). The chemistry and antioxidant properties of tocopherols and tocotrienols. *Lipids* **31**, 671–701.

Kaufmann, H. P. and Fiedler, H. (1938). Untersuehungen fiber die Brauehbarkeit des Maekey-Schmutzes in den Laugen Fette u. Seif. 46: 292–299 cited in McGregor, D. I. 1980. *In* "Analytical Chemistry of Rapeseed and Its Products, a Symposium" (J. K. Daun, D. I. McGregor and E. E. McGregor, eds.), pp. 59–66. Winnipeg, Canada, 5–6 May 1980.

Kaushik, N. and Agnihotri, A. (1997). Evaluation of improved method for determination of rapeseed-mustard FAMES by GC. *Chromatographia* **44**, 97–99.

Kaushik, N. and Agnihotri, A. (1999). High performance liquid chromatographic method for separation and quantification of intact glucosinolates. *Chromatographia* **49**(5/6), 281–284.

Kaushik, N. and Agnihotri, A. (2000). GLC analysis of Indian rapeseed-mustard to study the variability of fatty acid composition. *Biochemical Society Transactions* **28**, 581–583.

Kaushik, N. and Agnihotri, A. (2003). Fatty acid variability in Indian mustard (*Brassica juncea* L.) lines generated through inter-and/or intra specific hybridization. *In* "Proceedings of the 11th International Rapeseed Congress," pp. 274–276. The Royal Veterinary and Agricultural University, Copenhagen, Denmark, 6–10th July.

Khalatkar, A. S., Rakow, G. and Downey, R. K. (1991). Selection for quality and disease resistance in *Brassica juncea* cv. Pusa bold. *In* "Proceedings of the Eighth International Rapeseed Congress," p. 198. Saskatoon, Canada, 9–11 July.

Kirk, J. T. O. and Hurlstone, C. J. (1983). Variation and inheritance of erucic acid content in. *Brassica juncea. Zeitschrift fur Pflanzenzuchtung* **90**(4), 331–338.

Kirk, J. T. O. and Oram, R. N. (1981). Isolation of erucic acid free lines of *Brassica juncea*: Indian mustard now a potential oilseed crop in Australia. *Journal of Australian Institute of Agricultural Science* **47**, 51–52.

Knowles, P. F. (1968). Registration of UC–1safflower. *Crop Science* **8**, 641.

Kolovrat, O. (1988). Use of a modified palladium test in winter swede rape breeding. *Rostlinna- Vyroba* **34**(6), 667–672.

Kondra, Z. P. (1967). Some environmental effects on and genetics of the isothiocyanate and oxazolidinethione content of seed meal of *Brassica napus* L. and *B. campestris* L. Ph.D. Thesis, Forage and Oilseeds Sections, Saskatoon Research Station, Canada.

Kondra, Z. P. and Stefansson, B. R. (1970). Inheritance of the major glucosinolates of rapeseed (*Brassica napus*) meal. *Canadian Journal of Plant Science* **50**, 643–647.

Kondra, Z. P. and Thomas, P. M. (1975). Inheritance of oleic, linoleic and linolenic acids in seed oil of rapeseed (*Brassica napus*). *Canadian Journal of Plant Science* **55**, 205–210.

Kott, L. S. (1995). Production of mutants using the rapeseed doubled haploid system. *In* "Proceedings International Symposium on the Use of Induced Mutations and Molecular Techniques for Crop Improvement Organized by IAEA and FAO," pp. 505–515, Vienna, 19th–23rd June.

Kramer, J. K., Sauer, F. D., Farnworth, E. R., Stevenson, D. and Rock, G. A. (1998). Haematological and lipid changes in new born piglets fed milk-replacer diets containing erucic acid. *Lipids* **33**, 1–10.

Krzymanski, J. and Downey, R. K. (1969). Inheritance of fatty acid composition in winter forms of rapeseed (*Brassica napus*). *Canadian Journal of Plant Science* **49**, 313–319.

Krzymanski, J., Pietka, T., Krotka, K. and Michalski, K. (1995). Glucosinolate content in F1 hybrids of polish double-zero winter swede rape. *Rosliny-Oleiste* **16**(1), 12–24.

Kumar, P., Yadava, T. P. and Dhawan, K. (1982). Combining ability analysis for oil content in Indian mustard. *Indian Journal of Agricultural Science* **52**, 219–221.

Kumar, S., Singh, A. K., Kumar, M., Yadav, S. K., Chauhan, J. S. and Kumar, P. R. (2003). Standardization of near-infrared spectroscopy (NIRS) for determination of seed oil and protein content in rapeseed-mustard. *Journal of Food Science and Technology* **40**, 306–309.

Kumar, S., Yadav, S. K., Chauhan, I. S., Singh, A. K., Khan, N. A. and Kumar, P. R. (2004). Total glucosinolate estimation by complex formation between glucosinolate and tetrachloropalladate (II) using ELISA reader. *Journal of Food Science and Technology* **41**, 63–65.

Kuntzon, D. S., Thomson, G. A., Radke, S. E., Johnson, W. B., Knaff, V. C. and Kridl, J. C. (1992). Modification of Brassica seed oil by antisense expression of a stearoyl-acid carrier protein desaturase gene. *Proceedings of the National Academy of Sciences, USA* **89**, 2624–2628.

Leckband, G., Rades, H., Frauen, M. and Friedt, W. (2003). *NAPUS* 2000—a research programme for the improvement of the whole rapeseed. *In* "Proceedings of the 11th International Rapeseed Congress," pp. 209–211. The Royal Veterinary and Agricultural University, Copenhagen, Denmark, 6–10th July.

Lein, K. A. (1970). Quantitative Bestimmungsmethoden for samenglucosinolate in *Brassica* Arten and Ihre Anwendung in der Zuchtung Von Glucosinolatarамen Raps. *Zeitschrift fuer Pflanzenzuechtung* **63**, 137–154.

Lepage, G. and Roy, C. C. (1986). Direct transesterification of all classes of lipids in a one-step reaction. *Journal of Lipid Research* **27**, 114–120.

Leung, J., Fenton, T. W., Mueller, M. R. and Clandinin, A. (1979). Condensed tannins of rapeseed meal. *Journal of Food Science* **44**, 1313–1316.

Li, D. F. and Qui, H. L. (1987). Oil crop of China. *Chinese Academy Press Beijing* **4**, 7–11.

Lichter, R., De Groot, E., Fiebig, D., Schweiger, R. and Gland, A. (1988). Glucosinolates determined by HPLC in the seeds of microspore-derived homozygous lines of rapeseed (*Brassica napus* L.). *Plant Breeding* **100**, 209–221.

Liu, D. F. and Liu, H. L. (1989). Inheritance to erucic acid and ecosinoic acids in *B. juncea*. *Hereditas* **11**(5), 17–20.

Liu, H., Han, J. and Hu, X. (1991). Studies on the inheritance of seed-coat colour and other related characters of yellow-seeded *Brassica napus*. *In* "Proceedings of the Eighth International Congress," Vol. 5, pp. 1438–1444. Saskatoon, Canada.

Love, H. K., Rakow, G., Raney, J. P. and Downey, R. K. (1990). Development of low glucosinolate Mustard. *Canadian Journal of Plant Science* **70**, 419–424.

Love, H. K., Rakow, G., Raney, J. P. and Downey, R. K. (1991). Breeding improvements towards canola quality *Brassica juncea*. *In* "Proceedings of the Eighth International Rapeseed Congress," pp. 164–169. Saskatoon, Canada.

Luhs, W. and Friedt, W. (1995). Breeding high-erucic acid rapeseed by means of *Brassica napus* resynthesis. *In* "Proceedings of the Ninth International Rapeseed Congress," Vol. 2: D18, pp. 449–451. Cambridge, UK4–7 July.

Luhs, W., Weier, D., Marwede, V., Frauen, M., Lekband, G., Becker, H. C., Frentzen, M. and Friedt, W. (2003). Breeding of oilseed rape (*Brassica napus* L.) for modified tocopherol composition- synergy of conventional and modern approaches. *In* "Proceedings of 11th International Rapeseed

Congress," Vol. 1, pp. 194–197. The Royal Veterinary and Agricultural University, Copenhagen, Denmark 6–10th July.

Luthy, B. and Matile, P. (1984). The mustard oil bomb: Rectified analysis of the sub cellular organisation of the myosin's system. *Biochemistry and Physiology Pflanzen* **179**, 5–12.

Magrath, R. and Mithen, R. (1993). Maternal effects on the expression of individual aliphatic glucosinolates in seeds and seedlings of *Brassica napus*. *Plant Breeding* **111**, 249–252.

Magrath, R., Herron, C., Giamoustaris, A. and Mitten, R. (1993). The inheritance of aliphatic glucosinolates in *Brassica napus*. *Plant Breeding* **111**, 55–72.

Malabat, C., Atterby, H., Chaudhary, Q., Renard, M. and Gueguen, J. (2003). Genetic variability of rapeseed protein composition. *In* "Proceedings of the 11th International Rapeseed Congress," pp. 205–208. The Royal Veterinary and Agricultural University, Copenhagen, Denmark, 6–10th July.

Malik, R. S. (1990). Prospects for *Brassica carinata* as an oilseed crop in India. *Experimental Agriculture* **26**, 125–129.

Malode, S. N., Swamy, R. V. and Khalatkar, A. S. (1995). Introgression of 'OO' quality characters in *Brassica juncea* cv. Pusa bold. *In* "Proceedings of the Ninth International Rapeseed Congress," pp. 431–438. Cambridge, UK, 4–7 July 1995.

Marquard, R. (1990). Untersuchungen über den Einflu*β* von Sorte und Standort auf den Tocopherolgehalt verschiedener Pflanzenöle. Fat. Science Technology **92**, 452–455.

Matile, P. (1980). Die Senfolbombe: Zur komparumentierung des myrosinase systems. *Biochemistry and Physiology Pflanzen* **175**, 722–731 (cited in: Mitten, R. (1992). Leaf glucosinolate profiles and their relationship to pests and disease resistance in oilseed rape. *Euphytica* **63**, 71–142).

McGregor, D. I. (1974). A rapid and sensitive spot test for linolenic acid levels in rapeseed. *Canadian Journal of Plant Science* **54**, 211–213.

McGregor, D. I. (1977). A rapid and simple method of screening rapeseed and mustard seed for erucic acid content. *Canadian Journal of Plant Science* **57**, 133–142.

McGregor, D. I. (1985). Determination of glucosinolate in *Brassica* seed. *Cruciferae Newsletter* **10**, 132–136.

McGregor, D. I. and Downey, R. K. (1975). A rapid and simple assay for identifying low glucosinolate rapeseed. *Canadian Journal of Plant Science* **55**, 191–196.

McMillan, M., Spinks, E. A. and Fenwick, G. R. (1986). Preliminary observations on the effect of dietary Brussels sprouts on thyroid function. *Human Toxicology* **5**, 15–19.

Mika, V., Tillman, P., Koprna, R., Nerusil, P. and Kucera, V. (2003). Fast prediction of quality parameters in whole seeds of oilseed rape (*Brassica napus* L.). *Plant and Soil Environment* **49**(4), 141–145.

Milford, G. F. J., Fieldsend, J. K., Porter, A. J. R., Rawlinson, C. J., Evans, E. J. and Bilsborrow, P. E. (1989). Changes in glucosinolate concentration during the vegetative growth of single and double low cultivars of winter oil seed rape. *Aspects of Applied Biology* **23**, 83–90.

Miller, R. W., Van Etten, C. H., Mc Grew, C. E., Wolf, I. A. and Jones, Q. (1962). Amino acid composition of seed meals from 41 species of Cruciferae. *Journal of Agriculture and Food Chemistry* **10**, 426–430.

Minson, D. J. (1997). Ruminants: The protein producer. *Biologist* **44**, 463–464.

Mitra, R. and Bhatia, C. R. (1979). Bioenergetic considerations in the improvement of oil content and quality in oilseed crop. *Theoratical and Applied Genetics* **54**, 41–47.

Mitten, R. (1992). Leaf glucosinolate profiles and their relationship to pests and disease resistance in oilseed rape. *Euphytica* **63**, 71–83.

Moller, P., Rehman, M. H., Stolen, O. and Sorenson, H. (1985). Inheritance of fatty acids and glucosinolates in oilseed rape. *In* "Possibilities for Improvement of Rape Adapted for the Growth Conditions in the Production and Utilization of Cruciferous Crop," pp. 286–300. Nijhoff/Junk, Dordrecht.

Monpara, B. A. and Jaisani, B. G. (2000). Oral presentation. *In* "National Seminar on Oilseed and Oils Research and Development Needs in the Millennium." Hyderabad, India, February 2–4 (cited in Chauhan, J. S., Tyagi, M. K., Kumar, P. R., Tyagi, P., Singh, M. and Kumar, S. (2000). Breeding for oil and seed meal quality in rapeseed-mustard in India: A review. *Agriculture Reviews* **23**(2), 71–92).

Mou, T. M. and Liu, H. L. (1988). Genetic studies on the total content of glucosinolates in seeds of rape (*Brassica napus* L.). *Cruciferae Newsletter* **13**, 52–53.

Mou, T. M. and Liu, H. L. (1990). Genetic analysis of the glucosinolates content in seeds of rape (*Brassica napus* L.). *Acta Agronomica sinica* **16**(2), 97–105.

Murphy, D. J. (1995). The use of conventional and molecular genetics to produce new diversity in seed oil composition for the use of plant breeders- progress, problems and future prospects. *Euphytica* **85**, 433–440.

Murphy, D. J. (1996). Engineering oil production in rapeseed and other oil crops. *Trends in Biotechnology* **14**(6), 206–213.

Murphy, D. J. (1999). Production of novel oils in plants. *Current Opinions in Biotechnology* **10**, 175–180.

Nabloussi, A., Fernandez-Martinez, J. M., De Haro, A. and Velasco, L. (2003). Novel seed oil types in Ethiopian mustard. *In* "Proceedings of the 11th International Rapeseed Congress," Vol. 1, p. 193. The Royal Veterinary and Agricultural University, Copenhagen, Denmark, 6–10th July.

Newton, I. S. (1998). Long chain polyunsaturated fatty acids—the new frontier for nutrition. *Lipid Technology* **10**, 77–81.

Oram, R. N. and Kirk, J. T. O. (1993). Induction of mutations for higher seed quality in Indian mustard. *In* "Proceedings of the 10th Australian Plant Breeding Conference," pp. 187–191. Gold Coast, Queensland.

Oram, R. N., Walton, G., Marcroft, S., Potter, T. D., Burton, W. A., Castleman, G. H., Easton, A. A. and Kirk, J. T. O. (1997). Progress in breeding canola-grade Indian mustard. *In* "Proceedings of the 11th Australian Research Assembly on Brassicas," pp. 79–83. Agriculture WA, Perth.

Oram, R. N., Salisbury, P. A., Krick, J. T. O. and Burton, W. A. (1999). Development of early flowering, canola grade *Brassica juncea* germplasm. *In* "Proceedings of the 10th International Rapeseed Congress." Canberra, Australia, 26–29 September 1999.

Pallot, T. N., Leong, A. S., Allen, J. A., Golder, T. M., Greenwood, C. F. and Golebiowski, T. (1999). Precision of fatty acid analysis using near-infrared spectroscopy of whole seed *Brassica*. *In* "Proceedings of the 10th International Rapeseed Congress." Canberra, Australia; 1999: proceedings CD, September 26–29.

Palmer, C. E., Keller, W. A. and Arnison, P. G. (1996). Experimental haploidy in Brassica species. *In* "*In Vitro* Haploid Production in Higher Plants" (S. M. Jain, S. K. Sapory and R. E. Veilleux, eds.), pp. 143–172. Kluwer, Dordrecht, Netherlands.

Palmer, M. V., Yeung, S. P. and Sang, J. P. (1987). Glucosinolate content of seedlings, tissue cultures, and regenerant plants of B. *juncea* (Indian mustard). *Journal of Agriculture and Food Chemistry* **35**, 262–265.

Persson, C. (1985). High palmitic acid content in summer turnip rape (*Brassica Campestris* var. *annua* L.). *Cruciferae Newsletter* **10**, 137.

Persson, S. (1974). A method for determination of glucosinolates in rapeseed as TMS-derivatives. *In* "Proceedings of the Fourth International Rapeseed Conference," pp. 381–386. Giessen, West GermanyJune 4–8.

Potts, D. and Males, D. R. (1999). Inheritance of fatty acid composition in *Brassica juncea*. *In* "Proceedings of the 10th International Rapeseed Congress," Canberra, Australia, 26–29th September.

Potts, D. A., Rakow, G. W. and Males, D. R. (1999). Canola-quality *Brassica juncea*, a new oilseed crop for the Canadian prairies. *In* "Proceedings of the Xth GCIRC International Rapeseed Congress." Canberra, Australia, 26–29 September 1999.

Prabhu, H. R. (2000). Lipid peroxidation in culinary oils subjected to thermal stress. *Indian Journal of Clinical Biochemistry* **15**, 1–5.

Prakash, S., Kumar, P. R., Sethi, M., Singh, C. and Tandon, R. K. (2000). Mustard oil—the ultimate edible oil. *The Botanica* **50**, 94–101.

Prem, D. (2006). Induction of genetic variability for agro-morphological and biochemical traits in Indian mustard [*Brassica juncea* (L.) Czern & Coss] through mutagenesis in conjugation with doubled haploid technology. Ph.D. thesis TERI university, New Delhi.

Prem, D., Sarkar, G., Gupta, K. and Agnihotri, A. (2006). Haploid mutagenesis in Indian mustard [*Brassica juncea* (L.) Czern and Coss] for generation of novel agronomic and biochemical variability. *In* "Proceedings International Conference on Haploids in Higher Plants III." Vienna, Austria, February 12–15.

Rahman, M. H. and Poulsen, M. H. (1995). Inheritance of glucosinolates in oilseed rape. *In* "Proceedings of the Ninth International Rapeseed Congress," Cambridge, UK, 4–7 July.

Rahman, M. H., Joersbo, M. and Poulsen, M. H. (2001). Development of yellow-seeded *Brassica napus* of double low quality. *Plant Breeding* **120**, 473–475.

Rakow, G. (1973). Selection for content of linoleic and linolenic acid in rapeseed after mutagenic treatment. *Zeitschrift fuer Pflanzenzuechtung* **69**, 62–68 [cited in: Banga, S. K. (1996). Breeding for oil and meal quality. *In* "Oilseed and vegetable Brassicas: Indian perspective" (V. L. Chopra and S. Prakash, eds.), Chapter 11, pp. 234–249].

Rakow, G. (1995). Developments in the breeding of oil in other *Brassica* species. *In* "Proceedings of the Ninth International rapeseed congress," Vol. 2, pp. 401–406. Cambridge, UK.

Rakow, G. and Raney, J. P. (1995). Field performance of canola quality *Brassica juncea*. *In* "Proceedings Ninth International Rapeseed Congress Cambridge," Vol. 2, p. 428. UK.

Rakow, G. and Raney, J. P. (2003). Present status and future perspectives of breeding for seed quality in *Brassica* oilseed crops. *In* "Proceedings of the 11th International Rapeseed Congress," pp. 181–185. The Royal Veterinary and Agricultural University, Copenhagen, Denmark, 6–10th July.

Raney, J. P., Olson, T. V., Rakow, G. and Ripley, V. L. (2003a). *Brassica juncea* with a canola fatty acid composition from an interspecific cross with *Brassica napus*. *In* "Proceedings of the 11th International Rapeseed Congress," pp. 281–283. The Royal Veterinary and Agricultural University, Copenhagen, Denmark, 6–10th July.

Raney, J. P., Olson, T. V., Rakow, G. and Ripley, V. L. (2003b). Selection of near zero aliphatic glucosinolate *Brassica juncea* from an interspecific cross with *B. napus*. *In* "Proceedings of the 11th International Rapeseed Congress," pp. 284–286. The Royal Veterinary and Agricultural University, Copenhagen, Denmark, 6–10th July.

Raut, R. N. (1996). *Brassica carinata*- A new oilseed for higher production and sustainability *Second International Crop Science Congress*, New Delhi, Abstract, p. 196.

Razin, A. and Cedar, H. (1991). DNA methylation and gene expression. *Microbiology Reviews* **55**, 451–458.

Reinhardt, T. C. and Röbbelen, G. (1991). Quantitative analysis of fatty acids in intact rapeseed by Near Infrared Reflectance Spectroscopy. *In* "Proceedings of the Eighth International Rapeseed Congress," pp. 1380–1384. Saskatoon, Canada.

Renarid, S. and McGregor, L. (1976). Antithrombogenic effects of erucic acid poor rapeseed oils in the rats. *Review of Forage Crops* **23**, 393–396.

Robbelen, G. (1980). Breeding for low contents of glucosinoltes in rapeseed. *In* "Production and Utilization of Proteins in Oilseed Crops" (E. S. Bunting, ed.), pp. 91–106. Nijkoff-junk.

Rucker, B. and Robbelen, G. (1994). Inheritance of total and individual glucosinolate contents in seeds of winter oilseed rape (*Brassica napus* L.). *Plant Breeding* **113**, 206–216.

Sang, J. P. and Truscott, R. J. W. (1984). Liquid chromatographic determination of glucosinolates in rapeseed as desulfoglucosinolates. *Journal of Association of Official Analytical Chemist* **67**, 829–833.

Sauer, F. D. and Kramer, J. K. G. (1983). The problems associated with the feeding of high erucic acid rapeseed oils and some fish oils to experimental animals. *In* "High and Low Erucic Acid Rapeseed Oils" (J. K. G. Kramer, F. D. Sauer and W. J. Figden, eds.), pp. 254–292. Academic Press, Toronto, Canada.

Saxena, S. (2006). Golden mustard: Biotechnological approach for addressing vitamin A deficiency. *In* "Presentations at Third Workshop on Agricultural Biotechnology for Global Public Goods." National Acadamy of Agricultural Sciences and US National Acadamy, Chinnai, India, October 4–6th.

Scarth, R. and McVetty, P. B. E. (1999). Designer oil canola—a review of new food-grade Brassica oils with focus on high oleic, low linolenic types. "Proceedings of the Xth International Rapeseed Congress." Canberra, Australia, September 26–29, 1999.

Scarth, R., McVetty, P. B. E., Rimmer, S. R. and Stefansson, B. R. (1988). Stellar low linolenic–high linoleic acid summer rape. *Canadian Journal of Plant Science* **68**, 509–511.

Schilling, W. and Friedt, W. (1992). Breeding 'OO' rapeseed (*Brassica napus*) with differential glucosinolate content in the leaves. *In* "Proceedings Eighth International Rapeseed Congress." pp. 250–255. Saskatoon, Canada.

Schnug, E. and Haneklaus, S. (1988). Theoretical principles for the indirect determination of the total glucosinolate content in rapeseed and meal quantifying the sulphur concentration via X-ray fluorescence (X-RF method). *Journal of the Science of Food and Agriculture* **45**, 243–254.

Schwetka, A. (1982). Inheritance of seed color in turnip rape (*Brassica campestris* L.). *Theoratical and Applied Genetics* **62**, 161–169.

Shenk, J. and Westerhaus, M. O. (1993). Analysis of Agriculture and Food Products by Near-Iinfrared Reflectance Spectroscopy. pp. 41–51. *Infrasoft International*, Port Matilda, USA.

Shewmaker, C. K., Sheehy, J. A., Daley, M., Colburn, S. and Ke, D. Y. (1999). Seed-specific over expression of phytoene synthase: Increase in carotenoids and other metabolic effects. *Plant Journal* **20**, 401–412.

Shirzadegan, M. (1986). Inheritance of seed colour in *Brassica napus* L. *Zeitschrift fuer Pflanzenzuchtung* **96**, 140–146.

Shirzadegan, M. and Robbelen, G. (1985). Influence of seed colour and hull proportion on the quality properties of seeds in *Brassica napus* L. *Fette Seifen Anstrichm* **87**, 235–237.

Shpota, V. I. and Podkolzina, V. E. (1986). Producing erucic acid free Indian mustard forms. *Selektsiya I Semenovodsto* **4**, 19–21.

Siebel, J. and Pauls, K. P. (1989). Inheritance pattern of erucic acid content in populations of *Brassica napus* microspore-derived spontaneous diploids. *Theoratical and Applied Genetics* **77**, 489–494.

Simonne, A. H., Simonne, E. H., Eitenmiller, R. R., Mills, H. A. and Cresman, C. P. (1997). Could the Dumas method replace the Kjeldahl digestion for nitrogen and crude protein determinations in foods? *Journal of Science Food and Agriculture* **73**, 39–45.

Singh, H. and Yashpal, M. (1991). Genetic component and combining ability analysis of oil content in Indian mustard. *Journal of Oilseed Research* **8**, 117–120.

Singh, N. K., Kaushik, N., Sarkar, G. and Agnihotri, A. (1996). Variation and association analysis among major fatty acids in Rapeseed-mustard. *Cruciferae Newsletter* **18**, 88–89.

Smith, N. A., Singh, S. P., Wang, M. B., Stoutjesdijk, P. A., Green, A. G. and Waterhouse, P. M. (2000). Total silencing by intron-spliced hairpin RNAs. *Nature* **407**, 319–320.

Sodhi, Y. S., Mukhopadhyaya, A., Aruumugam, N., Verma, J. K., Gupta, V. and Petal, D. (2002). Genetic analysis of total glucosinolate in crosses involving a high glucosinolate Indian variety and a low glucosinolate line of *B. juncea*. *Plant Breeding* **121**, 508–511.

Spasibionek, S., Krzymanski, J. and Bartkowiak-Broda, I. (2003). Mutants of *Brassica napus* with changed fatty acid composition. *In* "Proceedings of the 11th International Rapeseed Congress," Vol. 1, pp. 221–224. The Royal Veterinary and Agricultural University, Copenhagen, Denmark, 6–10th July.

Spinks, E. A., Sones, K. and Fenwick, G. R. (1984). The quantitative analysis of glucosinolates in cruciferous vegetables, oilseeds and forage crops using high performance liquid chromatography. *Fette Seifen Anstrichmittel* **86**, 228–231.

Stahl, E. (1969). "Thin Layer Chromatography—A Laboratory Handbook." Springer, New York.

Stefansson, B. R. (1983). The development of improved rapeseed cultivars. *In* "High and Low Erucic Acid Rapeseed Oils: Production Usage Chemistry and Toxicological Evaluation" (J. K. G. Kramer, F. D. Sauer and W. J. Pigden, eds.), pp. 143–159. Academic Press, Don Mills.

Stefansson, B. R. and Hougen, F. W. (1964). Selection of rape plants (*Brassica napus*) with seed oil practically free from erucic acid. *Canadian Journal of Plant Science* **44**, 359–364.

Stefansson, B. R., Hougen, F. W. and Downey, R. K. (1961). Note on the isolation of rape plants with seed oil free from erucic acid. *Canadian Journal of Plant Science* **41**, 218–219.

Stiepel, M. (1926). Geschichte der chinesischen Philosophie; *Bd. 1*Das klassische Zeitalzer bis zur Han-Dynastie [cited in McGregor, D. I. (1980). Analytical Chemistry of rapeseed and its products. *In* "Proceedings of a Symposium," pp. 59–66. Winnipeg, Canada, 5–6 Ma 1980].

Stominski, B. A. and Campbell, L. D. (1987). Gas chromatographic determination of Indole glucosinolates—a re-examination. *Journal of the Science of Food and Agriculture* **40**, 131–143.

Stoutjesdijk, P. A., Hurlstone, C., Singh, S. P. and Green, A. G. (1999). Genetic manipulation for altered oil quality in *Brassicas*. *In* "Proceedings of the Xth

International Rapeseed Congress.'' Canberra, Australia. September 26–29, 1999.

Stringam, G. R. and McGregor, D. I. (1980). Inheritance and fatty acid composition of a yellow-embryo mutant in turnip rape (*Brassica campestris* L.). *Canadian Jounal of Plant Science* **60,** 97–102.

Stringam, G. R., Mc Gregor, D. I. and Pawlowski, S. H. (1974). Chemical and morphological characteristics associated with seed colour in rapeseed. *In* ''Proceedings of the Fourth International Science Congress,'' pp. 99–108. Giessen, West Germany.

TERI (2004). Plant biodiversity: Chemical characterization of glucosinolate content. Report No.1999BB64. Project Report NATP submitted to World Bank through the Indian Council of Agricultural Reasearch (http://www.teriin.org/reports/rep206/rep206.htm).

Teutonico, R. A. and Osborn, T. C. (1994). Mapping of RFLP and quantitative trait loci in *Brassica rapa* and comparison to the linkage maps of *B. napus, B. oleracea* and *Arabidopsis thaliana. Theoretical and Applied Genetics* **89**, 885–894.

Thangstad, O. P., Evjen, K. and Bones, A. (1991). Immunogold-EM localization of myrosinase in *Brassiceae. Protoplasma* **161**, 85–93.

Thies, W. (1971). Rapid and simple analysis of the fatty acid composition in individual rape cotyledons. I. Methods of gas and paper chromatography. *Zeitschrift fuer Pflanzenzuchtung* **65**, 181–202.

Thies, W. (1980). Analysis of glucosinolates via 'on-column' desulfation. *In* ''Analytical Chemistry of Rapeseed and Its Products'' (J. Daun, D. I. McGregor and E. E. McGregor, eds.). The Canola Council of Canada, Winnipeg.

Theis, W. (1983). Complex formation between glucosinolates and tetra chloropalladate (II) and its utilization in plant breeding. *Fette Seifen Anstrichm* **84**, 338–342.

Tholen, J. T., Shiefeng, S., Truscott, R. J. W. and Roger, J. W. (1989). The thymol method for glucosinolate determination. *Journal of the Science of Food and Agriculture* **49**, 157–165.

Tholen, J. T., Buzza, G., McGregor, D. I. and Truscott, R. J. W. (1993). Measuring of the glucosinolate content in rapeseed using the trubluglu meter. *Plant Breeding* **110**, 137–143.

Tiwari, P. N., Gambhir, P. N. and Rajan, T. S. (1974). Rapid and non-destructive determination of seed oil by pulsed NMR technique. *Journal of American Oil Chemist Society* **51**, 104–109.

Truscott, R. J. W. and Shen, S. (1987). Thymol method for glucosinolate estimation. *Cruciferae Newsletter* **9**, 116–118.

Underhill, E. W. and Kirkland, D. F. (1971). Gas chromatography of trimethylsilyl derivatives of glucosinolates. *Journal of Chromatography* **57**, 47–54.

Uppstrom, B. (1983). Glucosinolate pattern in different growth stages of high and low glucosinolate varieties of *Brassica napus. Sveriges utsadesforening Tidskrift* **93**, 331–336.

Vageeshbabu, H. S. and Chopra, V. L. (1997). Genetic and biotechnological approaches for reducing glucosinolates from rapeseed mustard Meal. *Journal of Plant Biochemistry and Biotechnology* **6**(2), 53–62.

Van Deynze, A. E. and Pauls, K. P. (1994). Seed colour assessment in *Brassica napus* using a near-infrared reflectance spectrometer adapted for visible light measurement. *Euphytica* **76**, 45–51.

Van Etten, C. H., McGrew, C. E. and Daxenbichler, M. E. (1974). Glucosinolate determination in cruciferous seeds and meals by measurement of enzymatically released glucose. *Journal of Agriculture and Food Chemistry* **22**, 483–487.

Velasco, L. and Mollers, C. (2002). Non-destructive assessment of protein content in single seed of rapeseed (*Brassica napus* L.) by near-infrared reflectance spectroscopy. *Euphytica* **123**, 89–93.

Velasco, L., Fernández-Martínez, J. M. and De Haro, A. (1995a). Applicability of NIRS for estimating multiple seed quality components in Ethiopian mustard. *In* "Proceedings of the Ninth International Rapeseed Congress," pp. 867–869. Cambridge, UK.

Velasco, L., Fernandez Martinez, J. M. and De Haro, A. (1997a). Determination of the fatty acid composition of the oil in intact-seed mustard by near-infrared reflectance spectroscopy. *Journal of American Oil Chemist Society* **74**, 1595–1602.

Velasco, L., Fernandez-Martinez, J. M. and De Haro, A. (1997b). Induced variability for C18 unsaturated fatty acids in Ethiopian mustard. *Canadian Journal of Plant Science* **77**, 91–95.

Velasco, L., Fernandez-Martinez, J. M., DeHaro, A. and Del Rio, M. (1995b). Obtainment of EMS-induced mutants in Ethiopian mustard (*Brassica carinata* Braun). *In* "Proceedings of the Ninth International Rapeseed Congress," D14.

Velasco, L., Matthaus, B. and Moellers, C. (1998). Non-destructive assessment of sinapic acid esters in *Brassica* species: I. Analysis by near-infrared reflectance spectroscopy. *Crop Science* **38**, 1645–1650.

Velasco, L., Goffman Fernando, D. and Becker Heiko, C. (1999). Development of calibration equations to predict oil content and fatty acid composition in *Brassicaceae* germplasm by near-infrared reflectance spectroscopy. *Journal of American Oil Chemist Society* **76**, 25–30.

Vera, C. L. and Woods, D. L. (1982). Isolation of independent gene pairs at two loci for seed coat color in *Brassica juncea*. *Canadian Journal of Plant Science* **62**, 47–50.

Vera, C. L., Woods, D. L. and Downey, R. K. (1979). Inheritance of seed coat color in *Brassica juncea*. *Canadian Journal of Plant Science* **59**, 635–637.

Vermorel, M., Heaney, R. K. and Fenwick, G. R. (1986). Nutritive value of rapeseed meals: Effects of individual Glucosinolates. *Journal of Agriculture and Food Science* **37**, 1197–1202.

Voelkar, T. A., Hayes, T. R., Crammer, A. M., Turner, J. C. and Davies, H. M. (1996). Genetic Engineering of a quantitative trait: Metabolic and genetic parameters influencing the accumulation of laurate in rapeseed. *Plant Journal* **9**, 229–241.

Voelker, T. A., Worell, A. C., Anderson, L., Bleibaum, J., Fan, C., Hawkins, D. J., Radke, S. E. and Davies, H. M. (1992). Fatty acid biosynthesis redirected to medium chains in transgenic oilseed plants. *Science* **257**, 72–74.

Wallsgrove, R., Benett, R., Keddle, G., Barlet, E. and Ludwig-Miller, J. (1999). Glucosinolate biosynthesis and pest disease interactions. *In* "Proceedings of the 10th International Rapeseed Congress," Canberra, Australia.

Wetter, L. R. (1955). The determination of mustard oils in rapeseed meal. *Canadian Journal of Biochemistry and Physiology* **33**, 980–984.

Wetter, L. R. (1957). The estimation of substituted thiooxazolidines in rapeseed meals. *Canadian Journal of Biochemistry and Physiology* **35**, 293–297.

Wetter, L. R. and Youngs, C. G. (1976). A thiourea-UV assay for total glucosinolate content in rapeseed meal. *Journal of American Oil Chemist Society* **53**, 162–164.

Worthington, R. F. and Hammons, R. O. (1971). Genotypic variation in fatty acid composition and stability of *Arachis hypogia* L. *Oil Oleagineux* **26**, 695–700.

Youngs, C. G. and Wetter, L. R. (1967). Microdetermination of the major individual isothiocyanates and oxazolidinethiones in rapeseed. *Journal of American Oil Chemist Society* **44**, 551–554.

Zhao, F. J., Evans, E. J., Bilsborrow, P. E. and Syers, J. K. (1993a). Influence of sulphur and nitrogen on seed yield and quality of low glucosinolate oilseed rape (*Brassica napus* L.). *Journal of Science of Food and Agriculture* **63**, 29–37.

Zhao, F. J., Evans, E. J., Bilsborrow, P. E. and Syers, J. K. (1993b). Sulphur uptake and distribution in double and single low varieties of oilseed (*Brassica napus* L.). *Plant and Soil* **150**, 69–76.

Zhou, Y. M. and Liu, H. L. (1987). Inheritance of fatty acid in rapeseed. (*Brassica napus*). *Acta Agronomica Sinica* **13**(1), 1–10.

Zukalova, H. and Vasak, J. (2002). The role and effect of Glucosinolates of *Brassica* species—a review. *Rostlinna Vyroba* **48**(4), 175–180.

Development and Practical Use of DNA Markers

KATARZYNA MIKOLAJCZYK

Plant Breeding and Acclimatization Institute, Research Division in Poznan, Strzeszynska 36, 60-479 Poznan, Poland

ABSTRACT

Double-low (00) varieties of oilseed rape (*Brassica napus* L.), characterized by no erucic acid in seed oil and a very low glucosinolates content in seed meal make one of the major plant oil sources nowadays. Further improvement of quality traits is very

Advances in Botanical Research, Vol. 45
Incorporating Advances in Plant Pathology

0065-2296/07 $35.00
DOI: 10.1016/S0065-2296(07)45004-2

important for establishing a strong position of rapeseed in the oil crops market and the main selection goals concern: increasing seed yield, broadening of genetic diversity, modification of seed oil composition and natural antioxidants content, and improvement of seed meal quality. Taking into account that *B. napus* is an amphiploid, phenotypical and statistical analyses applied during selection are often time-consuming and difficult. However, application of DNA markers that enable genotyping at early developmental levels, independently from changing environment conditions, could make the selection process time- and cost-effective. Development of genetic markers based on the most modern methods of molecular biology and biotechnology seems to be one of the main topics of scientific research to be applied for quality breeding. DNA markers can be developed as a result of genetic mapping and quantitative trait locus analyses as well as gene tagging and searching for important differences in nucleotide sequences. Genomic, proteomic, and metabolic studies enable insight into molecular mechanisms and biochemical pathways and also help in identification of crucial genes as well as genomic regions involved in expression of traits of interest. Up till now, several DNA markers have been developed and applied for oilseed rape quality breeding, that is, for monitoring of cytoplasmic male sterility systems as well as restorer genes, pathogenes and abiotic stress resistance, dwarf genotypes, and seed oil content. Genetic markers are also used for prediction of hybrid performance, gene-flow, and transgene assessment, and new challenges arise concerning development of reliable methods for accessing population diversity as well as testing of candidate varieties with the use of proper genetic markers.

I. INTRODUCTION

Oilseed rape (*Brassica napus* L.) is one of the most important oil crops of the moderate climate regions. Development of double-low varieties, that is, with no erucic acid in seed oil and with a very low glucosinolate (GC) content in seed meal and their introduction into commercial production (Downey and Rakow, 1987), has made rapeseed one of the main plant oil sources nowadays and there is a constant increase in oilseed rape seed production throughout the world (Bartkowiak-Broda *et al.*, 2005). Further improvements in both the oil composition and seed meal quality are crucial for establishing rapeseed as competitive crop with other oil-yielding crops. Therefore, the main selection goals in oilseed rape breeding for better productivity and quality are hybrid breeding, resistance for insects and diseases, breeding of dwarf genotypes that are resistant to lodging, improvement of the quality of seed meal by lowering fiber content—breeding for yellow-seeded genotypes, breeding of varieties differentiated in respect of seed oil composition, due to the food- and nonfood use of rapeseed oil, and broadening of genetic diversity of oilseed rape cultivars throughout wide crosses, resynthesis, mutagenesis, and recombination as well as genetic transformation.

Real progress in rapeseed quality breeding has been achieved with the use of classical recombinant pedigree-breeding methods accompanied by selection based on biometric and biochemical traits and statistical analyses. However, invention and development of DNA fingerprinting methods enabled development of DNA markers which can be used for monitoring of specific traits independently from changing environment conditions and at any developmental stages. Marker-assisted selection (MAS) has become a powerful tool in quality breeding and together with development of doubled haploid (DH) lines (Cegielska-Taras *et al.*, 2002) makes the breeding process more effective and cheaper (Barret *et al.*, 1999). DNA (genetic) markers can be also applied for allele genotyping, tagging of genes, development of genetic maps and localizing quantitative trait loci (QTLs), identifying cultivars, establishing genetic distance (GD) among cultivars, gene-flow monitoring, as well as transgene identification. Finally, all biotechnological approaches in quality breeding are accompanied by the use of DNA markers. Therefore, research on designing of specific genetic markers is undertaken at many laboratories worldwide. In this chapter, some molecular aspects of development and application of DNA markers for important agronomic traits of oilseed rape are presented.

II. DEVELOPMENT AND USE OF DNA MARKERS

There are several types of DNA markers which reflect the progress in molecular biology methods and techniques. First of all, restriction fragment length polymorphism (RFLP) markers (Botstein *et al.*, 1980) have been used for the construction of genetic linkage maps for *B. oleracea* (Slocum *et al.*, 1990), *B. rapa* (Song *et al.*, 1991), and *B. napus* (Landry *et al.*, 1991). The discovery of polymerase chain reaction (PCR) (Mullis and Faloona, 1987) and PCR-based techniques led to the development of various other markers, viz., RAPD (random amplified polymorphic DNA) (Williams *et al.*, 1990), SSRs (simple-sequence repeats) (Rafalski and Tingey, 1993), AFLP (amplified fragments length polymorphism) (Vos *et al.*, 1995), as well as SNP (single nucleotide polymorphism) (Haff and Smirnov, 1997), ACGMs (amplified consensus genetic markers) (Brunel *et al.*, 1999), SRAP (sequence-related amplified polymorphism) (Li and Quiros, 2001), and SINE (short-interspersed element)—crop-specific transposable element markers (Prieto *et al.*, 2005).

RFLP markers were applied as genetic markers by Botstein *et al.* (1980) and were widely used for genetic analyses in plant breeding (Tanksley *et al.*, 1989). They were applied for construction of genetic maps and tagging of genes as well as of QTLs. The following application of these markers in breeding programs made it possible to introgress many traits into new genetic background. The linkage map in *B. napus* comprising 120 markers was first developed by Landry *et al.* (1991) with the use of a progeny of a cross of two spring rapeseed varieties. Furthermore, genetic maps constructed with the use of RFLP codominant markers made a new tool for comparative investigations of the structure and evolution of plant genomes (Tanksley *et al.*, 1992; Whitkus *et al.*, 1992). Uzunova *et al.* (1995) developed genetic map of a population derived from DH lines produced from F_1 hybrid between an old winter rapeseed landrace characterized by high GC and erucic acid content and a winter double-low variety, using RFLP markers. This genetic map revealed four QTLs for seed GC content. By comparative analysis of RFLP marker rearrangements on genetic maps, reciprocal transposition as the effect of plant crossing as well as mapping the mosaic of ancestral genotypes while pedigree breeding were performed (Osborn *et al.*, 2003; Sharpe and Lydiate, 2003). The study on genome duplication patterns using RFLP markers revealed that loci in *B. napus* usually occur in homeologous pairs (one from *B. rapa* and one from *B. oleracea*) (Parkin *et al.*, 2003). Moreover, RFLP analysis was also used to investigate the genetic diversity and relationships of 83 *B. napus* accessions with 43 genomic clones, which revealed 161 polymorphic fragments (Diers and Osborn, 1994). The RFLP data analysis led to the categorization of three major groups of cultivars, that is the first group—only spring accessions, the second—mostly winter accessions, and the third—rutabagas and oilseed rape accessions from China and Japan, indicating that winter and spring cultivars of oilseed rape belong to genetically distinct groups (Diers and Osborn, 1994).

Invention of RAPD method (Williams *et al.*, 1990) is rapid and easy to perform and it has opened new possibilities for large-scale screening of bulked segregant individuals (Michelmore *et al.*, 1991). A taxonomic data on the species relationships within the U triangle (U, 1935) was generated with the use of RAPD markers (Demeke *et al.*, 1992). RAPD primers were applied for determining genetic relationship between three *B. napus* breeding lines (Halldén *et al.*, 1994) along with the use of RFLP probes. Fifty RFLP and 92 RAPD primers were used for analysis, which revealed more than 500 bands for RFLP- and more than 400 for RAPD method. The results obtained through both the markers from the three lines were consistent with the known pedigree data. Bootstrap analyses revealed that the application of about 30 probes would be sufficient to detect those relationships. In brief, the RAPD markers have the same resolving power as RFLP markers and the RAPD method is simpler, quicker, cheaper, and requires small amount of DNA (Halldén *et al.*, 1994).

Microsatellites or SSRs consist of short tandem nucleotide repeat motifs densely interspersed in genomes, which are mostly $(GA/TC)_n$ in plants (Lagerkrantz *et al.*, 1993). The variation in high frequency of repeat unit numbers was observed at many microsatellite loci, which can be analyzed by PCR with the use of specific primer pairs for particular sequences flanking individual loci (Weber and May, 1989), thus making SSRs highly polymorphic and codominant genetic markers (Uzunova and Ecke, 1999). SSR markers (Grist *et al.*, 1993) are useful for genetic mapping as well as for establishing genetic identity, relationship among genotypes, and genome structure (Kresovich *et al.*, 1995; Ribaut *et al.*, 1997). Pairs of primers flanking microsatellites from *Brassica* (Plieske and Struss, 2001a; Suwabe *et al.*, 2002) are available at http://www.brassica.info/ssr/SSRinfo.htm. Moreover, A and C genome-specific SSR loci were identified by Saal *et al.* (2001) and it was shown that in most cases a single primer pair detected two loci—single locus from each genome. Thus, the identified genome-specific microsatellite loci may be used as markers to identify somatic and resynthesized hybrids, aneuploid, translocation, and substitution lines in A and C genome species (Saal *et al.*, 2001). SSR markers have been placed on genetic maps of rapeseed (Rudolph *et al.*, 1999; Uzunova and Ecke, 1999).

AFLP markers (Vos *et al.*, 1995) were applied for molecular characterization of novel resynthesized rapeseed lines and analysis of their genetic diversity in comparison with spring rapeseed cultivars (Seyis *et al.*, 2003) since they are widely used for genetic mapping. A consensus genetic map of *B. napus* was constructed comprising 540 markers, of which most were RFLPs (58), RAPDs (269), and AFLPs (195) as well as 9 isozymes, 3 specific PCRs, 5 SSR markers, and the dwarf *Bzh* gene (Lombard and Delourme, 2001).

A method for tagging gene sequences by ACGMs was developed by Brunel *et al.* (1999). This method was based on the 87% conservation of the coding nucleotide sequences between *Arabidopsis* and *Brassica* (Cavell *et al.*, 1998), which was later applied for designing primer pairs for PCR amplification of particular genes. First six consensus markers for floral meristem control, *FAD7* chloroplast linoleate desaturase, *FAD3* cytosolic linoleate desaturase, and alcohol dehydrogenase were presented and degenerate primers allowed the amplification of fragments sized from 400 to 2300 bp (Brunel *et al.*, 1999). In addition, 32 ACGMs of *Arabidopsis* and *Brassica* were reported by Fourman *et al.* (2002) for genes linked to agronomic traits, as fatty acids biosynthesis pathway, plant defense mechanisms, disease resistance flowering time, and floral morphology. The ACGM markers were also applied for construction of a gene map for comparative study as well as tagging genes for developing of allele-specific DNA markers (Brunel *et al.*, 1999).

DNA marker system used for map construction, gene tagging, genomic, and cDNA fingerprinting as well as map-based cloning was developed by Li and Quiros (2001) and designed as SRAP. This method was aimed at the amplification of open reading frames (ORFs) and revealed that 20% of the SRAP markers were found codominant but they were evenly distributed in major linkage groups of *B. oleracea*. The forward primer used for PCR amplification consisted of a 14-base long core sequence, of which the first 10 bases at the 5′ end was a "filler" of no specific sequence, followed by the sequence CCGG and 3 variable and selective nucleotides at the 3′ end. The reverse primer was 18-base long and consisted of the same elements, except one difference, that is AATT. The primers did not form hairpins or other secondary structures but had GC content of 40–50%, whereas filler sequences were different from each other by 10 or 11 bases. Both primers were radiolabeled and the PCR products were separated by denaturing acrylamide gels and detected through autoradiography. *BoGLS-ALK* GC desaturation gene was tagged by using the same technique (Li and Quiros, 2001).

Newly developed allele-specific markers for direct identification of SNPs (Haff and Smirnov, 1997; Syvänen, 1999, 2005) have provided a new powerful tool for monitoring mutants in breeding process. Detection of SNP includes PCR amplification of a target sequence, fluorescent labeled primers, and microsequencing and detection of polymorphism. A- and C-allele-specific SNP markers were developed for the high oleic and low linolenic mutant of oilseed rape (Hu *et al.*, 2003).

Other kind of DNA markers were developed by Prieto *et al.* (2005) for monitoring of gene flow from *B. napus* to its relative—wild radish (*Raphanus raphanistrum*, RrRr, $2n = 18$). In this method, SINEs, transposable elements of class I, which are irreversibly inserted into genome (Okada, 1991) were identified and analyzed for both *B. napus* and *R. raphanistrum* by using sequence-specific amplified polymorphism (SSAP). An AFLP-like technique (Ellis *et al.*, 1998) led to the detection of individual SINE insertions as a band on a sequencing gel. *B. napus*-specific bands present at high frequency were further analyzed and converted into a single locus marker that could be revealed by PCR. Obtained SINE markers could be used for retracing gene flow from *B. napus* and *R. raphanistrum* (Prieto *et al.*, 2005).

III. MOLECULAR ASPECTS AND APPLICATION OF DNA MARKERS IN BREEDING FOR IMPORTANT AGRONOMIC TRAITS OF OILSEED RAPE SYSTEMS OF CYTOPLASMIC MALE STERILITY

A. CYTOPLASMIC MALE STERILITY

Cytoplasmic male sterility (CMS), a maternally inherited trait in which a plant is unable to produce functional pollen, has been observed in over 150 plant species and is often associated with chimeric mitochondrial (mt) ORFs, transcripts of which interfere with mt function and pollen development (Schnable and Wise, 1998). The CMS trait is encoded by mt genome characterized by mtDNA rearrangements. CMS system is used for commercial hybrid seed production in conjunction with nuclear fertility restorer (*Rf*) genes. There are two general categories of CMS systems—native or endogenous. In both the systems, the nucleus and the sterility inducing cytoplasm occur naturally in the same species, whereas in alloplasmic, male sterility results from the introduction of the nucleus of one species into a cytoplasm of another species (Brown *et al.*, 1998) ensuring thereby total cross-pollination, which is not the case in endogenous systems. In *B. napus*, two endogenous systems exist—*nap* and *pol* (Jean *et al.*, 1997; L'Homme *et al.*, 1997). Most *B. napus* varieties contain the *nap* cytoplasm but are male-fertile because they possess a restorer gene for *nap* CMS. The *nap* cytoplasm confers male sterility on a few exceptional varieties that lack restorer gene such as the cultivar "Bronowski" (Li *et al.*, 1998). The *pol* cytoplasm was originally identified in a Polish oilseed rape cultivar (Fu, 1981). Most *B. napus* varieties lack a restorer gene for *pol* CMS and, hence, are sterilized by *pol* cytoplasm. Physical mapping revealed differences between the male-sterile *pol* and male-fertile *cam* mt genome that are confined to a rearranged region around the *atp6* gene (L'Homme and Brown, 1993) and result led to the discovery of the CMS-associated *orf224/atp6* locus (Singh and Brown, 1991), which is present only in the *pol* cytoplasm. Additional evidence for the involvement of this locus in the CMS phenotype is associated with *orf224/atp6* transcripts, which are differentially processed in plants that carry restorer genes (Jean *et al.*, 1997; Singh *et al.*, 1996). The predicted ORF224 protein exhibits 79% sequence similarity to the predicted protein of the *orf222* portion of the CMS-associated *ofr222/nad5c/orf139* region in the male-sterile *nap* cytoplasm (L'Homme *et al.*, 1997) and transcripts of the region are less abundant and are qualitatively different in restored and nonrestored (i.e., sterile) plants. RFLP mapping showed that restorer genes for *nap* and *pol* CMS systems map to the same chromosomal position. Thus, in contrast to CMS in other species, different forms of *Brassica* endogenous CMS are restored by alleles of a single locus and the restoration properties of these alleles reflect their involvement in the modification of transcripts of corresponding CMS-associated mtDNA regions (Li *et al.*, 1998). In the *pol* CMS system, 11 DNA markers (10 RFLP and 1 RAPD) linked directly to the *Rfp1* restorer gene locus have been reported (Jean *et al.*, 1997). In addition, to perform map-based cloning of the *Rfp* gene, the *pol* cytoplasm and *Rfp* was transferred from *B. napus* to *B. rapa*. As a result a DH *pol* cytoplasm *B. campestris* population was generated that segregated for the *Rfp* gene (Formanova *et al.*, 2006). Genomic library was constructed with the use of DNA obtained from one of the *Rf* gene carrying DH line selfed progeny. A clone containing *Rfp*-linked RFLP marker was isolated for further isolation of the *Rfp* gene (Formanova *et al.*, 2006).

For the production of oilseed rape F_1 hybrids on a commercial scale, an effective system of pollination control is required, an alloplasmic *ogura* radish CMS system is widely used because it completely ensures cross-pollination (Bartkowiak-Broda *et al.*, 1979). This system was originally found in radish (*R. sativus* L.) by Ogura (1968) and later on transferred to *B. oleracea* and *B. napus* by interspecific crosses (Bannerot *et al.*, 1974). Male-sterile *B. napus* cybrids have also been produced through protoplast fusion (Pelletier *et al.*, 1983) to generate male-sterile lines without any defects (Pelletier *et al.*, 1987). *Ogura* CMS oilseed rape plants are phenotypically characterized by flowers with underdeveloped anthers (Fig. 1) and on molecular level, it is a result of expression *orf138* present in male sterile, whereas it is absent in male-fertile normal plant mt genome *orfB* locus, as revealed by physical mapping studies (Bonhomme *et al.*, 1992; Krishnasamy and Makaroff, 1993). Unlike the CMS-associated genes in the *pol* and *nap* cytoplasms, the accumulation of *orf138*-specific transcripts has so far not correlated with the presence or absence of the *Rfo* restorer. Nevertheless, restoration was associated with a reduced accumulation of the 19-kDa ORF138 mt membrane protein (Grelon *et al.*, 1994). Primers specific for 5′ and 3′ ends of the *orf138* nucleotide sequences (Krishnasamy and Makaroff, 1993) were used for PCR identification of the *ogura* CMS cytoplasm, while the fusion experiments of leaf protoplasts from fertile cabbage and cold-tolerant *ogura* CMS broccoli lines (Sigareva and Earle, 1997) and one 0.5-kb band were obtained on agarose gel for CMS cybrids and no bands for fertile

Fig. 1. Flower of a male-sterile *ogura* CMS plant with characteristic, underdeveloped anthers (kindly provided by professor I. Bartkowiak-Broda, Plant Breeding and Acclimatization Institute in Poznan, Poland). (See color Plate 1.)

cybrids. The *orf138*-specific pair of primers was applied for PCR monitoring of *ogura* CMS cytoplasm among *B. napus* cultivars (Mikolajczyk *et al.*, 1998). A sequence-characterized amplified region (SCAR) marker specific for *B. napus ogura* CMS cytoplasm was obtained after adjustment of PCR reaction conditions (Sigareva and Earle, 1997). The PCR analysis clearly indicated one 0.5-kb band on agarose gel for *ogura* CMS cytoplasm, and no bands for fertile cytoplasms (Fig. 2) make this SCAR marker a very useful tool for routine identification of *ogura* CMS-breeding lines before flowering as well as for monitoring of restored hybrids (Mikolajczyk *et al.*, 1998).

The other alloplasmic system is *kosena* radish CMS-Rf (kos CMS), transferred from *R. sativus* to *B. napus* by donor–recipient protoplast fusion (Sakai and Imamura, 1992; Sakai *et al.*, 1996). The *kos* CMS is encoded by the mtDNA *orf125* (Iwabuchi *et al.*, 1999) encoding a 17-kDa protein of a sequence homologous to that of *orf138*, except for two-amino acid substitution and a 39-bp deletion in the *orf138* coding region. Similarly, the accumulation of ORF125 is associated with the CMS phenotype in *B. napus* (Grelon *et al.*, 1994;

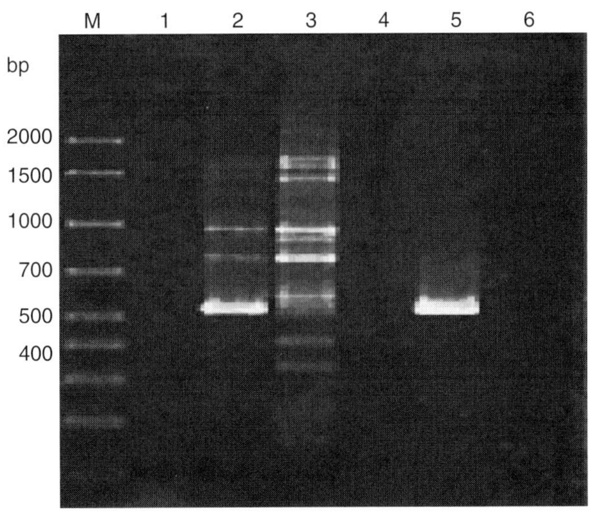

Fig. 2. 1.4% agarose gel electrophoresis of PCR products obtained with the use of primers for a SCAR marker specific for *orf138* mtDNA coding for *ogura* CMS. Lanes 1 and 4: control reactions without a template DNA; 2 and 5: PCR products obtained with the use of DNA from male-sterile *ogura* CMS plants, as a template; 3 and 6: PCR products obtained with the use of DNA from male-fertile plants, as a template; 1–3: PCR conditions as described by Sigareva and Earle (1997); 4–6: adjusted PCR conditions (Mikolajczyk *et al.*, 1998); M: molecular size marker—*Eco*RI and *Hin*dIII digested λ phage DNA (in bp).

Iwabuchi *et al.*, 1999) and the *Rf* restorer gene regulates protein expression at the translational level (Koizuka *et al.*, 1998).

The *tour* CMS system in *B. napus* was developed through the transfer of male-sterile cytoplasm from *B. juncea* (Rawat and Anand, 1979). The *Rf* genes for *tour* CMS were discovered in spring rapeseed germplasm (Banga *et al.*, 1994) and two AFLP markers linked to them were identified as a useful tool for screening a large germplasm breeding population for developing maintainer and restorer gene pools (Janeja *et al.*, 2003a).

A new alloplasmic male-sterile system in *B. napus* was developed by backcross substitution of its nucleus into the cytoplasm of a wild and weedy crucifer—*Enarthrocarpus lyratus* ($2n = 20$; E^1E^1) (Janeja *et al.*, 2003b). In this system, male sterility was complete, stable, and expressed in small flowers with rudimentary anthers. The required fertility restorer gene (*Rf*) has been introgressed from the cytoplasm donor species. Inheritance studies revealed a monogenic dominant control of fertility restoration. Bulked segregant analysis with 215 RAPD primers led to identification of putative primers associated with fertility restoration. Cosegregation analysis of eight of them with *Rf* gene revealed two markers, $OPK15_{700}$ and $OPZ06_{1300}$, flanking the *Rf* locus on either side at a distance of 8.2 and 2.5 cM, respectively. These DNA markers made a useful tool in MAS for improving the commercial potential of this CMS-fertility restorer system for hybrid seed production program in oilseed rape. The other closely linked markers were $OPZ06_{1400}$, $OPD09_{1000}$, and $OPB19_{1100}$ at distances of 5.2, 7.8, and 13.5 cM for *Rf*, respectively (Janeja *et al.*, 2003b).

B. *Rfo* RESTORER GENE FOR CMS *Ogu*-INRA

To produce the hybrid seeds by using CMS system, fertility restorer genes are required, which occur in native CMS-restorer systems. Since no restorer gene for alloplasmic radish *ogura* CMS system was available in *B. napus* (Rousselle, 1982), it was introduced from radish (*R. sativus* L.) through intergeneric crosses between a CMS line of *B. napus* and *Raphanobrassica* (*R. sativus* × *B. napus* amphidiploid, $2n = 56$) (Heyn, 1976). The fully restored plants with one dominant restorer gene were obtained for the best CMS cybrids (Pelletier *et al.*, 1987). Several experimental approaches were undertaken to search for the radish introgression into *B. napus* genome. Cosegregation studies of isozyme markers and male fertility restoration revealed that radish *Rfo* restorer gene for *ogura* CMS radish cytoplasm was introduced into oilseed rape genome along with an isozyme phosphoglucoisomerase-2 (Pgi-2) marker (Delourme and Eber, 1992). Tight linkage was found between the *Rfo* restorer gene and the radish Pgi-2 isozyme marker. Despite the fact that some improved lines with low GC content may lost, the marker (Delourme *et al.*, 1995) proved very useful as a codominant one and was applied for rapid screening of restored oilseed rape plants segregating populations and selection of homozygous restorer forms (Bartkowiak-Broda *et al.*, 2005). Genetic marker analysis was used to estimate the length of the *Pgi-2-Rfo* segment and to construct the physical map of chromosomal area. A comparison of the *Pgi-2* alleles in two progenitors (*B. rapa* L., AA and *B. oleracea* L., CC) of oilseed rape (AACC) indicated that the introgression may have taken place in the C genome of *B. napus* (Chèvre *et al.*, 1995) through homeologous recombination since the introgressed *Pgi-2* allele of radish has replaced the *Pgi-2* allele of oilseed rape (Delourme and Eber, 1992). In addition, the radish introgression was localized on the *B. napus* genetic map constructed from the cross Darmor. bzh. × Yudal and further saturated with RAPD markers. It was concluded that the introgression was mapped on the DY15 linkage group, and was not distal (Delourme *et al.*, 1998). Due to the fact that mapping of the *Rfo* gene region in rapeseed was hindered by the impaired meiotic recombination of the radish introgression with the homeologous rapeseed chromosome (Delourme *et al.*, 1998), characterization of the radish introgression was done with the use of the *Arabidopsis* genome sequence and radish genetic mapping (Giancola *et al.*, 2003). Radish genes linked to *Rfo* restorer gene were detected in the restored rapeseed genome by application of ACGMs (Brunel *et al.*, 1999), indicating several breaks of colinearity between radish and *Arabidopsis* genomes. In addition, precise genetic map around the *Rfo* gene in a segregating radish population was also constructed to determine the relative position of markers. A schematic diagram of the radish introgression in rapeseed indicated the integration of 50 cM of the radish genome in restored rapeseed and that orthologous *Rfo* gene was carried by *Arabidopsis* chromosome 1 (Giancola *et al.*, 2003). One of the main goals in the *Ogu*-INRA CMS system in quality breeding was to obtain double-low restorer lines with a shorter introgression and a good agronomic value including low seed GC content and a good yielding capacity. The development of low GC restorer lines might occur through the deletion of the part of the introgression, for example, one of such lines lost the radish *Pgi-2* allele expression without recovering that of the rapeseed *Pgi-2* allele and at the same time showing a defect in the meiotic transmission of the *Rfo* restorer gene and a very poor agronomic value (Primard-Brisset *et al.*, 2005). Nonspontaneous recombination was forced between that *Rfo*-carrying introgression and the rapeseed homologous chromosome from a low GC *B. napus* line in order to recover the lost C genome information. Just prior to meiosis, γ-ray irradiation was applied for inducing chromosome breakage of restorer plants and then pollen of the plants was used for crossing with low GC CMS plants. One of the F_2 families showed improved behavior for vigor, transmission rate of *Rfo*, and female fertility. It was characterized by a unique combination of molecular markers in comparison with other oilseed rape restorers, which indicated that the recombination event allowed the recovery of *B. oleracea* genetic information that was originally replaced by the radish introgression in the original restorers, which resulted in a duplicated region (originating from radish and *B. oleracea*) on the chromosome carrying the introgression in the analyzed F_2 plant family (Primard-Brisset *et al.*, 2005). Several DNA markers linked to the *Rfo* gene were detected. Analysis with the use of RAPD method performed on two F_2 populations and a population of segregating DHs lines using bulk segregant analysis (BSA) approach (Michelmore *et al.*, 1991) revealed six primers producing repeatable polymorphisms between paired bulks, each bulk corresponding to homozygous restored and male-sterile plants of three segregating populations. DNA polymorphisms generated by four of these primers, that is OPC02, OPD02, OPF06, and OPG02, were found to be completely linked to the restorer gene with the polymorphic DNA fragments associated either with the fertility restorer allele or with the sterility maintainer allele (Delourme *et al.*, 1994). In order to apply the RAPD markers for MAS of restorer oilseed rape lines as well as restored hybrids, three of the above RAPD primers—OPC02, OPD02, and OPG02 were used for analysis, which revealed that among the Polish breeding materials only the RAPD $OPC02_{1150}$ marker was specific for fertility *Rfo* restorer gene (Fig. 3) (Mikolajczyk *et al.*, 1998). Application of AFLP method led to identification of a polymorphic fragment specific for restored plant DNA, which was converted to a sequence tagged site (STS) marker—STS190 (Murayama *et al.*, 2002). Further investigation with the use of RT-PCR expression analysis method showed that the fragment was expressed specifically for restored genotypes as well as it was a part of a 633-bp ORF homologous to the genes of several putative disease resistance proteins of *Arabidopsis thaliana* (Murayama *et al.*, 2002). In order to better understand the molecular mechanism of fertility restoration, the radish *Rfo* restorer gene was cloned and analyzed with the use of map-based cloning approach relying on syntheny between radish and *Arabidopsis*. Radish genetic populations were generated by self-fertilization of individuals that possessed the *ogura* cytoplasm and were heterozygous for *Rfo*. They were used for mapping studies with the use of RFLP probes (Brown *et al.*, 2003) to find RFLP markers linked more closely to *Rfo* gene. The markers were applied for identification of synthetic *Arabidopsis* BACs' contigs and finally used as probes for screening of a BAC genomic library constructed from genomic DNA of radish plant homozygous for *Rfo*-associated alleles, in order to clone and sequence the radish *Rfo* region. Radish BAC and cosmid clones showing complete genetic linkage to *Rfo* were sequenced. In total, sequence of over 270 kb, with a minimum fivefold coverage of the genomic

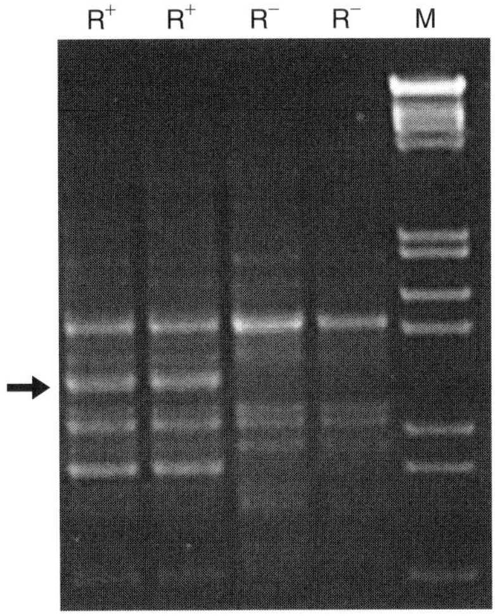

Fig. 3. 1.8% agarose gel electrophoresis of RAPD PCR with the use of the Operon Technologies OPC02 primer; R^+ and R^-: plants with or without *Rfo* restorer gene; M: molecular size marker—*Eco*RI and *Hin*dIII digested λ phage DNA (in base pairs); arrow indicates polymorphic $OPC02_{1150}$ marker totally linked to *Rfo* restorer gene for CMS *ogura* adapted from Mikolajczyk *et al.* (1998).

region, was assembled and annotated, and 43 genes were predicted. Although Wu *et al.* (2001) reported the occurrence of a gene for a chlorophyll *a*/*b*-binding protein at the *ogura* restorer locus in *B. napus*, none of the genes in the 270-kb sequence encodes such a protein. Plant transformation experiments revealed that one of the constructs containing the predicted genes caused recovery of male fertility after transformation of *ogura* CMS *B. napus* plants. Those male-fertile plants showed no phenotypic alterations apart from the change in floral phenotype from *ogura* CMS to the completely fertile restored phenotype (Brown *et al.*, 2003). The transgenically restored flowers were indistinguishable from normal male-fertile *B. napus* flowers. Only a single predicted open reading frame, designated by authors as g26, was located in the Bgl-5 radish DNA. 5′ and 3′ RACE analyses with the use of corresponding primers specific for predicted exons nucleotide sequences were performed in order to determine the structure of *Rfo* mRNA. A single product was obtained from fertile plant RNA using both 5′ and 3′ RACE and no product was obtained from sterile plant DNA. The polypeptide encoded by the g26 mRNA was 687 amino acids in length. The N-terminus was predicted to function as a mt targeting presequence, with the presequence cleavage site predicted at amino acid 51. Similar to the protein encoded by the Petunia *Rf* gene (Bentolila *et al.*, 2002), g26p contained 16 repeats of a 35-AA consensus sequence, the PPR motif (Small and Peeters, 2000). RT-PCR applied for expression analysis of the *Rfo* gene revealed products from flower, leaf, and cotyledon, but not root RNA of homozygous fertile radish plants (Brown *et al.*, 2003). Levels of ORF138 were markedly lower in flowers and leaves, but not in roots of restored versus CMS *ogura* radish plants indicating that *Rfo* like most other nuclear restorer genes acted by downregulating the expression of a CMS-associated mt gene (Krishnasamy and Makaroff, 1994). It was suggested that *Rfo* acted posttranslationally in *B. napus* by selectively targeting ORF138 for proteolysis (Bellaoui *et al.*, 1999). This further suggested that the *Rfo* protein might specifically associate with ORF138 protein rather than the orf138 transcript or mtDNA gene region (Brown *et al.*, 2003). Research on the genetics of fertility restoration of *kosena* radish CMS revealed that genetically this system was the same as that of the *ogura* CMS-Rf-system. Two dominant alleles, *Rf1* and *Rf2* that act complimentary are must for complete fertility restoration in radish, whereas in *B. napus kosena* CMS cybrid, only *Rf1* allele restores the fertility. It is found to be associated with accumulation of ORF125 mt encoded protein (Koizuka *et al.*, 2000). The gene that restores fertility to the radish *kosena* CMS cytoplasm has been sequenced and characterized (Koizuka *et al.*, 2003) which revealed that protein encoded by this gene was identical to that of the *Rfo* protein. Thus, although there were differences in the mt CMS determinants between the radish *kosena* and *ogura* cytoplasms, fertility restoration of both systems could be conferred by genes encoding the same polypeptide (Brown *et al.*, 2003). Radish *kosena* CMS plants are found to express an allele of that gene possessing four substituted amino acids in the second and third repeats of the PPR. It is suggested that the domains formed by these repeats in ORF687 were essential for fertility restoration. Protein levels of the *kosena* CMS-associated mt protein ORF125 are considerably reduced in plants in which fertility was restored, although mRNA expression was normal. It was suggested that ORF687 directly or indirectly lowered the level of ORF125, thus causing the restoration of fertility in CMS plants (Koizuka *et al.*, 2003).

C. PREDICTION OF HYBRID PERFORMANCE WITH THE USE OF GENETIC MARKERS

Heterosis is a term used for describing the common phenomenon in the organic world in which the performance of a hybrid is greater than either parent in different aspects. It exists in higher plants for many measurable characters such as plant height, stem diameter, leaf size, plant biomass production, and seed yield (Liu *et al.*, 2002). Hybrid cultivars of oilseed rape (*B. napus* L.) have been successfully used in agriculture for many years due to their significant yield increase of 30–60% over the mid-parents (Grant and Beversdorf, 1985; Sernyk and Stefansson, 1983). Even stronger heterosis has been observed in the interspecific hybrids among *Brassica* species. Since the heterosis in a hybrid is contributed by genetic complementation between divergent parents, it was assumed that the GD between the parents affects the amount of heterosis expressed in an F_1 hybrid (Frei *et al.*, 1986; Griffing and Linstrom, 1954). Schwartz (1960) and Frei *et al.* (1986) proposed that heterozygosity in isozymes and allozymes could be used for prediction of heterosis in crops. Studies on GD revealed by morphological and molecular markers showed their relationship with hybrid performance in oilseed rape. However, some of the results from the different types of research are contradictory with respect to the relationship between GD and heterosis. It has been found that not all polymorphic DNA fragments contribute to heterosis due to the considerable number of fragments that are either located in noncoding regions or have no association with agronomically important traits. It has, therefore, been proposed that the discrimination between general and special heterosity might be useful for heterosis prediction (Zhang *et al.*, 1996). In order to search for the molecular basis of heterosis of interspecific hybrids between *B. napus* and *B. campestris*, RFLP markers have been used in the diallel analysis for agronomic traits (Liu *et al.*, 2002); however, the results obtained were not satisfactory for prediction of a general model. As such, the information obtained so far on the genetic diversity of the parental lines does not appear to be a reliable basis for predicting F_1 yield and heterosis, thereby indicating the need to develop specific strategies for identification of parental lines with a high level of combining ability (Yu *et al.*, 2005).

D. GENE-FLOW AND TRANSGENE ASSESSMENT

Genetic modifications of crop plants with the use of biotechnological methods, and especially creation of transgenic forms, have led to possible problems concerning difficulties in eliminating the transgenic forms while crop rotation; genes flow through cross-pollination within plants of the same species growing in other fields; introgression of a transgene by hybridization with wild, weedy relatives of the transgenic crop; a horizontal transfer to microorganisms, as well as modification in the cropping system, which can give the opportunity to change selective pressure on a population interacting with the crop (Reboud, 2003). In order to cope with the problems, various studies have been performed by using various methods, isozymes, transgene markers, as well as RAPD and AFLP markers to search for a gene flow from transgenic *B. napus* genotypes to closely related weedy species, which hybridized spontaneously with oilseed rape, and especially—*B. rapa* and

R. raphanistrum (Chévre *et al.*, 1998; Jørgensen and Andersen, 1994). The introgression in a mixed population of *B. napus* and *B. rapa* has revealed that the plants had an outcrossing rate at least 80% (Hansen *et al.*, 2003). The other method of monitoring the movement of transgenes by tagging them with green fluorescent protein (GFP) was invented by Halfhill *et al.* (2001). GFP was cloned from the jellyfish *Aequorea victoria* (Prasher *et al.*, 1992) and when coupled to another transgene, GFP indicated the presence of transgenic material by whole-plant fluorescence without tissue sampling and laboratory analysis (Leffel *et al.*, 1997). *B. napus* cv. Westar was transformed with two GFP constructs, GFP alone and linked to a synthetic *Bacillus thuringiensis* (Bt) endotoxin gene. GFP fluorescence was macroscopically detectable throughout the entire life cycle of rapeseed. However, the intensity of F_1 hybrid lines was lower than homozygous parent. Nevertheless, GFP could be used as a tool for monitoring of transgene flow (Halfhill *et al.*, 2001).

E. TESTING OF CANDIDATE VARIETY

The current testing system of a candidate variety based mainly on standard morphological analyses, such as of fatty acids profile by GLC method, HPLC analysis of glucosynolates content, as well as starch–gel electrophoresis of cotyledon isozymes, may be insufficient to reveal the variety distinctness, uniformity, and stability. Biochemical and morphological markers detect only a limited degree of polymorphism, being dependent on changing environmental conditions and developmental stages of plants. In addition, the presence of many types of hybrid varieties, which are often genetically modified, makes the situation more complex (Tommasini *et al.*, 2003). Therefore, genetic markers which are independent from environmental conditions and developmental stage can constitute a powerful tool for identifying and monitoring of new and existing varieties to protect intellectual property of plant breeders. RAPD and RFLP markers have been used for *B. napus* fingerprinting. However, due to the limitations of the methods, their practical applications are limited. Therefore, the application of codominant SSRs markers combined with fluorescence-based automated technology has been used for testing of distinctness and uniformity oilseed rape varieties (Tommasini *et al.*, 2003). Three multiplex SSR sets, each composed of five markers, have been developed and tested to determine the extent of diversity within and between different varieties using a fluorescence-based semiautomated detection technology. As a result, varieties have been separated in three groups corresponding to winter, spring, and forage types. It was concluded that it should be possible to select a set of markers for a suitable separation. The results obtained from the technology indicate no significant correlation between SSR and morphological data. However, GD measured by SSRs showed correlation to pedigree. SSR method can be applied for prescreening or grouping of existing and candidate varieties, which therefore, in turn, reduce the number of varieties to be grown for morphological and biochemical tests, thus complementing the current variety testing system in *B. napus* (Tommasini *et al.*, 2003).

F. PATHOGENS AND ABIOTIC STRESS RESISTANCE

Stem canker of crucifers, also called the blackleg disease and caused by *Leptosphaeria maculans* (Desm.) Ces. & de Not. (anamorph *Phoma lingam* Tode ex Fr.), a heterothallic haploid dothideomycete fungus, is the most damaging disease of oilseed *Brassica* crops, including oilseed rapes. In the autumn, *L. maculans* produces pseudothecia following sexual outcrosses on infected debris from the previous growing season. After being discharged by the ascospores, pseudothecia infect leaves of the growing plants, which lead to the development of leaf lesions (phoma leaf spots) that contain the asexual fruiting bodies, pycnidia. Then, in winter and early spring, during a symptomless biotrophic phase, the fungus grows systematically in intracellular species down the petiole toward the crown of the stem base. In the late spring and early summer in Europe, *L. maculans* again causes necrosis of the crown tissues which may result in the plant lodging prior to harvest (Hammond and Lewis, 1986; Hammond *et al.*, 1985). Polygenic quantitative resistance (Pilet *et al.*, 1998, 2001) or specific resistance involves major resistance genes in *B. napus* (Hall, 1992; West *et al.*, 2001). In *L. maculans*, nine avirulence genes (*AvrLm1*: nine genes) (Ansan-Melayah *et al.*, 1997; Balesdent *et al.*, 2006) and the corresponding nine major resistance genes (*Rlm1* to *Rlm9*) have been identified in *Brassica* sp. (Delourme *et al.*, 2004) on the basis of gene-for-gene reactions (Flor, 1971). The *Rlm* resistance genes control the disease until the corresponding avirulent allele (*AvrLm*) dominated the pathogen population (Ansan-Melayah *et al.*, 1997; Balesdent *et al.*, 2006; Rouxel *et al.*, 2003). However, extensive breeding of plants possessing only a single effective *Rlm* gene leads to the loss of the resistance in a few years, due to the adaptation of the *L. maculans* field populations (Gout *et al.*, 2006). For evaluation of genetic variation of *L. maculans* field population in France, minisatellite markers have been used and it is assumed that genetic structure of populations from Western Europe is similar to that of Australia (Gout *et al.*, 2006). All *Brassica* species containing the B genome, that is *B. nigra* L. Koch (BB, $2n = 16$), *B. juncea* L. Czer (AABB, $2n = 36$), and *B. carinata* A. Br. (BBCC, $2n = 34$), carry complete resistance which remains throughout the life of the plant (Rimmer and van den Berg, 1992). The introduction of resistance genes in the oilseed rape genome has been attempted either by sexual crosses (Chèvre *et al.*, 1996; Roy, 1978; Sacristan and Gerdemann, 1986; Struss *et al.*, 1996) or by asymmetric protoplast fusion (Sjödin and Glimelius, 1989). Analyses of addition lines containing one B chromosome in the genetic background of an oilseed rape (Chèvre *et al.*, 1996; Struss *et al.*, 1991, 1996) indicated a mono- or oligogenic control of this resistance. Stable *B. napus–B. juncea* recombinant lines resistant to blackleg (*L. maculans*) were selected following identification of chromosomal and genomic origin of the interspecific introgression of *B. juncea*, which conferred blackleg resistance at the cotyledon stage in *B. napus*. In addition, three RAPD markers totally linked to the introgression have been identified, namely OPI01-HaeIII, $OPG02_{800}$, and $OPT01_{800}$ (Chèvre *et al.*, 1997) and found to be localized on the DY17 linkage group (Barret *et al.*, 1998a) of the *B. napus* genetic map (Foisset *et al.*, 1996). Moreover, the estimated size of the substituted *B. napus* fragment was 39 cM and the resistance was introgressed in rapeseed genome by homologous recombination (Barret *et al.*, 1998a). Inheritance studies revealed that unlike the oligo- or polygenically inherited resistance genes of *B. napus* known so far, the B genome resistance genes of the recombinant lines behaved monogenically dominant (Plieske *et al.*, 1998). However, the various described lines seem to harbor independent genes, suggesting that each gene is effective on its own (Saal and Struss, 2005). Moreover, four RFLP and three AFLP have been found to be linked to genes of the B genome of *Brassica* mediating resistance against *P. lingam* in oilseed rape (Plieske and Struss, 2001b). One each in RFLP and AFLP markers have been converted into three PCR-based STS markers: one of dominant and two of codominant inheritance. The STS markers used for mapping studies indicated syntheny and homology between the B-genome resistance gene donors and *B. napus* in the region of the resistance genes. The location of the resistance genes in the B genome/*B. napus* recombinant lines is estimated more likely on the A genome (Plieske and Struss, 2001b). In order to obtain cultivars with enhanced durability of resistance, it was desirable to combine independently acting genes of resistance (pyramidization). This process could be accelerated by using tightly linked molecular markers. The striking similarity of the cloned resistance genes (R genes) had enabled researchers to identify R gene-like sequences—resistance gene analogues (RGAs) or resistance gene homologues (RGHs), which might provide a source for candidate genes in mapping experiments. RGAs had been mapped in various important crop species, including soybean (Kanazin *et al.*, 1996; Yu *et al.*, 1996), corn (Collins *et al.*, 1998), potato (Leister *et al.*, 1996), and oilseed rape (Fourmann *et al.*, 2001; Joyeux *et al.*, 1999). Recessive resistance locus r_j *lm2* (Saal *et al.*, 2004) derived from the B genome of *B. juncea* was introgressed into *B. napus*. RGA and RAPD markers linked to the locus have been identified and converted into SCARs to facilitate marker-assisted backcross breeding (Saal and Struss, 2005).

Sclerotinia sclerotiorum (Lib.) de Bary is a fungal pathogen of plants (Purdy, 1979), infecting leaves, stems (*Sclerotinia* stem rot), and pods at different developmental stages and causing significant devastations in *B. napus*. Although chemical methods have been used to control the disease, yet resistant cultivars may be the most economical, effective, and ecological way to cope with this problem. Genetic studies have shown that the resistance for *Sclerotinia* can be either monogenic or polygenic, depending on the plant species (Baswana *et al.*, 1991). The genetic basis of partial resistance for *Sclerotinia* was searched in rapeseed population families derived from the cross between a male sterility restorer and a partially resistant line by the use of RFLP, AFLP, SSR, and RAPD markers. Six QTLs were detected, of which three were found to be associated with leaf resistance at the seedling stage and three other with disease resistance at the mature plant stage (Zhao and Meng, 2003).

G. YELLOW-SEEDED GENOTYPES

Presently grown cultivars of *B. napus* are black-seeded, and the development of yellow-seeded forms has been proposed as a means to improve rapeseed meal quality by reducing fiber content and increasing protein and oil content (Shirzadegan and Röbbelen, 1985). Since no yellow-seeded forms exist in natural germplasm of *B. napus*, resynthesis of *B. napus* through crosses between yellow-seeded *B. rapa* and yellow- or brown-seeded *B. alboglabra* has been used to create yellow-seeded *B. napus* (Chen *et al.*, 1988). In China, yellow-seeded *B. chinensis* has been used for developing yellow-seeded *B. napus* (Liu, 1983) and a first yellow-seeded cultivar, HUA-yellow No. 1, was registered for commercial production in 1990 (Liu *et al.*, 1991). In addition, yellow-seeded *B. napus* has also been developed from the interspecific crosses with yellow-seeded *B. juncea* and *B. carinata* by Rashid *et al.* (1994). In Canada, a number of yellow-seeded germplasm lines have been developed from complex interspecific crosses involving the black-seeded *B. napus* cv. Regent and yellow-seeded germplasm from *B. rapa* ssp. yellow sarson, *B. carinata*, and *B. juncea*. Pedigree selection was used to isolate yellow-seeded plants in segregating generations after crosses that resulted in identification of a yellow-seeded *B. napus* strain, YN90-1016, in 1990. As a result of hybridization of this line with black-seeded germplasm, agronomically improved and high oil content yellow-seeded *B. napus* lines were developed (Rakow *et al.*, 1999). In Poland, yellow-seeded winter oilseed rape has also been developed as a result of crossing of a spontaneous mutant of a double-low winter oilseed rape line characterized by brighter seed coat with a yellow-seeded spring oilseed rape line obtained from Agriculture Research Station, Saskatoon, Canada in 1979. Yellow-seeded forms have been improved in respect of quality features by crossing with black-seeded, double-low lines/varieties, and these have been stabilized for seed coat color by selection and inbreeding, as well as development of DH lines (Piotrowska *et al.*, 2003). Study on the inheritance of seed coat color in *B. napus* (Shirzadegan, 1986; Van Deynze and Pauls, 1994) led to the conclusion that the black seed coat color in *B. napus* is controlled by three dominant genes and that yellow seeds are produced only when all three loci were in homozygous recessive condition. In addition, an RFLP marker linked to seed coat color in *B. napus* (Van Deynze *et al.*, 1995) has also been identified. Sommers *et al.* (2001) reported that in *B. napus*, a single major gene (pigment 1) and two additional genes controlled seed color and that these genes are flanked by eight, one, and two RAPD markers, respectively. Liu *et al.* (2005) reported a single gene control for partial dominance of yellow seed color, and two RAPD and eight AFLP markers are found at the close vicinity of the seed coat color locus as revealed by the segregation analysis of Chinese breeding materials. The partial dominance and single gene control together with the available molecular markers would greatly facilitate the further breeding of yellow-seeded hybrid varieties (Liu *et al.*, 2005). On the basis of availability of many reports on different yellow-seeded genes existing in different plant materials, it may not be surprising that both dominant yellow-seeded and recessive types have been found in *Brassica* (Liu *et al.*, 2005).

H. DWARF GENOTYPES

Susceptibility to lodging is an important problem in oilseed rape crops which may lead to a significant yield decrease and reduce the final crop dry weight and harvest index (Islam and Evan, 1994). Lodging problems might increase due to the orientation of rapeseed breeding toward the development of hybrid varieties, as hybrid vigor can lead to a height gain of more than 20 cm in size. Oilseed rape breeding for lodging resistance is hampered since this trait is under complex genetic control and its expression is influenced by environmental conditions. A dwarf mutant was obtained by Foisset *et al.* (1995) and genetic analysis of this character revealed that it is under the control of a single mutated gene (*Bzh*) with an additive effect, the dwarf plant being *bzh*/*bzh*. With the use of the random genetic mapping approach, RFLP and RAPD markers linked to *Bzh* gene were identified as well as the linkage group containing *Bzh* was extended. Among the eight RAPD markers the nearest flanking marker, named $OPM07_{730}$, was linked at 0.8 ± 0.7 cM (Foisset *et al.*, 1995) and it was converted into a more convenient SCAR marker (Barret *et al.*, 1998b). Another dwarf oilseed rape mutant, named NDF-1, was obtained by Wang *et al.* (2004) and it was applied for hybrid-breeding programs due to its lodging resistance and promissing seed weight value. Genetic mapping revealed that dwarfism is under the control of a major *ndf1* gene with a mainly additive effect and nonsignificant dominance effect (Wang *et al.*, 2004).

IV. SEED OIL CONTENT

Winter cultivars contain 42–49% of dry seed weight, whereas spring cultivars contain about 40% (Krzymanski, 1984). The oil quality mainly depends on specific fatty acid composition (Rakow and Raney, 2003; Scarth and McVetty, 1999; Thelen and Ohlrogge, 2002). Seeds of double-low winter cultivars of oilseed rape contain on average 4–6% of total saturated ($C_{16:0}$ palmitic and $C_{18:0}$ stearic) fatty acids, 0–2% of $C_{22:1}$ erucic acid, 56–68% of $C_{18:1}$ oleic acid, 18–22% of $C_{18:2}$ linoleic acid, and 10–13% of $C_{18:3}$linolenic acid (Krzymanski, 1970). Double-low varieties meet current customer requirements of rapeseed oil as a food product for human consumption (Spasibionek, 2006). For other purposes such as meal and for oil in industrial and technological use, quality changes are needed.

Rapeseed oil is mainly used for human consumption as salad and frying oils as well as margarines and shortenings. In addition, it is applied for industrial and technical purposes, that is, for production of cosmetics, pharmaceuticals, candles, soaps, detergents, coatings, polymers, drying oils, surfactant, lubricant, as well as for biofuel biocomponents production (Altin *et al.*, 2001; McDonell *et al.*, 1999; Töpfer *et al.*, 1995). Differential fatty acid compositions are required due to various uses of rapeseed oil (Friedt and Lühs, 1999; Mikolajczyk and Bartkowiak-Broda, 2003). One of the main advantages of rapeseed oil as a nutrition product is the presence of polyunsaturated linoleic- and linolenic fatty acids at 2:1 quantity ratio, which makes it a valuable source of essential for human health exogenic fatty acids (Fitzpatrick and Scarth, 1998; Leckband *et al.*, 2002; Scarth and McVetty, 1999; Simopoulos, 2000). However, this characteristic could be a disadvantage for industrial and technology use. Polyunsaturated fatty acids cause flexibility and oxidative rancidity of the oil. Moreover, when used for commercial frying, such oil requires partial hydrogenation. Given the health risk associated with hydrogenated oils as well as the possibilities of economical losses due to the instability of the oils characterized by the high level of polyunsaturated fatty acids, it would be beneficial to breed the rapeseed cultivars of lowered linolenic acid content (Scarth and McVetty, 1999).

For industrial and technical applications, over 80% and 75% content of erucic- or oleic acids, respectively, and at the same time 3% and less of linolenic acid would be required. Keeping this in view, breeding for changed fatty acids composition in rapeseed is being done by organizations worldwide (Rakow and Raney, 2003; Scarth and McVetty, 1999; Spasibionek, 2006). Conventional recombinant-breeding strategy accompanied by mutagenesis as well as wide crossing and biotechnological methods have given the opportunity to change fatty acid composition, though logistical problem such as how to breed newer cultivars characterized by differentiated fatty acid contents without cross-contaminations still remains a concern.

Different approaches have been followed for introducing the low linolenic acid trait into rapeseed genotypes. However, the breeding process is complicated because this trait has a complex genetic inheritance (Pleines and Freidt, 1989), being highly influenced by the environment (Bartkowiak-Broda and Krzymanski, 1983). Therefore, DNA markers appear as an accurate and environment independent tool for application in breeding of the low linolenic oilseed rape (Snowdon and Friedt, 2004).

Searching for DNA markers linked to linolenic acid content was done on F_2 segregating populations obtained from a cross between a spring cultivar Topas and an R4 low linolenic mutant line of cultivar Oro (Rakow, 1973; Röbbelen and Nitsch, 1975) with the use of 8 RFLP and 34 RAPD primers (Tanhuanpää *et al.*, 1995), accompanied by biochemical analysis of seed fatty acids content. Variance analysis revealed one RAPD marker, named 25a, significantly associated with 8% of linolenic acid content. The marker was then applied for analysis of segregating population of inbred- as well as DHs lines obtained after crossing of a double-low winter oilseed rape line PN1775/02 and spring low linolenic Canadian varieties, Stellar (Scarth *et al.*, 1988) and Apollo (Scarth *et al.*, 1994) and it proved very useful for identifying the double-low type genotypes (Fig. 4) (Mikolajczyk *et al.*, 1999, 2004). RAPD marker $OPK01_{1100}$ proved to be specific for linolenic acid content of 00-type while analyzing the F_2 population obtained after crossing of the German rapeseed cultivar Duplo and a low linolenic line resulting from the cross with Mutant Oro (Hu *et al.*, 1995). Other RAPD marker found associated to linolenic acid content was conversed to a codominant SCAR marker. In addition, two RAPD markers were found to be associated with oleic/linoleic acid content (Hu *et al.*, 1999). Tanhuanpää *et al.* (1995) reported two markers generated by OPK01 primer, which were associated with linolenic acid concentration. However, this primer did not reveal polymorphism in respect of linolenic acid concentration when applied for analysis of DH line population developed from F_1 hybrid obtained from a cross of the low linolenic variety Stellar with double-low Drakkar line (Jourdren *et al.*, 1996a).

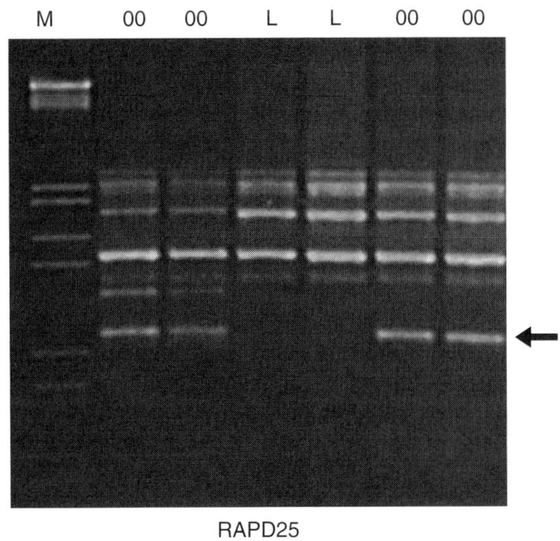

Fig. 4. 1.4% agarose gel electrophoresis of PCR with the use of the RAPD25 primer (Tanhuanpää *et al.*, 1995) associated with about 10% content of linolenic acid in seed oil; M: molecular size marker—*Eco*RI and *Hin*dIII digested λ phage DNA (in base pairs); arrow indicates polymorphic band associated with the normal content of linolenic acid in seed oil adapted from Mikolajczyk *et al.* (1999).

Two oilseed rape lines were crossed: low linolenic Stellar and Drakkar (9% C 18:3) and DH lines were obtained from F_1 hybrid plant microspores. RAPD method was applied for identifying markers linked to the linolenic acid content and two linkage groups—L1 and L2 of six markers (72.7 and 75.6 cM) were determined (Jourdren *et al.*, 1996b). Each corresponded to a major QTL, which explained 24% and 30.7% of the total phenotypic variation of the linolenic acid content. On the first linkage group, $B07_{100}$, $M20_{1180}$, $I06_{650}$, $P05_{1420}$, $L13_{1150}$, and $M14_{350}$ RAPD markers were identified and on the second, $P08_{650}$, $D08_{1310}$, $R13_{1600}$, $R12_{1580}$, $X14_{1720}$, and $X14_{700}$. $P05_{1420}$ and $P08_{650}$ were also polymorphic for linolenic acid content of double-low plants in Apollo, Stellar as well as for the low linolenic winter rapeseed mutant line developed by Spasibionek (2006) (Fig. 5) (Mikolajczyk *et al.*, 1999, 2004). With the use of RFLP markers, a QTL was identified near an RFLP locus detected by a cDNA clone encoding *fad3* desaturase implicating that the low linolenic acid content of Stellar might be due to a mutation in the gene (Thormann *et al.*, 1996). Sommers *et al.* (1998) reported a total of 16 RAPD markers associated with linolenic acid content and distributed over 3 linkage groups, as a result of mapping of a DHs population derived from a cross between the low linolenic Apollo line and the YN90-1016 high linolenic line. Three QTLs on three linkage groups were identified which accounted for 73.5% of the variation in linolenic acid levels in *B. rapa*

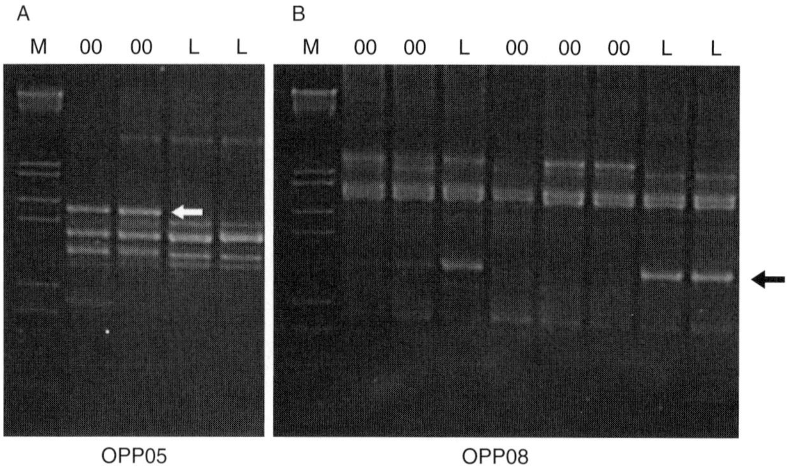

Fig. 5. 1.4% agarose gel electrophoresis of RAPD PCR with the use of the Operon Technologies primers: (A) OPP05 and (B) OPP08; 0: double-low type rapeseed plants (about 10% of linolenic acid in seed oil), L: low linolenic genotypes of rapeseed oil (less than 2% of linolenic acid in seed oil); M: molecular size marker—*Eco*RI and *Hin*dIII digested λ phage DNA (in base pairs); arrows indicate polymorphic bands associated with the normal- (0) or low linolenic acid content (L) in seed oil adapted from Mikolajczyk *et al.* (1999).

(Tanhuanpää and Schulman, 2002). Tanhuanpää *et al.* (1998) identified a QTL for oleic acid in a spring *B. rapa* and a positioned and sequenced *fad2* gene associated with desaturation of oleic acid. Allele-specific markers for *fad2* were developed using 3′ mismatch (Tanhuanpää *et al.*, 1998). Bulks were obtained from segregating population of DH lines developed from F_1 hybrid plant after a cross between a Canadian line characterized by the high oleic- (79.2%) and low linolenic (1.6%) acid content (HOLL), and a DH line with 67.8% oleic acid and 7.5% linolenic acid. RAPD markers exhibiting a strong association with oleic and linolenic acid were identified and conversed into more reliable and effective SCAR markers (Javidfar *et al.*, 2006). Due to the fact that some RAPD markers were not polymorphic on different genetic backgrounds, attempts were performed to develop allele-specific markers for the mutated low linolenic genotypes. On the basis of biochemical analyses, it was established that the *fad3* gene encoding the microsomal Δ^{15} desaturase was associated with the induced variation in linolenic acid content (Jourdren *et al.*, 1996a). Map-based cloning and sequencing of the *fad3* gene of *A. thaliana* (Arondel *et al.*, 1992) enabled designing of primers specific for the microsomal *fad3* gene and to apply them for cloning, sequencing and analysis of the orthologous *B. napus* gene. With the use of the primers, *fad3* alleles of nonmutated and the low linolenic mutant plants were PCR amplified and a polymorphism between the low- and the high linolenic acid genotypes was established. In addition, two independent mutations were done for low linolenic acid content (Jourdren *et al.*, 1996b). A and C alleles of *fad3* gene which map the two L1 and L2 loci controlling linolenic acid content in oilseed rape seeds were obtained from Stellar and Drakkar lines, sequenced, and compared.Two mutations were identified and a specific codominant PCR test, that is restriction-site generating PCR (RG-PCR), was developed for detecting each point mutation (Barret *et al.*, 1999). Chemically induced high-oleic and low linolenic mutant and a nonmutated variety Quantum were used for cloning and sequencing of *fad2* and *fad3* alleles. And, in each of them, single nucleotide mutations were detected and allele-specific SNP markers were developed to be applied for MAS during trait introgression and breeding (Hu *et al.*, 2003). In addition, *B. napus* extraplastid FAD3 linoleate desaturase was characterized by expression in *Saccharomyces cerevisiae* by Reed *et al.* (2000). Finally, new mutants of winter rapeseed double-low breeding line (*B. napus*), characterized by high oleic (over 76%) and low linolenic (about 2.8%) content, were obtained by Spasibionek (2006) with the use of a chemical mutagen (EMS). Interestingly, the morphology of the mutant plants, especially of the LL mutant, differed from the nonmutated ones (Fig. 6). DH population was developed from F_1 hybrid which was obtained after crossing of nonmutated double-low line and the LL mutant inbred line. Presently, allele-specific markers for the LL mutant genotype are in the process of development (Mikolajczyk *et al.*, unpublished data) in order to obtain an effective tool for MAS in further breeding programs.

Due to industrial use of rapeseed oil, there is an increased interest in the high erucic acid (HEA) varieties (50–60%). The levels of erucic acid can range from 1% to 60% (Krzymanski and Downey, 1969) and the trait is controlled by two genes with multiple alleles acting in an additive manner (Kondra and Stefansson, 1965). Two regions controlling nearly all phenotypic variation in erucic acid concentration have been identified out of segregating DH lines obtained from a cross between high erucic and high linolenic *B. napus* cv. Major and a DH line developed from Stellar (Thormann *et al.*, 1996). In another study, 14 RAPD markers linked to the E1 and E2 loci have been identified and two genes localized which do not contribute uniformly to the erucic acid (C22:1) level (Jourdren *et al.*, 1996c). A key component of very long fatty acids synthesis elongase complex is the β-ketoacyl-CoA synthase, which in *Arabidopsis* is encoded by the *FAE1* gene. Barret *et al.* (1998c) isolated and analyzed two sequences homologous to the *FAE1* gene from a *B. napus* immature embryo cDNA library. The two genes cosegregated with

Fig. 6. Morphology of winter oilseed rape plants: (A) double-low genotype (00), characterized by about 65% of oleic acid and 10% of linolenic acid in seed oil; (B) high oleic mutant (HOMut), characterized by over 75% of oleic acid in seed oil; (C) low linolenic mutant (LLMut) genotype, characterized by less than 2% of linolenic acid in seed oil (kindly provided by Dr. S. Spasibionek, Institute of Plant Breeding and Acclimatization in Poznan, Poland). (See color Plate 2.)

the E1 and E2 loci governing erucic acid content in *B. napus*, and it was established that *BN-FAE1* gene corresponded to unique *B. rapa* gene and the *BN-FAE2* gene to the unique *B. oleracea* gene (Fourmann *et al.*, 1998). Although changes in nucleotide sequences between the high- and low erucic genotypes employ genetic markers for selection of specific genotypes, further increase of erucic acid content could be reached effectively by genetic transformation techniques. In order to search for regions responsible for the HEA trait, heterologous expression experiments were performed, and it was shown that one-amino acid substitution, that is serine residue at position 282 of the HEA sequences instead of phenylalanine in LEA *B. napus* cv. Westar, leads to disturbing of FAE activity (Katavic *et al.*, 2002).

V. CONCLUSION

Present market demands, concerning raw material obtained from seeds of oilseed rape, require several specific products of different quality, depending on their application. Therefore, breeding programs are oriented on specific and effective methods of genotype identification. Common phenotype analyses based on biochemical methods as well as biometrical measures and observations while selection may be insufficient because plant phenotypes are often influenced by changing environmental conditions and different developmental stages. It is not the case with the use of DNA markers for selection programs because genome analysis is independent from those factors. DNA markers can be developed as a result of genetic mapping and QTL analyses as well as gene tagging and searching for important differences in nucleotide sequences. A progress in searching for the complex, amphidiploid *B. napus* genome is possible by constantly developing and utilizing methods of molecular biology and biotechnology. Several kinds of DNA markers have been thus developed and applied for routine analyses while selection. Development of DH lines, along with the use of markers, makes the selection process more simple and effective, due to elimination of heterozygous forms. It is important that DNA marker should be allele-specific and codominant, easy to perform, inexpensive, and not too complicated. In such cases, selection would be more effective and less laborious. However, new challenges arise concerning development of reliable methods for accessing population diversity as well as testing of candidate varieties with the use of proper genetic markers.

ACKNOWLEDGMENTS

I would like to thank Professor Iwona Bartkowiak-Broda, Plant Breeding and Acclimatization Institute in Poznan, for critically reading this chapter and his helpful remarks.

REFERENCES

Altin, R., Cetinkaya, S. and Yűcesu, H. S. (2001). The potential of using vegetable oil fuels as fuel for diesel engines. *Energy Conversion and Management* **42**, 529–538.

Ansan-Melayah, D., Rouxel, T., Bertrandy, J., Letarnec, B., Mendes-Pereira, E. and Balesdent, M. H. (1997). Field efficiency of *Brassica napus* specific resistance correlates with *Leptosphaeria maculans* population structure. *European Journal of Plant Pathology* **103**, 835–841.

Arondel, V., Lemieux, B., Hwanh, I., Gibson, S., Goodman, H. M. and Somerville, C. R. (1992). Map-based cloning of a gene controlling omega-3 fatty acid desaturation in *Arabidopsis*. *Science* **258**, 1353–1354.

Balesdent, M. H., Louvard, K., Pinochet, X. and Rouxel, T. (2006). A large-scale survey of *Leptosphaeria maculans* occurring on oilseed rape in France. *European Journal of Plant Pathology* **114**, 53–65.

Banga, S. S., Banga, S. K. and Sandha, G. S. (1994). Hybrids in oilseed rape based on *tour* cytoplasm. *Cruciferae Newsletter* **16**, 73–74.

Baannerot, H., Boulidard, L., Cauderon, Y. and Tempe, J. (1974). Cytoplasmic male sterility transfer from *Raphanus* to *Brassica*. *In* "Proceedings of Eucarpia Meeting Cruciferae Vegetable Crop Sect., 25, 52–54. Dundee, Scotland.

Barret, P., Guérif, J., Reynoird, J. P., Delourme, R., Eber, F., Renard, M. and Chèvre, A. M. (1998a). Selection of stable *Brassica napus-Brassica juncea* recombinant lines resistant to blackleg (*Leptosphaeria maculans*). 2. A 'to and fro' strategy to localise and characterise interspecific introgressions on *B. napus* genome. *Theoretical and Applied Genetics* **96**, 1097–1103.

Barret, P., Delourme, R., Foisset, N. and Renard, M. (1998b). Development of a SCAR (sequence characterised amplified region) marker for molecular tagging of the dwarf *BREIZH* (*Bzh*) gene in *Brassica napus L. Theoretical and Applied Genetics* **97**, 828–833.

Barret, P., Delourme, R., Renard, M., Domergue, F., Lessire, R., Delseny, M. and Roscoe, T. J. (1998c). A rapeseed FAE1 gene is linked to the E1 locus associated with variation in the content of erucic acid. *Theoretical and Applied Genetics* **96**, 177–186.

Barret, P., Delourme, R., Brunel, D., Jourdren, C., Horvais, R. and Renard, M. (1999). Low linolenic acid level in rapeseed can be easily assessed through the detection of two single base substitution in *fad3* genes. *In* "Proceedings of the 10th International Rapeseed Congress." Canberra, Australia, 26–29.09.1999, CD ROM.

Bartkowiak-Broda, I. and Krzymanski, J. (1983). Inheritance of C18 fatty acid composition in seed oil of zero erucic winter rape *Brassica napus* L. *In* "Proceedings of Sixth International Rapeseed Conference," Vol. 1, pp. 477–482. Paris, France.

Bartkowiak-Broda, I., Rouselle, P. and Renard, M. (1979). Investigations of two kinds of cytoplasmic male sterility in rapeseed (*Brassica napus*, L.). *Genetica Polonica* **4**, 487–497.

Bartkowiak-Broda, I., Mikolajczyk, K., Spasibionek, S. and Cegielska-Taras, T. (2005). Genetic and breeding research timing at increasing the value of rapeseed oil as a source of renewable energy. *In* "Alternative Plants for Sustainable Agriculture" (S. Jezowski, K. M. Wojciechowicz and E. Zenkteler, eds.), pp. 129–139, 2006. Institute of Plant Genetics PAS, Poznan, Poland.

Baswana, K. S., Rastogi, K. B. and Sharma, P. P. (1991). Inheritance of stalk rot resistance in cauliflower (*Brassica oleracea* var. *Botrytis* L.). *Euphytica* **57**, 93–96.

Bellaoui, M., Grelon, M., Pelletier, G. and Budar, F. (1999). The restorer *Rfo* gene acts post-translationally on the stability of the ORF138 Ogura CMS-associated protein in reproductive tissues of rapeseed cybrids. *Plant Molecular Biology* **40**, 893–902.

Bentolila, S., Alfonso, A. A. and Hanson, M. R. (2002). A pentatricopeptide repeat-containing gene restores fertility to cytoplasmic male-sterile plants. *Proceedings of the National Academy of Sciences of the United States of America* **99**, 10887–10892.

Bonhomme, S., Budar, F., Lancelin, D., Small, I., Defrance, M.-C. and Pelletier, G. (1992). Sequence and transcript analysis of the *Nco2.5* Ogura-specific fragment correlated with cytoplasmic male sterility in *Brassica* hybrids. *Molecular & General Genetics* **235**, 340–348.

Botstein, D., White, R. L., Skolnick, M. and Davis, R. V. (1980). Construction of a genetic map in a man using restriction fragment length polymorphisms. *American Journal of Human Genetics* **32**, 314–331.

Brown, G. G., Domaj, M., DuPauw, M., Jean, M., Li, X. Q. and Landry, B. S. (1998). Molecular analysis of *Brassica* CMS and its application to hybrid seed production. *In* "Proceedings of International Symposium on Brassicas" (Thomas Grégoire and Antonio A. Monteiro, eds.) Acta Horticulture (ISHS) **459**, 265–274.

Brown, G. G., Formanova, N., Jin, H., Wargachuk, R., Dendy, C., Prashant, P., Laforest, M., Zhang, J., Cheung, W. Y. and Landry, B. S. (2003). The radish *Rfo* restorer gene of Ogura cytoplasmic male sterility encodes a protein with multiple pentartricopeptide repeats. *The Plant Journal* **35**, 262–272.

Brunel, D., Froger, N. and Pelletier, G. (1999). Development of amplified consensus genetic markers (ACGM) in *Brassica napus* from known *Arabidopsis thaliana* sequences of known biological function. *Genome* **42**, 378–402.

Cavell, A. C., Lydiate, D. J., Parkin, I. A. P., Dean, C. and Trick, M. (1998). Colinearity between a 30-centimorgan segment of *Arabidopsis thaliana* chromosome 4 and duplicated regions within the *Brassica napus* genome. *Genome* **41**, 62–69.

Cegielska-Taras, T., Tykarska, T., Szala, L., Kuraś, L. and Krzymański, J. (2002). Direct plant development from microspore-derived embryos of winter oilseed rape *Brassica napus L.* ssp. *Oleifera* (DC.) Metzger. *Euphytica* **124**(3), 341–347.

Chen, B. Y., Heneen, W. K. and Jönsson, R. (1988). Resynthesis of *Brassica napus* L. through interspecific hybridization between, *B. alboglabra* Bailey and, *B. campestris L.* with special emphasis on seed colour. *Plant Breeding* **101**, 52–59.

Chèvre, A. M., Delourme, R., Eber, F., Margale, E., Quiros, C. F. and Arus, P. (1995). Genetic analysis and nomenclature for seven isozyme systems in *Brassica nigra*, *B. oleracea* and, *B. campestris*. *Plant Breeding* **114**, 473–480.

Chèvre, A. M., Eber, F., This, P., Barret, P., Languy, X., Brun, H., Delseny, M. and Renard, M. (1996). Characterization of *Brassica nigra* chromosomes and of blackleg resistance in, *B. napus-B. nigra* addition lines. *Plant Breeding* **115**, 113–118.

Chèvre, A. M., Barret, P., Eber, F., Dupuy, P., Brun, H., Tanguy, X. and Renard, M. (1997). Selection of stable *Brassica napus-B. juncea* recombinant lines resistant to blackleg (*Leptosphaeria maculans*). 1. Identification of molecular markers, chromosomal and genomic origin of the introgression. *Theoretical and Applied Genetics* **95**, 1104–1111.

Chévre, A. M., Eber, F., Baranger, A., Hureau, G., Barret, P. and Renard, M. (1998). Characterization of backcross generations obtained under field conditions from oilseed rape–wild radish F1 interspecific hybrids: An assessment of transgene dispersal. *Theoretical and Applied Genetics* **97**, 90–98.

Collins, N. C., Webb, C. A., Seah, S., Ellis, J. G., Hulbert, S. H. and Pryor, A. (1998). The isolation and mapping of disease resistance gene analogs in maize. *Molecular Plant-Microbe Interactions* **11**, 968–978.

Delourme, R. and Eber, F. (1992). Linkage between an isozyme marker and a restorer gene in radish cytoplasmic male sterility of rapeseed (*Brassica napus* L.). *Theoretical and Applied Genetics* **85**, 222–228.

Delourme, R., Bouchereau, A., Hubert, N., Renard, M. and Landry, B. S. (1994). Identification of RAPD markers linked to a fertility restorer gene for the *Ogura* radish cytoplasmic male sterility of rapeseed (*Brassica napus* L.). *Theoretical and Applied Genetics* **88**, 741–748.

Delourme, R., Eber, F. and Renard, M. (1995). Breeding double-low restorer lines in radish cytoplasmic male-sterility of rapeseed (*Brassica napus L.*). *In* "Proceeding of the Ninth International Rapeseed Congress," pp. 6–8. Cambridge, UK.

Delourme, R., Foisset, N., Horvais, R., Barret, P., Champagne, G., Cheung, W. Y., Landry, B. S. and Renard, M. (1998). Characterisation of the radish introgression carrying the *Rfo* restorer gene for the *Ogu*-INRA cytoplasmic male sterility in rapeseed (*Brassica napus L.*). *Theoretical and Applied Genetics* **97**, 129–134.

Delourme, R., Pilet-Nayel, M. L., Archipiano, M., Horvais, R., Tanguy, X., Rouxel, T., Brun, H., Renard, M. and Balesdent, M. H. (2004). A cluster of major specific resistance genes to *Leptosphaeria maculans* in *Brassica napus*. *Phytopathology* **94**, 578–583.

Demeke, T., Adams, R. P. and Chibbar, R. (1992). Potential taxonomic use of random amplified polymorphic DNA (RAPD): A case study in *Brassica*. *Theoretical and Applied Genetics* **84**, 990–994.

Diers, B. W. and Osborn, T. C. (1994). Genetic diversity of oilseed *Brassica napus* germ plasm based on restriction fragment length polymorphisms. *Theoretical and Applied Genetics* **88**, 662–668.

Downey, R. K. and Rakow, G. (1987). Rapeseed and mustard. *In* "Principles of Cultivar Development, Vol. 2, Crop Species" (W. R. Fehr, ed.), pp. 437–486. Macmillan Publishing Company, New York.

Ellis, T. H., Poyser, S. J., Knox, M. R., Vershinin, A. V. and Ambrose, M. J. (1998). Polymorphism of insertion sites of Ty1-copia class retrotransposons and its use for linkage and diversity analysis in pea. *Molecular & General Genetics* **260**, 9–19.

Fitzpatrick, K. and Scarth, R. (1998). Improving the health and nutritional value of seed oils. *P.B.I. Bulletin*. NRC-CRC. January, 15–19.

Flor, H. H. (1971). Current status of the sene-for-fene concept. *Annual Review of Phytopathology* **9**, 275–296.

Foisset, N., Delourme, R., Barret, P. and Renard, M. (1995). Molecular tagging of the dwarf *BREIZH* (*Bzh*) gene in *Brassica napus*. *Theoretical and Applied Genetics* **91**, 756–761.

Foisset, N., Delourme, R., Barret, P., Hubert, N., Landry, B. C. and Renard, M. (1996). Molecular-mapping analysis in *Brassica napus* using isozyme, RAPD and RFLP markers on a doubled-haploid progeny. *Theoretical and Applied Genetics* **93**, 1017–1025.

Formanova, N., Li, X. Q., Ferrie, A. M. R., DePauw, M., Keller, W. A., Landry, B. and Brown, G. G. (2006). Towards positional cloning in *Brassica napus*: Generation and analysis of doubled haploid, *B. rapa* possessing the, *B. napus pol* CMS and *Rfp* nuclear restorer gene. *Plant Molecular Biology* **61**, 269–281.

Fourmann, M., Barret, P., Renard, M., Pelletier, G., Delourme, R. and Brunel, D. (1998). The two genes homologous to *Arabidopsis FAE1* co-segregate with the two loci govering erucic acid content in *Brassica napus*. *Theoretical and Applied Genetics* **96**, 852–858.

Fourmann, M., Charlot, F., Froger, N., Delourme, R. and Brunel, D. (2001). Expression, mapping and genetic variability of *Brassica napus* disease resistance analogues. *Genome* **44**, 1083–1099.

Fourman, M., Barret, P., Froger, N., Baron, C., Charlot, F., Delourme, R. and Brunel, D. (2002). From *Arabidopsis thaliana* to *Brassica napus*: Development of amplified consensus genetic markers (ACGM) for construction of a gene map. *Theoretical and Applied Genetics* **105**, 1196–1206.

Frei, O. M., Stuber, C. W. and Goodman, M. M. (1986). Use of allozymes as genetic markers for predicting performance in maize single cross hybrids. *Crop Science* **26**, 37–42.

Friedt, W. and Lühs, W. (1999). Breeding of rapeseed (*Brassica napus*) for modified seed quality—synergy of conventional and moderm approaches. *In* "Proceedings of the 10th International Rapeseed Congress." Canberra, Australia, 26–29.09.1999, CD ROM.

Fu, T. D. (1981). Production and research in the People's Republic of China. *Eucarpia Cruciferae Newsletter* **6**, 6–7.

Giancola, S., Marbadour, S., Desloire, S., Clouet, V., Falentin-Guyomarc'h, H., Laloui, W., Falentin, C., Pelletier, G., Renard, M., Bendahmane, A., Delourme, R. and Budar, F. (2003). Characterization of a radish introgression carrying the Ogura fertility restorer gene *Rfo* in rapeseed, using the *Arabidopsis* genome sequence and radish genetic mapping. *Theoretical and Applied Genetics* **107**, 1442–1451.

Gout, L., Eckert, M., Rouxel, T. and Balesdent, M. H. (2006). Genetic variability and distribution of mating type alleles in field populations of *Leptosphaeria maculans* from France. *Applied and Environmental Microbiology* **72**(1), 185–191.

Grant, I. and Beversdorf, W. D. (1985). Heterosis and combining ability estimates in spring oilseed rape (*Brassica napus L.*). *Canadian Journal of Genetics and Cytology* **27**, 472–478.

Grelon, M., Budar, F., Bonhomme, S. and Pelletier, G. (1994). Ogura cytoplasmic male-sterility (CMS)-associated *orf138* is translated into a mitochondrial membrane polypeptide in male-sterile *Brassica* cybrids. *Molecular & General Genetics* **243**, 540–547.

Griffing, B. and Linstrom, E. W. (1954). A study of the combining ability of corn inbreds having varying proportions of corn belt and non-corn belt germplasm. *Agronomy Journal* **46**, 545–552.

Grist, S. A., Firgaira, F. A. and Morley, A. A. (1993). Dinucleotide repeat polymorphisms isolated by the polymerase chain reaction. *BioTechniques* **15**, 304–309.

Haff, L. H. and Smirnov, I. P. (1997). Single nucleotide polymorphism identification assays using a thermostable DNA polymerase and delayed extraction MALDI-TOF mass spectrometry. *Genome Research* **7**, 378–388.

Halfhill, M. D., Richards, H. A., Mabon, S. A. and Stewart, C. N., Jr. (2001). Expression of GFP and Bt transgenes in *Brassica napus* and hybridization with *Brassica rapa*. *Theoretical and Applied Genetics* **103**, 659–667.

Hall, R. (1992). Epidemiology of blackleg of oilseed rape. *Canadian Journal of Plant Pathology* **14**, 46–55.

Halldén, C., Nilsson, N.-O., Rading, I. M. and Säll, T. (1994). Evaluation of RFLP and RAPD markers in a comparison of *Brassica napus* breeding lines. *Theoretical and Applied Genetics* **88**, 123–128.

Hammond, K. E. and Lewis, B. G. (1986). The timing and sequence of events leading to stem canker disease in populations of *Brassicas napus* var. *oleifera* in the field. *Plant Pathology* **35**, 551–564.

Hammond, K. E., Lewis, B. G. and Musa, T. M. (1985). A systemic pathway in the infection of oilseed rape plants by *Leptospheria maculans*. *Plant Pathology* **34**, 557–565.

Hansen, L. B., Siegismund, H. R. and Jørgensen, R. B. (2003). Progressive introgression between *Brassica napus* (oilseed rape) and, *B. rapa*. *Heredity* **91**, 276–283.

Heyn, F. W. (1976). Transfer of restorer genes from *Raphanus* to cytoplasmic male-sterile *Brassica napus*. *Cruciferae Newsletter* **1**, 15–16.

Hu, J., Quiros, C., Arus, P., Struss, D. and Röbbelen, G. (1995). Mapping of a gene determining linolenic acid concentration in rapeseed with DNA-based markers. *Theoretical and Applied Genetics* **90**, 258–262.

Hu, J., Li, G., Struss, D. and Quiros, C. F. (1999). SCAR and RAPD markers associated with 18-carbon fatty acids in rapeseed, *Brassica napus*. *Plant Breeding* **118**, 145–150.

Hu, X., Sullivan, M. L., Gupta, M. and Thompson, S. A. (2003). Cloning of fad2 and fad3 genes and development of gene-specific markers for high oleic and low linolenic acids in Canola (*Brassica napus L.*). *In* "Proceedings of the 11th International Rapeseed Congress," pp. 186–189. Copenhagen, Denmark, 6–10 July 2003.

Islam, N. and Evan, E. J. (1994). Influence of lodging and nitrogen rate on the yield and yield attributes of oilseed rape (*Brassica napus* L.). *Theoretical and Applied Genetics* **88**, 530–534.

Iwabuchi, M., Koizuka, N., Fujimoto, H., Sakai, T. and Imamura, J. (1999). Identification and expression of the kosena radish (*Raphanus sativus* L. cv. Kosena) homologue of the ogura radish CMS-associated gene, *orf138*. *Plant Molecular Biology* **39**, 183–188.

Janeja, H. S., Banga, S. S. and Lashmikumaran, M. (2003a). Identification of AFLP markers linked to fertility restorer genes for tournefortii cytoplasmic male-sterility system in *Brassica napus*. *Theoretical and Applied Genetics* **107**, 148–154.

Janeja, H. S., Banga, S. K., Bhaskar, P. B. and Banga, S. S. (2003b). Alloplasmic male sterile *Brassica napus* with *Enarthrocarpus lyratus* cytoplasm: Introgression and molecular mapping of an, *E. lyratus* chromosome segment carrying a fertility restoring gene. *Genome* **46**, 792–797.

Javidfar, F., Ripley, V. L., Roslinsky, V., Zeiali, H. and Abdmishani, C. (2006). Identification of molecular markers associated with oleic and linolenic acid in spring oilseed rape (*Brassica napus* L.). *Plant Breeding* **125**, 65–71.

Jean, M., Brown, G. G. and Landry, B. S. (1997). Genetic mapping of nuclear fertility restorer genes for the 'Polima' cytoplasmic male sterility in canola (*Brassica napus* L.) using DNA markers. *Theoretical and Applied Genetics* **95**, 321–328.

Jourdren, C., Barret, P., Brunel, D., Delourme, R. and Renard, M. (1996a). Specific molecular marker of the genes controlling linolenic acid content in rapeseed. *Theoretical and Applied Genetics* **93**, 512–518.

Jourdren, C., Barret, P., Horvais, R., Delourme, R. and Renard, M. (1996b). Identification of RAPD markers linked to linolenic acid genes in rapeseed. *Euphytica* **90**, 351–357.

Jourdren, C., Barret, P., Horvais, R., Foisset, N., Delourme, R. and Renard, M. (1996c). Identification of RAPD markers linked to the loci controlling erucic acid level in rapeseed. *Molecular Breeding* **2**, 61–71.

Jørgensen, R. B. and Andersen, B. (1994). Spontaneous hybridization between oilseed rape (*Brassica napus*) and weedy, *B. campestris* (Brassicaceae): A risk of growing genetically modified oilseed rape. *American Journal of Botany* **81**, 1620–1626.

Joyeux, A., Fortin, M. G., Mayerhofer, R. and Good, A. G. (1999). Genetic mapping of plant disease resistance homologues using a minimal *Brassica napus* population. *Genome* **42**, 735–743.

Kanazin, V., Marek, L. F. and Shoemaker, R. C. (1996). Resistance gene analogs are conserved and clustered in soybean. *Proceedings of the National Academy of Sciences of the United States of America* **93**, 11746–11750.

Katavic, V., Mietkiewska, E., Barton, D. L., Giblin, E. M., Reed, D. W. and Taylor, D. C. (2002). Restoring enzyme activity in nonfunctional low erucic acid *Brassica napus* fatty acid elongase 1 by a single amino acid substitution. *European Journal of Biochemistry* **269**, 5625–5631FEBS 2002.

Koizuka, N., Fujimoto, H., Sakai, T. and Imamura, J. (1998). Translational control of ORF125 expression by a radish fertility-restoration gene in *Brassica napus*. *In* "Plant Mitochondria: From Gene to Function" (I. M. Møller, P. Gardestorm, K. Glimelius and E. Glaser, eds.), pp. 83–86. Blackhuys Publishers, Leiden, The Netherlands.

Koizuka, N., Imai, R., Iwabuchi, M., Sakai, T. and Imamura, J. (2000). Genetic analysis of fertility restoration and accumulation of ORF125 mitochondrial protein in the kosena radish (*Raphanus sativus* cv. Kosena) and a *Brassica napus* restorer line. *Theoretical and Applied Genetics* **100**, 949–955.

Koizuka, N., Imai, R., Fujimoto, H., Hayakawa, T., Kimura, Y., Kohno-Murase, J., Sakai, T., Kawasaki, S. and Imamura, J. (2003). Genetic characterization of a pentatricopeptide repeat protein gene, *orf687*, that restores fertility in the cytoplasmic male-sterile Kosen radish. *The Plant Journal* **34**(4), 407–415.

Kondra, Z. P. and Stefansson, B. R. (1965). Inheritance of erucic and eicosenoic acid content of rapeseed oil (*Brassica napus*). *Canadian Journal of Genetics and Cytology* **7**, 505–510.

Kresovich, S., Szewc-McFadden, A. K., Bliek, S. M. and McFerson, J. R. (1995). Abundance and characterization of simple-sequence repeats (SSRs) isolated from a size-fractionated genomic library of *Brassica napus* L. (rapeseed). *Theoretical and Applied Genetics* **91**, 206–211.

Krishnasamy, S. and Makaroff, C. A. (1993). Characterization of the radish mitochondrial *ofrB* locus: Possible relationship with male sterility in Ogura radish. *Current Genetics* **24**, 156–163.

Krishnasamy, S. and Makaroff, C. A. (1994). Organ-specific reduction in the abundance of a mitochondrial protein accompanies fertility restoration in cytoplasmic male-sterile radish. *Plant Molecular Biology* **26**, 935–946.

Krzymanski, J. (1970). Genetyczne mozliwości ulepszania skladu chemicznego nasion rzepaku ozimego. (Genetic possibilities of improving of the winter oilseed rape seeds chemical content (In Polish).) *Hodowla Roslin Aklimatyzacja i Nasiennictwo* **14**(2), 95–133.

Krzymanski, J. (1984). Hodowlane mozliwosci ulepszania zawartosci oleju i bialka w nasionach rzepaku. (Breeding possibilities for improvement of oil- and proteins content in oilseed rape seeds (in Polish).) Wyniki badan nad rzepakiem ozimym 1983. IHAR Radzikow, pp. 104–111.

Krzymanski, J. and Downey, R. K. (1969). Inheritance of oleic, linoleic and linolenic acids in seed oil of rapeseed, *Brassica napus*. *Canadian Journal of Plant Science* **55**, 205–210.

Lagerkrantz, U., Ellegren, H. and Andersson, L. (1993). The abundance of various polymorphic microsatellite motifs differ between plants and vertebrates. *Nucleic Acids Research* **21**, 1111–1115.

Landry, B. S., Hubert, N., Etoh, T., Harada, J. J. and Lincoln, S. E. (1991). A genetic map for *Brassica napus* based on restriction fragment length polymorphisms detected with expressed DNA sequences. *Genome* **34**, 543–552.

Leckband, G., Frauen, M. and Friedt, W. (2002). NAPUS 2000. Rapeseed (*Brassica napus*) breeding for improvement human nutrition. *Food Research International* **35**, 273–278.

Leffel, S. M., Mabon, S. A. and Stewart, C. N., Jr. (1997). Applications of green fluorescent protein in plants. *BioTechniques* **23**, 912–918.

Leister, D., Ballvora, A., Salamini, F. and Gebhardt, C. (1996). A PCR-based approach for isolating pathogen resistance genes from potato with potential wide application in plants. *Nature Genetics* **14**, 421–429.

L'Homme, Y. and Brown, G. G. (1993). Organizational differences between cytoplasmic male sterile and male fertile *Brassica* mitochondrial genome are confined to a single transposed locus. *Nucleic Acids Research* **21**, 1903–1909.

L'Homme, Y., Stahl, R. J., Li, X. Q., Hameed, A. and Brown, G. G. (1997). *Brassica nap* cytoplasmic male sterility is associated with expression of a mt DNA region containing a chimeric gene similar to the *pol* CMS-associated *orf224* gene. *Current Genetics* **31**, 325–335.

Li, G. and Quiros, C. F. (2001). Sequence-related amplified polymorphism (SRAP), a new marker system based on a simple PCR reaction: Its application to mapping and gene tagging in *Brassica*. *Theoretical and Applied Genetics* **103**, 455–461.

Li, X.-Q., Jean, M., Landry, B. S. and Brown, G. G. (1998). Restorer genes for different forms of *Brassica* cytoplasmic male sterility map to a single nuclear locus that modifies transcripts of several mitochondrial genes. *Proceedings of the National Academy of Sciences of the United States of America* **95**, 10032–10037.

Liu, H. L. (1983). Studies on the breeding of yellow seeded *Brassica napus* L.. *In* "Proceedings of the Sixth International Rapeseed Congress," Vol. 1, pp. 637–641. Paris, France. Groupe Consultatif International de Recherche Sur la Colza, Paris, France.

Liu, H. L., Han, J. X. and Hu, X. J. (1991). Studies on the inheritance of seed coat colour and other related characteristics of yellow seeded *Brassica napus*. *In* "Proceedings of the Eighth International Rapeseed congress" (D. I. McGregor, ed.), Vol. 5, pp. 1438–1444. Saskatoon, Sask. Groupe Consultatif International de Recherche sur la Colza, Paris, France.

Liu, R., Qian, W. and Meng, J. (2002). Association of RFLP markers and biomass heterosis in trigenomic hybrids of oilseed rape (*Brassica napus* ×, *B. campestris*). *Theoretical and Applied Genetics* **105**, 1050–1057.

Liu, Z., Fu, T., Tu, J. and Chen, B. (2005). Inheritance of seed colour and identification of RAPD and AFLP markers linked to the seed colour gene in rapeseed (*Brassica napus* L.). *Theoretical and Applied Genetics* **110**, 303–310.

Lombard, V. and Delourme, R. (2001). A consensus linkage map for rapeseed (*Brassica napus* L.): Construction and integration of three individual maps from DH populations. *Theoretical and Applied Genetics* **103**, 491–507.

McDonell, K., Ward, S., Leahy, J. J. and McNulty, P. (1999). Properties of rapeseed oil for use as a diesel fuel extender. *Journal of the American Oil Chemists' Society* **76**(5), 539–543.

Michelmore, R. W., Paran, I. and Kesseli, R. V. (1991). Identification of markers linked to disease resistance genes by bulked segregant analysis: A rapid method to detect markers in specific genomic regions using segregating populations. *Proceedings of the National Academy of Sciences of the United States of America* **88**, 9828–9832.

Mikolajczyk, K. and Bartkowiak-Broda, I. (2003). Markery DNA w hodowli jakosciowej rzepaku ozimego (*Brassica napus* L.) w aspekcie modyfikacji zawartosci kwasow tluszczowych. [DNA markers in rapeseed (*Brassica nalus* L.) breeding with respect to fatty acids content modification (in Polish).]. *Rosliny Oleiste Oilseed Crops* **XXIV**, 33–49.

Mikolajczyk, K., Matuszczak, M., Pietka, T., Bartkowiak-Broda, I. and Krzymanski, J. (1998). Zastosowanie markerow DNA do badan skladnikow mieszancow.

[The use of DNA markers for hybrid components analysis (in Polish).]. *Rosliny Oleiste Oilseed Crops* **XIX**(2), 463–471.

Mikolajczyk, K., Spasibionek, K. and Krzymanski, J. (1999). Poszukiwanie markerow DNA sprzezonych z cecha obnizonej zawartosci kwasu linolenowego w materialach hodowlanych rzepaku ozimego. [Search for DNA markers linked to the low linolenic acid content in winter oilseed rape breeding materials (in Polish).]. *Rosliny Oleiste Oilseed Crops* **XX**(2), 414–421.

Mikolajczyk, K., Spasibionek, S., Cegielska-Taras, T., Nowakowska, J. and Bartkowiak-Broda, I. (2004). Analysis of oilseed rape (*Brassica napus* L.) doubled haploid lines differentiated in respect of linolenoc acid content with the use of RAPD method (in Polish). *In* "Genetyka w ulepszaniu roslin uzytkowych (Genetics in Crop Improvement)" (P. Krajewski, Z. Zwierzykowski and P. Kachlicki, eds.), Vol. 11, pp. 111–116. Institute of Plant Genetics of the Polish Academy of Sciences, Dissertations and Monographs.

Mullis, K. and Faloona, F. (1987). Specific synthesis of DNA *in vitro* via a polymerase-catalysed chain reaction. *Methods in Enzymology* **44**, 224–249.

Murayama, S., Habuchi, T., Yamagishi, H. and Terach, T. (2002). Identification of a sequence-tagged site (STS) marker linked to a restorer gene for Ogura cytoplasmic male sterility in radish (*Raphanus sativus* L.) by non-radioactive AFLP analysis. *Euphytica* **129**, 61–68.

Ogura, H. (1968). Studies on the new male sterility in Japanese radish, with special references to the utilisation of this sterility towards the practical raising of hybrid seeds. *Memoirs of the Faculty of Agriculture, Kagoshima University* **6**, 39–78.

Okada, N. (1991). SINEs. *Current Opinion in Genetics & Development* **1**, 498–504.

Osborn, T. C., Butrulle, D. V., Sharpe, A. G., Pickering, K. J., Parkin, I. A. P., Parker, J. S. and Lydiate, D. J. (2003). Detection and effects of a homeologous reciprocal transposition in *Brassica napus*. *Genetics* **165**, 1569–1577.

Parkin, I. A. P., Sharpe, A. G. and Lydiate, D. J. (2003). Patterns of genome duplication within the *Brassica napus* genome. *Genome* **46**, 291–303.

Pelletier, G., Primard, C., Vedel, F., Chetrit, P., Remy, R., Rousselle, P. and Renard, M. (1983). Intergeneric cytoplasm hybridization in *Cruciferae* by protoplast fusion. *Molecular & General Genetics* **191**, 244–250.

Pelletier, G., Primard, C., Vedel, F., Chetrit, P., Renard, M., Pellan-Delourme, R. and Mesquida, J. (1987). Molecular phenotype and genetic characterization of mitochondrial recombinants in rapeseed. *In* "Proceedings of the Seventh International Rapeseed Conference," pp. 113–118. Poznan, Poland.

Pilet, M. L., Delourme, R., Foisset, N. and Renard, M. (1998). Identification of loci contributing to quantitative field resistance to blackleg disease, causal agent *Leptosphaeria maculans* (Desm.) Ces. Et de Not., in Winter rapeseed (*Brassica napus* L.). *Theoretical and Applied Genetics* **96**, 23–30.

Pilet, M. L., Duplan, G., Archipiano, H., Barret, P., Baron, C., Horvais, R., Tanguy, X., Lucas, M. O., Renard, M. and Delourme, R. (2001). Stability of QTL for field resistance to blackleg across two genetic backgrounds in oilseed rape. *Crop Science* **41**, 197–205.

Piotrowska, A., Krzymanski, J., Bartkowiak-Broda, I. and Krotka, K. (2003). Characteristic of yellow-seeded lines of winter oilseed rape. *In* "Proceedings of the 11th International Rapeseed Congress," pp. 247–249. Copenhagen, Denmark, 6–10 July 2003. BP5.**12**.

Pleines, S. and Freidt, W. (1989). Genetic control of linolenic acid concentration in seed oil of rapeseed (*Brassica napus* L.). *Theoretical and Applied Genetics* **78**, 793–797.

Plieske, J. and Struss, D. (2001a). Microsatellite markers for genomic analysis in *Brassica*. I. Development and abundance in *Brassica* species. *Theoretical and Applied Genetics* **102**, 689–694.

Plieske, J. and Struss, D. (2001b). STS markers linked to *Phoma* resistance genes of the *Brassica* B-genome revealed sequence homology between *Brassica nigra* and *Brassica napus*. *Theoretical and Applied Genetics* **102**(4), 483–488.

Plieske, J., Struss, D. and Röbbelen, G. (1998). Inheritance of resistance derived from the B-genome of *Brassica* against *Phoma lingam* in rapeseed and the development of molecular markers. *Theoretical and Applied Genetics* **97**, 929–936.

Prasher, D. C., Eckenrode, V. K., Ward, W. W., Prendergast, F. G. and Cormier, M. J. (1992). Primary structure of the *Aequorea victoria* green fluorescent protein. *Gene* **111**, 229–233.

Prieto, J. L., Pouilly, N., Jenczewski, E., Deragon, J. M. and Chèvre, A. M. (2005). Development of crop-specific transposable element (SINE) markers for studying gene flow from oilseed rape to wild radish. *Theoretical and Applied Genetics* **111**, 446–455.

Primard-Brisset, C., Pouoard, J. P., Horvais, R., Eber, F., Pelletier, G., Renard, M. and Delourme, R. (2005). A new recombined double low restorer line for the *Ogu*-INRA cms in rapeseed (*Brassica napus* L.). *Theoretical and Applied Genetics* **111**, 736–746.

Purdy, L. H. (1979). *Sclerotinia sclerotiorum*: History, diseases and symptomatology, host range, geographic distribution and impact. *Phytopathology* **69**, 875–890.

Rafalski, J. A. and Tingey, S. V. (1993). Genetic diagnostic in plant breeding—RAPDs, microsatellites and machines. *Trends in Genetics* **9**, 275–280.

Rakow, G. (1973). Selektion auf Linol- und Linolensäuregehalt in Rapssamen nach mutagener Behandlung. *Zeitschrift für Pflanzenzüchtung* **69**, 62–82.

Rakow, G. and Raney, J. P. (2003). Present status and future perspectives of breeding for seed quality in *Brassica* oilseed crops. *In* "Proceedings of the 11th International Rapeseed Congress," pp. 181–185. Copenhagen, Denmark, 6–10 July 2003. BO5.3.

Rakow, G., Relf-Eckstein, J., Raney, P. and Gugel, R. (1999). Development of high yielding, disease resistant, yellow-seeded *Brassica napus*. *In* "Proceedings of the 10th International Rapeseed Congress." Canberra, Australia, Session C07, 26–29 September 1999. CD-ROM, Group Consultif International de Recherche sur la Colza, Paris, France.

Rashid, A., Rakow, G. and Downey, R. K. (1994). Development of yellow-seeded *Brassica napus* L. through interspecific crosses. *Plant Breeding* **112**, 127–134.

Rawat, D. S. and Anand, I. J. (1979). Male sterility in Indian mustard. *Indian Journal of Genetics & Plant Breeding* **39**, 412–414.

Reboud, X. (2003). Effect of a gap on gene flow between otherwise adjacent transgenic *Brassica napus* crops. *Theoretical and Applied Genetics* **106**, 1048–1058.

Reed, D. W., Schäfer, U. A. and Covello, P. S. (2000). Characterization of the *Brassica napus* extraplastidal linoleate desaturase by expression in *Saccharomyces cerevisiae*. *Plant Physiology* **122**, 715–720.

Ribaut, J.-M., Hu, X., Hoisington, D. and González-de-León, D. (1997). Use of STS and SSRs as rapid and reliable preselection tools in a marker-assisted selection-backcross scheme. *Plant Molecular Biology Reporter* **15**(2), 154–162.

Rimmer, S. R. and van den Berg, C. G. J. (1992). Resistance of oilseed *Brassica* ssp. to blackleg caused by *Leptosphaeria maculans*. *Canadian Journal of Plant Pathology* **14**, 56–66.

Röbbelen, G. and Nitsch, A. (1975). Genetical and physiological investigations on mutants for polynoic fatty acids in rapeseed, *Brassica napus* L. *Zeitschrift für Pflanzenzüchtung* **75**, 93–105.

Rousselle, P. (1982). Premiers résultats d'un programme d'introducion de l'androstérilité "Ogura" du radis chez le colza. *Agronomie* **2**, 859–864.

Rouxel, T., Penaud, A., Pinochet, X., Brun, H., Gout, L., Delourme, R., Schmit, J. and Balesdent, M. H. (2003). A 10-year survey of populations of *Leptosphaeria maculans* in France indicates a rapid adaptation toward the *Rlm1* resistance gene of oilseed rape. *European Journal of Plant Pathology* **109**, 871–881.

Roy, N. N. (1978). A study on disease variation in the population of an interspecific cross of *Brassica juncea,* L. ×, *B. napus* L. *Euphytica* **27**, 145–149.

Rudolph, B., Uzunova, M. I. and Ecke, W. (1999). Development and genetic mapping of microsatellite markers in rapeseed. *In* "Proceedings of the 10th International Rapeseed Congress." Canberra, Australia, 26–29.09.1999, CD ROM.

Saal, B. and Struss, D. (2005). RGA- and RAPD-derived SCAR markers for a *Brassica* B-genome introgression conferring resistance to blackleg in oilseed rape. *Theoretical and Applied Genetics* **111**, 281–290.

Saal, B., Plieske, J., Hu, J., Quiros, C. F. and Struss, D. (2001). Microsatellite markers for genome analysis in *Brassica*. II. Assignment of rapeseed microsatellites to the A and C genomes and genetic mapping in *Brassica oleracea*, L. *Theoretical and Applied Genetics* **102**, 695–699.

Saal, B., Brun, H., Glais, I. and Struss, D. (2004). Identification of a *Brassica juncea*-derived recessive gene conferring resistance to *Leptosphaeria maculans* in oilseed rape. *Plant Breeding* **123**, 505–511.

Sacristan, M. D. and Gerdemann, M. (1986). Different behavior of *Brassica juncea* and, *B. carinata* as sources of *Phoma lingam* resistance in experiments of interspecific transfer to, *B. napus*. *Plant Breeding* **97**, 304–314.

Sakai, T. and Imamura, J. (1992). Alteration of mitochondrial genomes containing *atpA* genes in the sexual progeny of cybrids between *Raphanus sativus* cms line and *Brassica napus* cv. Westar. *Theoretical and Applied Genetics* **84**, 923–929.

Sakai, T., Liu, H. J., Iwabuchi, M., Kohno-Murase, J. and Imamura, J. (1996). Introduction of a gene from fertility restored radish (*Raphanus sativus*) into *Brassica napus* by fusion of X-irradiated protoplasts from a radish restorer line and iodoacetamide-treated protoplasts from a cytoplasmic male-sterile cybrid of, *B. napus*. *Theoretical and Applied Genetics* **93**, 373–379.

Scarth, R. and McVetty, P. B. E. (1999). Designer oil Canola—a review of a new-food grade *Brassica* oils with focus on high oleic, low linolenic types. *In* "Proceedings of the 10th International Rapeseed Congress." Canberra, Australia, 26–29.09.1999, CD ROM.

Scarth, R., McVetty, P. B. E., Rimmer, S. R. and Stefansson, B. R. (1988). 'Stellar' low linolenic-high linoleic acid summer rape. *Canadian Journal of Plant Science* **68**, 509–511.

Scarth, R., Rimmer, S. R. and McVetty, P. B. E. (1994). 'Apollo' low linolenic acid summer rape. *Canadian Journal of Plant Science* **75**, 203–204.

Schnable, P. S. and Wise, R. P. (1998). The molecular basis of cytoplasmic male sterility and fertility restoration. *Trends in Plant Science* **3**(5), 175–180.

Schwartz, D. (1960). Genetic studies on mutant enzymes in maize: Synthesis of hybrid enzymes by heterozygotes. *Proceedings of the National Academy of Sciences of the United States of America* **46**, 1210–1215.

Sernyk, J. L. and Stefansson, B. R. (1983). Heterosis in summer rape (*B. napus* L.). *Canadian Journal of Plant Science* **63**, 407–413.

Seyis, F., Snowdon, R. J., Lühs, W. and Friedt, W. (2003). Molecular characterization of novel resynthesized rapeseed (*Brassica napus*) lines and analysis of their genetic diversity in comparison with spring rapeseed cultivars. *Plant Breeding* **122**, 473–478.

Sharpe, A. G. and Lydiate, D. J. (2003). Mapping the mosaic of ancestral genotypes in a cultivar of oilseed rape (*Brassica napus*) selected via pedigree breeding. *Genome* **46**, 461–468.

Shirzadegan, M. (1986). Inheritance of seed colour in *Brassica napus* L. *Zeitschrift für Pflanzenzüchtung* **96**, 140–146.

Shirzadegan, M. and Röbbelen, D. (1985). Influence of seed colour and hull proportions on quality properties of seeds in *Brassica napus* L. *Fette Seifen Anstrichmittel* **87**, 235–237.

Sigareva, M. A. and Earle, E. D. (1997). Direct transfer of a cold-tolerant Ogura male-sterile cytoplasm into cabbage (*Brassica oleracea* ssp. *capitata*) via protoplast fusion. *Theoretical and Applied Genetics* **94**, 213–220.

Simopoulos, A. P. (2000). Human requirement for n-3 polyunsaturated fatty acids. *Poultry Science* **79**, 961–970.

Singh, M. and Brown, G. G. (1991). Suppression of cytoplasmic male sterility by nuclear genes alters expression of a novel mitochondrial region. *Plant Cell* **3**, 1349–1362.

Singh, M., Hamel, N., Menassa, R., Li, X. Q., Young, B., Jean, M., Landry, B. S. and Brown, G. G. (1996). Nuclear genes associated with a single brassica CMS restorer locus influence transcripts of three different mitochondrial gene regions. *Genetics* **143**, 505–516.

Sjödin, C. and Glimelius, K. (1989). *Brassica naponigra*, a somatic hybrid resistant to *Phoma lingam*. *Theoretical and Applied Genetics* **77**, 651–656.

Slocum, M. K., Figdore, S. S., Kennard, W. C., Suzuki, J. Y. and Osborn, T. C. (1990). Linkage arrangement of restriction fragment length polymorphisms in *Brassica oleracea*. *Theoretical and Applied Genetics* **80**, 57–64.

Small, I. D. and Peeters, N. (2000). The PPR motif—a TPR-related motif prevalent in plant organellar proteins. *Trends in Biochemical Sciences* **25**, 46–47.

Snowdon, R. J. and Friedt, W. (2004). Molecular markers in *Brassica* oilseed breeding: Current status and future possibilities. *Plant Breeding* **123**, 1–8.

Sommers, D. J., Friesen, K. R. D. and Rakow, G. (1998). Marker assisted selection of low linolenic acid in oilseed species. *Theoretical and Applied Genetics* **96**, 897–903.

Sommers, D. J., Rakow, G., Prabhu, V. K. and Friesen, K. R. D. (2001). Identification of a major gene and RAPD markers for yellow seed coat colour in *Brassica napus*. *Genome* **44**, 1077–1082.

Song, K. M., Suzuki, J. Y., Slocum, M. K., Williams, P. H. and Osborn, T. C. (1991). A linkage map of *Brassica rapa* (syn. *campestris*) based on restriction fragment length polymorphisms. *Theoretical and Applied Genetics* **82**, 296–304.

Spasibionek, S. (2006). New mutants of winter rapeseed (*Brassica napus* L.) with changed fatty acid composition. *Plant Breeding* **125**, 259–267.

Struss, D., Bellin, U. and Röbbelen, G. (1991). Development of B-genome chromosome addition lines of, *B. napus* using different interspecific *Brassica* hybrids. *Plant Breeding* **106**, 204–214.

Struss, D., Quiros, C. F., Plieske, J. and Röbbelen, G. (1996). Construction of *Brassica* B-genome synteny groups based on chromosomes extracted from three different sources by phenotypic, isozyme and molecular markers. *Theoretical and Applied Genetics* **93**, 1026–1032.

Suwabe, K., Iketani, H., Nunome, T. and Kage, T. (2002). Isolation and characterization of microsatellites in *Brassica rapa*, L. *Theoretical and Applied Genetics* **104**, 1092–1098.

Syvänen, A.-C. (1999). From gels to chips: Minisequencing primer extension for analysis of point mutations and single nucleotide polymorphisms. *Human Mutation* **13**, 1–10.

Syvänen, A.-C. (2005). Toward genome-wide SNP genotyping. *Nature Genetic Supplement* **37**, S5–S10.

Tanhuanpää, P. K. and Schulman, A. (2002). Mapping of genes affecting linolenic acid content in *Brassica rapa* ssp. *oleifera*. *Molecular Breeding* **10**, 51–62.

Tanhuanpää, P. K., Vilkki, J. P. and Vilkki, H. J. (1995). Association of RAPD marker with linolenic acid concentration in the seed oil of rapeseed (*Brassica napus* L.). *Genome* **38**, 414–416.

Tanhuanpää, P. K., Vilkki, J. P. and Vihinen, M. (1998). Mapping and cloning of *FAD2* gene to develop allele-specific PCR for oleic acid in spring turnip rape (*Brassica rapa* ssp. *oleifera*). *Molecular Breeding* **4**, 543–550.

Tanksley, S. D., Young, N. D., Paterson, A. H. and Bonierbale, M. W. (1989). RFLP mapping in plant breeding: New tools for an old science. *BioTechnology* **7**, 257–264.

Tanksley, S. D., Ganal, M. W., Prince, J. P., de Vincente, M. C., Bonierbale, M. W., Broun, P., Fulton, T. M., Giovannoni, J. J., Grandillo, S., Martin, G. B., Messeguer, R., Miller, J. C., *et al.* (1992). High density molecular linkage maps of the tomato and potato genomes. *Genetics* **132**, 1141–1160.

Thelen, J. J. and Ohlrogge, J. B. (2002). Metabolic engineering of fatty acid biosynthesis in plants. *Metabolic Engineering* **4**, 12–21.

Thormann, C. E., Romero, J., Mantet, J. and Osborn, T. C. (1996). Mapping loci controlling the concentration of erucic and linolenic acids in seed oil of *Brassica napus* L. *Theoretical and Applied Genetics* **93**, 282–286.

Tommasini, L., Batley, J., Arnold, G. M., Cooke, R. J., Donini, P., Lee, D., Law, J. R., Lowe, C., Moule, C., Trick, M. and Edwards, K. J. (2003). The development of multiplex simple sequence repeat (SSR) markers to complement distinctness, uniformity and stability testing of rape (*Brassica napus* L.) varieties. *Theoretical and Applied Genetics* **106**, 1091–1101.

Töpfer, R., Martini, N. and Schell, J. (1995). Modification of plant lipid synthesis. *Science* **268**, 681–686.

U, N. (1935). Genome analysis in *Brassica* with spezial reference to the experimental formation of, *B. napus* peculiar mode of fertilization. *Japanese Journal of Botany* **7**, 389–452.

Uzunova, M., Ecke, W., Wiessleder, K. and Röbbelen, G. (1995). Mapping the genome of rapeseed (*Brassica napus* L.). I. Construction of an RFLP linkage map and localization of QTLs for seed glucosinolate content. *Theoretical and Applied Genetics* **90**, 194–204.

Uzunova, M. I. and Ecke, W. (1999). Abundance, polymorphism and genetic mapping of microsatellites in oilseed rape (*Brassica napus* L.). *Plant Breeding* **118**, 323–326.

Van Deynze, A. E. and Pauls, K. P. (1994). The inheritance of seed colour and vernalization requirement in *Brassica napus* using doubled haploid populations. *Euphytica* **74**, 77–83.

Van Deynze, A. E., Landry, B. S. and Pauls, K. P. (1995). The identification of restriction fragments length polymorphisms linked to seed colour genes in *Brassica napus*. *Genome* **38**, 534–542.

Vos, P., Hogers, R., Sleeker, M., Reijans, M., Lee, T., Homes, M., Freiters, A., Pot, J., Peleman, J., Kuiper, M. and Zabeau, M. (1995). AFLP: A new concept for DNA fingerprinting. *Nucleic Acids Research* **23**, 4407–4414.

Wang, M. L., Zhao, Y., Chen, F. and Yiu, X. C. (2004). Inheritance and potentials of a mutated dwarfing gene *ndf1* in *Brassica napus*. *Plant Breeding* **123**(5), 449–453.

Weber, J. L. and May, M. (1989). Abundant class of human DNA polymorphisms which can be types using the polymerase chain reaction. *American Journal of Human Genetics* **44**, 388–396.

West, J. S., Kharbanda, P. D., Barbetti, M. J. and Fitt, B. D. L. (2001). Epidemiology and management of *Leptosphaeria maculans* (phoma stem canker) on oilseed rape in Australia, Canada and Europe. *Plant Pathology* **50**, 10–27.

Whitkus, R., Doebley, J. and Lee, M. (1992). Comparative genome mapping of sorghum and maize. *Genetics* **132**, 1119–1130.

Williams, J. G. K., Kubelik, A. R., Livak, K. J., Rafalski, J. A. and Tingey, S. V. (1990). DNA polymorphisms amplified by arbitrary primers are useful as genetic markers. *Nucleic Acids Research* **18**, 6531–6535.

Wu, Y., Tulsieram, L. and Rothstein, S. J. (2001). Identification and characterization of a putative light-harvesting chlorophyll a/b binding protein gene encoded at a fertility restorer locus for the Ogura CMS in *Brassica napus* L. *Theoretical and Applied Genetics* **102**, 759–766.

Yu, C. Y., Hu, S. W., Zhao, H. X., Guo, A. G. and Sun, G. L. (2005). Genetic distances revealed by morphological characters, isozymes, proteins and RAPD markers and their relationships with hybrid performance in oilseed rape (*Brassica napus* L.). *Theoretical and Applied Genetics* **110**, 511–518.

Yu, Y. G., Buss, G. R. and Saghai Maroof, M. A. (1996). Isolation of a superfamily of candidate disease-resistance genes in soybean based on a conserved nucleotide-binding site. *Proceedings of the National Academy of Sciences of the United States of America* **93**, 11751–11756.

Zhang, Q. F., Zhou, Z. Q., Yang, G. P., Xu, C. G. and Liu, K. D. (1996). Molecular marker heterozygosity and the hybrid performance in *indica* and *japonica* rice. *Theoretical and Applied Genetics* **93**, 1218–1224.

Zhao, J. and Meng, J. (2003). Genetic analysis of loci associated with partial resistance to *Sclerotinia sclerotiorum* in rapeseed (*Brassica napus* L.). *Theoretical and Applied Genetics* **106**, 759–764.

Self-Incompatibility

RYO FUJIMOTO AND TAKESHI NISHIO

Graduate School of Agricultural Science,
Tohoku University, Japan

ABSTRACT

Self-incompatibility system has been most intensively studied in *Brassica* because of the advantage as plant materials and the importance as crop species. The determinants of self-recognition specificity in the stigma and the pollen are SRK and SP11/SCR, respectively. The genes of SRK and SP11 have many alleles in a species and are closely linked with each other in the *S*-locus complex. A set of *SRK* and *SP11* alleles are referred to as *S* haplotype. Sequence analyses of many alleles of *SRK* and *SP11* in *Brassica* species revealed pairs of *S* haplotypes having highly similar *SRK* and *SP11* alleles between different species. The interspecific pair of *S* haplotypes is considered to be derived from the same ancestral *S* haplotype and to have

Advances in Botanical Research, Vol. 45
Incorporating Advances in Plant Pathology

0065-2296/07 $35.00
DOI: 10.1016/S0065-2296(07)45005-4

maintained the same recognition specificity. The structure of the *S*-locus complex is highly variable, and this high structural polymorphism is considered to be important for suppressing the recombination between *SRK* and *SP11*, which may result in the breakdown of self-incompatibility. The first candidate protein identified as the female *S* determinant was SLG, which is abundant on the stigma surface, but the function of SLG in self-incompatibility is unclear. The dominance relationships of the *SRK* alleles are different from those of the *SP11* alleles, and dominance order of *S* haplotypes is nonlinear. Recently, methylation of the promoter region of SP11 has been reported to be responsible for transcriptional suppression of recessive *SP11*. Several mutations in *SP11* and *SRK* causing self-compatibility have been identified, and other genes controlling self-incompatibility have also been reported. These molecular studies can contribute to the development of the F_1 hybrid breeding using the self-incompatibility system in *Brassica* crops.

I. INTRODUCTION

Many species in the genus *Brassica* have a self-incompatibility system, which is controlled by a single *S* locus with multiple alleles (Bateman, 1955). The determinants of self-recognition specificity in the stigma and the pollen have been discovered in the past 15 years. The female determinant is a membrane-spanning serine threonine kinase, named *S* receptor kinase (SRK) (Stein *et al.*, 1991), which has an extracellular domain (*S* domain), a transmembrane domain, and an intracellular domain (kinase domain). The male determinant is a small cysteine-rich protein, named SP11/SCR (*S*-locus protein 11/*S*-locus cysteine rich) (SP11 hereafter) (Schopfer *et al.*, 1999; Suzuki *et al.*, 1999). SP11 binds SRK in an allele-specific manner (Kachroo *et al.*, 2001; Takayama *et al.*, 2001). It is generally thought that binding between a ligand, that is SP11, and a receptor, that is SRK, induces reactions that lead to self-pollen rejection. The first candidate protein identified as the female *S* determinant was *S*-locus glycoprotein (SLG). SLG is a secreted glycoprotein, which is abundant on the stigma surface (Nasrallah *et al.*, 1988). The high degree of sequence similarity between SLG and the *S* domain of SRK led to the identification of SRK (Stein *et al.*, 1991).

SRK, *SP11*, and *SLG* are closely linked to each other in the *S* locus, and the alleles of these three genes are transmitted to progeny as one set. Therefore, a set of alleles of *SRK*, *SP11*, and *SLG* is termed *S* haplotype (Nasrallah and Nasrallah, 1993). About 50 and 30 *S* haplotypes have been identified in *Brassica oleracea* and *B. rapa*, respectively (Nou *et al.*, 1993; Ockendon, 2000). Here, we designate the S_1 haplotype of *B. rapa*, *B. oleracea*, and *Raphanus sativus* as *BrS-1*, *BoS-1*, and *RsS-1*, respectively, and alleles of *SRK*, *SP11*, and *SLG* of *BrS-1* as *BrSRK-1*, *BrSP111*, and *BrSLG-1*, respectively.

II. SEQUENCE DIVERSITY OF MULTIPLE ALLELES OF THE *S*-LOCUS GENES

Nucleotide sequences of *SLG*, the first exon of *SRK* encoding the *S* domain, and *SP11* of many *S* haplotypes have been determined in *B. rapa*, *B. oleracea*, and *R. sativus*. On the basis of nucleotide sequences of the *S*-locus genes, *S* haplotypes are classified into two groups, class I and class II (Nasrallah *et al.*, 1991). Nucleotide sequence analysis of many class-I *SLG* alleles in *B. rapa* and *B. oleracea* identified an interspecific pair of *S* haplotypes having high-sequence similarity, that is *BoS-14* and *BrS-25* (Kusaba *et al.*, 1997). Between *BrS-8* and *BrS-46*, which have a highly similar part in *SLG*, the third to seventh exons of *SRK* encoding the kinase domain are also highly similar (Kusaba and Nishio, 1999), suggesting that recombination occurred between these *S* haplotypes, although recombination is considered to be suppressed in the *S* locus. In phylogenetic analysis of the nucleotide sequences of *SLG* alleles in *B. rapa*, *B. oleracea*, and *R. sativus*, the *SLG* alleles in *R. sativus* did not form a species-specific cluster, suggesting that diversification of the *SLG* alleles occurred before diversification between *Brassica* and *Raphanus* (Sakamoto *et al.*, 1998).

Deduced amino acid sequences of SP11 in *B. rapa* are much variable except for eight conserved cysteine residues (Watanabe *et al.*, 2000). Determining the nucleotide sequences of *SP11* alleles and the first exon of *SRK* of many class-I *S* haplotypes in *B. oleracea* and *B. rapa*, Sato *et al.* (2002) revealed that several interspecific pairs of *S* haplotypes having similar *SRK* alleles have highly similar *SP11* alleles. One intergeneric pair, which has high nucleotide sequences of both *SRK* and *SP11* between *B. rapa* and *R. sativus*, was also found (Okamoto *et al.*, 2004).

The *SP11* alleles of class-II *S* haplotypes have relatively narrow nucleotide sequence variation (Shiba *et al.*, 2002). Nucleotide sequence similarities of *SP11* and the first exon of *SRK* between the class-II *S* haplotypes are higher than those between class-I *S* haplotypes, suggesting that class-II *S* haplotype diversification occurred more recently than that of the class-I *S* haplotypes.

III. RECOGNITION SPECIFICITIES OF INTERSPECIFIC PAIRS OF *S* HAPLOTYPES

The same recognition specificity between *BrS-46* and *BoS-7* has been demonstrated by pollination tests using interspecific hybrids, which have no interspecific incompatibility with *B. rapa* and *B. oleracea* (Kimura *et al.*, 2002). Pollination tests by three different methods, that is use of interspecific hybrids, use of transgenic plants, and bioassay of recombinant SP11 proteins, indicated that four class-I interspecific pairs between *B. rapa* and *B. oleracea* have the same recognition specificities (Sato *et al.*, 2003). Using *B. rapa* transgenic plants having an *RsSP11–21* transgene, one intergeneric pair, *BrS-9* and *RsS-21*, whose amino acid identities of the *S* domain of SRK and SP11 are 89.5% and 91.0%, respectively (Okamoto *et al.*, 2004), was revealed to have the same recognition specificity (Sato *et al.*, 2004). On the other hand, three interspecific pairs in class-II *S* haplotypes, that is *BrS-40/BoS-5*, *BrS-44/BoS-2b*, and *BrS-60/BoS-15*, were found to have slightly different, completely different, and the same recognition specificities, respectively (Sato *et al.*, 2003, 2006). The *S* haplotypes in an interspecific pair or an intergeneric pair are considered to be derived from the same ancestral *S* haplotypes, with the same recognition specificity being kept after speciation. However, interspecific pairs of class-II *S* haplotypes may tend to change the recognition specificities (Sato *et al.*, 2006).

The important regions for the recognition specificities of SRK and SP11 have been investigated by comparing amino acid sequences of SRK and SP11 in interspecific pairs. Amino acid substitutions of SRK between interspecific pairs are few in HVRs, although the HVRs are highly variable among different *S* haplotypes, suggesting that the HVRs in SRK are important regions for recognition specificities (Sato *et al.*, 2003). Revealing variations in regions other than the HVRs of SRK between different *S* haplotypes, Sainudiin *et al.* (2005) have indicated that other regions are also responsible for determining the recognition specificity of SRK. Among the six regions of SP11, which are divided on the basis of the conserved cysteine residues, Regions III, V, and VI of class-I SP11s have few amino acid differences between the interspecific pairs having the same recognition specificities, suggesting the possibility that these domains participate in the recognition specificity of SP11 (Sato *et al.*, 2003). By domain swapping and alanine-scanning mutagenesis, a domain between the fifth cysteine and the sixth cysteine of SP11 [named Region V by Sato *et al.* (2003)] has been revealed to determine the recognition specificity of SP11 (Chookajorn *et al.*, 2004). Recognition specificity was changed by replacing Region III and Region V between different two *S* haplotypes, also suggesting the importance of these domains in determining the recognition specificity of SP11 (Sato *et al.*, 2004). In hypotheses on generation of a new *S* haplotype proposed by these two studies (Chookajorn *et al.*, 2004; Sato *et al.*, 2004), it was proposed that a new *S* haplotype was generated through some intermediates. Maintaining recognition specificity of an original *S* haplotype, the recognition specificities of the intermediates may have been gradually altered, finally resulting in the generation of a new *S* haplotype, which has lost the recognition specificity of the original *S* haplotype.

IV. GENOME STRUCTURE OF THE *S* LOCUS

The genome structure of the *S* locus has been investigated in some *S* haplotypes of *B. rapa* and *B. napus*. Comparison of genome structures of the *S* locus between several class-I *S* haplotypes has indicated that gene placement, distance between *SP11*, *SRK*, and *SLG*, and the orientation of these genes at the *S* locus are different between *S* haplotypes (Brugiere *et al.*, 2000). Recombination between *SRK* and *SP11*, which results in the breakdown of self-incompatibility, seldom occurs, and recombination suppression is considered to be mainly due to the heteromorphism of the *S* locus. Other *S*-locus-linked genes have also been found in the *S* locus (Nasrallah, 2000). Sequence polymorphism of the outside regions from *SP6* and *SLL2* between different *S* haplotypes is lower than that of the intergeneric region between *SP6* and *SLL2* (Cui *et al.*, 1999; Shiba *et al.*, 2003). Comparison between class-I *S* haplotypes and a class-II *S* haplotype has indicated that the genome structure of the *S* locus of a class-II *S* haplotype is similar to that of class-I *S* haplotypes, but that the order of *SRK* and *SLG* in the class-II *S* haplotype is reverse to that in the class-I haplotypes (Fukai *et al.*, 2003) (Fig. 1). The order of *SRK* and *SP11* is completely conserved among four class-II *S* haplotypes, although the region between *SP11* and *SRK* is highly variable (Kakizaki *et al.*, 2006).

An *S* haplotype in *B. oleracea* has more than 200 kb between *SRK* and *SLG* (Nasrallah, 2000). Fragment sizes in *B. oleracea* detected by Southern

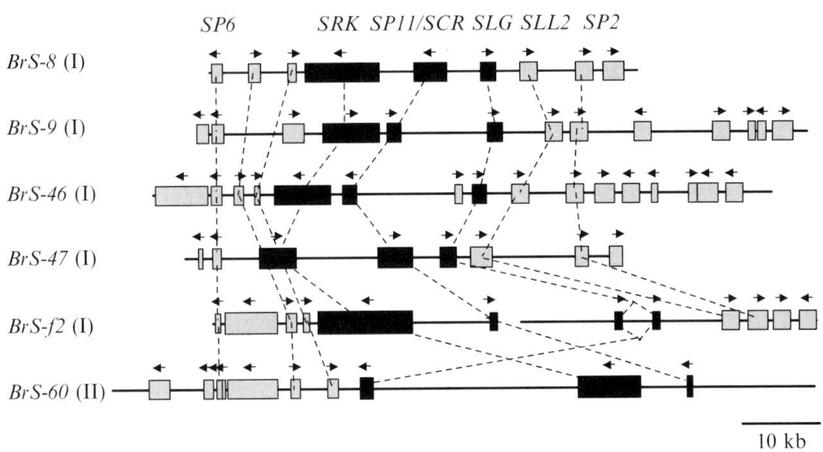

Fig. 1. Structural polymorphism of the *S* locus in *B. rapa*. Boxes are genes in the *S* locus. Arrows indicate the direction of transcription. I, class-I *S* haplotype; II, class-II *S* haplotype.

blot analysis using pulsed-field gel electrophoresis have been found to be much larger than those in *B. rapa*, suggesting that the *S*-locus region in *B. oleracea* is larger than that in *B. rapa* and that the sizes of the *S* locus are relatively constant in *B. rapa* but not in *B. oleracea* (Kimura *et al.*, 2002; Suzuki *et al.*, 2000a). More detailed analysis was performed by determining the nucleotide sequences of the *S* locus in six *S* haplotypes, which are two class-I interspecific pairs and one class-II pair between *B. rapa* and *B. oleracea*. Comparison of the structure of the *S* locus in these three interspecific pairs clearly demonstrated that the *B. oleracea S* locus is larger than the *B. rapa S* locus, and revealed more retrotransposon-like sequences, termed *STFs* (*S*-locus retrotransposon families), in the *S*-locus of *B. oleracea* than in that of *B. rapa* (Fujimoto *et al.*, 2006a). Most *STFs* are considered to have been inserted after speciation of *B. rapa* and *B. oleracea* (Fig. 2). Although the genome size of *B. oleracea* (ca 600 Mb) is not so different from that of *B. rapa* (ca 500 Mb), the distances between *SP11* and *SRK* in *B. oleracea* are more than twice those in *B. rapa*. The reason for the existence of many more retrotransposons in the *B. oleracea S* locus than in the *B. rapa S* locus is of interest.

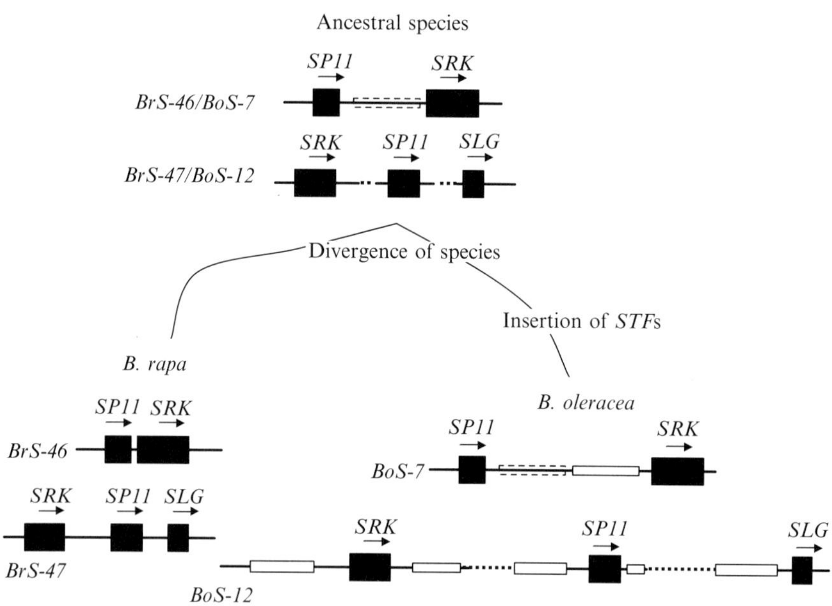

Fig. 2. Diversification of the *S*-locus structure between *B. rapa* and *B. oleracea*. Insertions of many retrotransposon-like sequences (*STFs*) are detected in *B. oleracea*. White boxes indicate *STFs*.

V. FUNCTION OF SLG IN SELF-INCOMPATIBILITY

SLG is located near *SRK*, and the sequence of *SLG* is similar to that of the *S* domain of *SRK*. It has been suggested that *SLG* arose from duplication of the *S* domain of *SRK* (Tantikanjana *et al.*, 1993). Although most of the *S* haplotypes in *B. rapa* and *B. oleracea* have the *SLG* gene, some *S* haplotypes in *B. rapa* and *B. oleracea* lack the *SLG* gene despite being self-incompatible (Okazaki *et al.*, 1999; Sato *et al.*, 2002; Suzuki *et al.*, 2003). Self-incompatible *S* haplotypes that have a frameshift mutation or a nonsense mutation in the *SLG* genes have also been found (Suzuki *et al.*, 2000b). These findings indicate that SLG is not essential for the reaction of self-incompatibility.

In phylogenetic analysis of SLG and the *S* domain of SRK in *B. rapa* and *B. oleracea*, SLG and the *S* domain of SRK in the same *S* haplotypes are clustered in the same clades, suggesting that gene conversion events between *SLG* and the *S* domain of *SRK* in the same *S* haplotypes may have occurred many times (Nishio and Kusaba, 2000). Sato *et al.* (2002) have reported examples of gene conversion events from *SLG* to *SRK* and from *SRK* to *SLG*. New *S* haplotypes may have been generated by genomic rearrangements due to mutations and recombination events. The gene conversion from *SLG* to the *S* domain of *SRK* may also have contributed to the generation of new *S* haplotypes. It has been speculated that the significance of the *SLG* gene is that it is a reservoir of sequence variation to generate new *S* haplotypes (Miege *et al.*, 2001; Nishio and Kusaba, 2000).

It has been reported that coexpression of *SLG* with *SRK* stabilizes the SRK protein in transgenic plants (Dixit *et al.*, 2000) and that the introduction of an *SLG* gene together with *SRK* strengthens self-incompatibility (Takasaki *et al.*, 2000). SLG can be considered to be an enhancer of self-incompatibility. On the other hand, introduction of an *SLG* gene with *SRK* did not strengthen self-incompatibility (Silva *et al.*, 2001). The function of SLG in self-incompatibility is still controversial.

VI. DOMINANCE RELATIONSHIP OF *S* HAPLOTYPES

Because self-incompatibility in *Brassica* is sporophytically controlled, there are dominance relationships of *S* haplotypes in the stigma and the pollen (Thompson and Taylor, 1966). Codominance is common and observed more frequently in the stigma than in the pollen. The dominance relationships are different between the stigma and the pollen, and dominance order of *S* haplotypes is nonlinear (Hatakeyama *et al.*, 1998; Thompson and Taylor, 1966).

The dominance relationship in the stigma is considered to be determined by the SRK protein itself, but it is not clear whether the *S* domain or the kinase domain is more important for determining the dominance relationship of *SRK*. Two models proposed by Hatakeyama *et al.* (2001) are competition of an SRK-mediated signaling pathway and posttranscriptional modification of SRK. The former model suggests the importance of the kinase domain in determining the dominance relationships of the *SRK* alleles. *BrS-46* is dominant over *BrS-54* in the stigma, but there are only two different amino acid residues in the kinase domain between *BrS-46* and *BrS-54* (Fujimoto *et al.*, 2006c). Amino acid sequence similarity of the kinase domain between *BrS-8* and *BrS-46* is also high, with only one amino acid substitution (Kusaba and Nishio, 1999). Amino acid sequence identities of the *S* domain of SRK between these three *S* haplotypes were less than 85.3%. The kinase domain sequences of *B. oleracea S* haplotypes in the interspecific pair with *BrS-8* and *BrS-46* have more differences between them, that is eight amino acid substitutions. These *S* haplotypes would be interesting materials for the study of the dominance relationship of *SRK*.

In pollen, class-I *S* haplotypes are generally dominant over class-II *S* haplotypes (Nasrallah *et al.*, 1991). Expression analysis using *in situ* hybridization has indicated that the dominance relationship in the pollen is controlled at the mRNA level (Kusaba *et al.*, 2002; Shiba *et al.*, 2002). This transcriptional suppression of recessive SP11 has been reported to be controlled by methylation of the promoter region of *SP11* (Shiba *et al.*, 2006). The promoter region of a recessive *SP11* allele is *de novo* methylated in tapetum cells before initiation of *SP11* transcription, and most of methylated sites are asymmetric sites. Interestingly, expression of a dominant *SP11* allele is not necessary for the suppression of gene expression of recessive *SP11* alleles (Fujimoto *et al.*, 2006b). Although *de novo* methylation of the *SP11* promoter is known to suppress the expression of *SP11*, enzyme participating in *de novo* methylation and the mechanism of recognition of dominant and recessive *SP11* alleles have not been revealed.

VII. SELF-COMPATIBILITY CAUSED BY MUTATIONS IN THE *S* LOCUS

Most plants in *B. rapa* and *B. oleracea* are self-incompatible, and there are few self-compatible lines obtained by spontaneous mutations, suggesting an advantage of self-incompatibility in these species. One self-compatible line which has been studied at the molecular level is a *BoS-f1* homozygote in Chinese kale, *B. oleracea* var. *alboglabra*. In the *BoS-f1* homozygote, the

S domain and the transmembrane domain of *SRK* are deleted, and *SRK* is not expressed (Nasrallah *et al.*, 1994). The original *S* haplotype of *BoS-f1* is considered to be *BoS-28* because of high nucleotide sequence identities of *SLG* and the kinase domain of *SRK* between them (Fujimoto *et al.*, 2006b). The *BoS-f1* homozygote has no sequence homologous with *BoSP11-28* and the reported *SP11* sequences, suggesting that *SP11* is also deleted. Sequences of *SLG* and the kinase domain of *SRK* of three independent lines in *B. oleracea* var. *alboglabra*, O-405, BO479, and A-12 are 100% identical, suggesting that all these self-compatible lines in *B. oleracea* var. *alboglabra* have *BoS-f1* in common.

"Yellow sarson" is a self-compatible oilseed cultivar in India. The self-compatibility of "yellow sarson" is controlled by two loci, *S* and *M*, the *M* locus being independent of the *S* locus (Hinata *et al.*, 1983). The loss of *SRK* expression has been reported to be one of the causes of self-compatibility of "yellow sarson" having *BrS-f2* (Nasrallah *et al.*, 1994; Watanabe *et al.*, 1997). Recently, we have found that the original *S* haplotype of *BrS-f2* is *BrS-54* and that neither *SRK* nor *SP11* is expressed. Insertion of a retrotransposon in *SRK* and deletion of a promoter region of *SP11* are inferred to be the causes of the lack of expression of these genes (Fujimoto *et al.*, 2006b).

We obtained self-compatible plants from a self-pollinated population of an F_1 hybrid cultivar in Chinese cabbage having *BrS-46/BrS-54*. Pollination tests indicated that this self-compatibility is linked to *BrS-54* and that the recognition function of the stigma is lost. The *SRK* allele of this self-compatible plant, named *BrSRK-54f*, is normally transcribed and translated. The nucleotide sequence of the first exon of *BrSRK-54f* is different from that of *BrSRK-54*, but the same as that of *BrSLG-54f* and *BrSLG-54*, suggesting that gene conversion from *SLG* to *SRK* has occurred and resulted in the loss of the recognition specificity of *BrSRK-54* (Fujimoto *et al.*, 2006c).

Although monogenomic species in *Brassica* are generally self-incompatible, digenomic species such as *B. napus*, *B. juncea*, and *B. carinata* are self-compatible, suggesting that these digenomic species have lost the function of self-incompatibility. It has been reported that the self-compatible line of *B. napus*, "Westar," has a class-I *SRK-A10* allele having a frameshift mutation due to a 1-bp deletion and a class-II *SRK* allele encoding a functional protein of BoSRK-15 (Goring *et al.*, 1993; Haffani *et al.*, 2004). Failure to produce an active *BoSP11-15*, the presence of a nonfunctional *BoSP11-15* gene, and the recessiveness of *BoSP11-15* to the class-I *S* haplotype, that is *S-A10*, are three possible explanations for the self-compatibility of "Westar" (Haffani *et al.*, 2004). The stigmas of *BrS-60* having the same recognition specificity as *BoS-15* do not reject the pollen grains of allotetraploids produced by interspecific hybridization between homozygotes of *BrS-46* and *BoS-15*, indicating that

BoSP11-15 expression is suppressed in the allotetraploids of *BrS-46/BoS-15* (Kimura *et al.*, 2002). This result suggests the possible suppression of *BoSP11-15* expression by the dominance relationship with a class-I *S* haplotype in "Westar." Further analyses are required to elucidate the mechanisms of compatibility of digenomic *Brassica* species.

VIII. OTHER GENES PARTICIPATING IN *BRASSICA* SELF-INCOMPATIBILITY

In the self-incompatibility reaction, the *S* haplotype-specific interaction between SRK and SP11 initially occurs and pollen tube elongation is finally inhibited. The process leading to pollen tube rejection from *S* haplotype-specific interaction is not well understood, but some genes participating in the self-incompatibility reaction after interaction of SRK and SP11 have been identified.

One gene is *ARC1* (armadillo-repeat-containing 1), which has been isolated by the yeast-two-hybrid screen of proteins interacting with the kinase domain of SRK (Gu *et al.*, 1998). The suppression of *ARC1* expression by the antisense method weakens self-incompatibility, indicating that ARC1 is a component of the pollen rejection mechanism of self-incompatibility (Stone *et al.*, 1999). ARC1 has a U-box motif and E3 ubiquitin ligase activity, suggesting that certain proteins participating in self-incompatibility reaction in the pistil are degraded by ubiquitination (Stone *et al.*, 2003).

Genes of thioredoxin-h protein 1 (THL1) and THL2 have also been isolated by the yeast-two-hybrid screen (Bower *et al.*, 1996). *THL1* and *THL2* belonging to the thioredoxin-h family are expressed ubiquitously in *B. napus*. An *in vitro* experiment has revealed that THL1 inhibits the autophosphorylation activity of SRK, suggesting that THL1 and THL2 prevent the constitutive activation of SRK, that is, constitutive expression of the pollen rejection trait (Cabrillac *et al.*, 2001). The suppression of both *THL1* and *THL2* results in a constitutive but weak self-incompatibility reaction, leading to the hypothesis that THL1 and THL2 are negative regulators of the autophosphorylation activity of SRK (Haffani *et al.*, 2004).

The third gene is *MLPK* (*M*-locus protein kinase). The *M* locus controlling self-compatibility of "yellow sarson" is independent of the *S* locus and epistatic to the *S*-locus gene (Hinata *et al.*, 1983). *MLPK* has been isolated as a candidate gene of *M* by map-based cloning. MLPK of "yellow sarson" has one amino acid substitution of a single nucleotide change that leads to the loss of the autophosphorylation activity (Murase *et al.*, 2004). MLPK belongs to a subfamily of receptor-like cytoplasmic kinase (RLCK), and it has been speculated that MLPK forms a complex with SRK and inhibits pollen tube germination or elongation, although it is not yet clear how MLPK functions in pollen tube rejection.

IX. F_1 HYBRID BREEDING USING THE SELF-INCOMPATIBILITY SYSTEM

Many cultivars of Brassicaceae vegetables, for example cabbage, broccoli, cauliflower, Chinese cabbage, turnip, and radish, are F_1 hybrids and their seeds are produced using the self-incompatibility system or male sterility. As for the yield of F_1 hybrid seeds, F_1 hybrid breeding using the self-incompatibility system is much superior to that using male sterility. However, instability of self-incompatibility influenced by environmental factors, such as high temperature, sometimes results in production of low-quality seeds containing a high percentage of selfed seeds. Strong and stable self-incompatibility is required for F_1 hybrid breeding. Although no *S* haplotypes providing strong self-incompatibility are known, class-II *S* haplotypes are known to exhibit a weak self-incompatibility phenotype (Nasrallah and Nasrallah, 1993). Among class-II *S* haplotypes, *BoS-15* has been found to show the weakest incompatibility, perhaps due to the low affinity between SP11 and SRK (Sato *et al.*, 2006). If SLG stabilizes SRK and enhances self-incompatibility, the *S* haplotypes lacking SLG may result in a weaker self-incompatibility phenotype. The strength of self-incompatibility is controlled by genetic background (Ruffio-Chable *et al.*, 1997) and may also be influenced by interaction of *S* haplotypes and the genetic background.

In F_1 hybrid breeding, development of inbred lines is an important breeding process. Inbred lines are commonly obtained by bud pollination, but such pollination is quite laborious. Therefore, various methods have been developed for overcoming self-incompatibility. Among these methods, carbon dioxide treatment of self-pollinated flowers (Nakanishi and Hinata, 1973) in a greenhouse or a plastic house is an effective labor-saving method for seed production of inbred lines. The yield of selfed seeds by bud pollination is generally low in lines having strong self-incompatibility. The effect of genotype on the response to carbon dioxide gas is also known (Nakanishi and Hinata, 1973; Niikura and Matsuura, 2000).

A simple method for identification of the *S* haplotype is required in F_1 hybrid breeding using the self-incompatibility system because test crossing requires considerable time and labor. PCR-RFLP analysis using primers having conserved *SLG* sequences has been successfully used for identification of *S* haplotypes (Brace *et al.*, 1993; Nishio *et al.*, 1994). However, there are

S haplotypes lacking the *SLG* alleles, that is *BoS-24*, *BrS-32*, *BrS-33*, and *BrS-36* (Okazaki *et al.*, 1999; Sato *et al.*, 2002). Some *S* haplotypes have sequence variation in *SLG*, for example *BoS-2* and *BoS-13* (Kusaba *et al.*, 2000; Miege *et al.*, 2001), and some have duplication of *SLG*, for example *BoS-15* (Cabrillac *et al.*, 1999; Fujimoto *et al.*, 2006a). PCR-RFLP of *SRK* sequences can provide supplementary data for the identification of *S* haplotypes (Nishio *et al.*, 1997). Dot blot analysis using the *SP11* alleles as probes (Fujimoto and Nishio, 2003) is a simple and reliable method of *S* haplotype identification. Since many samples can be analyzed at once without electrophoresis, this method is highly suitable for practical plant breeding. However, there are still many *S* haplotypes for which the *SP11* and *SRK* sequences have not been determined in *B. rapa* and *B. oleracea*.

X. CONCLUSION

The collection of many *S* haplotypes and the development of *S* tester lines with determined nucleotide sequences of the *S*-locus genes are important for F_1 hybrid breeding using the self-incompatibility system as well as for the molecular genetic study of self-incompatibility.

REFERENCES

Bateman, A. J. (1955). Self-incompatibility systems in angiosperms. III. *Cruciferae*. *Heredity* **9**, 52–68.

Bower, M. S., Matias, D. D., Fernandes-Carvalho, E., Mazzurco, M., Gu, T., Rothstein, S. J. and Goring, D. R. (1996). Two members of the thioredoxin-h family interact with the kinase domain of a *Brassica S* locus receptor kinase. *The Plant Cell* **8**, 1641–1650.

Brace, J., Ockendon, D. J. and King, G. J. (1993). Development of a method for the identification of *S* alleles in *Brassica oleracea* based on digestion of PCR-amplified DNA with restriction endonucleases. *Sex Plant Reproduction* **6**, 133–138.

Brugiere, N., Rothstein, S. J. and Cui, Y. (2000). Molecular mechanisms of self-recognition in *Brassica* self-incompatibility. *Trends Plant Science* **5**, 432–438.

Cabrillac, D., Delorme, V., Garin, J., Ruffio-Chable, V., Giranton, J. L., Dumas, C., Gaude, T. and Cock, J. M. (1999). The S_{15} self-incompatibility haplotype in *Brassica oleracea* includes three *S* gene family members expressed in stigmas. *The Plant Cell* **11**, 971–986.

Cabrillac, D., Cock, J. M., Dumas, C. and Gaude, T. (2001). The *S*-locus receptor kinase is inhibited by thioredoxins and activated by pollen coat proteins. *Nature* **410**, 220–223.

Chookajorn, T., Kachroo, A., Ripoll, D. R., Clark, A. G. and Nasrallah, J. B. (2004). Specificity determinants and diversification of the *Brassica* self-incompatibility pollen ligand. *Proceedings of the National Academy of Sciences of the United States of America* **101**, 911–917.

Cui, Y., Brugiere, N., Jackman, L., Bi, Y. M. and Rothstein, S. J. (1999). Structural and transcriptional comparative analysis of the *S* locus regions in two self-incompatible *Brassica napus* lines. *The Plant Cell* **11**, 2217–2231.

Dixit, R., Nasrallah, M. E. and Nasrallah, J. B. (2000). Post-transcriptional maturation of the *S* receptor kinase of *Brassica* correlates with co-expression of the *S*-locus glycoprotein in the stigmas of two *Brassica* strains and in transgenic tobacco plants. *Plant Physiology* **124**, 297–311.

Fujimoto, R. and Nishio, T. (2003). Identification of *S* haplotypes in *Brassica* by dot-blot analysis of *SP11* alleles. *Theoretical and Applied Genetics* **106**, 1433–1437.

Fujimoto, R., Okazaki, K., Fukai, E., Kusaba, M. and Nishio, T. (2006a). Comparison of the genome structure of the self-incompatibility (*S*) locus in interspecific pairs of *S* haplotypes. *Genetics* **173**, 1157–1167.

Fujimoto, R., Sugimura, T., Fukai, E. and Nishio, T. (2006b). Suppression of gene expression of a recessive *SP11/SCR* allele by an untranscribed *SP11/SCR* allele in *Brassica* self-incompatibility. *Plant Molecular Biology* **61**, 577–587.

Fujimoto, R., Sugimura, T. and Nishio, T. (2006c). Gene conversion from *SLG* to *SRK* resulting in self-compatibility in *Brassica rapa*. *FEBS Letters* **580**, 425–430.

Fukai, E., Fujimoto, R. and Nishio, T. (2003). Genomic organization of the *S* core region and the *S* flanking regions of a class-II *S* haplotype in *Brassica rapa*. *Molecular Genetics and Genomics* **269**, 361–369.

Goring, D. R., Glavin, T. L., Schafer, U. and Rothstein, S. J. (1993). An *S* receptor kinase gene in self-compatible *Brassica napus* has a 1-bp deletion. *The Plant Cell* **5**, 531–539.

Gu, T., Mazzurco, M., Sulaman, W., Matias, D. D. and Goring, D. R. (1998). Binding of an arm repeat protein to the kinase domain of the *S*-locus receptor kinase. *Proceedings of the National Academy of Sciences of the United States of America* **95**, 382–387.

Haffani, Y. Z., Gaude, T., Cock, J. M. and Goring, D. R. (2004). Antisense suppression of thioredoxin *h* mRNA in *Brassica napus* cv. Westar pistils causes a low level constitutive pollen rejection response. *Plant Molecular Biology* **55**, 619–630.

Hatakeyama, K., Takasaki, T., Watanabe, M. and Hinata, K. (1998). Molecular characterization of *S* locus genes, *SLG* and *SRK* in a pollen-recessive self-incompatibility haplotype of *Brassica rapa* L. *Genetics* **149**, 1587–1597.

Hatakeyama, K., Takasaki, T., Suzuki, G., Nishio, T., Watanabe, M., Isogai, A. and Hinata, K. (2001). The *S* receptor kinase gene determines dominance relationships in stigma expression of self-incompatibility in *Brassica*. *Plant Journal* **26**, 69–76.

Hinata, K., Okazaki, K. and Nishio, T. (1983). Gene analysis of self-compatibility in *Brassica campestris* var. *yellow sarson* (a case of recessive epistatic modifier). *Proceedings of the Sixth International Rapeseed Conference* **1**, 354–359.

Kachroo, A., Schopfer, C. R., Nasrallah, M. E. and Nasrallah, J. B. (2001). Allele-specific receptor-ligand interactions in *Brassica* self-incompatibility. *Science* **293**, 1824–1826.

Kakizaki, T., Takada, Y., Fujioka, T., Suzuki, G., Satta, Y., Shiba, H., Isogai, A., Takayama, S. and Watanabe, M. (2006). Comparative analysis of the *S*-intergenic region in class-II *S* haplotypes of self-incompatible *Brassica rapa* (syn. *campestris*). *Genes & Genetic Systems* **81**, 63–67.

Kimura, R., Sato, K., Fujimoto, R. and Nishio, T. (2002). Recognition specificity of self-incompatibility maintained after the divergence of *Brassica oleracea* and *Brassica rapa*. *Plant Journal* **29**, 215–223.

Kusaba, M. and Nishio, T. (1999). Comparative analysis of *S* haplotypes with very similar *SLG* alleles in *Brassica rapa* and *Brassica oleracea*. *Plant Journal* **17**, 83–91.

Kusaba, M., Nishio, T., Satta, Y., Hinata, K. and Ockendon, D. J. (1997). Striking sequence similarity in inter- and intra-specific comparisons of class I *SLG* alleles from *Brassica oleracea* and *Brassica campestris*: Implications for the evolution and recognition mechanism. *Proceedings of the National Academy of Sciences of the United States of America* **94**, 7673–7678.

Kusaba, M., Matsushita, M., Okazaki, K., Satta, Y. and Nishio, T. (2000). Sequence and structural diversity of the *S* locus genes from different lines with the same self-recognition specificities in *Brassica oleracea*. *Genetics* **154**, 413–420.

Kusaba, M., Tung, C. W., Nasrallah, M. E. and Nasrallah, J. B. (2002). Monoallelic expression and dominance interactions in anthers of self-incompatible *Arabidopsis lylata*. *Plant Physiology* **128**, 17–20.

Miege, C., Ruffio-Chable, V., Schierup, M. H., Cabrillac, D., Dumas, C., Gaude, T. and Cock, J. M. (2001). Intrahaplotype polymorphism at the *Brassica S* locus. *Genetics* **159**, 811–822.

Murase, K., Shiba, H., Iwano, M., Che, F. S., Watanabe, M., Isogai, A. and Takayama, S. (2004). A. membrane-anchored protein kinase involved in *Brassica* self-incompatibility signaling. *Science* **303**, 1516–1519.

Nakanishi, T. and Hinata, K. (1973). An effective time for CO_2 gas treatment in overcoming self-incompatibility in *Brassica*. *Plant Cell Physiology* **14**, 873–879.

Nasrallah, J. B. (2000). Cell-cell signaling in the self-incompatibility response. *Current Opinion in Plant Biology* **3**, 368–373.

Nasrallah, J. B. and Nasrallah, M. E. (1993). Pollen-stigma signaling in the sporophytic self-incompatibility response. *The Plant Cell* **5**, 1325–1335.

Nasrallah, J. B., Yu, S. M. and Nasrallah, M. E. (1988). Self-incompatibility genes of *Brassica oleracea*: Expression, isolation, and structure. *Proceedings of the National Academy of Sciences of the United States of America* **85**, 5551–5555.

Nasrallah, J. B., Nishio, T. and Nasrallah, M. E. (1991). The self-incompatibility genes of *Brassica* expression and use in genetic ablation of floral tissue. *Annual Review of Plant Physiology and Plant Molecular Biology* **42**, 393–422.

Nasrallah, J. B., Rundle, S. J. and Nasrallah, M. E. (1994). Genetic evidence for the requirement of the *Brassica S*-locus receptor kinase gene in the self-incompatibility response. *Plant Journal* **5**, 373–384.

Niikura, S. and Matsuura, S. (2000). Genetic analysis of the reaction level of self-incompatibility to a 4% CO_2 gas treatment in the radish (*Raphanus sativus* L.). *Theoretical and Applied Genetics* **101**, 1189–1193.

Nishio, T. and Kusaba, M. (2000). Sequence diversity of *SLG* and *SRK* in *Brassica oleracea* L. *Annals of Botany* **85**(Suppl. A), 141–146.

Nishio, T., Sakamoto, K. and Yamaguchi, J. (1994). PCR-RFLP of the *S*-locus for the identification of breeding lines in cruciferous vegetables. *Plant Cell Reports* **13**, 546–550.

Nishio, T., Kusaba, M., Sakamoto, K. and Ockendon, D. J. (1997). Polymorphism of the kinase domain of the *S*-locus receptor kinase gene (*SRK*) in *Brassica oleracea* L. *Theoretical and Applied Genetics* **95**, 335–342.

Nou, I. S., Watanabe, M., Isogai, A. and Hinata, K. (1993). Comparison of *S*-alleles and *S*-glycoproteins between two wild populations of *Brassica campestris* in Turkey and Japan. *Sexual Plant Reproduction* **6**, 79–86.

Ockendon, D. J. (2000). The *S*-allele collection of *Brassica oleracea*. *Acta Horticulture* **539**, 25–30.

Okamoto, S., Sato, Y., Sakamoto, K. and Nishio, T. (2004). Distribution of similar self-incompatibility (*S*) haplotypes in different genera, *Raphanus* and *Brassica*. *Sexual Plant Reproduction* **17**, 33–39.

Okazaki, K., Kusaba, M., Ockendon, D. J. and Nishio, T. (1999). Characterization of *S* tester lines in *Brassica oleracea*: Polymorphism of restriction fragment length of *SLG* homologues and isoelectric points of *S*-locus glycoproteins. *Theoretical and Applied Genetics* **98**, 1329–1334.

Ruffio-Chable, V., Herve, Y., Dumas, C. and Gaude, T. (1997). Distribution of *S*-haplotypes and its relationship with self-incompatibility in *Brassica oleracea*. Part 1. In inbred lines of cauliflower (*B. oleracea* var botrytis). *Theoretical and Applied Genetics* **94**, 338–346.

Sainudiin, R., Wong, W. S., Yogeeswaran, K., Nasrallah, J. B., Yang, Z. and Nielsen, R. (2005). Detecting site-specific physicochemical selective pressures: Applications to the class I HLA of the human major histocompatibility complex and the SRK of the plant sporophytic self-incompatibility system. *Journal of Molecular Evolution* **60**, 315–326.

Sakamoto, K., Kusaba, M. and Nishio, T. (1998). Polymorphism of the *S*-locus glycoprotein gene (*SLG*) and the *S*-locus related gene (*SLR1*) in *Raphanus sativus* L. and self-incompatible ornamental plants in the Brassicaceae. *Molecular & General Genetics* **258**, 397–403.

Sato, K., Nishio, T., Kimura, R., Kusaba, M., Suzuki, T., Hatakeyama, K., Ockendon, D. J. and Satta, Y. (2002). Co-evolution of the *S*-locus genes, *SRKSLG*, and *SP11/SCR*, in *Brassica oleracea* and *B. rapa*. *Genetics* **162**, 931–940.

Sato, Y., Fujimoto, R., Toriyama, K. and Nishio, T. (2003). Commonality of self-recognition specificity of *S* haplotypes between *Brassica oleracea* and *Brassica rapa*. *Plant Molecular Biology* **52**, 617–626.

Sato, Y., Okamoto, S. and Nishio, T. (2004). Diversification and alteration of recognition specificity of the pollen ligand SP11/SCR in self-incompatibility of *Brassica* and *Raphanus*. *The Plant Cell* **16**, 3230–3241.

Sato, Y., Sato, K. and Nishio, T. (2006). Interspecific pairs of class II *S* haplotypes having different recognition specificities between *Brassica oleracea* and *Brassica rapa*. *Plant Cell Physiology* **47**, 340–345.

Schopfer, C. R., Nasrallah, M. E. and Nasrallah, J. B. (1999). The male determinant of self-incompatibility in *Brassica*. *Science* **286**, 1697–1700.

Shiba, H., Iwano, M., Entani, T., Ishimoto, K., Shimosato, H., Che, F. S., Satta, Y., Ito, A., Takada, Y., Watanabe, M., Isogai, A. and Takayama, S. (2002). The dominance of alleles controlling self-incompatibility in *Brassica* pollen is regulated at the RNA level. *The Plant Cell* **14**, 491–504.

Shiba, H., Kenmochi, M., Sugihara, M., Iwano, M., Kawasaki, S., Suzuki, G., Watanabe, M., Isogai, A. and Takayama, S. (2003). Genomic organization of the *S*-locus region of *Brassica*. *Bioscience Biotechnology and Biochemistry* **67**, 622–626.

Shiba, H., Kakizaki, T., Iwano, M., Tarutani, Y., Watanabe, M., Isogai, A. and Takayama, S. (2006). Dominance relationships between self-incompatibility alleles controlled by DNA methylation. *Nature Genetics* **38**, 297–299.

Silva, N. F., Stone, S. L., Christie, L. N., Sulaman, W., Nazarian, K. A. P., Burnett, L. A., Arnoldo, M. A., Rothstein, S. J. and Goring, D. R. (2001). Expression of the *S* receptor kinase in self-compatible *Brassica napus* cv. Westar leads to the allele-specific rejection of self-incompatible *Brassica napus* pollen. *Molecular & General Genetics* **265**, 552–559.

Stein, J. C., Howlett, B., Boyes, D. C., Nasrallah, M. E. and Nasrallah, J. B. (1991). Molecular cloning of a putative receptor protein kinase gene encoded at the self-incompatibility locus of *Brassica oleracea*. *Proceedings of the National Academy of Sciences of the United States of America* **88**, 8816–8820.

Stone, S. L., Arnoldo, M. and Goring, D. R. (1999). A breakdown of *Brassica* self-incompatibility in ARC1 antisense transgenic plants. *Science* **286**, 1729–1731.

Stone, S. L., Anderson, E. M., Mullen, R. T. and Goring, D. R. (2003). ARC1 is an E3 ubiquitin ligase and promotes the ubiquitination of proteins during the rejection of self-incompatible *Brassica* pollen. *The Plant Cell* **15**, 885–898.

Suzuki, G., Kai, N., Hirose, T., Fukui, K., Nishio, T., Takayama, S., Isogai, A., Watanabe, M. and Hinata, K. (1999). Genomic organization of the *S* locus: Identification and characterization of genes in *SLG*/*SRK* region of S^9 haplotype of *Brassica campestris* (syn. *rapa*). *Genetics* **153**, 391–400.

Suzuki, G., Watanabe, M and Nishio, T. (2000a). Physical distances between *S*-locus genes in various *S*. haplotypes of *Brassica rapa* and B. *oleracea*. *Theoretical and Applied Genetics* **101**, 80–85.

Suzuki, G., Kakizaki, T., Takada, Y., Shiba, H., Takayama, S., Isogai, A. and Watanabe, M. (2003). The *S* haplotypes lacking *SLG* in the genome of *Brassica rapa*. *Plant Cell Reports* **21**, 911–915.

Suzuki, T., Kusaba, M., Matsushita, M., Okazaki, K. and Nishio, T. (2000b). Characterization of *Brassica* *S*-haplotypes lacking *S*-locus glycoprotein. *FEBS Letters* **482**, 102–108.

Takasaki, T., Hatakeyama, K., Suzuki, G., Watanabe, M., Isogai, A. and Hinata, K. (2000). The *S* receptor kinase determines self-incompatibility in *Brassica* stigma. *Nature* **403**, 913–916.

Takayama, S., Shimosato, H., Shiba, H., Funato, M., Che, F. S., Watanabe, M., Iwano, M. and Isogai, A. (2001). Direct ligand-receptor complex interaction controls *Brassica* self-incompatibility. *Nature* **413**, 534–538.

Tantikanjana, T., Nasrallah, M. E., Stein, J. C., Chen, C. H. and Nasrallah, J. B. (1993). An alternative transcript of the *S* locus glycoprotein gene in a class II pollen-recessive self-incompatibility haplotype of *Brassica oleracea* encodes a membrane-anchored protein. *The Plant Cell* **5**, 657–666.

Thompson, K. F. and Taylor, J. P. (1966). Non-linear dominance relationships between *S*alleles. *Heredity* **21**, 345–362.

Watanabe, M., Ono, T., Hatakeyama, K., Takayama, S., Isogai, A. and Hinata, K. (1997). Molecular characterization of *SLG* and *S*-related genes in a self-compatible *Brassica campestris* L. var. *yellow sarson*. *Sexual Plant Reproduction* **10**, 332–340.

Watanabe, M., Ito, A., Takada, Y., Ninomiya, C., Kakizaki, T., Takahata, Y., Hatakeyama, K., Hinata, K., Suzuki, G., Takasaki, T., Satta, Y. Shiba, H., *et al.* (2000). Highly divergent sequences of the pollen self-incompatibility (*S*) gene in class-I *S* haplotypes of *Brassica campestris* (syn. *rapa*) L. *FEBS Letters* **473**, 139–144.

Fingerprinting of Oilseed Rape Cultivars

VLADISLAV ČURN AND JANA ŽALUDOVÁ

Faculty of Agriculture, University of South Bohemia, Czech Republic

ABSTRACT

Although oilseed rape was domesticated relatively recently, breeding of *Brassica napus* and other oilseed Brassicas has been very intensive in the last few decades. However, it has focused only on some of the most important agronomic/economic traits. Because of the convergent nature of this process in all areas of rape breeding, the genetic basis of new cultivars has been decreasing. It is, therefore, very difficult to characterize rape cultivars on the basis of phenotypic characters. Likewise,

Advances in Botanical Research, Vol. 45
Incorporating Advances in Plant Pathology

0065-2296/07 $35.00
DOI: 10.1016/S0065-2296(07)45006-6

registration and protection of new rape cultivars rely on a relatively small number of morphological characters, and as the number of cultivars increases, the ability to distinguish them on a morphological basis alone becomes more difficult. New descriptors, like molecular markers, are required to maintain the efficiency of registration testing. To complement morphological traits, isozyme analysis has been used for cultivar identification and was included in DUS testing. Isozyme analysis allows a more direct observation of the genotype of a cultivar, but the number of isozyme markers is limited. It seems that DUS testing would benefit from the use of DNA markers that have been shown to be more rapid and cost-effective, and their several types have been used to assess genetic diversity in the genus *Brassica*. Molecular markers in general can also be used as potential techniques for cultivar identification. In comparison with morphological traits, molecular markers have many advantages and these have been successfully applied in registration activities like cultivar identification and controls of seed purity of hybrid varieties. A relatively high number of different molecular (protein and nucleic acids) techniques and approaches are available for plant genotyping purposes. The marker systems differ in information content, number of scorable polymorphisms, degree of automation and labor, and financial costs. Principles and advantages of particular marker techniques, application of molecular markers in rapeseed cultivar identification, and in Brassicas-breeding programs are discussed in this chapter.

I. INTRODUCTION

The range of genetic variability of cultivated crops comparing to the whole germplasm diversity of particular crop plant is relatively low. Although oilseed rape (*Brassica napus* L.) belongs to the group of lately domesticated crops (Soltis *et al.*, 1998), in the last few decades, breeding of *B. napus* (canola, oilseed rape, rapeseed) and other oilseed Brassicas has been very intensive and it focused only on several most important agronomic/economic traits. And because of convergent character of this process in all areas of rape breeding, the genetic base of new cultivars has been decreasing; it is therefore very difficult to characterize rape cultivars as well as cultivars of many crops on the basis of phenotypic characters (Nielsen, 1985). Likewise, registration and protection of rape cultivars rely on a relatively small number of morphological characters and as the number of cultivars increases, the ability to distinguish them on a morphological basis alone becomes more difficult (Lombard *et al.*, 2000). New descriptors like molecular markers are required to maintain the efficiency of registration testing.

Although various morphological and quality traits, and yielding characteristics are currently explored for cultivar protection (ISTA and UPOV directions), new markers are being developed to maintain the efficacy of registration and distinctness uniformity stability (DUS) testing, which guarantees the quality of new cultivars for farmers and traders. Moreover, it is also important to protect intellectual property of the plant breeders (plant breeders' rights), which encourages the continuous development of new cultivars.

In that context, plant breeders' rights have evolved since the introduction of the concept of essential derivation into the International Union for the Protection of new Varieties of plants (UPOV) convention of 1991. An essentially derived cultivar is a cultivar that is derived from and that confirms to the genotype of the initial cultivar and is clearly distinguishable from the initial cultivar by phenotypic characteristics such as a single simply inherited trait. This phenomenon has gathered momentum since the development of techniques like gene manipulations and GMO technology, which make it possible to introduce a single gene directly into an existing genotype/cultivar (Ondřej and Drobník, 2002). For most species, DUS testing only relies on the comparison of morphological traits between candidate cultivars and all the previously registered cultivars (the reference collection). This protocol has a number of limitations and is too expensive and time consuming. For economic reasons, DUS testing requires minimal replication (two locations and two replicates per location in EU), which leads to unreliable estimates of the interactions between genotypes and environment. These interactions can be important for the morphological traits used because their expression is influenced by environmental conditions. Secondly, candidate varieties are compared with the reference collection which increases in size year after year. To complement morphological traits, isozyme analysis has been used for cultivar identification and has been included in DUS testing. In the 1980s, mainly isozymes and storage proteins had been tested as markers for cultivar characterization of various crops, including oilseed rape (Čurn and Sáková, 1997; Gupta and Röbbelen, 1986; Mündges *et al.*, 1990). Isozyme analysis allows a more direct observation of the genotype of a cultivar, but the number of isozyme markers is limited. However, a disadvantage of these biochemical markers in rape cultivars seems to be their relatively low level of polymorphism (Lee *et al.*, 1996), probably as the result of the genetic similarity of modern cultivars. Isozymes may be suitable for differentiation of *B. napus* from other Brassicas (*B. oleracea, B. rapa*, and so on), but for the identification of individual oilseed rape cultivars, it is necessary to use additional marker systems for precise cultivar description (Čurn and Sáková, 1997).

DUS testing would benefit from the use of DNA markers that have been shown to be more rapid and cost-effective. Molecular markers in general can be reliably used as potential techniques for cultivar identification. In comparison with morphological traits, molecular markers have many advantages. Their expression is independent of environmental conditions, a lengthy survey of plant growth methods is not needed and the potential number of markers is nearly unlimited in relation to isozymes. Molecular markers have been successfully applied in registration activities like cultivar identification (Mailer *et al.*, 1994) or controls of seed purity of hybrid varieties (Marshall *et al.*, 1994). Restriction fragment length polymorphism (RFLP) analysis as the first molecular marker system has been shown to be a valuable tool for detecting patterns of DNA polymorphism among and within *Brassica* species (Dos Santos *et al.*, 1994; Figdore *et al.*, 1988; Hallden *et al.*, 1994). However, the procedure is laborious, expensive and automation is difficult. Also, only a few loci are detected per assay.

The newer DNA marker systems are based on PCR technology, and for this reason are more suitable for routine cultivar identification, require small amount of DNA, and are generally fast and simple. Random amplified polymorphic DNA (RAPD) analysis allows large number of markers to be assayed inexpensively using PCR and oligonucleotide primers of arbitrary sequence (Williams *et al.*, 1990). RAPD analysis has been widely used for detection of genetic polymorphisms in *Brassica* species, especially in the 1990 (Jain *et al.*, 1994; Kresovich *et al.*, 1992; Mailer *et al.*, 1994). Microsatellites or simple sequence repeats (SSRs) are codominant, highly polymorphic PCR-based markers, and may be expected to be very powerful in cultivar discrimination. Although the development of locus-specific oligonucleotide primers is time-consuming and expensive, a range of specific primer pairs for crop Brassicas has been developed. These primers may prove valuable for cultivar identification (Kresovich *et al.*, 1995; Szewc-McFadden *et al.*, 1996; Uzunova and Ecke, 1999). Alternatively, amplified fragment length polymorphism (AFLP) analysis can be employed. AFLP analysis is a technique by which selected fragments from a digestion of total plant DNA are amplified by PCR (Vos *et al.*, 1995), and it seems to be a powerful technique for cultivar identification (Powell *et al.*, 1996). It provides a large number of markers in a single analysis without requiring sequence information for their development (Vos *et al.*, 1995). AFLPs have been successfully applied for intraspecific and interspecific genetic diversity analyses and for genetic linkage mapping (Weising *et al.*, 2005). As compared with RFLP, for which polymorphism among cultivars is low (Diers and Osborn, 1994), AFLPs have very attractive qualities for DUS testing in rapeseed. Results of AFLP analysis as a tool for oilseed rape cultivar identification are the most promising ones as compared to other available methods (Das *et al.*, 1999; Lombard *et al.*, 2000). This is due to the high multiplex ratio, which is the number of information points analyzed per experiment (Powell *et al.*, 1996).

II. FINGERPRINTING TOOLS IN OILSEED BRASSICAS CULTIVAR IDENTIFICATION

A number of different molecular techniques markers and approaches are available for plant genotyping purposes (Table I). The marker systems differ in information content, number of scorable polymorphisms, degree of

TABLE I
Examples of Some Important Traits for Which Genetic Marker Loci Have Been Used to Map Loci or Clone Genes in Brassica Oilseeds

Character	Specific trait	Molecular marker	Selected references
Morphological traits	Pod length	RAPD	Hawkins *et al.*, 2005
	Plant height (dwarf locus)	SCAR	Barret *et al.*, 1998; Foisset *et al.*, 1995
	Seed color	RAPD	Somers *et al.*, 2001
		RAPD, AFLP	Zhi-Wen *et al.*, 2005
	Flowering time		Ferreira *et al.*, 1995c
	Petalless flower		Fray *et al.*, 1997
Male sterility	"Ogura" fertility restorer	RFLP, RAPD	Delourme *et al.*, 1998
		RAPD	Delourme *et al.*, 1994; Hansen *et al.*, 1997
	"Polima" fertility restorer	RFLP, RAPD	Jean *et al.*, 1997, 1998
	tournefortii fertility restorer	AFLP	Janeja *et al.*, 2003
	CMS from *Arabidopsis thaliana*	RFLP	Leino *et al.*, 2003
	Recessive genic male sterility	AFLP	Ke *et al.*, 2004, 2005
	Dominant genic male sterility	AFLP	Lu *et al.*, 2004; Song *et al.*, 2006
Self-incompatibility			
Abiotic stress	Cold tolerance, winter hardiness	RFLP, AFLP	Kole *et al.*, 2002a
Disease resistance	Blackleg (*Leptosphaeria maculans*)	RAPD	Chevre *et al.*, 1997; Delourme *et al.*, 2004
		RAPD, RFLP	Plieske *et al.*, 1998
		SCAR, RGA (resistance gene analogue)	Saal and Struss, 2005
		STS	Dion *et al.*, 1995; Ferreira *et al.*, 1995a; Plieske and Struss, 2001
		RAPD, RFLP	Pilet *et al.*, 1998a,b, 2001
	White rust (*Albugo candida*)		Ferreira *et al.*, 1995b; Kole *et al.*, 2002b
	Turnip mosaic virus (TuMV)	AFLP, SSR	Hughes *et al.*, 2003
		RFLP	Walsh *et al.*, 1999
	Turnip yellows virus (TuYV)	AFLP	Dreyer *et al.*, 2001
	Sclerotinia sclerotiorum		Zhao and Meng, 2003
Oil content/quality	Erucic acid content	RFLP	Ecke *et al.*, 1995
		RAPD	Rajcan *et al.*, 1999

(*continues*)

TABLE I (*continued*)

Character	Specific trait	Molecular marker	Selected references
	Glucosinolate content	RFLP	Uzunova *et al.*, 1995
	Oleic acid content	SCAR, RAPD	Hu *et al.*, 1999
		AFLP	Schierholt *et al.*, 2000
		RAPD, ISSR, SCAR	Javidfar *et al.*, 2006
	Linoleic acid content	SCAR, RAPD	Hu *et al.*, 1999
	Linolenic acid content	RAPD	Hu *et al.*, 1995
		RFLP, RAPD	Tanhuanpää *et al.*, 1995
		RAPD	Jourdren *et al.*, 1996; Somers *et al.*, 1998
	Seed glucosinulates	RFLP	Howell *et al.*, 2003; Uzunova *et al.*, 1995

automation and labor, and financial costs. In following parts, the short descriptions of basic techniques used for the purposes of cultivar identification and description are given.

III. BIOCHEMICAL MARKERS, ANALYSIS OF ISOZYMES, AND STORAGE PROTEINS

Isozyme analysis has become a useful tool for various applications in plant genetics and breeding. Isozyme markers have been specifically applied in genetic and taxonomic studies and in breeding programs of the family *Brassicaceae*. In *B. napus*, isozymes have been used in taxonomic studies (Truco and Arús, 1987; Vaughan and Waite, 1967) and to identify and confirm interspecific hybrids (Sundberg and Glimelius, 1986, 1991). Enzyme polymorphism detected by gel electrophoresis has been proposed as a powerful tool for the testing of purity of F_1 hybrids (Nijenhuis, 1971; Wills and Wiseman, 1980; Wills *et al.*, 1979; Woods and Thurman, 1976). Isozymes have also been reported as biochemical genetic markers used for characterization and cultivar identification in Brassicas (Arús *et al.*, 1985; Mündges *et al.*, 1990).

Isozymes as biochemical markers have some advantages in plant genetic studies and breeding programs, such as the alleles at most isozyme loci are codominant, and cause no deleterious changes in plant phenotype. This codominance also allows homozygotes to be distinguished from heterozygotes. The equipment and materials needed for screening the isozyme-banding patterns (zymograms) of plants are relatively cheaper, and it is possible to screen large numbers of plants rapidly. Further since only small sample amount of plant tissue is needed, the integrity of the whole plant can be maintained (Moore and Collins, 1983).

The genus *Brassica* consists of diploid and amphidiploid species and is characterized by high genetic and morphological variation. In *Brassica* vegetable crops, the breeding of F_1 hybrid varieties is now quite widespread, and for testing the purity of F_1 hybrid seed lots the isoenzyme analysis has become a useful tool (Arús, 1983; Arús and Shields, 1983; Nijenhuis, 1971).

Isoenzyme and protein polymorphisms detected by gel electrophoresis have been used in the family Brassicaceae in phylogenetic and taxonomic studies (Arús *et al.*, 1991; Denford and Vaughan, 1977; Truco and Arús, 1987; Vaughan and Denford, 1968; Vaughan and Waite, 1967; Vaughan *et al.*, 1966; Yadava *et al.*, 1979), as markers for verification the hybrid character of somatic hybrids (Forsberg *et al.*, 1994; Sjödin and Glimelius, 1989; Sundberg and Glimelius, 1991) and for characterization of disomic addition lines (Chevre *et al.*, 1991). One of the first applications of isoenzyme electrophoresis in plant breeding and seed production was study of ACP isozymes in F_1 hybrid varieties of Brussels sprouts cabbage (Nijenhuis, 1971). The applications, viz., purity testing of F_1 hybrids and testing the homogeneity of seed lots, have been proposed as a powerful tool in plant breeding (Arús *et al.*, 1982; Gasanova, 1990; Luhový and Reňáková, 1989; Wills and Wiseman, 1980; Wills *et al.*, 1979; Woods and Thurman, 1976). In last 20 years, isozymes and storage proteins have been used as genetic biochemical markers for characterization and cultivar identification in oilseed rape (Arús *et al.*, 1985; Čurn and Sáková, 1997; Gupta and Röbbelen, 1986; Mündges *et al.*, 1990), and testing the homogeneity of androgenetic lines of oilseed rape (Čurn, 1995).

IV. MOLECULAR TECHNIQUES, DETECTION OF POLYMORPHISM ON THE NUCLEIC ACIDS LEVEL

A. SEQUENCING

Polymorphism at the DNA level can be studied by numerous approaches, and acquisition of information about primary sequence of nucleotides on particular DNA fragment through DNA sequencing is one of these approaches (Sambrook and Russel, 2001). DNA sequencing provides highly robust, the most reproducible and informative, data set. Modern technologies of DNA sequencing also overcome its limitations. Methodology and applications of DNA sequencing are widely reviewed by Alphey (1997) and Hillis *et al.* (1996).

B. RESTRICTION FRAGMENT LENGTH POLYMORPHISM

RFLP is based on the variations in the length of DNA fragments produced by a specific restriction endonuclease from genomic DNAs of two or more individuals of a species (Kahl, 2004). The basis of RFLP technique is digestion of genomic (or chloroplast, mitochondrial) DNA by specific enzymes, known as restriction endonucleases. These enzymes are a part of specific restriction/methylation system produced by a variety of prokaryotes, and their main function in prokaryotic cells is to recognize foreign DNA molecules and their cutting in specific palyndromic motives, mostly consisting of four, five, or six bases. Restriction digestion of DNA molecules leads to specific set of DNA fragments of well-defined length. Because of high specificity of DNA digestion, these sets of restriction fragments are highly reproducible. These populations of DNA fragments are electrophoresed, Southern-blotted, and position of specific fragments is visualized after hybridization with radioactive or nonradioactive labeled probe. Probe-based methods like RFLP were the first widely reported means of revealing DNA sequence variations in a diverse range of organisms, including cultivars of main crops (Ainsworth and Sharp, 1989; Karp *et al.*, 1996; Weising *et al.*, 1995). RFLP analysis utilizes genomic or cDNA clones as probes following cutting (restriction) of the target DNA with a specific enzyme. This is a very effective way of revealing differences between varieties and the large number of restriction enzyme/probe combinations available makes RFLP analysis a powerful tool. This is also robust and repeatable in different laboratories. RFLPs are relatively small in size and are codominant in nature. For identification, the use of multiple copy probes, either random or of known derivation, can be advantageous (Lee *et al.*, 1996). Cooke (1995) reported analyses with repetitive oligonucleotide as probes, with the $(GATA)_n$ repeats proving to be widely applicable. Utilization of multilocus probes which recognize tandemly repeated DNA motifs such mini- or microsatellites, then resulted in patters of markers which are called as DNA fingerprints (Jeffreys *et al.*, 1985). These fingerprints or DNA profiles are highly variable, are individual specific, and can be used as markers system for cultivar/genotype identification. However, there are some difficulties associated with this approach, including the need for relatively large amounts of good quality DNA, the source and public availability of probes, the time taken for results to be produced, and the fact that radioactive labeling of probes is widely used to reveal the DNA profiles, which together undoubtedly place a limitation on its routine use (Botstein *et al.*, 1980; Tanksley *et al.*, 1989). At present, especially for purposes of DNA profiling based on microsatellite locus analyses, RFLP approach is replaced by PCR-based approaches.

C. PCR AND OTHER PCR-BASED TECHNIQUES

Amplification technologies based on polymerase chain reaction (PCR) address some of these difficulties and offer apparent advantages over RFLP. PCR is a method that allows the production of millions of copies of particular segments of DNA after the enzymatic *in vitro* amplification of particular DNA regions, thus facilitating their analysis (Innis *et al.*, 1990; Mullis *et al.*, 1994). For more than 20 year since the times of PCR invention (Saiki *et al.*, 1985, 1988), this technique revolutionized the analysis methodology in plant molecular biology. PCR assay usually consists of three steps—denaturation, primer annealing, and polymerization. The selectivity of PCR reaction is directly determined by primers. Primers are short single-stranded oligonucleotides with the sequence complementary to template sequence. Primer design also influences the character of resulted amplified DNA fragments (patterns of DNA profiles). Till now, numerous PCR-based techniques have been developed, viz., PCR with arbitrary primers (RAPD, DAF, AP-PCR), PCR with specific primers, and many other modification of standard PCR approach (for review see Caetano-Anollés and Gresshoff, 1997; Weising *et al.*, 2005).

V. AMPLIFIED FRAGMENT LENGTH POLYMORPHISM

The AFLP technique combines components of RFLP analysis with PCR technology (Vos and Kuiper, 1997; Vos *et al.*, 1995). Total genomic DNA is digested with a pair of restriction enzymes, normally frequent and a rare cutter. Adaptors of known sequence (addition contains one to three arbitrary bases) are then ligated to the DNA fragments. Primers complementary to the adaptors are used to amplify the restriction fragments. The PCR-amplified fragments can then be separated by gel electrophoresis followed by visualizing the banding patterns. A range of enzymes and primers are available to manipulate the complexity of AFLP fingerprints to suit application. Care is needed in selection of primers with selective bases. AFLP profiles require no prior DNA sequence information and the number and nature of fragments amplified are altered by the choice of primer pair. The technique also has the advantage of sampling many loci simultaneously and, in addition, is more robust than arbitrary priming techniques such as RAPD, since more stringent conditions are used in the PCR. AFLP has predominantly been applied in genetic mapping studies (Ballvora *et al.*, 1995; Becker *et al.*, 1995; Meksem *et al.*, 1995; Van Eck *et al.*, 1995). The technique is semiquantitative because the intensity of AFLP bands can be used to determine the zygosity (Van Eck *et al.*, 1995).

AFLP has proven to be a competent tool in bacterial taxonomic studies (Huys *et al.*, 1996; Janssen *et al.*, 1996), and it has also been used for the identification of fungal isolates (Mueller *et al.*, 1996). The use of AFLP to analyze cultivars of various species, including cereals, potatoes, sunflowers, Brassicas, beans, and lentils, has been reported (Cooke and Reeves, 1998; Law *et al.*, 1998; Lombard *et al.*, 2000; Sobotka *et al.*, 2004).

A. SSR or Microsatellites

Microsatellites are tandemly repeated DNA sequences, usually with a repeat unit of 2–4 bp (Morgante and Olivieri, 1993). The di-, tri-, or tetranucleotide repeats are arranged in tandem arrays consisting of 5–50 copies. They have codominant character and allow distinguishing homo- and heterozygotes. For microsatellites also, the high level of polymorphism is typical and they are site polymorphic due to a variable number of tandem repeats. SSRs are abundant in plants, occurring on average every 6–7 kb (Cardle *et al.*, 2000). These repeat motifs are flanked by conserved nucleotide sequences from which forward and reverse primers can be designed to PCR-amplify the DNA section, containing the SSR. SSR alleles, amplified products of variable length, can be separated by gel electrophoresis and visualized by silver-staining, autoradiography (if primers are radioactively labeled), or via automation (if primers are fluorescently labeled). SSR analysis is amenable to automation and multiplexing, and allows genotyping to be performed on large numbers of lines and multiple loci to be analyzed simultaneously. In many species, multiple alleles have been shown to exist for some microsatellites, due to variations in the copy number of this repeat unit. The markers resulting from the analyses are called microsatellite markers, viz., SSR, STMS (sequence-tagged microsatellite sites), and SSLP (simple sequence length polymorphism). Because of presence of microsatellites in chloroplast DNA, specific primers for analysis of chloroplast microsatellites are available and these markers are known as cpSSRs. Many other modifications of these analyses are known nowadays, depending on primer design like DAMD, ISSR, MP-PCR, AMP-PCR, and RAMP. There are also methods that use various combinations of repeat-specific primers as IRAP, REMAP (Kalendar *et al.*, 1999), or combining repeat-specific and AFLP primers in technique S-SAP (Waugh *et al.*, 1997). These approaches and techniques are reviewed in Weising *et al.* (2005) or Caetano-Anollés and Gresshoff (1997).

The potential of SSR analysis for plant breeding and cultivar-related work in general has generated much research activity and as a result there are now microsatellites available for several crop species, including oilseed rape

(Cooke and Reeves, 1998; Karp *et al.*, 1997; Weising *et al.*, 1998.). Methods have been published for high throughput, nonradioactive detection of microsatellite alleles (Lagoda *et al.*, 1998; van den Berg, 1997). The data from SSR analyses are generally easy to score and a further significant advantage is that microsatellites can be multiplexed so that a number of markers (loci) can be evaluated in a single PCR and separated on the same gel. This is not only more efficient but also suitable for automation, which enhances the data-gathering process as well as improves cost-effectiveness. Methods of SSR analyses, based on the differential fluorescent labeling of microsatellite primers and separation of multiplexed products using automated DNA sequences, are being increasingly used for crop plant genotyping (Mitchell *et al.*, 1997). The use of multiple array capillary electrophoresis for the rapid (<60 min) separation of multiplexed microsatellite products has also been demonstrated (Mansfield *et al.*, 1998). Such approaches, although requiring expensive capital investment, have considerable potential for very efficient and automated profiling, database production, and cultivar verification, and their wider use will prove difficult to resist.

SSR was also modified using different design of primers. ISSR technique is such a modification of microsatellite analysis (Kantety *et al.*, 1995; Zietkiewics *et al.*, 1994). Pattern of ISSR markers is a consequence of PCR amplification of internal SSR using anchored primers without exact knowledge of their sequence. Hantula *et al.* (1996), Charters *et al.* (1996), and Zietkiewics *et al.* (1994) reported this method to be more reliable and repeatable than RAPD analysis and providing higher polymorphism. This technique is more rapid, less time- and labor-consuming, and economically more viable than AFLP or SSR. However, the character of ISSR markers is dominant although some authors reported presence of codominant markers too (Fisher *et al.*, 1996).

B. SINGLE NUCLEOTIDE POLYMORPHISM

Single nucleotide polymorphisms (SNPs) are marker basically not used for DNA fingerprinting. However, due to their diallelic character and stability, these are a powerful tool for QTL mapping and molecular marker-assisted selection approach in plant breeding. SNPs are defined as single nucleotide position substitutions and are the most common forms of DNA sequence polymorphism. In *Arabidopsis* they are reported to occur once every 3.3 kb (Mohler and Schwarz, 2004). SNP polymorphism may be detected using several methodological approaches, as PCR-RFLP (CAPS technique), allele-specific PCR, SSCP, Sanger dideoxysequencing, minisequencing and pyrosequencing, heteroduplex analysis, and allele-specific probe hybridization.

VI. MOLECULAR MARKER TECHNIQUES: APPLICATION IN CULTIVAR IDENTIFICATION IN OILSEED RAPE

Utilization of isozymes and storage proteins for identification of cultivars, seed lots, and F_1 hybrids are mentioned above. So applications of molecular markers will be discussed here. In the last 20 years, we have seen a remarkable shift in applications of different methods of DNA-profiling for cultivar identification. RFLP analysis exploiting random, multilocus, or highly specific probes still plays very important role as an effective method for specific purposes like construction of linkage maps, detection of specific DNA sequences, genes or alleles, and in taxonomic studies (Kraic, 1994). However, for its limitations in many applications, this method is replaced by PCR-based approaches.

Several studies were focused on application of molecular markers and techniques in taxonomic studies. Very complex assessment of relationships in 74 different taxa in the family Brassicaceae on the basis of restriction polymorphism of cpDNA was reported by Warwick and Black (1991, 1993). The subtribe *Brassicinae* on the basis of these studies was divided into two clearly distinguishable groups: *rapa–oleracea and nigra* (Warwick *et al.*, 1992). Phylogenetic and taxonomic problems in the family Brassicaceae were studied not only on the basis of restriction pattern of cpDNA but also on the basis of RFLP analysis of total mitochondrial genomic DNA. On the cytodemic level (higher taxonomic unit defined with respect to basal chromosomal number and crossability of species), analyses of mitochondrial and chloroplast DNA provide comparable results (Pradhan *et al.*, 1992). Restriction analyses of mtDNA also allowed determining phylogenetic relationships between diploid and allotetraploid or autotetraploid species like *Diplotaxis viminea* and allotetraploid *D. muralis* or *B. maurorum* and autotetraploid *B. cossoneana*. RFLP approach was also employed to study evolutionary relations in different *Brassica* species (Pradhan *et al.*, 1992; Song *et al.*, 1988a) and relationships on subspecies levels in cultivated *Brassica* plants (Song *et al.*, 1988b, 1990). RFLP and isozyme markers were also utilized as tools for evaluation of genetic diversity in populations of *B. campestris* and its subspecies (McGrath and Quiros, 1992).

Several researchers also adopted the RAPD-PCR approach to these (Kresovich *et al.*, 1992; Thormann *et al.*, 1994). Unfortunately, both techniques are marked by different characters and properties and in direct comparison, RFLP and RAPD markers provide relatively distinct or dissimilar results. RFLP and RAPD analyses give comparable results on intraspecific level, and also the level of genetic resolution of RAPD markers is comparable to RFLPs, for example in analyses of breeding lines of *B. oleracea* and *B. napus* (Dos Santos *et al.*, 1994; Hallden *et al.*, 1994). But the differences can be seen on interspecies level and RAPD markers are less suitable marker systems than RFLPs, or SSRs/AFLPs (Sobotka *et al.*, 2004; Thormann *et al.*, 1994). RAPD markers were successfully employed for genetic fingerprinting in several *Brassica* crops, in broccoli and cauliflower (Hu and Quiros, 1991), *Brassica* species (Demeke *et al.*, 1992; Jain *et al.*, 1994), oilseed rape (Hallden *et al.*, 1994; Mailer *et al.*, 1994; Sáková and Čurn, 1998), and for parentage analyses and genetic mapping (Reiter *et al.*, 1992; Scott *et al.*, 1992; Uzunova *et al.*, 1995). RAPDs were also used as markers to distinguish different *Brassica* species (*B. nigra, B. juncea, B. hirta*, and *B. rapa*), 6 *B. napus* cultivars and 10 *B. napus* doubled haploid lines which have been produced via microspore cultures. Some RAPD primers attained high level of polymorphism and resolution and allowed to detect distinguishable sublines in DH lines. The highest level of polymorphism (71.43%) was expressed in *Brassica* species, remarkably low level was between registered cultivars (Sáková and Čurn, 1998). The DNA template as molecular markers are also utilized also for genome characterization of resynthesized oilseed rape plants (Poulsen *et al.*, 1993), interspecific hybrids, and transgenic plants (Hájková *et al.*, 2005; Kerlan *et al.*, 1993).

Recently, new approaches have been introduced for the fingerprinting of *Brassica* cultivars such as APLP and SSR. Although these markers are widely used in some other crops, in Brassicas their utilization is not so frequent. RAPD, AFLP, SSR, and RFLP are often similar from the point of view of genetic similarity (Powell *et al.*, 1996; Thormann *et al.*, 1994). Among all mentioned techniques, AFLPs are the most efficient because of their capacity to reveal many polymorphic bands per assay. However, they do not offer the highest level of polymorphism (Powell *et al.*, 1996). Lombard *et al.* (2000) found that only 2 combinations of AFLP primers from total number of 17 tested could distinguish 83 oilseed rape cultivars. Also Sobotka *et al.* (2004) reported higher level of polymorphism and higher ability of resolution of AFLPs comparing to RAPDs or SSRs. AFLP markers are used in last few years for molecular characterization and genetic diversity evaluation of *B. napus* cultivars (Seyis *et al.*, 2003). Another application of this technique is detection of hybrids between *B. napus* and its relatives (Hansen *et al.*, 2001, 2003), and transgenic oilseed rape and wild relatives (Warwick *et al.*, 2003). Čurn *et al.* (2002) and Sobotka *et al.* (2004) gave comparison of RAPD and AFLP, and showed that AFLP methodology detected polymorphism more efficiently than RAPD approach. In this study, AFLPs yielded 7.3 polymorphic bands per primer(s) but RAPDs only 0.45. Das *et al.* (1999) evaluated genetic relationship among nine cultivars of *B. rapa*. They reported that level of polymorphism in both RAPD and AFLP approaches was considerably higher, but the number of polymorphic bands per AFLP assay was 5.6-fold higher compared to RAPD (42.6 vs 7.6).

The number of polymorphic loci per assay is important for cultivar identification. The new registered cultivars need not display polymorphism in analysis with established cultivars, and it may be then necessary to test other primers. The low reproducibility of RAPDs is well known (Jones *et al.*, 1997). Automated fluorescence dye-labeled AFLP techniques offer significant improvements over radioactive labeling methods by increasing the scoring accuracy and the typing efficiency (Sobotka *et al.*, 2004). However, this technique requires special equipments and is more expensive in comparison with RAPDs and SSRs, although the automated DNA sequencer is now the standard equipment of most molecular biology laboratories. Automated fluorescence dye-labeled AFLP method allows detecting higher number of bands with higher sensitivity and resolution.

The last discussed marker systems are microsatellites. SSRs have usually revealed the highest level of polymorphism of all DNA markers tested (Maughan *et al.*, 1995; Powell *et al.*, 1996; Salimath *et al.*, 1995). A range of specific SSR primers has been developed for Brassica and related species (Kresovich *et al.*, 1995; Szewc-McFadden *et al.*, 1996; Uzunova and Ecke, 1999), and it is also possible to use known primer from related species like *Arabidopsis* (Westman and Kresovich, 1998). Uzunova and Ecke (1999) have used 11 sequence-specific SSR primer pairs for genetic analysis of 31 conventional oilseed rape varieties and only 1 tested primer pair came up to expectations. Kresovich *et al.* (1995) and Szewc-McFadden *et al.* (1996) also reported that limited degree of polymorphism was detected by SSR in *B. napus*. Reports by Čurn *et al.* (2002) and Sobotka *et al.* (2004) also confirmed limited availability of this marker system for oilseed rape cultivar identification. They found only five microsatellites to be polymorphic, but except Bn12A only two different alleles were detected (Sobotka *et al.*, 2004).

Interesting modification of standard procedure of microsatellite polymorphism detection is the usage of 5′-anchored SSR primer. Anchored SSR primers are complementary to genomic microsatellites and contain short oligonucleotide "anchor" sequences that ensure the primers anneal to either the 5′ or 3′ end of genomic repeats. The advantage is detection of greater number of loci assayed per reaction. Charters *et al.* (1996) distinguished all 20 analyzed oilseed rape cultivars using one pair of 5′-anchored SSR primers, and each primer pair was able to distinguish only 16 cultivars. Specific primers combined with 5′-anchored primers detect higher level of polymorphism than only single specific pairs (Varghese *et al.*, 2000).

VII. CONCLUSION

Identification of registered cultivars, study and characterization of diversity of germplasm, and verification of purity of breeding lines and seed lots are the main areas of utilization of molecular markers in rapeseed breeding programs. Therefore, better knowledge of distribution of donor genes, genetic disposition of agriculturally/economically important traits, and accessibility of genetic resources, selection, and marker systems play an important role in more extensive employment of molecular marker techniques and approaches in plant breeding.

REFERENCES

Ainsworth, C. C. and Sharp, P. J. (1989). The potential role of DNA probes in plant variety protection. *Plant Varieties & Seeds* **2**, 27–34.

Alphey, L. (1997). "DNA Sequencing (Introduction to Biotechniques)." Bios Scientific Publishers, Oxford, UK.

Arús, P. (1983). Genetic purity of commercial seed lots. *In* "Isozymes in Plant Genetics a Breeding, Part A" (S. D. Tanksley and T. J. Orton, eds.), pp. 415–423. Elsevier Science Publication, Amsterdam, Oxford, New York.

Arús, P. and Shields, C. R. (1983). Cole crops (*Brassica oleracea* L.). *In* "Isozymes in Plant Genetics and Breeding, Part B" (S. D. Tanksley and T. J. Orton, eds.), pp. 339–350. Elsevier Science Publication, Amsterdam, Oxford, New York.

Arús, P., Tanksley, S. D., Orton, T. J. and Jones, R. A. (1982). Electrophoretic variation as a tool for determining seed purity and for breeding hybrid varieties of *Brassica oleracea*. *Euphytica* **31**, 417–428.

Arús, P., Shields, C. R. and Orton, T. J. (1985). Application of isozyme electrophoresis for purity testing a cultivar identification of F_1 hybrids of *Brassica oleracea*. *Euphytica* **34**, 651–657.

Arús, P., Chevre, A. M., Delourme, R., Eber, F., Kerlan, M. C., Margale, E. and Quiros, C. F. (1991). Isoenzyme nomenclature for eight enzyme systems in three *Brassica* species. *In* "Proceedings of the Eighth International Rapeseed Congress." Saskatoon, July 9–10, 1991.

Ballvora, A., Hesselbach, J., Niewöhner, J., Leister, D., Salamini, F. and Gebhardt, C. (1995). Marker enrichment and high resolution map of the segment of potato chromosome VII harbouring the nematode resistance gene *Gro1*. *Molecular & General Genetics* **249**, 82–90.

Barret, P., Delourme, R., Foisset, N. and Renard, M. (1998). Development of a SCAR (sequence characterised amplified region) marker for molecular tagging of the dwarf BREIZH (*Bzh*) gene in *Brassica napus* L. *Theoretical and Applied Genetics* **97**, 828–833.

Becker, J., Vos, P., Kuiper, M., Salamini, F. and Heun, M. (1995). Combined mapping of AFLP and RFLP markers in barley. *Molecular & General Genetics* **249**, 65–73.

Botstein, D., White, R. L., Skolnick, M. and Davis, R. W. (1980). Construction of a genetic linkage map in man using restriction fragment length polymorphisms. *American Journal of Human Genetics* **32**, 314–331.

Caetano-Anollés, G. and Gresshoff, P. M. (1997). "DNA Markers Protocols, Applications, and Overviews." Wiley-Liss Inc., New York.

Cardle, L., Ramsay, L. and Milbourne, D. (2000). Computational and experimental characterization of physically clustered simple sequence repeats in plants. *Genetics* **156**, 847–854.

Charters, Y. M., Robertson, A., Wilkinson, M. J. and Ramsay, G. (1996). PCR analysis of oilseed rape cultivars (*Brassica napus* L. ssp. *oleifera*) using 5′-anchored simple sequence repeats (SSR) primers. *Theoretical and Applied Genetics* **92**, 442–447.

Chevre, A. M., This, P., Eber, F., Deschamps, M., Renard, M., Delseny, M. and Quiros, C. F. (1991). Characterization of disomic addition lines *Brassica napus–Brassica nigra* by isozyme, fatty acid, an RFLP markers. *Theoretical and Applied Genetics* **81**, 43–49.

Chevre, A. M., Barret, P., Eber, F., Dupuy, P., Brun, H., Tanguy, X. and Renard, M. (1997). Selection of stable *Brassica napus-B. Juncea* recombinant lines resistant lines resistant to blackleg (*Leptosphaeria maculans*). 1. Identification of molecular markers, chromosomal and genomic origin of the introgression. *Theoretical and Applied Genetics* **95**, 1104–1111.

Cooke, R. J. (1995). Gel electrophoresis for the identification of plant varieties. *Journal of Chromatography* **698**, 281–299.

Cooke, R. J. and Reeves, J. C. (1998). Cultivar identification general discussion and review of new methods. *In* "Encyclopedia of Seed Production of World Crops" (A. F. Kelly and R. A. T. George, eds.). John Wiley & Sons, Chichester, UK.

Čurn, V. (1995). Acid phosphatase and leucine aminopeptidase isozymes as biochemical markers of homogeneity in oil seed rape androgenetic lines. *Plant Growth Regulation* **16**, 59–63.

Čurn, V. and Sáková, L. (1997). Využití biochemických markerů ve šlechtění řepky a dalších brukvovitých plodin. *Genetika a Šlechtění* **33**, 281–305.

Čurn, V., Ovesná, J., Sáková, L. and Sobotka, R. (2002). Identification of oilseed rape cultivars using AFLP markers. *Journal of Central European Agriculture* **3**, 285–292.

Das, S., Rajagopal, J., Bhatia, S., Srivastava, P. S. and Lakshmikumaran, M. (1999). Assessment of genetic variation within *Brassica campestris* cultivars using amplified fragment length polymorphism and random amplification of polymorphic DNA markers. *Journal of Biosciences* **24**, 433–440.

Delourme, R., Bouchereau, A., Hubert, N., Renard, M. and Landry, B. S. (1994). Identification of RAPD markers linked to fertility restorer gene for the *Ogura* radish cytoplasmic male sterility of rapeseed (*Brassica napus* L.). *Theoretical and Applied Genetics* **88**, 741–748.

Delourme, R., Foisset, N., Horvais, R., Barret, P., Champagne, G., Cheung, W. Y., Landry, B. S. and Renard, M. (1998). Characterisation of the radish introgression carrying the *Rfo* restorer gene for the *Ogu*-INRA cytoplasmic male sterility in rapeseed (*Brassica napus* L.). *Theoretical and Applied Genetics* **97**, 129–134.

Delourme, R., Pilet-Nayel, M. L., Archipiano, M., Horvais, R., Tanguy, X., Rouxel, T., Brun, H., Renard, M. and Balesdent, M. H. (2004). A cluster of major specific resistance genes to *Leptosphaeria maculans* in *Brassica napus*. *Phytopathology* **94**, 578–583.

Demeke, T., Adams, R. P. and Chibbar, R. (1992). Potential taxonomic use of random amplified polymorphic DNA (RAPD): A case study in *Brassica*. *Theoretical and Applied Genetics* **84**, 990–994.

Denford, K. E. and Vaughan, J. G. (1977). A comparative study of certain seed isoenzyme in the ten chromosome complex of *Brassica campestris* and its allies. *Annals of Botany* **41**, 411–418.

Diers, B. W. and Osborn, T. C. (1994). Genetic diversity of oilseed *Brassica napus* germplasm based on restriction fragment length polymorphisms. *Theoretical and Applied Genetics* **88**, 662–668.

Dion, Y., Gugel, R. K., Rakow, G. F. W., Seguin-Swartz, G. and Landry, B. S. (1995). RFLP mapping of resistance to the blackleg disease [causal agent, *Leptosphaeria maculans* (Desm.) Ces. et de Not. in canola (*Brassica napus* L). *Theoretical and Applied Genetics* **91**, 1190–1194.

Dos Santos, J. B., Nienhuis, J., Skroch, P., Tivang, J. and Slocum, M. K. (1994). Comparison of RAPD and RFLP genetic markers in determining genetic similarity among *Brassica oleracea* L. genotype. *Theoretical and Applied Genetics* **87**, 909–915.

Dreyer, F., Graichen, K. and Jung, C. (2001). A major quantitative trait locus for resistance to Turnip Yellows Virus (TuYV, syn. Beet western yellows virus, BWYV) in rapeseed. *Plant Breeding* **120**, 457–462.

Ecke, W., Uzunova, M. and Weißleder, K. (1995). Mapping the genome of rapeseed (*Brassica napus* L.). II. Localization of genes controlling erucic acid synthesis and seed oil content. *Theoretical and Applied Genetics* **91**, 972–977.

Ferreira, M. E., Rimmer, S. R., Williams, P. H. and Osborn, T. C. (1995a). Mapping loci controlling *Brassica napus* resistance to *Leptosphaeria maculans* under different screening conditions. *Phytopathology* **85**, 213–217.

Ferreira, M. E., Williams, P. H. and Osborn, T. C. (1995b). Mapping of a locus controlling resistance to *Albugo candida* in *Brassica napus* using molecular markers. *Phytopathology* **85**, 218–220.

Ferreira, M. E., Satagopan, J., Yandell, B. S., Williams, P. H. and Osborn, T. C. (1995c). Mapping loci controlling vernalisation requirement and flowering time in *Brassica napus*. *Theoretical and Applied Genetics* **90**, 727–732.

Figdore, S. S., Kennard, W. C., Song, K. M., Slocum, M. K. and Osborn, T. C. (1988). Assessment of degree of restriction fragment length polymorphism in *Brassica*. *Theoretical and Applied Genetics* **75**, 833–840.

Fisher, P. J., Gardner, R. C. and Richardson, T. E. (1996). Single locus microsatellites isolated using 5′ anchored PCR. *Nucleic Acids Research* **24**, 4369–4371.

Foisset, N., Delourme, R., Barret, P. and Renard, M. (1995). Molecular tagging of the dwarf BREIZH (Bzh) gene in *Brassica napus*. *Theoretical and Applied Genetics* **91**, 756–761.

Forsberg, J., Landgren, M. and Glimelius, K. (1994). Fertile somatic hybrids between *Brassica napus* and *Arabidopsis thaliana*. *Plant Science* **95**, 213–223.

Fray, M. J., Puangsomlee, P., Goodrich, J., Coupland, G., Evans, E. J., Arthur, A. E. and Lydiate, D. J. (1997). The genetics of stamenoid petal production in oilseed rape (*Brassica napus*) and equivalent variation in *Arabidopsis thaliana*. *Theoretical and Applied Genetics* **94**, 731–736.

Gasanova, N. D. (1990). Evaluation of cabbage genetic purity by isoenzyme phenotypes. (Оценка генетической чистотъı капустъı по изоферментнъım фенотипам). *Research Bulletin N. I. Vavilov Institute Plant Industry* **197**, 24–25.

Gupta, S. K. and Röbbelen, G. (1986). Identification of rapeseed (*Brassica napus*) cultivars by electrophoresis. *Plant Breeding* **96**, 363–370.

Hájková, P., Hrubý, J., Pernová, E., Čurn, V. and Žaludová, J. (2005). Monitorig pěstebních ploch, přenos a detekce transgenů geneticky modifikované řepky olejky (*Rassica napus* L. var. *napus*). *Sborník vědeckých prací VÚP Troubsko* **15**, 93–100.

Hallden, C., Nilsson, N.-O., Rading, T. M. and Säll, T. (1994). Evaluation of RFLP and RAPD markers in a comparison of *Brassica napus* breeding lines. *Theoretical and Applied Genetics* **88**, 123–128.

Hansen, L. B., Siegismund, H. R. and Jorgensen, R. B. (2001). Introgression between oilseed rape (*Brassica napus* L.) and its weedy relative *B. rapa* L. in a natural population. *Genetic Resources and Crop Evolution* **48**, 621–627.

Hansen, L. B., Siegismund, H. R. and Jorgensen, R. B. (2003). Progressive introgression between *Brassica napus* (oilseed rape) and *B. rapa*. *Heredity* **91**, 276–283.

Hansen, M., Halldén, C., Nilsson, N. O. and Säll, T. (1997). Marker-assisted selection of restored male-fertile *Brassica napus* plants using a set of dominant RAPD markers. *Molecular Breeding* **3**, 449–456.

Hantula, J., Dusabenygasani, M. and Hamelin, R. C. (1996). Random amplified microsatellites (RAMS)–a novel method for characterizing genetic variation within fungi. *European Journal of Forest Pathology* **26**, 159–166.

Hawkins, G. P., Zhang, X., Thiagarjan, M. R., Corrigan, L. M. and Stringam, G. R. (2005). Identification of RAPD markers linked to pod length in *Brassica napus* L. canola. *Canadian Journal of Plant Science* **85**, 803–808.

Hillis, D. M., Moritz, C. and Mable, B. K. (1996). "Molecular Systematics." Sinauer Associates, Sunderland, MA.

Howell, P. M., Sharpe, A. G. and Lydiate, D. J. (2003). Homoeologous loci control the accumulation of seed glucosinolates in oilseed rape (*Brassica napus*). *Genome* **46**, 454–460.

Hu, J. and Quiros, C. F. (1991). Identification of broccoli and cauliflower cultivars with RAPD markers. *Plant Cell Reports* **10**, 505–511.

Hu, J., Quiros, C., Arus, P., Struss, D. and Robbelen, G. (1995). Mapping of a gene determining linolenic acid concentration in rapeseed with DNA-based markers. *Theoretical and Applied Genetics* **90**, 258–262.

Hu, J., Li, G., Struss, D. and Quiros, C. F. (1999). SCAR and RAPD markers associated with 18-carbon acids in rapeseed *Brassica napus*. *Plant Breeding* **118**, 145–150.

Hughes, S. L., Hunter, P. J., Sharpe, A. G., Kearsey, M. J., Lydiate, D. J. and Walsh, J. A. (2003). Genetic mapping of the novel *Turnip mosaic virus* resistance gene *TuRB03* in *Brassica napus*. *Theoretical and Applied Genetics* **107**, 1169–1173.

Huys, G., Coopman, R., Janssen, P. and Kersters, K. (1996). High-resolution genotypic analysis of the genus Aeromonas by AFLP fingerprinting. *International Journal of Systematic Bacteriology* **46**, 572–580.

Innis, M. A., Gelfand, D. H., Sninsky, J. J. and White, T. J. (1990). "PCR Protocols A Guide to Methods and Applications." Academic Press, San Diego.

Jain, A., Bhatia, S., Banga, S. S., Prakash, S. and Lakshmikumaran, M. (1994). Potential use of random amplified polymorphic DNA (RAPD) technique to study the genetic diversity in Indian mustard (*Brassica juncea*) and its relationship to heterosis. *Theoretical and Applied Genetics* **8**, 116–122.

Janeja, H. S., Banga, S. S. and Lakshmikumaran, M. (2003). Identification of AFLP markers linked to fertility restorer genes for tournefortii cytoplasmic male-sterility system in Brassica napus. *Theoretical and Applied Genetics* **107**, 148–154.

Janssen, P., Coopman, R., Huys, G., Swings, J., Marjo, Bleeker, Vos, P., Zabeau, M. and Kersters, K. (1996). Evaluation of the DNA fingerprinting method AFLP as a new tool in bacterial taxonomy. *Microbiology* **142**, 1881–1893.

Javidfar, F., Ripley, V. L., Roslinsky, V., Zeinali, H. and Abdmishani, C. (2006). Identification of molecular markers associated with oleic and linolenic acid in spring oilseed rape (*Brassica napus*). *Plant Breeding* **125**, 65–71.

Jean, M., Brown, G. G. and Landry, B. S. (1997). Genetic mapping of nuclear fertility restorer genes for the 'Polima' cytoplasmic male sterility in canola (*Brassica napus* L.) using DNA markers. *Theoretical and Applied Genetics* **95**, 321–328.

Jean, M., Brown, G. G. and Landry, B. S. (1998). Targeted mapping approaches to identify DNA markers linked to the Rfpl restorer gene for the 'Polima' CMS of canola (*Brassica napus* L.). *Theoretical and Applied Genetics* **97**, 431–438.

Jeffreys, A. J., Wilson, V. and Thein, S. L. (1985). Individualspecific 'fingerprints' of human DNA. *Nature* **316**, 76–79.

Jones, C. J., Edwards, K. J., Castaglione, S., Winfield, M. O., Sala, F., vandeWiel, C., Bredemeijer, G., Vosman, B., Matthes, M., Daly, A., Brettschneider, R., Bettini, P., *et al.* (1997). Reproducibility testing of RAPD, AFLP and SSR markers in plants by a network of European laboratories. *Molecular Breeding* **3**, 381–390.

Jourdren, C., Barret, P., Horvais, R., Delourme, R. and Renard, M. (1996). Identification of, RAPD markers linked to linolenic acid genes in rapeseed. *Euphytica* **90**, 351–357.

Kahl, G. (2004). "The Dictionary of Gene, Technology Genomics, Transcriptomics, Proteomics." Wiley-VCH, Weinheim.

Kalendar, R., Grob, T., Regina, M., Suoniemi, A. and Schulman, A. H. (1999). IRAP and REMAP two new retrotransposon-based DNA fingerprinting techniques. *Theoretical and Applied Genetics* **98**, 704–711.

Kantety, R. V., Zeng, X., Bennetzen, J. L. and Zehr, B. E. (1995). Assessment of genetic diversity in dent and popcorn (*Zea mays* L.) inbred lines using inter-simple sequence repeat (ISSR) amplification. *Molecular Breeding* **1**, 365–373.

Karp, A., Seberg, O. and Buiatti, M. (1996). Molecular techniques in the assessment of botanical diversity. *Annals of Botany* **78**, 143–149.

Karp, A., Edwards, K. J., Bruford, M., Funk, S., Vosman, B., Morgante, M., Seberg, O., Kremer, A., Boursot, P., Arctander, P., Tautz, D. and Hewitt, G. M. (1997). Molecular technologies for biodiversity evaluation opportunities and challenges. *Nature Biotechnology* **15**, 625–628.

Ke, L., Sun, Y., Liu, P. and Yang, G. (2004). Identification of AFLP fragments linked to one recessive genic male sterility (RGMS) in rapeseed (*Brassica napus* L.) and conversion to SCAR markers for marker-aided selection. *Euphytica* **138**, 163–168.

Ke, L. P., Sun, Y. Q., Hong, D. F., Liu, P. W. and Yang, G. S. (2005). Identification of AFLP markers linked to one recessive genic male sterility gene in oilseed rape, *Brassica napus*. *Plant Breeding* **124**, 367–370.

Kerlan, M. C., Chevre, A. M. and Eber, F. (1993). Interspecific hybrids between a transgenic rapeseed (*Brassica napus*) and related species cytogenetical characterization and detection of the transgene. *Genome* **36**, 1099–1106.

Kole, C., Thormann, C. E., Karlsson, B. H., Palta, J. P., Gaffney, P., Yandell, B. and Osborn, T. C. (2002a). Comparative mapping of loci controlling winter survival and related traits in oilseed *Brassica rapa* and *B. napus*. *Molecular Breeding* **9**, 201–210.

Kole, C., Williams, P. H., Rimmer, S. R. and Osborn, T. C. (2002b). Linkage mapping of genes controlling resistance to white rust (*Albugo candida*) in *Brassica rapa* (syn. *campestris*) and comparative mapping to *Brassica napus* and *Arabidopsis thaliana*. *Genome* **45**, 22–27.

Kraic, J. (1994). Identifikácia, vzájomné rozlišovanie odrôd a charaktezizácia niektorých vlastností odrod pomocou detekcie DNA polymorfismu. Pol'nohospodárske skúšobníctvo **2**, 15–16.

Kresovich, S., Szewc-McFadden, A. K., Bliek, S. M. and McFerson, J. R. (1995). Abundance and characterization of simple-sequence repeats (SSRs) isolates from a size-fractionated genomic library of *Brassica napus* L. (rapeseed). *Theoretical and Applied Genetics* **91**, 206–211.

Kresovich, S., Williams, J. G. K., McFerson, J. R., Routman, E. J. and Schaal, B. A. (1992). Characterization of genetic identities and relationships of *Brassica oleracea* L. via a random amplified polymorphic DNA assay. *Theoretical and Applied Genetics* **85**, 190–196.

Lagoda, P. J. L., Noyer, J. L., Dambier, D., Baurens, F. C., Grapin, A. and Lanaud, C. (1998). Sequence tagged microsatellite site (STMS) markers in the *Musaceae*. *Molecular Ecology* **7**, 657–666.

Law, J. R., Domini, P., Koebner, R. M. D., Reeves, J. C. and Cooke, R. J. (1998). DNA profiling and plant variety registration. III The statistical assessment of distinctness in wheat using amplified fragment length polymorphisms. *Euphytica* **102**, 335–342.

Lee, D., Reeves, J. C. and Cooke, R. J. (1996). DNA profiling and plant variety registration 2. Restriction fragment length polymorphisms in varieties of oilseed rape. *Plant Varieties and Seeds* **9**, 181–190.

Leino, M., Teixeira, R., Landgren, M. and Glimelius, K. (2003). *Brassica napus* lines with rearranged *Arabidopsis* mitochondria display CMS and a range of developmental aberrations. *Theoretical and Applied Genetics* **106**, 1156–1163.

Lombard, V., Baril, C. P., Dubreuil, P., Blouet, F. and Zhang, D. (2000). Genetic relationships and fingerprinting of rapeseed cultivars by AFLP: Consequences for varietal registration. *Crop Science* **40**, 1417–1425.

Lu, G. Y., Yang, G. S. and Fu, T. D. (2004). Molecular mapping of dominant genic male sterility gene *Ms* in rapeseed (*Brassica napus*). *Plant Breeding* **123**, 262–265.

Luhový, M. and Reňáková, M. (1989). Rychlé stanovení hybridnosti růžičkové kapusty elektroforézou isoenzymů kyselé fosfatázy. *Bull. Výzk. Šlecht. Úst. Zelin., Olomouc* **33**, 61–66.

Mailer, R. J., Scarth, R. and Fristensky, B. (1994). Discrimination among cultivars of rapeseed (*Brassica napus* L.) using DNA polymorphisms amplified from arbitrary primers. *Theoretical and Applied Genetics* **87**, 697–704.

Mansfield, E. S., Robertson, J. M., Vainer, M., Isenberg, A. R., Frazier, R. R., Ferguson, K., Chow, S., Harris, D. W., Barker, D. L., Gill, P. D., Budowle, B. and McCord, B. R. (1998). Analysis of multiplexed short tandem repeat (STR) systems using capillary array electrophoresis. *Electrophoresis* **19**, 101–107.

Marshall, P., Marchand, M. C., Lisieczko, Z. and Landry, B. S. (1994). A simple method to estimate the percentage of hybridity in canola (*Brassica napus*) F1 hybrids. *Theoretical and Applied Genetics* **89**, 853–858.

Maughan, P. J., Maroof, M. A. S. and Buss, G. R. (1995). Microsatellite and amplified sequence length polymorphisms in cultivated and wild soybean. *Genome* **38**, 715–723.

McGrath, J. M. and Quiros, C. F. (1992). Genetic diversity at isozyme and RFLP loci in *Brassica campestris* as related to crop type and geographical origin. *Theoretical and Applied Genetics* **83**, 783–790.

Meksem, K., Leister, D., Peleman, J., Zabeau, M., Salamini, F. and Gebhardt, C. (1995). A high-resolution map of the vicinity of the *R1* locus on chromosome V of potato based on RFLP and AFLP markers. *Molecular & General Genetics* **249**, 74–81.

Mitchell, S. E., Kresovich, S., Jester, C. A., Hernandez, C. J. and Szewc-McFadden, A.K. (1997). Application of multiplex PCR and fluorescence-based semi-automated allele sizing technology for genotyping plant genetic resources. *Crop Science* **37**, 617–624.

Mohler, V. and Schwarz, G. (2004). Genotyping tools in plant breeding from restriction fragment length polymorphisms to single nucleotide polymorphisms. *In* "Biotechnology in Agriculture and Forestry, Vol. 55. Molecular Marker Systems" (H. Lörz and G. Wenzel, eds.), pp. 24–38. Springer-Verlag, Berlin, Heidelberg.

Moore, G. A. and Collins, G. B. (1983). New challenges confronting plant breeders. *In* "Isozymes in Plant Genetics and Breeding, Part A" (S. D. Tanksley and T. J. Orton, eds.), pp. 25–58. Elsevier Science Publication, Amsterdam, Oxford, New York.

Morgante, M. and Olivieri, A. M. (1993). PCR-amplified microsatellites as markers in plant genetics. *Plant Journal* **3**, 175–182.

Mueller, U. G., Lipari, S. E. and Milgroom, M. M. (1996). Amplified fragment length polymorphism (AFLP) fingerprinting of fungi cultured by the fungus-growing ant *Cyphomyrmex minutus*. *Molecular Ecology* **5**, 119–122.

Mullis, K. B., Ferre, F. and Gibbs, R. A. (1994). "The Polymerase Chain Reaction." Birkhauser, Boston.

Mündges, H., Köhler, W. and Friedt, W. (1990). Identification of rapeseed cultivars (*Brassica napus*) by starch gel electrophoresis of enzymes. *Euphytica* **45**, 179–187.

Nielsen, G. (1985). The use of isozymes as probes to identify and label plant varieties and cultivars. *Isozymes: Current Topics in Biological and Medical Research* **12**, 1–32.

Nijenhuis, B. T. E. (1971). Estimation of the proportion of inbred seed in Brussels sprouts hybrid seed by acid phosphatase isoenzyme analysis. *Euphytica* **20**, 498–507.

Ondřej, M. and Drobník, J. (2002). "Transgenose rostlin." Academia, Praha.

Pilet, M. L., Delourme, R., Foiset, N. and Renard, M. (1998a). Identification of loci contributing to quantitative resistance to blackleg disease, causial agent *Leptosphaeria maculans* (Desm.) Ces. et de Not., in winter rapeseed (*Brassica napus* L). *Theoretical and Applied Genetics* **96**, 23–30.

Pilet, M. L., Delourme, R., Foiset, N. and Renard, M. (1998b). Identification of QTL involved in field resistance to light leaf spot (*Pyrenopeziza brassicae*) and blackleg resistance (*Leptosphaeria maculans*) in winterseed rape (*Brassica napus* L.). *Theoretical and Applied Genetics* **97**, 398–406.

Pilet, M. L., Duplan, G., Archipiano, M., Barret, P., Baron, C., Horvais, R., Tanguy, X., Lucas, O. M., Renard, M. and Delourme, R. (2001). Stability of QTL for field resistance to blackleg across two genetic backgrounds in oil rape. *Crop Science* **41**, 197–205.

Plieske, J. and Struss, D. (2001). STS markers linked to *Phoma* resistance genes of the *Brassica* B-genome revealed sequence homology between *Brassica nigre* and *Brassica napus*. *Theoretical and Applied Genetics* **102**, 483–488.

Plieske, J., Struss, D. and Röbbelen, G. (1998). Inheritance of resistance derived from the B-genome of *Brassica* against *Phoma lingam* in rapeseed and the development of molecular markers. *Theoretical and Applied Genetics* **97**, 929–936.

Poulsen, G. B., Kahl, G. and Weising, K. (1993). Oligonucleotide fingerprinting of resynthesized *Brassica napus*. *Euphytica* **70**, 53–59.

Powell, W., Morgante, M., Andre, C. H., Hanafey, M., Vogel, J., Tingey, S. and Rafalski, A. (1996). The comparison of RFLP, RAPD, AFLP and SSR (microsatellite) markers for germplasm analysis. *Molecular Breeding* **2**, 225–238.

Pradhan, A. K., Prakash, S., Mukhopadhyay, A. and Pental, D. (1992). Phylogeny of *Brassica* and allied genera based on variation in chloroplast and mitochondrial DNA patterns molecular and taxonomic classifications are incongruous. *Theoretical and Applied Genetics* **85**, 331–340.

Rajcan, I., Kasha, K. J., Kott, L. S. and Beversdorf, W. D. (1999). Detection of molecular markers associated with linolenic and erucic acid levels in spring rapeseed (*Brassica napus* L.). *Euphytica* **105**, 173–181.

Reiter, R. S., Williams, J. G. K., Feldmann, K. A., Rafalski, J. A., Tingey, S. D. and Scolnik, P. A. (1992). Global and local genome mapping in *Arabidopsis thaliana* by using recombinant inbred lines and random amplified polymorphic DNAs. *Proceedings of the National Academy of Sciences of the United States of America* **89**, 1477–1481.

Saal, B. and Struss, D. (2005). RGA- and RAPD-derived SCAR markers for a *Brassica* B-genome introgression conferring resistance to blackleg in oilseed rape. *Theoretical and Applied Genetics* **111**, 281–290.

Saiki, R. K., Scharf, S., Faloona, F., Mullis, K. B., Horn, G. T., Erlich, H. A. and Arnheim, N. (1985). Enzymatic amplification of β-globin genomic sequences and restriction site analysis for diagnosis of sickle-cell anemia. *Science* **230**, 1350–1354.

Saiki, R. K., Geffand, D. H., Stoffel, S., Scharf, S. J., Higuchi, R., Horn, G. T., Mullis, K. B. and Erlich, H. A. (1988). Primer-directed enzymatic amplification of DNA with thermostable DNA-polymerase. *Science* **239**, 487–491.

Sáková, L. and Čurn, V. (1998). Identifikace a klasifikace vybraných odrůd brukvovitých plodin a dihaploidních linií řepky pomocí RAPD markerů. *Czech Journal of Genetics and Plant Breeding* **34**, 61–67.

Salimath, S. S., Olivieri, A. C., Godwin, I. D. and Bennetzen, J. L. (1995). Assessment of genome origins and genetic diversity in the genus *Eleusine* with DNA markers. *Genome* **38**, 757–763.

Sambrook, J. and Russel, D. W. (2001). "Molecular Cloning a Laboratory Manual." Cold Spring Harbor Laboratory Press, Cold Spring Harbor, NY.

Scott, M., Haymes, M. and Williams, S. M. (1992). Parentage analysis using RAPD PCR. *Nucleic Acids Research* **20**, 5493.

Seyis, F., Snowdon, R. J., Lush, W. and Friedt, W. (2003). Molecular characterization of novel resynthesized rapeseed (*Brassica napus*) lines and analysis of their genetic diversity in comparison with spring rapeseed cultivars. *Plant Breeding* **122**, 473–478.

Schierholt, A., Becker, H. C. and Ecke, W. (2000). Mapping a high oleic acid mutation in winter oilseed rape (*Brassica napus* L.). *Theoretical and Applied Genetics* **101**, 897–901.

Sjödin, C. and Glimelius, K. (1989). *Brassica naponigra*, a somatic hybrid resistant to *Phoma lingam*. *Theoretical and Applied Genetics* **77**, 651–656.

Sobotka, R., Dolanská, L., Čurn, V. and Ovesná, J. (2004). Fluorescence-based AFLPs occur as the most suitable marker system for oilseed rape cultivar identification. *Journal of Applied Genetics* **45**, 161–173.

Soltis, D. E., Soltis, P. S. and Doyle, J. J. (1998). "Molecular Systematics of Plants II DNA Sequencing." Kluwer Academic Publishers, Norwell, MA.

Somers, D. J., Friesen, K. R. D. and Rakow, G. (1998). Identification of molecular markers associated with linoleic acid desaturation in *Brassica napus*. *Theoretical and Applied Genetics* **96**, 897–903.

Somers, D. J., Rakow, G., Prabhu, V. K. and Friesen, K. R. D. (2001). Identification of a major gene and RAPD markers for yellow seed coat colour in *Brassica napus*. *Genome* **44**, 1077–1082.

Song, K. M., Osborn, T. C. and Williams, P. H. (1988a). *Brassica* taxonomy based on nuclear restriction fragment length polymorphisms (RFLPs). *Theoretical and Applied Genetics* **76**, 593–600.

Song, K. M., Osborn, T. C. and Williams, P. H. (1988b). *Brassica* taxonomy based on nuclear restriction fragment length polymorphisms (RFLPs). *Theoretical and Applied Genetics* **75**, 784–794.

Song, K. M., Osborn, T. C. and Williams, P. H. (1990). *Brassica* taxonomy based on nuclear restriction fragment length polymorphisms (RFLPs). *Theoretical and Applied Genetics* **79**, 497–506.

Song, L. Q., Fu, T. D., Tu, J. X., Ma, C. Z. and Zang, G. S. (2006). Molecular validation of allele inheritance for dominant genic male sterility gene in *Brassica napus* L. *Theoretical and Applied Genetics* **113**, 55–62.

Sundberg, E. and Glimelius, K. (1986). A method for production of interspecific hybrids within *Brassiceae* via somatic hybridization, using resynthesis of *Brassica napus* as a model. *Plant Science* **43**, 155–162.

Sundberg, E. and Glimelius, K. (1991). Production of cybrid plants within *Brassicaceae* by fusing protoplasts and plasmolytically induced cytoplasts. *Plant Science* **79**, 205–216.

Szewc-McFadden, A. K., Kresovich, S., Bliek, S. M., Mitchell, S. E. and McFerson, J. R. (1996). Identification of polymorphic, conserved simple sequence repeats (SSRs) in cultivated *Brassica* species. *Theoretical and Applied Genetics* **93**, 534–538.

Tanhuanpää, P. K., Vilkki, J. P. and Vilkki, H. J. (1995). Association of an RAPD marker with linolenic acid concentration in the seed oil of rapeseed (*Brassica napus* L.). *Genome* **38**, 414–416.

Tanksley, S. D., Young, N. D., Paterson, A. H. and Bonierbale, M. W. (1989). RFLP mapping in plant breeding New tools for an old science. *BioTechnology* **7**, 257–264.

Thormann, C. E., Ferreira, M. E., Camargo, L. E. A., Tivang, J. G. and Osborn, T. C. (1994). Comparison of RFLP and RAPD markers to estimating genetic relationships within and among cruciferous species. *Theoretical and Applied Genetics* **88**, 973–980.

Truco, M. J. and Arús, P. (1987). Comparative study on the isozymes of *Brassica campestris, B. oleracea* a *B. napus*. *Cruciferae Newsletter* **12**, 18–19.

Uzunova, M. I. and Ecke, W. (1999). Abundance, polymorphism and genetic mapping of microsatellites in oilseed rape (*B. napus* L.). *Plant Breeding* **118**, 323–326.

Uzunova, M., Ecke, W., Weissleder, K. and Röbbelen, G. (1995). Mapping the genome of rapeseed (*Brassica napus* L.) I. Construction of an RFLP linkage map and localization of QTLs for seed glucosinolate content. *Theoretical and Applied Genetics* **90**, 194–204.

Varghese, J. P., Rudolph, B., Uzunova, M. I. and Ecke, W. (2000). Use of 5′-anchored primers for the enhanced recovery of specific microsatellite markers in Brassica napus L. *Theoretical and Applied Genetics* **101**, 115–119.

van den Berg, B. M. (1997). Horizontal ultrathin-layer multi-zonal electrophoresis of DNA: An efficient tool for large-scale polymerase chain reaction (PCR) fragment analysis. *Electrophoresis* **18**, 2861–2864.

Van Eck, H. J., Rouppe van der, Voort, J., Draaistra, J., Van Zandvoort, P., Van Enckevort, E., Segers, B., Peleman, J., Jacobsen, E., Helder, J. and Bakker, J. (1995). The inheritance and chromosomal localization of AFLP markers in a non-inbred potato offspring. *Molecular Breeding* **1**, 397–410.

Vaughan, J. G. and Denford, K. E. (1968). An acrylamide gel electrophoretic study of the seed proteins of *Brassica* and *Sinapis* species, with special reference to their taxonomic value. *Journal of Experimental Botany* **19**, 724–732.

Vaughan, J. G. and Waite, A. (1967). Comparative electrophoretic studies of the seed proteins of certain species of *Brassica* and *Sinapis*. *Journal of Experimental Botany* **18**, 100–109.

Vaughan, J. G., Waite, A., Boulter, D. and Waiters, S. (1966). Comparative studies of the seed proteins of *Brassica campestris, Brassica oleracea*, and *Brassica nigra*. *Journal of Experimental Botany* **17**, 332–343.

Vos, P. and Kuiper, M. (1997). AFLP analysis. *In* "DNA Markers Protocols, Applications, and Overviews" (G. Caetano-Anolles and P. M. Gresshoff, eds.), pp. 115–132. Wiley-Liss Inc., New York.

Vos, P., Hogers, R., Bleeker, M., Rijans, M., Van de, Lee, T., Hornes, M., Frijters, A., Pot, J., Peleman, J., Kuiper, M. and Zabeau, M. (1995). AFLP a new technique for DNA fingerprinting. *Nucleic Acids Research* **23**, 4407–4414.

Walsh, J. A., Sharpe, A. G., Jenner, C. E. and Lydiate, D. J. (1999). Characterisation of resistance to turnip mosaic virus in oilseed rape (*Brassica napus*) and genetic mapping of *TuRB01*. *Theoretical and Applied Genetics* **99**, 1149–1154.

Warwick, S. I. and Black, L. D. (1991). Molecular systematics of *Brassica* and allied genera (subtribe Brassicinae, Brassicacaea)–chloroplast genome and cytodeme congruence. *Theoretical and Applied Genetics* **82**, 81–92.

Warwick, S. I. and Black, L. D. (1993). Molecular relationships in subtribe Brassicinae (Cruciferae, tribe Brassiceae). *Canadian Journal of Botany* **71**, 906–918.

Warwick, S. I., Black, L. D. and Aguinagalde, I. (1992). Molecular systematics of *Brassica* and allied genera (Subtribe Brassicinae, Brassiceae)–chloroplast DNA variation in the genus *Diplotaxis*. *Theoretical and Applied Genetics* **83**, 839–850.

Warwick, S. I., Simard, M. J., Legere, A., Beckie, H. J., Braun, L., Zhu, B., Mason, P., Seguin-Swartz, G. and Stewart, C. N. (2003). Hybridization between transgenic *Brassica napus* L. and its wild relatives *Brassica rapa* L., *Raphanus raphanistrum* L., *Sinapis arvensis* L., and *Erucastrum gallicum* (Willd) OE Schulz. *Theoretical and Applied Genetics* **107**, 528–539.

Waugh, R., Mclean, K., Flavell, A. J., Pearce, S. R., Kumar, A., Thomas, B. B. T. and Powell, W. (1997). Genetic distribution of *Bare-1*-like retrotransposable elements in the barley genome revealed by sequence-specific amplification polymorphisms (S-SAP). *Molecular & General Genetics* **253**, 687–694.

Weising, K., Nybom, H., Wolff, K. and Meyer, W. (1995). "DNA Fingerprinting in Plants and Fungi." CRC Press, Boca Raton, FL.

Weising, K., Winter, P., Hüttel, B. and Kahl, G. (1998). Microsatellites markers for molecular breeding. *Journal of Crop Production* **1**, 113–143.

Weising, K., Nybom, H., Wolff, K. and Kahl, G. (2005). "DNA Fingerprinting in Plants. Principles, Methods, and Applications." CRC Press, Boca Raton, FL.

Westman, A. L. and Kresovich, S. (1998). The potential for cross-taxa simple-sequence repeats (SSR) amplification between *Arabidopsis thaliana* L. and crop brassicas. *Theoretical and Applied Genetics* **96**, 272–281.

Williams, J. G. K., Kubelik, A. R., Livak, K. J., Rafalski, J. A. and Tingey, S. V. (1990). DNA polymorphisms amplified by arbitrary primers are useful as genetic markers. *Nucleic Acids Research* **18**, 6531–6535.

Wills, A. B. and Wiseman, E. M. (1980). Acid phosphatase isoenzymes of *Brassica oleracea* seedlings and their applications to sib testing in F_1 hybrids. *The Annals of Applied Biology* **94**, 137–142.

Wills, A. B., Fyfe, S. K. and Wiseman, E. M. (1979). Testing F_1 hybrids of *Brassica oleracea* for sibs by seed isoenzyme analysis. *The Annals of Applied Biology* **91**, 263–270.

Woods, S. and Thurman, D. A. (1976). The use of seed acid phosphatases in the determination of the purity of F_1 hybrid Brussels sprout seed. *Euphytica* **25**, 707–712.

Yadava, J. S., Chowdhury, J. B., Kakar, S. N. and Nainawatee, H. S. (1979). Comparative electrophoretic studies of proteins and enzymes of some *Brassica* species. *Theoretical and Applied Genetics* **54**, 89–91.

Zhao, J. and Meng, J. (2003). Genetic analysis of loci associated with parential resistance to *Sclerotinia sclerotiorum* in rapeseed (*Brassica napus* L.). *Theoretical and Applied Genetics* **106**, 759–764.

Zhi-Wen, L., Ting-Dong, F., Jin-Xing, T. and Bao-Yuan, C. (2005). Inheritance of seed colour and identification of RAPD and AFLP markers linked to the seed colour gene in rapeseed (*Brassica napus* L.). *Theoretical and Applied Genetics* **110**, 303–310.

Zietkiewics, E., Rafalski, A. and Labuda, D. (1994). Genome fingerprint by sequence repeat (SSR)-anchored polymerase chain reaction amplification. *Genomics* **20**, 176–183.

Haploid and Doubled Haploid Technology

L. XU, U. NAJEEB, G. X. TANG, H. H. GU, G. Q. ZHANG,
Y. HE AND W. J. ZHOU

Institute of Crop Science, College of Agriculture and Biotechnology, Zhejiang University, China

ABSTRACT

The microspore culture technique has its wide applications in plant genetic research and breeding programmes in oilseed *Brassicas* due to its relative simplicity, efficiency in haploid and doubled haploid production, mutation and germplasm generation, and gene transformation. Various factors could influence microspore embryogenesis and haploid production including donor plant genotype, donor plant physiology, microspore developmental stage, culture conditions, culture environment and pretreatments. Stress is also an essential component during embryogenesis induction in microspore culture. Efficient plant regeneration from microspores mostly occurs through direct embryogenesis ensuring minimal occurrence of cytogenetic abnormalities. Appropriate stress conditions such as chilling, partial desiccation, cotyledon excision, and successive subculture of microspore-derived embryos could promote plant development in oilseed rape. Medium renovation, phytohormones and plant

Advances in Botanical Research, Vol. 45
Incorporating Advances in Plant Pathology

0065-2296/07 $35.00
DOI: 10.1016/S0065-2296(07)45007-8

growth regulators, and chromosome doubling agents such as colchicine treatment also affect plant regeneration in *Brassica* species. Compared to colchicine treatments of microspore-derived embryos and plants, immediate colchicine treatment of isolated microspores results in high embryogenesis and diploidisation and low chimeric percentages. The ploidy level of microspore-derived plants of *Brassica* species could be estimated by different methods at various stages. Mutation breeding techniques are widely used in plant breeding for producing useful mutants and variants. Microspore culture also provides an ideal method for mutation because the mutated traits can be fixed in homozygous condition by chromosome doubling, which can enforce to obtain target mutation traits efficiently. Ultraviolet irradiation, mutagenic agents ethyl methane sulphonate and sodium azide could be applied to isolated microspores and the derived embryos of rapeseed. Utilization of microspore-derived embryos for production of desired traits such as the altered fatty acids, disease resistance and glucosinolate compositions through mutagenesis and selection is advancing and also discussed.

I. INTRODUCTION

The microspore culture technique has its potential applications in plant genetic research and breeding programs due to its relative simplicity and efficiency in haploid production (Palmer *et al.*, 1996a). Since the first report of isolated microspore culture in *Brassica napus* by Lichter (1982), there has been remarkable progress in developing this system (Chuong and Beversdorf, 1985; Polsoni *et al.*, 1988; Swanson *et al.*, 1987), and the techniques have now been used to generate haploid and doubled haploid (DH) plants in many *Brassica* species (Ferrie *et al.*, 1995; Gu *et al.*, 2003b,c, 2004b,c; Zhang *et al.*, 2003, 2004; Zhou *et al.*, 2002a,b).

In the last two decades, remarkable progress in isolated microspore culture technology has been made in all major *Brassica* species, especially in oilseed rape (*B. napus*) (Palmer *et al.*, 1996a). Comprehensive utilization of DH production system has been involved in *Brassica*-breeding programs as well as in gene transfer, biochemical and physiological studies, and other manipulations (Palmer *et al.*, 1996b). However, all these applications largely depend on efficient microspore embryogenesis and embryo development. Anther culture and microspore culture are nowadays widely used in many *Brassica* species (Keller and Armstrong, 1982; Maheshwari *et al.*, 1980).

In angiosperms, the occurrence of haploid plants was first reported in *Datura stramonium* (Blakeslee *et al.*, 1922). Thereafter, it was noted that spontaneous haploidy occurs in many other species. In *Brassica*, spontaneous haploids have been noted in number of species like *B. napus, B. carinata, B. campestris*, and *B. oleracea* (Komatsu, 1936; Kuriyama and Watanabe, 1955; Morinaga and Fukushima, 1933; Olsson and Hagberg, 1955; Ramanujam, 1941;

Stringam and Downey, 1973; Thompson, 1956, 1969). It has been suggested that haploid plants can be found in all varieties of spring and winter oilseed rapes that occur naturally (Thompson, 1974), although Banga and Labana (1986) reported only 10 spontaneous haploids in 76 lines examined. New cultivars and breeding lines have been developed using these haploids (Thompson, 1974; Wang *et al.*, 2002; Wu *et al.*, 1999; Zhang *et al.*, 2004).

Following the initial success in the production of pollen-derived plants from cultured anthers of *D. innoxia* (Guha and Maheshewari, 1964, 1966), attempts were made to culture anthers from *Brassica* species. The first report was by Kameya and Hinata (1970) who described callus and haploid plants from cultured anthers of *B. oleracea*. Later, microspore embryogenesis was reported in *B. napus* and *B. campestris* (Keller and Armstrong, 1975), and then haploids from several *Brassica* species were reported through anther culture (Jain *et al.*, 1980, 1989; Keller and Armstrong, 1978, 1979; Leelavathi *et al.*, 1984; Lichter, 1981; Sharma and Bhojwani, 1985).

Different factors are influencing the success of microspore embryogenesis and anther culture in *Brassica.* The most important of these is the genotype of cultural material. This phenomenon has been described in all species of *Brassica* where microspore and anther culture have been attempted (Arnison and Keller, 1990; Arnison *et al.*, 1990a,b; Baillie *et al.*, 1992; Burnett *et al.*, 1992; Ockendon, 1985; Thurling and Chay, 1984). The conditions under which plants are grown are also considered to be important for successful microspore culture. Pretreatment of isolated flower buds prior to microspore isolation and culture, such as low level of irradiations, ethanol, modified environment, and antimicrotubule agents like colchicine, have been known to enhance embryogenesis in *Brassica* (Klimaszewska and Keller, 1983; MacDonald *et al.*, 1988a,b; Pechan and Keller, 1989; Zaki and Dickinson, 1991). Medium composition like amount of carbohydrates used, nitrogen supply, plant growth regulators, and pH of the medium are also proved to be important in *Brassica* anther culture. Most commonly used formulation is the one developed by Lichter (1982) with various modifications (Gland *et al.*, 1988; Huang and Keller, 1989; Lichter, 1989; Zhou *et al.*, 2002a,b).

In the process of transferring the genes into oilseed rape, the conventional method is the consecutive selection of the hybrids, containing the herbicide resistant genes and other characters (such as yellow seeds, low glucosinolate, and high erucic acid). These putative pure lines are conventionally obtained through six to eight generations of self-pollinations and subsequent stepwise selections. Thus, the selection efficiency was extremely low, hard, and time-consuming. The microspore culture could be an ideal method of haploid breeding. By doubling the chromosomes (e.g., through colchicine treatment), one can get several DH-regenerated plants of not only pure heredity but also stable characters (Gu *et al.*, 2003b; Zhang *et al.*, 2004; Zhou *et al.*, 2002b). Using this method, the breeding process is accelerated and the selection efficiency enhanced (Gu *et al.*, 2003c, 2004a,b; Kott, 1998; Palmer *et al.*, 1996a). Adding mutagens or herbicides in microspore culture medium, the regenerated mutants with herbicide resistant genes have been obtained (Zhang and Takahata, 1999). Also, the conventionally bred herbicide resistant oilseed rape presents some risks such as gene drifting, biodiversity erosion, weeds resistance induction, and weed population dynamics (Momoh *et al.*, 2002; Pu, 2003).

Anther culture and microspore culture provide the opportunity to produce haploid embryos at high frequency in many species of *Brassica* and their commercial cultivars (Arnison and Keller, 1990; Lichter, 1989; Swanson, 1990). So, this procedure can be used in the development of homozygous lines for breeding programs (Wenzel, 1980). The chromosome doubling, that is either spontaneously or induced by chemicals, is among one of advantages of haploid embryo production and is extensively used in *Brassica*-breeding programs (Hoffmann *et al.*, 1982; Scarth *et al.*, 1991; Stringam and Thiagarajah, 1991). A DH population obtained from heterozygous starting material provides a broad spectrum of genetic recombinants from which selection can be made easily. For protoplast isolation and culture from haploid tissues of *Brassica*, the culture systems have been developed (Chuong *et al.*, 1987; Li and Hohlenbach, 1982; Thomas *et al.*, 1976). The advantage of this system is that it provides large number of cells for mutation and selection in a manner similar to isolated microspores. In *B. napus*, fatty acid composition of microspore-derived embryo is quite similar to that of normal embryo, and the total fatty acid per unit weight at early stages of development is comparable (Pomeroy *et al.*, 1991; Taylor *et al.*, 1990). Thus, embryos derived through microspore culture or haploid culture can be utilized for basic biochemical and physiological studies (Holbrook *et al.*, 1991; Johnson-Flanagan and Singh, 1993).

Utilization of microspore-derived embryos for production of desired traits such as the altered fatty acids, disease resistance, and glucosinolate compositions through mutagenesis and selection is advancing. For the storage of germplasm, haploid system presents an attractive alternative where germplasm in the form of microspores and embryos can be maintained by cryopreservation without loss of embryogenic or morphogenic potential. However, success is more likely if fungal toxins are used in these selections. Microspore population presents a rich source of genetic variation from which large number of genetic recombinations can be made that serves as a basis for cultivar improvement and development.

II. INCREASING MICROSPORE EMBRYOGENESIS

Haploid *Brassica* plants from seed occur naturally in low frequencies and have been used in breeding programs. However, the capacity to truly exploit *Brassica* haploids originates with the discovery of methods to induce embryos from anthers and isolated microspores *in vitro*. Since haploid plants were first obtained from microspore culture of *B. napus* (Lichter, 1982), there has been a rapid development in the application of this technique in rapeseed breeding and for the understanding of basic biochemical and physiological aspects of embryogenesis. Microspores provide a uniform, synchronous, and easily accessible population of cells for such studies (Kott, 1996; Palmer *et al.*, 1996b).

The microspore culture protocol used was developed in our laboratory (Zhou *et al.*, 2002b) with a few modifications for various *Brassica* species. The donor plants were grown under growth cabinet conditions with a 16-h photoperiod (~350 μE/m^2 s), a day/night temperature of 20/15°C and a high nutritive status. Vernalization of winter *B. napus* genotypes was conducted with a combination of 4°C and 8-h photoperiod for 8 weeks. The donor plants were grown in individual 3-liter pots, watered daily, and fertilized once a week with liquid NPK fertilizer. One week before microspore isolation, the temperature was adjusted to 12/10°C day/night regime. Approximately 10 plants were grown for each genotype (Zhou *et al.*, 2002a). For each treatment, 9–12 flower buds at the late uninucleate stage and early binucleate stage of microspore development were selected. The proper stage was determined from a combination of bud size (usually 2–3 mm in length for *B. rapa*, 3–4 mm for *B. napus*, and 5–6 mm for *B. oleracea*) and ratio of petal length to anther length (P/A, usually 0.5–0.75 for *B. rapa* and *B. napus*, 1.0–1.2 for *B. oleracea*), followed by the identification of microspore development under a dissecting microscope (Gu *et al.*, 2004a).

The flower buds were surface sterilized in 1% Gevonium (0.5% sodium hypochlorite) for 18 min prior to washing three times with sterile water. The buds were then macerated in cold B_5-13 medium (Gamborg *et al.*, 1968; can be provided by Duchefa Biochemie BV) with 13% sucrose, buffered at pH 6.0, and filtered through a 40-μm nylon filter into a centrifuge tube. The crude microspore suspension was centrifuged at 750 rpm for 5 min, and the microspores were resuspended in NLN-13 medium (Lichter, 1982; can be provided by Duchefa Biochemie BV) with 13% sucrose at pH 6.0. After adding activated charcoal (AC), the microspore suspension was then dispensed into 60-mm × 15-mm Petri dishes, 4 ml in each. A microspore density of about 2×10^4 per ml was used, that is, one flower bud equaled one Petri dish. The dishes were sealed with double layers of Parafilm, incubated at 30°C in the dark for 7 days (*B. napus*), or at 32°C for 2 days (*B. rapa* and *B. oleracea*), and then moved to 24°C still in the dark. Medium refreshing was conducted after 1 day of induction following microspore isolation. The suspension was collected and recentrifuged at 750 rpm for 5 min. The medium was then changed with the same amount of fresh NLN-13 medium, and cultured as described above (Gu *et al.*, 2003b, 2004a).

As soon as embryos were visible to the naked eye (about 9–15 days after isolation), cultures were transferred to a slow rotary shaker (45 rpm), still in darkness at 24°C. At the late torpedo stage, large embryos were transferred to solid MS regeneration medium (2% sucrose, half-strength macronutrients and 0.1-mg/liter GA_3, solidified with 0.9% agar, pH 5.8) in plastic growth containers. After an initial period of 10 days at 2°C, the cultures were incubated in a growth chamber with 24°C and 16-h day length and low light intensity (~60 $\mu E/m^2$ s) from fluorescent tubes. After about 30 days of microspore culture, embryos with normal shape (healthy-looking) at the cotyledon stage were counted. These embryos were then transferred to plastic growth containers containing solid MS-2 medium (Murashige and Skoog, 1962; 2% sucrose, 0.9% agar, pH 5.8). Approximately 3 weeks later, shoots developed from the embryos. They were then cut off from any callus or hypocotyl tissue and transferred to plastic growth vessels with fresh solid MS-2 medium for roots induction. Recalcitrant embryos could be trimmed by cutting to induce germination. After another 3 weeks, rooted plantlets were transferred to a soil–perlite mixture and were kept for 2 weeks in a nursing room with a temperature of 24°C, 16-h daylength, low light intensity, and high relative humidity. Gradual adaptation to glasshouse conditions followed (Zhang *et al.*, 2006; Zhou *et al.*, 2002a).

Various factors could influence microspore embryogenesis including donor plant genotype, donor plant physiology, microspore developmental stage, culture conditions, culture environment, and pretreatments (Dunwell, 1996). In *Brassica* microspore culture, a short heat shock treatment is required to stimulate microspore embryogenesis, although it may be replaced by other stresses such as low levels of γ-irradiation, ethanol, and colchicine (Pechan and Keller, 1989; Zhao *et al.*, 1996). Compared to heat shock stress, cold pretreatment is less frequently used in *Brassica* species in spite of its critical role in cereal microspore culture (Pechan and Smykal, 2001; Touraev *et al.*, 1997).

Genotype of the cultural material is one of the most important factors among various factors influencing success of microspore and anther culture in *Brassica* (Arnison and Keller, 1990; Arnison *et al.*, 1990a,b; Baillie *et al.*, 1992; Burnett *et al.*, 1992; Ockendon, 1985; Thurling and Chay, 1984) and even plant-to-plant variation within genotype is common (Ockendon, 1985; Phippen and Ockendon, 1990; Seguin-Swartz *et al.*, 1983). The situation is further complicated by genotype × environment interactions (Dunwell *et al.*, 1985; Roulund *et al.*, 1990; Thurling and Chay, 1984). In *B. oleracea* convar. *capitata* (L.) (Alef.), it was reported that field-grown plants were more responsive than those grown in greenhouse conditions (Roulund *et al.*, 1990). This shows that some genotypes may have specific environmental requirements in order to yield embryogenic microspores. In order to investigate the molecular events occurring during the embryo induction process, attempts were made by using two-dimensional gel electrophoresis of protein of DH plants previously characterized as responsive and unresponsive to culture (Fabijanski *et al.*, 1992). These experiments indicated that the initial phase of culture involves a typical heat shock response that was similar to the one previously documented in the other broccoli tissues (Fabijanski *et al.*, 1987).

Growth conditions are also important factor for successful microspore culture in *Brassica* and other species. It is well established that plants grown at low temperatures yield the most responsive microspores (Dunwell *et al.*, 1985). Although the temperature under which plants are grown is a critical factor, physiological state of donor plant is also important. In *B. napus*, higher frequencies of embryogenesis were obtained with older inflorescences than that with younger ones (Takahata *et al.*, 1991). It shows the influence of donor plant age on microspore embryogenesis. It is well established that embryogenesis is related to specific stages of microspore and pollen development (Fan *et al.*, 1988; Kott *et al.*, 1988a; Telmer *et al.*, 1993), and it varies from early uninucleate to early binucleate stage. The highest frequency of microspore embryogenesis was observed with microspore at late uninucleate stage in two genotypes of *B. napus* (Kott *et al.*, 1988a). Pretreatments of isolated flower buds prior to microspore isolation and culture have also been proved effective in *Brassica* species. These treatments include low levels of irradiations; ethanol, modified atmosphere, and antimicrotubule agents, for example colchicine, have been known to enhance embryogenesis in *Brassica* (Klimaszewska and Keller, 1983; MacDonald *et al.*, 1988a,b; Pechan and Keller, 1989; Zaki and Dickinson, 1991).

Colchicine can be applied practically at any stage during the microspore culture process, from isolated microspores to the regenerated plants. Colchicine added *in vitro* to the induction medium with freshly isolated rapeseed microspores within the first 3 days of culture reportedly produced improved embryogenesis and with no negative effect on embryo development (Iqbal *et al.*, 1994; Zaki and Dickinson, 1991). Short-term exposure of isolated microspores to colchicine increases the number of symmetric cell divisions and the frequency of embryogenesis in *B. napus* (Zhang *et al.*, 2003). The optimal time for such treatment seems to be the first 12–15 h after microspore isolation. Further methods are being explored involving the use of other antimicrotubule compounds such as trifluralin, oryzalin, amiprophos-methyl (APM), and pronamide, in addition to colchicine, to affect embryogenesis and chromosome doubling during the early stages of microspore culture. The combination of colchicine concentration and treatment duration is critical for embryogenesis and diploidization. Results of our experiments showed that an efficient embryogenesis and diploidization of haploid microspores of spring and winter *B. napus* could be achieved by treating them immediately with colchicine (Zhou *et al.*, 2002a,b). A high doubling efficiency of 83–91% is obtained from 500-mg/liter colchicine treatment for 15 h. In addition, at this level only few polyploid and chimeric plants are formed.

AC is often used in culture media due to its ability to modify the medium composition and sometimes can improve or regulate plant growth *in vitro*. It also increases embryogenesis from microspore in some plants when added to culture medium (Johansson, 1983), and in *B. napus*, Gland *et al.* (1988) reported that addition of AC into culture medium did not increase the number of embryos per flower bud but promoted the development of embryos to develop into whole plants. This positive effect is probably related to absorption of toxic metabolites which may be produced by nonembryogenic microspores (Kott *et al.*, 1988b). In nine morphotypes of *B. oleracea*, effect of addition of a 0.1-ml drop of AC on microspore culture embryogenesis was studied (Dias, 1999). Embryogenesis was greatly increased in all morphotypes by addition of 0.1-ml AC to the microspore culture media. However, the magnitude of response to the addition of AC varied with different plants and morphotypes.

Stress is an essential component during embryogenesis induction in microspore culture. Cold pretreatment has been used in cereal microspore culture but very seldom attempted in *Brassica* microspore culture. The effect of cold pretreatment on flower buds subjected to a liquid medium on microspore embryogenesis was investigated in spring and winter *B. napus*, as well as in *B. rapa* and *B. oleracea*. Cold pretreatment significantly enhanced microspore embryogenesis (by one- to sevenfold) compared to commonly used microspore culture protocol in *B. napus*, while it was less effective in *B. rapa* or even negative in *B. oleracea*. A significant enhancement of microspore embryogenesis by cold pretreatment of flower buds was reported in *B. napus* (Gu *et al.*, 2003a). The same promoting effect of cold pretreatment was also observed in *B. napus* (Lichter, 1982) and *B. oleracea* (white cabbage) (Osolnik *et al.*, 1993), and the more recently reported in *B. rapa* (Sato *et al.*, 2002).

The highest rates of germinated embryos (90%) and plantlets development (58.46%) are obtained by exposing microspore-derived embryos to chilling at 4°C. These results indicated that cold treatment not only enhanced microspore embryogenesis but also improved the germination and development of plants from microspore-derived embryos in *B. napus* (Gu *et al.*, 2004a; Zhou *et al.*, 2002b). Partial desiccation of the embryos can increase the rate of germinated embryos and plantlet development, and it is closely related to the duration of treatment used.

In *Brassica* microspore culture, cold pretreatment alone cannot stimulate microspore embryogenesis. It should be combined with other treatments such as heat shock so as to increase microspore embryogenesis (Pechan and Smykal, 2001). In ordinary protocols of *Brassica* microspore culture, if donor plants have been grown under optimal environmental conditions, where a low temperature is considered as a favorable and important factor for microspore embryogenesis, cold pretreatment of flower buds seems less important or has even been omitted (Palmer *et al.*, 1996a). However, cold pretreatment of flower buds could further enhance microspore embryogenesis and embryo development in *B. napus* even if donor plants have been grown under such a low-temperature condition which infers that cold pretreatment may have other functional effects on microspore embryogenesis. As far as the reason of beneficial effect on microspore embryogenesis by cold pretreatment of flower buds is concerned, many efforts have been made to explain this process.

In *B. rapa* as well as in other crops after cold pretreatment (Sato *et al.*, 2002; Sopory and Munshi, 1996), increased percentage of bicellular stage microspores with two equal nuclei was observed, which was supposed to be the necessary developmental stage for embryogenesis induction. It was suggested that the induction effect of osmotic stress under cold pretreatment conditions could be a possible reason for increased microspore embryogenesis potential. Proper culture osmoticum has proved to be essential to the stimulation of microspore embryogenesis. This osmotic requirement for microspore induction and development could be met by sucrose, but can also be substituted by polyethylene glycol (PEG) (Ilić-Grubor *et al.*, 1998). On the other hand, cold pretreatment could be substituted by osmotic stress for the induction of cereal microspore embryogenesis (Raina and Irfan, 1998; Roberts-Oehlschlager and Dunwell, 1990), which implies that osmotic stress rather than cold pretreatment may be the key factor affecting microspore embryogenesis. Prolonged cold pretreatment of flower buds showed similar promoting effect in microspore embryogenesis. Thus, flower buds or inflorescences for microspore culture can be stored for certain periods under low temperature.

III. EFFICIENT PLANT REGENERATION FROM MICROSPORE-DERIVED EMBRYOS

Plant regeneration from cultured anthers and microspores mostly occurs through direct embryogenesis and all the developmental stages characteristic of zygotic embryogenesis can be identified in the cultures (Palmer *et al.*, 1996a);

however, developmental process may proceed through secondary embryogenesis (Loh and Ingram, 1982; Loh *et al.*, 1983; Lott and Haube, 1983). Only few examples have been reported about callusing before plant regeneration occurred (Govil *et al.*, 1986; Sunderland, 1971), and few studies have been carried out on plantlet development from microspore-derived embryos in *Brassica* species. With *B. napus*, the frequencies of plantlet development varied from 1 to 47% (Chuong *et al.*, 1988; Kott and Beversdorf, 1990), while with *B. campestris* (*B. rapa*), the frequencies ranged from 5 to 20% (Baillie *et al.*, 1992; Burnett *et al.*, 1992). The effects of genotype, bud size, microspore stage, culture condition and different media on embryo production, and plantlet development have been also investigated (Ferrie and Keller, 1995; Gland-Zwerger, 1995).

Almost every culture has some older microspores that do not divide but do release some toxic elements that affect the rest of the potentially embryogenic microspores. Fletcher *et al.* (1998) and our results (Zhou *et al.*, 2002a) show that a change of induction medium 15–24 h after microspore isolation may be essential to allow normal embryo initiation and development. Poor embryogenesis and embryo germination are observed from ordinary microspore culture without change of induction medium and without *in vitro* colchicine treatment. Subculturing over an extended period (6–10 weeks) is required to induce germination and produce shoots, which originate from secondarily initiated embryos from hypocotyl epidermal cells. Obviously, this trimming procedure is quite labor intensive as well as expensive and should be avoided in a breeding program.

In terms of plant development from microspore-derived embryos, the stage at which embryos are transferred to solidified medium is critical for the plantlet formation (Niu *et al.*, 1999). Moreover, embryos transferred at the cotyledonary stage resulted in the highest frequency of plant regeneration from isolated microspores (Burnett *et al.*, 1992). It was reported that the culture medium was crucial to the plantlet development (Chuong and Beversdorf, 1985). Some reports showed that 1/2 MS (Murashige and Skoog, 1962) or 1/2 B5 (Gamborg *et al.*, 1968) media had a better effect on the plantlet formation than full MS or B5 media (Gland-Zwerger, 1995). In addition, the genotype of donor plant influenced the plantlet development, and the rate of plantlet formation reached from 9.5 to 37.5% (Mathias, 1998).

Cold pretreatment significantly enhanced microspore embryogenesis (by one- to sevenfold) compared to commonly used microspore culture protocol in *B. napus*, while it was less effective in *B. rapa* or even negative in *B. oleracea*. The appropriate duration of cold pretreatment was found to be 2–4 days, which stimulated the best microspore embryogenesis. Cold pretreatment was also able to promote embryo development including the

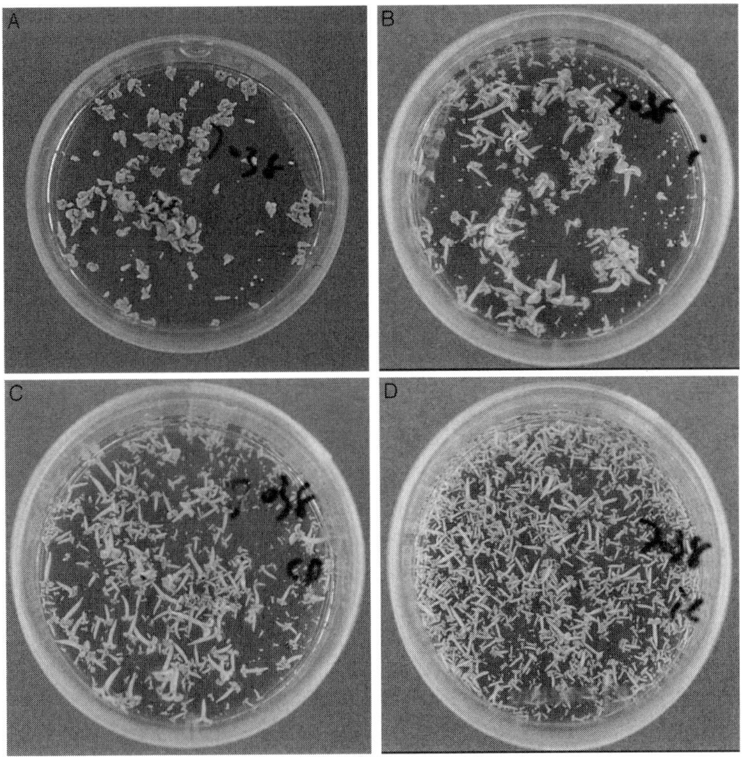

Fig. 1. Four-week-old embryos from isolated microspores of *B. napus* cv. Topas line 7038. (A) The control (direct isolation without cold pretreatment or medium refreshing). (B) Medium refreshed after 1 day of induction of isolated microspores (Medium refreshing was conducted after 1 day of induction following microspore isolation. The suspension was gathered and recentrifuged at 750 rpm for 5 min. The medium was then changed with the same amount of fresh NLN-13 medium, and cultured as before), without cold pretreatment. (C) Two days of cold pretreatment of flower buds before microspore isolation (cold pretreatment of flower buds was subjected to a liquid culture medium containing 13% sucrose. The dishes were sealed with double layers of Parafilm, and were then placed for 0, 1, 2, 3, and 4 days at 4°C in the dark. Microspore isolation and culture procedure was the same as before.), without medium refreshing. (D) Two days of cold pretreatment of flower buds before microspore isolation, followed by medium refreshing 1 day after microspore isolation. (See color Plate 3.)

improvement of embryo quality and acceleration of embryogenesis. When incorporating with medium refreshing, cold pretreatment could initiate the most microspore embryogenesis than any other treatment used. With further improvement, cold pretreatment method may have a positive effect in *Brassica*-breeding programs (Gu *et al.*, 2004a) (Fig. 1). The highest rates of germinated embryos (90.0%) and plantlet development (58.46%) are obtained by exposing microspore-derived embryos to chilling at 4 °C. These results indicated that cold treatment improved the germination of microspore-derived embryos and development of plants from embryos in *B. napus* (Zhang *et al.*, 2006).

Partial desiccation of the embryos can increase the rate of germinated embryos and plantlet development, and it is closely related to the duration of treatment used. The drying of the embryos for 1 or 2 days could significantly accelerate the embryo germination and development of plants from microspore-derived embryos. Prolonged partial desiccation of the embryos (i.e., 3 days) showed embryos that might suffer from nutrient starvation as indicated by the loss of embryo fresh mass. When drying for over 4 days, most embryos lost the viability and died (Zhang *et al.*, 2006). This interaction of partial desiccation and starvation stress requires further investigation.

It is known that the culture medium is an important factor affecting the plantlet development (Li *et al.*, 2005; Zhang *et al.*, 2005, 2006). Benzylaminopurine (BAP) is helpful for the growth and development of shoots *in vitro* of oilseed *Brassica* (Tang *et al.*, 2003). Zhang *et al.* (2006) evaluated the effects of chilling, partial desiccation, cotyledon excision, and successive subculture of microspore-derived embryos on plant development in oilseed rape (*B. napus* L.). The results showed that of the five media, all the genotypes showed the best response when the embryos were cultured on the half-strength (Murashige and Skoog, 1962) medium with 2.0-mg/liter BAP. Desiccation for 1 day also increased the embryo germination and plantlet development in all *B. napus* genotypes tested. Cutting the cotyledons of the embryos at late cotyledonary stage significantly increased the frequency of plantlet development. The highest rate of plantlet development was obtained from cultures of embryos sampled with size of less than 4.0 mm. The successive subculture further improved the embryo germination and development of plantlets from microspore-derived embryos.

The treatment of cotyledon excision of the embryos is another effective way to stimulate the embryo germination and development of plants from embryos. It was showed that cutting the cotyledons of the embryos at late cotyledonary stage significantly increased the frequency of plantlet development. Subculturing over an extended period is helpful to induce germination and produce shoots, which originate from secondarily initiated embryos from epidermal cells of hypocotyls. Meanwhile, it was observed that without cotyledon excision treatment, for the genotype ZJU436, the embryos expanded after transfer to the induction media but did not produce shoots. The results indicated that without cotyledons, nutrition in the embryos could concentrate on the process of plantlet development. This was the first report to remove the cotyledons from the microspore-derived embryos, with the

Fig. 2. (A) Large and healthy-looking embryos at late cotyledonary stage, regenerated from microspores of *B. napus* L., have a normal elongated root/shoot axis with two very conspicuous cotyledons surrounding the shoot apex. These embryos are also used for the treatment of cotyledon excision, that is, the half or all of both cotyledons were excised from the embryos (but the apical shoot meristem, elongated root/shoot axis, and the distal root meristem all remained), and then transferred to the solid induction medium (1/2 MS + 2.0-mg/liter BAP) with 2.0% sucrose and 0.9% agar. (B–D) Direct plant regeneration from microspore-derived embryos treated immediately with 500-mg/liter colchicine for 15 h in F_1 hybrid 5396 of winter *B. napus*. A 0.2% stock solution of colchicine in the induction medium (NLN-13 medium) was prepared and filter-sterilized using a 0.2-μm bacterial filter (Sartorius NML). Appropriate volumes of stock solution were added to the culture medium immediately after microspore isolation in order to give the desired colchicine concentration. After colchicine treatment in darkness at 30 °C for 6 and 15 h, the microspores were washed twice by centrifuging the microspore suspension with fresh induction medium. For further culture, the microspores were resuspended in fresh NLN-13 medium and cultured as before. After transferred to solid MS regeneration medium for 4 weeks (B); subcultured to solid MS medium for 2 weeks (C); and transferred to a soil–perlite mixture (D). (See color Plate 4.)

improved germination and development of plants from embryos (Zhang *et al.*, 2006) (Fig. 2). In addition, the cotyledon excised from the embryos could also be used for the quality selection at very early stage of breeding program in oilseed *Brassica*, as reported by Albercht *et al.* (1995).

Medium renovation, phytohormones and plant growth regulators, and chromosome-doubling agents, such as colchicine treatment, also affect plant development in *Brassica* species (Gu *et al.*, 2004c; Sato *et al.*, 2002; Zhang *et al.*, 2005; Zhou *et al.*, 2002b). High frequency of plant development could be obtained by abscisic acid (ABA) treatment of embryos (Hansen, 2000). These are areas that need to be further explored in order to achieve better plant development from microspore-derived embryos in *Brassica*-breeding programs.

IV. HIGH FREQUENCY PRODUCTION OF DH PLANTS

Haploid plant regeneration directly from the shoot apex of microspore-derived embryos is an important step in DH production. Direct and quick regeneration ensures minimal occurrence of cytogenetic abnormalities, which is extremely important in a breeding program (Fletcher *et al.*, 1998). Large and healthy-looking embryos obtained from colchicine-treated microspores, after transfer to solid regeneration medium and an initial period of low temperature (2°C) for about 10 days, germinate well at 24°C and can easily regenerate vigorous plants. These plants have complete roots, stems, leaves, and shoot apex, and are ready for transfer to the soil. The time period from microspore isolation to regenerated plants in soil is only 8–10 weeks, and more importantly, most of these plants can be DH plants already.

Although microspore culture is an effective technology for the production of DH parental lines required for producing F_1 hybrids of modern cultivars with desirable genetic recombinants, and plant regeneration from isolated microspores has been reported in *B. rapa* subsp. *oleifera* (Baillie *et al.*, 1992; Burnett *et al.*, 1992; Ferrie *et al.*, 1995; Guo and Pulli, 1996; Lichter, 1989), subsp. *pekinensis* (Kuginuki *et al.*, 1997; Sato *et al.*, 1989), subsp. *chinensis* (Cao *et al.*, 1994), and subsp. *parachinensis* (Wong *et al.*, 1996), the application of this technique for wide utilization as a breeding tool for these crosses still remains an obstacle. The less efficient production of DH plants is especially pronounced in *B. rapa* subsp. *chinensis*.

Plants regenerated from microspore-derived embryoids can be haploid, diploid, or polyploid. From rapeseed microspore culture, it is reported that 70–90% of regenerated plants are haploid (Charne *et al.*, 1988; Chen and Beversdorf, 1992). The most widely used method of chromosome doubling has been the use of the antimicrotubule agent colchicine. Colchicine treatment of isolated microspores increases the doubling efficiency of regenerated plants in *B. napus* (Chen *et al.*, 1994; Hansen and Andersen, 1996; Möllers *et al.*, 1994; Zhao *et al.*, 1996). The usual methods of chromosome doubling involve soaking roots or whole plants in a colchicine solution (Fletcher *et al.*, 1998), or culturing plantlets in colchicine-containing medium in the greenhouse (Mathias and Robbelen, 1991). Other alternatives are injecting colchicine into the secondary buds or applying a colchicine-soaked cotton swab to axillary buds (Gland, 1981; Lichter *et al.*, 1988). Such methods often result in the formation of chimeric plants with relatively small sectors of diploid tissue which will produce few selfed seeds. Consequently, an extra cycle of seed multiplication becomes necessary in the greenhouse before the first field planting.

Zhou *et al.* (2002a) reported that an efficient diploidization of haploid microspores of spring *B. napus* can be achieved by treating them immediately with colchicine. A high doubling efficiency of 83–91% is obtained from 500-mg/liter colchicine treatment for 15 h. In addition, at this level only few polyploid and chimeric plants are formed. These results are in good agreement with previous reports of the effects of such treatments. Concentrations of 50- to 500-mg/liter colchicine are required and treatment duration should be prolonged to 18–24 h in order to achieve a maximum diploidization of 90–94% (Chen *et al.*, 1994; Hansen and Andersen, 1996; Iqbal *et al.*, 1994; Möllers *et al.*, 1994).

Treatment duration of colchicine is another critical parameter for diploidization, and it is closely related to the concentrations used. Our experiment revealed that treatment duration of 30 h had less positive effects on embryogenesis and doubling efficiency, especially at higher colchicine concentrations (Zhou *et al.*, 2002a). However, even 42- to 72-h treatment duration is also very effective when applying at low concentration (10 mg/liter) of colchicine (Iqbal *et al.*, 1994; Möllers *et al.*, 1994; Zhao *et al.*, 1996).

Compared to colchicine treatments of microspore-derived plants and microspore-derived embryos, immediate colchicine treatment of isolated microspores results in high embryogenesis and diploidization and low chimeric percentages in *B. napus* (Chen *et al.*, 1994; Zhou and Hagberg, 2000). Plant treatment for chromosome doubling significantly delays plant growth and development. This process also reduces seed set so that an additional growth cycle is necessary to produce sufficient seed for field testing. In addition, plant colchicine treatment generates large amount of colchicine waste, which needs specialized handling and disposal. Therefore, immediate chromosome doubling by colchicine application of isolated microspores is safer, quicker, and more efficient.

We found that colchicine treatment during the first 15 h in microspore culture caused an increase in doubling efficiency, which was the highest (84.5% and 88.2%) with 500-mg/liter colchicine treatment. Application of colchicine for 6 h was less effective. High concentration (1000 mg/liter) of colchicine treatment generated significant numbers (4.7–9.5%) of polyploids, even treated for 6 h (5.5% and 9.5%). The study also revealed that changing of induction medium 15 h after microspore isolation induced higher spontaneous doubling efficiency, as compared with medium change 6 h after isolation (Zhou *et al.*, 2002b).

An improved protocol for production of DHs through microspore culture of *B. rapa* subsp. *chinensis* has been established by Gu *et al.* (2003b) (Fig. 3). Among other modifications, this protocol is mainly characterized by change of culture media and reduction of sucrose concentrations after the first 48 h of culture initiation, which results in a high frequency of microspore embryogenesis and good embryo quality. Almost every culture includes some unavoidable older microspores that do not divide but release some toxic compounds that affect the rest of the potentially embryogenic microspores adversely. A change of induction medium at the initial stage of microspore culture may be essential to get rid of any toxic elements released from the older microspores, thus allowing normal embryo initiation and development. Furthermore, different concentrations of sucrose are required at different stages of microspore development, which is also rather genotype dependent. It would be interesting to further investigate the response of *B. rapa* genotypes to the replacement of medium from NLN-13 (Lichter, 1982; 13% sucrose) to new NLN-13, as it works well with *B. napus* genotypes tested. Similar results are reported in *B. napus* (Hansen and Svinnset, 1993), *B. rapa* (Baillie *et al.*, 1992; Burnett *et al.*, 1992), and *B. juncea* (Lionneton *et al.*, 2001).

In the previous studies reported by Cao *et al.* (1994) and Wong *et al.* (1996) with the same variety of subsp. *chinensis*, the embryo production was quite low and the rate of plant regeneration was not reported. Gu *et al.* (2003b) reported that in *B. rapa* subsp. *chinensis*, a high frequency of spontaneous DHs obtained without any subculture or any treatment with chromosome-doubling agents. The percentage of spontaneous diploidization reported is more than 70% of regenerated plants compared to ~35% in the *B. napus* controls, which indicates that this modified method produces sufficient numbers of DH plants (Fig. 3).

Zhang *et al.* (2001) reported a similar percentage of spontaneous diploids in Chinese cabbage, but the rate of embryo development was not reported. Half of 45 genotypes produced 70% of DH plants. However, ploidy level was estimated at the flowering stage by flower morphology, pollen fertility, flower size, and seed set. This method is likely to misjudge some fertile plants like tetraploids for diploids and therefore overestimates the percentage of actual DH production. A high percentage of spontaneous diploids were found in cauliflower (56%) and broccoli (55–71%) (Chatelet *et al.*, 1999; Wang *et al.*, 1999). Such high spontaneous diploidization has not been reported in other *Brassica* species. In *B. rapa* subsp. *oleifera*, the percentage of spontaneous diploids reported is ~30% (Keller and Armstrong, 1975), compared to 5–50% in *B. napus* (Charne *et al.*, 1988; Chen and Beversdorf, 1992; Möllers *et al.*, 1994; Zhou *et al.*, 2002a,b) and 10–20% in *B. juncea* (Lionneton *et al.*, 2001), which is insufficient for exploitation in actual DH-breeding programs, and thus requires an additional step treatment with chromosome-doubling agents.

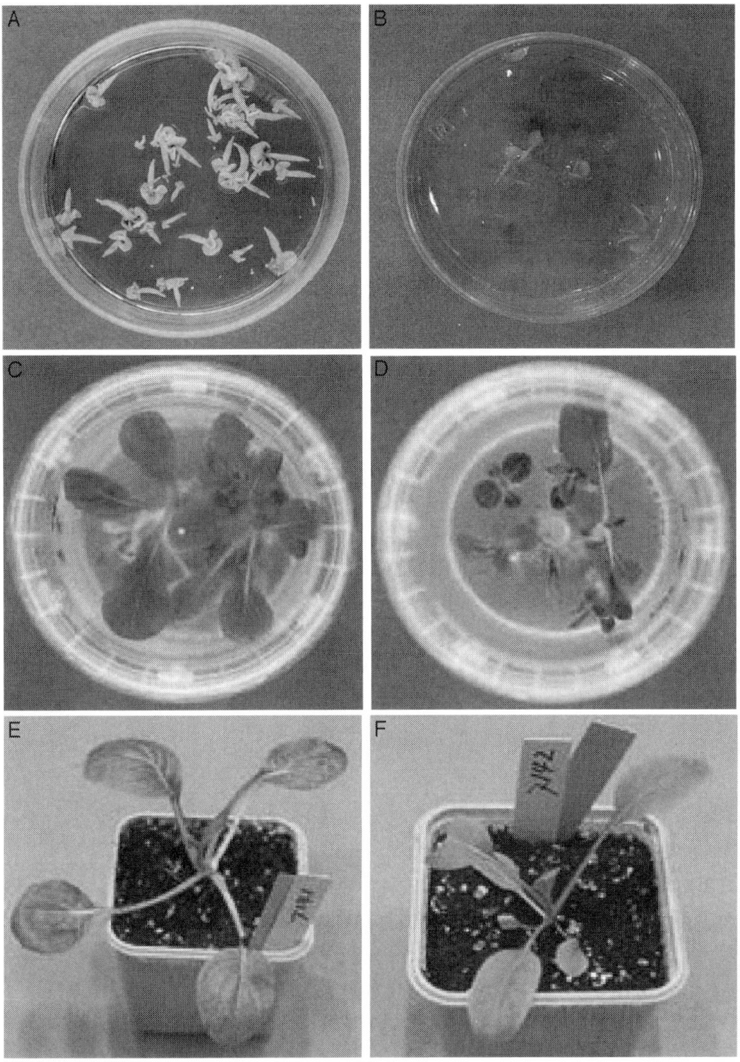

Fig. 3. High-frequency spontaneous production of DH plants in microspore cultures of *B. rapa* subsp. *chinensis* using an improved protocol primarily by replacement of culture media and reduction of sucrose concentrations from 17 to 10% after the first 48 h of culture induction. Large and healthy embryos of *B. rapa* subsp. *chinensis* genotypes 6986 (A) and 6987 (B) obtained in liquid NLN induction medium for 4 weeks, after the change of culture media and reduction of sucrose concentrations immediately after the microspore isolation. Direct plant regeneration from microspore-derived embryos of *B. rapa* subsp. *chinensis* genotypes 6986 (C) and 6987 (D) in the solid B5-2 regeneration medium (2% sucrose, 0.8% agar, pH 5.8), after cut free of any callus or hypocotyl tissue from the embryo-derived shoots. Normal rooted and DH androgenic plants of *B. rapa* subsp. *chinensis* genotypes 6986 (E) and 6987 (F) in a soil–perlite mixture, after kept for 2 weeks in a nursing room [24°C, 16-h photoperiod, low light intensity (40 $\mu E/m^2$ s), and high humidity]. (See color Plate 5.)

It is generally assumed that the occurrence of spontaneous diploids is primarily governed by genetic factors (Chen and Beversdorf, 1992; Keller and Armstrong, 1978). The basic diploid *Brassica* types such as *B. rapa* and *B. oleracea* might be more prone to spontaneous diploidization. For these types, a slightly higher percentage of early bicellular microspores are optimum for most embryogenesis in microspore cultures. On the contrary, for the synthetic amphidiploid *Brassica* types such as *B. napus* and *B. juncea*, predominantly late uninucleate microspores are needed for the microspore culture. But the real mechanism of spontaneous diploidization in microspore cultures still remains uncertain, and thus requires further investigation.

B. rapa subsp. *chinensis* is a very important vegetable, which is normally used for its leaves and stems. Gu *et al.* (2003b) reported that a certain number of tetraploids were obtained in two genotypes investigated. Tetraploid plants can develop into a very vigorous vegetative body, and have larger leaves and stems than those of diploid plants, which might be very useful (as some kinds of novel types) for vegetable breeders. Therefore, these tetraploid plants from isolated microspores could be directly exploited as new varieties in breeding programs.

V. DETERMINATION OF PLOIDY LEVEL

The ploidy level of microspore-derived plants of *Brassica* species could be estimated at various stages. For practical breeding purposes, the ploidy level analysis was generally determined at the flowering stage by flower morphology, pollen fertility, flower size, and seed set (Fletcher *et al.*, 1998; Hansen and Andersen, 1996; Zhou *et al.*, 2002c). By using both microspore-derived haploid plants and seed-germinated plants of *B. napus* as materials, the difference of stoma guard cell circumferences between them was significant and the stoma guard cell circumferences in the fifth leaf among these plants were relatively steady. Thus, the stoma guard cell circumferences of plants could be used as a subsidiary index to estimate the ploidy level of *B. napus* (Liu *et al.*, 2002). But these methods are likely to misjudge some fertile plants like tetraploids for diploids and therefore overestimate the percentage of actual DH production. So the precise ploidy level analysis should be conducted by a flow cytometer (FCM) such as the ploidy analyzer (PA) (Partec GmbH, Germany) (Zhou *et al.*, 2002a; see below).

FCM analysis of ploidy levels of microspore-derived plants in oilseed *Brassica* species was made from newly emerged young leaves of 3-week-old androgenic plants in soil (Zhou *et al.*, 2002a). A small portion of fresh leaf blades was placed in a 6-cm Petri dish in 0.6-ml detergent solution [100-ml deionized water, 0.1-mol/liter citric acid, 0.5-g Tween 20 (Serva 37470)], and was chopped with a sharp razor blade. One minute later, 2.4-ml staining solution [100-ml deionized water, 0.4-mol/liter Na_2HPO_4, 0.2-mg DAPI (Partec 7202)] was added to the chopped material. The suspension of cell nuclei and debris was filtered through a 40-μm nylon gauze and the filtrate was immediately analyzed with the FCM. The channel for the peak of the frequency curve representing DNA from diploid cells was calibrated by using leaf tissue from normal diploid rape that was used as a standard throughout the experiments. The DNA distribution curves were automatically analyzed by the FCM and recorded by a printer. The DNA peaks were compared to the standard peak and assigned to the ploidy level of haploid, diploid, triploid, tetraploid, and so on (Fig. 4).

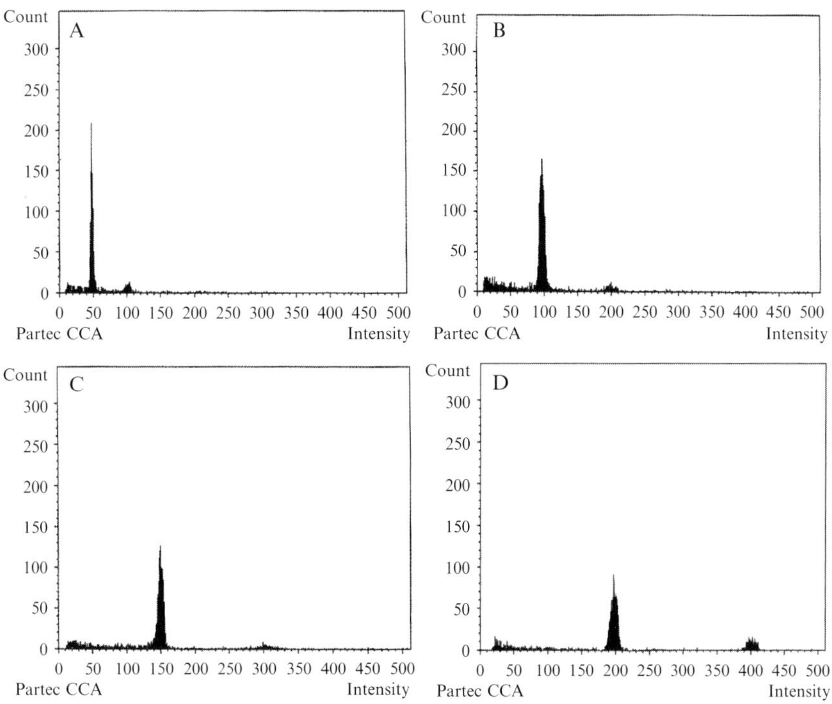

Fig. 4. FCM histograms of (A, 1*n*) haploids, (B, 2*n*) diploids, (C, 3*n*) triploids, and (D, 4n) tetraploids of microspore-derived plants in *B. napus* L. FCM analysis of ploidy levels of microspore-derived plants in oilseed was made from newly emerged young leaves of 3-week-old androgenic plants in soil. A small portion of fresh leaf blades was chopped in 0.6-ml detergent solution (100-ml deionized water, 0.1-mol/liter citric acid, 0.5-g Tween-20), and was then added 2.4-ml staining solution (100-ml deionized water, 0.4-mol/liter Na_2HPO_4, 0.2-mg DAPI). The suspension of cell nuclei and debris was filtered through a 40-μm nylon gauze and the filtrate was immediately analyzed with the FCM. The instrument was calibrated against normal diploid plant that was used as a standard throughout the experiments. The frequency curve of the sample DNA was compared to the standard peak and assigned to a ploidy level of haploid, diploid, triploid, tetraploid, and so on.

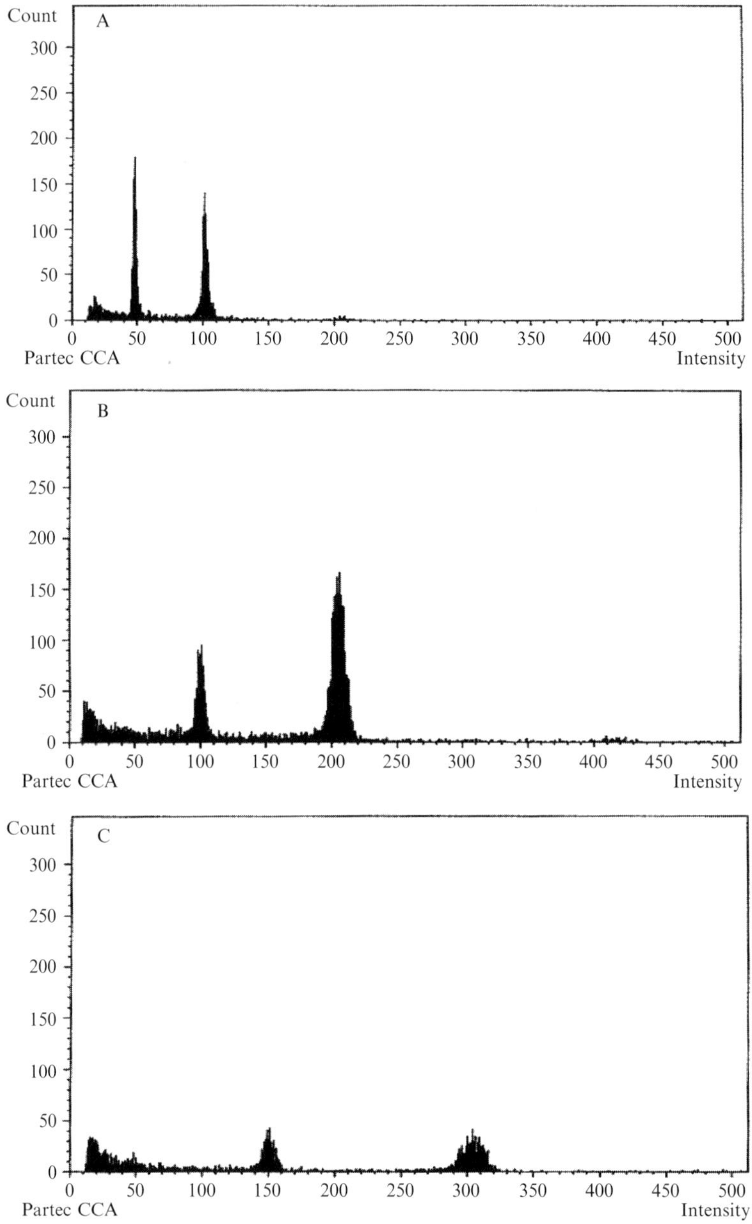

Fig. 5. FCM histograms of chimerics such as haploid plus diploid (A, $1n + 2n$), diploid plus tetraploid (B, $2n + 4n$), and triploid plus hexaploid (C, $3n + 6n$) of microspore-derived plants in *B. napus* L. FCM analysis of ploidy levels of microspore-derived plants in oilseed was made from newly emerged young leaves of 3-week-old androgenic plants in soil, in the same way as described in Fig. 4.

By using this method of FCM analysis, the ploidy level of microspore-derived plants could be well determined from fresh, newly emerged leaves by flow cytometric analysis at the much early stage (seedling stage). For instance, colchicine treatment during the first 15 h in microspore culture caused an increase in doubling efficiency, which was the highest (83.3% and 91.3%) with 500-mg/liter colchicine treatment. Application of colchicine for 30 h was less effective. High concentration (1000 mg/liter) of colchicine treatment generated significant numbers (7.3–14.3%) of polyploids and (few) chimerics, especially when treatment duration was 30 h (11.8% and 14.3%). The experiment also revealed that changing of induction medium 15 h after microspore isolation showed higher spontaneous doubling efficiency, as compared with medium change 30 h after isolation (Zhou *et al.*, 2002a). Furthermore, an unequivocal identification of haploids, diploids, triploids, and tetraploids (Fig. 4), as well as chimerics such as haploid plus diploid, diploid plus tetraploid, and triploid plus hexaploid of microspore-derived plants in *B. napus*, was possible from their discrete peaks of histograms [similar results see Zhou *et al.* (2002b)] (Fig. 5).

Gu *et al.* (2003b) also reported a large number of regenerated *B. rapa* plants that were relatively stable and consistent on ploidy level at that time, and they were haploids, diploids, and tetraploids. The results showed that more than 70% of regenerated plants in *B. rapa* genotypes were spontaneously DHs, less than 15% were haploids, and about 8.5% were tetraploids. *B. napus* genotypes had on an average of about 35% spontaneously DHs, about 50% haploids, and no tetraploids. Less than 10% of the plants tested in *B. rapa* were other polyploids, such as triploids, pentaploids, hexaploids, or chimerics with haploid and diploid cells and diploid plus tetraploid cells.

VI. MUTATION AND SELECTION FOR RAPESEED IMPROVEMENT

Mutation-breeding techniques have been widely used in genetic study and plant breeding for the production of useful mutants and variants. According to the statistic of mutation variety database (MVD) by FAO/IAEA, the number of mutated varieties registered online reached 2271, with 59 oil plants and rapeseed occupied only 15 (FAO/IAEA, 2003). Mutated varieties and mutated genes had accelerated agricultural production and economic development. Large number of mutation techniques *in vitro* were greatly developed in these years and widely used for improving quality and resistance of crop plants (Maluszynski *et al.*, 1995).

Microspore culture provides an ideal method for mutation because the mutated traits can be fixed in homozygous condition by chromosome doubling. In contrast, selection with diploid cells may allow recessive genes to remain undetected and to be carried unnoticed in the heterozygote. *In vitro* mutation technique to isolated microspores provides a novel approach in rapeseed mutation breeding, which can enforce to obtain target mutation traits efficiently and speed up the breeding process. The current progresses on *in vitro* mutagenesis and selection combined with microspore culture technology and mutant applications were summarized by Gu *et al.* (2003a).

There are two major mutation methods for isolated microspores, microspore-derived embryos and cotyledons of oilseed *Brassica in vitro*, one is the chemical mutation such as using mutagenic agents ethyl methane sulfonate (EMS) and sodium azide (NaN_3), and the other is physical mutation as ultraviolet (UV) and γ-irradiation (γ-radiation). In Brassicas, the efficient regeneration of plants from microspores provides a useful system for mutagenesis and selection. In addition, haploid suspension cell cultures of Brassicas and haploid embryos that are capable of sustained secondary embryogenesis are equally useful for such studies (Hoffmann, 1978; Ingram *et al.*, 1984; Loh and Lim, 1992; MacDonald *et al.*, 1988; Prabhudesai and Bhaskaran, 1991; Shu and Loh, 1987).

Breeding oilseed rape resistant to herbicides is now one of the most important objectives in breeding programs. The method of microspore and embryo culture, selection of the embryos resistant to the herbicide at the embryo stage, was used for herbicide-resistant breeding. Furthermore, the DH population was successfully produced by colchicine treatment. Its genotype was pure and the population could acquire stable inheritance. Through the physical and chemical mutations, the mutants with herbicide resistance were obtained in *B. napus* (Swanson *et al.*, 1988, 1989). Adding the mutagens or herbicides in microspore culture medium, the regenerated mutants with herbicide resistant genes have been obtained in Chinese cabbage (*B. campestris* subsp. *pekinensis*) (Zhang and Takahata, 1999). Resistant oilseed rape plants to glyphosate and haloxyfop were obtained after addition of these chemicals to the culture media. Culturing the embryos *in vitro* in the herbicide-containing media could produce these regenerated plants quickly and effectively. Moreover, the DH-regenerated plants can also be obtained by doubling the chromosomes (Xu *et al.*, 2005). These pure DH populations could be used to select new oilseed rape varieties directly or as the parents in the oilseed rape breeding. Compared with the conventional breeding method, the microspore culture approach greatly saves time and labor, enhances the selection efficiency, and offers a high potential in accelerating the breeding program of oilseed rape. Further studies are necessary for yield stability, quality, and resistance in the descendants of resistant plants, obtained from the microspore culture.

The fatty acid contents and components affect the quality of rapeseed greatly, including oil content, oleic acid, unsaturated fatty acid, linolenic acid, and so on. By chemical mutation, the oleic acid contents were enhanced from 60 to 85% and the linolenic acid contents were decreased from 10 to 3% (Kott, 1996). Through radiation mutation, the oleic acid contents were lifted from 47.1% to more than 50% and the highest was up to 70.4%, the linolenic acid contents were decreased to below 8% from the normal 11.4%, and the saturated fatty acid was also decreased to below 5% from the original 5.7% (Ferrie, 1999). According to the different uses of erucic acid (high erucic acid content has the special usage in industrial process), the mutants with different erucic acid contents were selected. Using UV treatment, Fletcher *et al.* (1998) had got the mutant with low erucic acid content (0.1%) from the initial 0.36%; however, Barro *et al.* (2003) enhanced the erucic acid from 42.8 to 49.5%. The seed oil content increased from parental 38.2 to 44.3% with about 16% increasing through being exposed to UV (Fletcher *et al.*, 1998).

The fodder from rapeseed with high glucosinolates was harmful to animals. Through the application of microspore mutation technology, the mutants with lower glucosinolates than parents' were obtained. In the *B. napus*, the lowest glucosinolate content was 16 μM compared with 99.6 μM of the parents after the microspore treated with UV. In the *B. carinata*, the average of parents' glucosinolates was 80.6 μM, after the microspore treated with UV, the average glucosinolate content of mutants was 37.5 μM, which was nearly half of the initial (Barro *et al.*, 2003).

Combining the rapeseed microspore induction and culturing together with pathogen selection, mutants with rapeseed *Alternaria brassicicola* resistance were obtained in *B. napus* (Ahmad *et al.*, 1991). Using the same method, mutants with *Sclerotinia sclerotiorum* resistance were selected in rapeseed (Liu *et al.*, 1997) and the mutants which were resistant to root rot disease were observed in Chinese cabbage (*B. campestris* subsp. *pekinensis*) (Zhang and Takahata, 1999). Some excellent agricultural characters can also be obtained through microspore mutation. Shi *et al.* (1995) reported that the mutants with the characters of long pod and dwarf in *B. napus* were obtained by using the chemical mutation of 0.2% and 0.25% EMS.

We have also carried out some mutation experiments *in vitro* in *B. napus*: UV irradiation, mutagenic agents EMS, and NaN_3 were applied to isolated microspores and microspore-derived embryos of four *B. napus* genotypes *in vitro* (He *et al.*, 2007a). In the UV light irradiation experiment, the isolated microspores and microspore-derived embryos were exposed for different time duration. Embryo yield showed a sharp decrease with increasing UV exposure. Embryo germination was also decreased when isolated microspores and microspore-derived embryos were treated by UV irradiation. Mutagenic treatments, NaN_3 and EMS were applied to the isolated microspores and embryos at early cotyledonary stage for various time intervals. When the isolated microspores were treated, chemical mutagens with low concentration promoted the embryo yield. With increasing of the mutagen concentrations and prolonged exposure time, embryo yield reduced gradually. When the concentration reached the maximum, no embryo was observed. Rate of embryo germination and plant regeneration decreased when the isolated microspores were treated by EMS. Many fertile microspore-derived embryos died after the applying of EMS. Embryo germination was also decreased with the treatment of EMS. But surprisingly, when 0.01% EMS treated each genotype for 5 h, the rate of embryo germination and plant regeneration all exceeded the control, among which the rate of embryo germination for genotype h58 reached 52.38%, which was twice over the control. Different mutagens had different effects. The application of NaN_3 had a promotive effect on embryogenesis and plant regeneration. When the isolated microspores were treated by 10-μM NaN_3 for 1 h, rate of plant regeneration of genotypes M9, h57, and h58 reached 11.11, 15.79, and 22.22%, respectively, which all exceeded over the controls. When 10-μM NaN_3 treated microspore-derived embryos of genotype h58 for 1 h, rate of plant regeneration reached 19.05%, which showed significant increase over the control. But when the concentration of NaN_3 reached 100 μM, no plant was regenerated in all four genotypes. So it is very important to use the appropriate concentration of NaN_3 in the *in vitro* mutagenesis.

UV irradiation, mutagenic agents EMS, and NaN_3 were also applied to cotyledons of two *B. napus* genotypes *in vitro* (He *et al.*, 2007b). Cotyledons were exposed in the UV radiation for different time intervals, and the results showed that suitable UV treatments strongly promoted the plant regeneration. When the cotyledons were exposed in the UV for 120 s and 60 s, the rate of callus induction and plant regeneration of the two cultivars reached the maximum respectively. When EMS was applied to cotyledons, the rate of callus induction and plant regeneration decreased with increasing of the mutagen concentrations and prolonging of the treatment time. When the treatment (1% EMS for 25 h) was applied to cotyledons, the rate of plant regeneration reduced to zero. NaN_3 significantly increased the callus induction with the increasing of the mutagen concentrations used. When cotyledons were treated 1000-μM NaN_3 for 5 h, the rate of callus induction of cv. N1 reached 89.29%, which was a significant increase over the control. However, 1000-μM NaN_3-treated cotyledons for 25 h completely impeded regeneration. Similarly, higher concentrations of mutagens (EMS and NaN_3) turned cotyledons yellow and some even deceased. With the further development of these protocols of *in vitro* mutagenesis on isolated microspores, microspore-derived embryos, and cotyledons of *B. napus*, some novel rapeseed germplasms or genotypes with unique agronomic and quality characters could be obtained, which are ultimately necessary for crop improvement and sustainable development.

VII. CONCLUSION

Anther and microspore culture techniques are mostly used for haploid production in Brassicas. These techniques have been widely utilized for breeding commercially important species. In the Brassicas, plant microspore culture and haploid breeding have several advantages such as DH breeding, mutation breeding, and gene transformation breeding, and play an important role in crop improvement and germplasm generation. With the development of microspore culture technology and its related breeding method, more and more international seed companies and breeding workers are using this technology.

Different factors are considered to be important for the successful haploid production, such as donor plant genotype, donor plant physiology, microspore developmental stage, culture conditions, culture environment, and pretreatments. Stress is also an essential component during embryogenesis induction in microspore culture, and such stresses including the medium refreshing, cold pretreatment, heat shocks, addition of colchicines, and AC to culture media are reported to promote the embryogenesis in Brassicas. Furthermore, proper culture osmoticum is also essential to the stimulation of microspore embryogenesis. This osmotic requirement for microspore induction and development could be met by sucrose, PEG or carrageenan oligosaccharides.

A significant enhancement of microspore embryogenesis by cold pretreatment of flower buds was reported in *B. napus* (Gu *et al.*, 2003a). Gland *et al.* (1988) reported that addition of AC into culture medium promoted the development of embryos to develop into whole plants in *B. napus*. Short-term exposure of isolated microspores to colchicine increases the number of symmetric cell divisions and the frequency of embryogenesis in *B. napus* (Zhang *et al.*, 2003).

Plant regeneration from microspores mostly occurs through direct embryogenesis. Successful embryogenesis after passing through all zygotic stages leads to the plant development. Factors like donor genotype, bud size, microspore stage, culture condition, different media, medium renovation, phytohormones, plant growth regulators, and chromosome-doubling agents such as colchicine treatment all affect plant regeneration greatly in *Brassica* species. Cold treatment not only enhanced microspore embryogenesis (Gu *et al.*, 2004a; Zhou *et al.*, 2002b) but also improved the germination and development of plants from embryos in *B. napus*. ABA treatment of embryos can also lift the frequency of plant development.

Appropriate stress conditions, such as cotyledon excision, chilling, partial desiccation, and successive subculture of microspore-derived embryos, are also considered to affect plant development in oilseed rape (*B. napus* L.). Cutting the cotyledons of the embryos at late cotyledonary stage can significantly increase the frequency of plantlet development. Subculturing over an extended period is helpful to induce germination and produce shoots, which originate from secondary initiated embryos from epidermal cells of hypocotyls.

Microspore culture is an effective technology for the production of DH parental lines required for producing F_1 hybrids of modern cultivars with desirable genetic recombinants, but the application of this technique for wide utilization as a routine breeding tool for these crosses still remains an obstacle because of the less efficient production of DH plants. Also, haploid plant regeneration directly and rapidly from the shoot apex of microspore-derived embryos is an important step in DH production. Direct and quick regeneration ensures minimal occurrence of cytogenetic abnormalities, which is extremely important in a breeding program (Fletcher *et al.*, 1998). The most widely used method of chromosome doubling is the use of the anti-microtubule agent colchicine. Colchicine treatment of isolated microspores increases the doubling efficiency of regenerated plants in *B. napus* obviously.

Compared to colchicine treatments of microspore-derived plants and microspore-derived embryos, immediate colchicine treatment of isolated microspores results in high embryogenesis and diploidization and low chimeric percentages in *B. napus* (Chen *et al.*, 1994; Zhou and Hagberg, 2000; Zhou *et al.*, 2002a). Large and healthy-looking embryos obtained from colchicine-treated microspores, after transfer to solid regeneration medium and an initial period of low temperature (2°C) for 10 days, germinate well at 24°C and can easily regenerate vigorous plants. Plant treatment for chromosome doubling significantly delays plant growth and development. This process also reduces seed set so that an additional growth cycle is necessary to produce sufficient seed for field testing. Therefore, immediate chromosome doubling by colchicine application of isolated microspores is safer, quicker, and more efficient.

An improved protocol for production of DHs through microspore culture of *B. rapa* subsp. *chinensis* has been established by Gu *et al.* (2003b). This protocol is mainly characterized by replacement of culture media and reduction of sucrose concentrations after the first 48 h of culture initiation, which results in a high frequency of microspore embryogenesis and good embryo quality.

Mutation-breeding techniques are widely used in plant breeding for producing useful mutants and variants. Microspore culture provides an ideal method for mutation because the mutated traits can be fixed in homozygous condition by chromosome doubling, which can enforce to obtain target mutation traits efficiently and accelerate the breeding process. In contrast, selection with diploid cells may allow recessive genes to remain undetected and to be carried unnoticed in the heterozygote. UV irradiation, mutagenic agents EMS, and NaN_3 were applied to isolated microspores, microspore-derived embryos, and cotyledons of rapeseed.

The method of microspore and embryo culture, selection of the embryos resistant to herbicide at the embryo stage after the mutations are used for herbicide-resistant breeding. The mutants with high oil contents and high oleic acid, low linolenic acid, and low erucic acid were obtained through mutations. Combining the rapeseed microspore induction and culturing together with pathogen selection, mutants with some rapeseed disease resistance were obtained in *B. napus*. The excellent agricultural characters can also be selected by using mutation techniques. Therefore, haploid microspore culture and DH technology have great potential for breeding crops due to their wide utilization and convenience, and at the same time they can be used as tools for more fundamental research in plant development and biotechnological processes.

ACKNOWLEDGMENTS

This work was supported by the Science and Technology Departments of Zhejiang Province (2004C22013, 2005C22004) and Hangzhou Municipality (2003132E32), the Zhejiang Natural Science Foundation of China (Y304162), the National Basic Research Program of China (2006CB101600), and the National Natural Science Foundation of China (30370852). I (W. J. Zhou, the corresponding author) am grateful to Per Hagberg (Svalöf Weibull AB, Sweden), and my other postdoc fellows and graduate students, past and present, for their contributions in this and related researches. The authors thank Dr. Raziuddin (a visiting professer in Zhejiang-University from NWFP Agricultural University, Peshawar-Pakistan) for critically reading the manuscript.

REFERENCES

Ahmad, I., Day, J. P., Macdonald, M. V. and Ingram, D. S. (1991). Haploid culture and UV mutagenesis in rapid-cycling *Brassica napus* for the generation of resistance to chlorsulfuron and *Alternaria brassicicola*. *Annals of Botany* **67**, 521–525.

Albercht, S. C., Mollers, C. and Robbelen, G. (1995). Selection *in vitro* for erucic-acid content in segregating populations of microspore-derived embryoids of *Brassica napus*. *Plant Breeding* **114**, 210–214.

Arnison, P. G. and Keller, W. A. (1990). A survey of the anther culture response of *Brassica oleracea* L. cultivars grown under field conditions. *Plant Breeding* **104**, 125–133.

Arnison, P. G., Donaldson, P., Ho, L. C. C. and Keller, W. A. (1990a). The influence of various physical parameters on anther culture of broccoli (*Brassica oleracea* var. *italica*). *Plant Cell Tissue and Organ Culture* **20**, 147–155.

Arnison, P. G., Donaldson, P., Jackson, A., Semple, C. and Keller, W. A. (1990b). Genotype specific response of cultured broccoli (*Brassica oleracea* var. *italica*) anthers to cytokinins. *Plant Cell Tissue and Organ Culture* **20**, 217–222.

Baillie, A. M. R., Epp, D. J., Hutcheson, D. and Keller, W. A. (1992). *In vitro* culture of isolated microspores and regeneration of plants in *Brassica campestris*. *Plant Cell Reports* **11**, 234–237.

Banga, S. S. and Labana, K. S. (1986). Spontaneous haploidy in *Brassica napus*. *Cruciferae Newsletter* **11**, 54–55.

Barro, F., Fernandez-Escobar, J., De la Vega, M. and Martin, A. (2003). Modification of glucosinolate and erucic acid contents in doubled haploid lines of *Brassica carinata* by UV treatment of isolated microspores. *Euphytica* **129**, 1–6.

Blakeslee, A. F., Belling, F., Farnham, M. E. and Bergner, A. D. (1922). A haploid mutant in the Jimson weed, *Datura stramonium*. *Science* **55**, 646–647.

Burnett, L., Yarrow, S. and Huang, B. (1992). Embryogenesis and plant regeneration from isolated microspores of *Brassica rapa* L. ssp. *oleifera*. *Plant Cell Reports* **11**, 215–218.

Cao, M. Q., Li, Y., Liu, F. and Dore, C. (1994). Embryogenesis and plant regeneration of pakchoi (*Brassica rapa* L. ssp. *chinensis*) via *in vitro* isolated microspore culture. *Plant Cell Reports* **13**, 447–450.

Charne, D. G., Pukacki, P., Kott, L. S. and Beversdorf, W. D. (1988). Embryogenesis following cryopreservation in isolated microspores of rapeseed, *Brassica napus* L. *Plant Cell Reports* **7**, 407–409.

Chatelet, P., Gindreau, K. and Herve, Y. (1999). Development and use of microspore culture applied to vegetable *Brassica oleracea* breeding. *In* "Anther and Pollen from Biology to Biotechnology" (C. Clement, E. Pacini and J. C. Audran, eds.), pp. 249–260. Springer-Verlag, Berlin, Heidelberg, NY.

Chen, J. L. and Beversdorf, W. D. (1992). Cryopreservation of isolated microspores of winter rapeseed (*Brassica napus* L.) for *in vitro* embryo production. *Plant Cell Tissue Organ Culture* **31**, 141–149.

Chen, Z. Z., Snyder, S., Fan, Z. G. and Loh, W. H. (1994). Efficient production of doubled haploid plants through chromosome doubling of isolated microspores in *Brassica napus*. *Plant Breeding* **113**, 217–221.

Chuong, P. V. and Beversdorf, W. D. (1985). High frequency embryogenesis through isolated microspore culture in *Brassica napus* L. and *B. carinata* Braun. *Plant Science* **39**, 219–226.

Chuong, P. V., Beversdorf, W. D. and Pauls, K. P. (1987). Plant regeneration from haploid stem peel protoplasts of *Brassica napus* L. *Journal of Plant Physiology* **130**, 57–65.

Chuong, P. V., Deslauriers, C., Ott, L. S. and Beversdorf, W. D. (1988). Effects of donor genotype and bud sampling on microspore culture of *Brassica napus*. *Canadian Journal of Botany* **66**, 1653–1657.

Dias, J. S. (1999). Effect of activated charcoal on *Brassica oleracea* microspore culture embryogenesis. *Euphytica* **108**, 65–69.

Dunwell, J. M. (1996). Microspore culture. *In* "*In Vitro* Haploid Production in Higher Plants" (S. M. Jain, S. K. Sopory and R. E. Veilleux, eds.), Vol. 1, pp. 205–216. Kluwer Academic Publishers, Dordrecht.

Dunwell, J. M., Cornish, L. M. and Decourcel, A. G. L. (1985). Influence of genotype, plant growth, temperature and anther incubation temperature on microspore embryo production in *Brassica napus* ssp. *Oleifera*. *Journal of Experimental Botany* **36**, 679–689.

FAO/IAEA (2003). MVD: Mutation varieties database http//www-mvd. iaea. org/ MVD/.

Fabijanski, S., Altosaar, I. and Arnison, P. G. (1987). Heat shock response of *Brassica oleracea* L. (broccoli). *Journal of Plant Physiology* **128**, 29–38.

Fabijanski, S., Altosaar, I. and Arnison, P. G. (1992). Heat shock response during anther culture of brossoli (*B. oleracea* L. (broccoli)). *Plant Cell Tissue Organ Culture* **26**, 203–212.

Fan, Z., Armstrong, K. C. and Keller, W. A. (1988). Development of microspores *in vivo* and *in vitro* in *Brassica napus* L. *Protoplasma* **147**, 191–199.

Ferrie, A. (1999). Combining microspores and mutagenesis. PBI Bulletin, pp. 17–19. National Research Council of Canada, 110 Gymnasium Place, Saskatoon, Saskatchewan, Canada.

Ferrie, A. M. R., Epp, D. J. and Keller, W. A. (1995). Evaluation of *Brassica rapa* L. genotypes for microspore culture response and identification of a highly embryogenic line. *Plant Cell Reports* **14**, 580–584.

Ferrie, A. M. R. and Keller, W. A. (1995). Development of methodology and applications of doubled haploids in *Brassica rapa*. *In* "GCIRC Proceedings of the Ninth International Rapeseed Congress," pp. 805–809. GCIRC, Cambridge, UK.

Fletcher, R., Coventry, J. and Kott, L. S. (1998). Doubled haploid technology for spring and winter *Brassica napus* (revised ed.). Technical Bulletin p. 42. OAC Publication, University of Guelph, Ontario, Canada.

Gamborg, O. L., Miller, R. A. and Ojima, K. (1968). Nutrient requirements of suspension cultures of soybean root cells. *Experimental Cell Research* **50**, 151–158.

Gland, A. (1981). Doubling chromosomes in interspecific hybrids by colchicine treatment. *Eucarpia Cruciferae Newsletter* **6**, 20–22.

Gland, A., Lichter, R. and Schweiger, H. G. (1988). Genetic and exogenous factors affecting embryogenesis in isolated microspore culture of *Brassica napus* L. *Journal of Plant Physiology* **132**, 613–617.

Gland-Zwerger, A. (1995). Culture conditions affecting induction and regeneration in isolated microspore cultures of different *Brassica* species. *In* "GCIRC Proceedings of the Ninth International Rapeseed Congress," pp. 799–801. GCIRC, Cambridge, UK.

Gu, H. H., Zhang, D. Q., Zhang, G. Q. and Zhou, W. J. (2003a). Advances on *in vitro* microspore technology for mutation breeding in rapeseed. *Acta Agriculturae Zhejiangenesis* **15**(5), 318–322.

Gu, H. H., Zhou, W. J. and Hagberg, P. (2003b). High frequency spontaneous production of doubled haploid plants from microspore culture in *Brassica rapa* ssp. *chinensis*. *Euphytica* **134**, 239–245.

Gu, H. H., Zhu, J. Q., Zhang, G. Q. and Zhou, W. J. (2003c). Microspore culture and ploidy identification of regenerated plant in Chinese flowering cabbage

(*Brassica rapa* ssp. *parachinensis*). *Journal of Agricultural Biotechnology* **11**, 572–576.

Gu, H. H., Hagberg, P. and Zhou, W. J. (2004a). Cold pretreatment enhances microspore embryogenesis in oilseed rape (*Brassica napus* L.). *Plant Growth Regulation* **42**, 137–143.

Gu, H. H., Tang, G. X., Zhang, G. Q. and Zhou, W. J. (2004b). Embryogenesis and plant regeneration from isolated microspores of winter cauliflower (*Brassica oleracea* var. *botrytis*). *Journal of Zhejiang University (Agricultural Life Science)* **30**, 34–38.

Gu, H. H., Zhang, D. Q. and Zhou, W. J. (2004c). Effects of medium renovation and colchicine treatment on embryogenesis of isolated microspores in *Brassica rapa* ssp. *chinensis*. *Acta Agronomica Sinica* **30**, 78–81.

Guha, S. and Maheshwari, S. C. (1964). *In vitro* production of embryos from anthers of Datura. *Nature (London)* **204**, 297.

Guha, S. and Maheshwari, S. C. (1966). Cell division and differentiation of embryos in the pollen grains of Datura *in vitro*. *Nature (London)* **212**, 97–98.

Guo, Y. D. and Pulli, S. (1996). High-frequency embryogenesis in *Brassica campestris* microspore culture. *Plant Cell Tissue Organ Culture* **46**, 219–225.

Govil, S., Babbar, S. B. and Gupta, S. C. (1986). Plant regeneration from *in vitro* cultured anthers of black mustard (*Brassica nigra* Koch). *Plant Breeding* **97**, 64–71.

Hansen, M. (2000). ABA treatment and desiccation of microspore-derived embryos of cabbage (*Brassica oleracea* ssp. *capitata* L.) improves plant development. *Journal of Plant Physiology* **156**, 164–167.

Hansen, M. and Svinnset, K. (1993). Microspore culture of Swede (*Brassica napus* ssp. *rapifera*) and the effects of fresh and conditioned media. *Plant Cell Reports* **12**, 496–500.

Hansen, N. J. P. and Andersen, S. B. (1996). *In vitro* chromosome doubling potential of colchicine, oryzalin, trifluralin and APM in *Brassica napus* microspore culture. *Euphytica* **88**, 159–164.

He, Y., Chen, S. B., Jin, Z. L., Wan, G. L., Xu, L., Tang, G. X. and Zhou, W. J. (2007a). Effects of mutagenic treatments of isolated microspores and microspore-derived embryos on embryogenesis and plant regeneration in oilseed rape (*Brassica napus* L.). *In* "GCIRC Proceedings of the 12th International Rapeseed Congress." Vol. II, pp. 69–72. GCIRC, Wuhan, China.

He, Y., Wan, G. L., Jin, Z. L., Xu, L., Tang, G. X. and Zhou, W. J. (2007b). Effects of mutagenic treatments of cotyledons on *in vitro* plant regeneration in oilseed rape (*Brassica napus* L.). *In* "GCIRC Proceedings of the 12th International Rapeseed Congress." Vol. II, pp. 54–57. GCIRC, Wuhan, China.

Hoffmann, F. (1978). Mutation and selection of haploid cell culture systems of rape and rye. *In* "Production of Natural Compounds by Cell Culture Methods" (A. W. Alfermann and E. Reinhord, eds.), pp. 319–329. Ges. und Strahlen und Umweltforsch, Munich.

Hoffmann, F., Thomas, E. and Wenzel, G. (1982). Anther culture as a breeding tool in rape. II. Progeny analysis of androgenic lines and induced mutants from haploid cultures. *Theoretical and Applied Genetics* **61**, 225–232.

Holbrook, L. A., Van Rooijen, G. J. H., Wilen, R. W. and Moloney, M. M. (1991). Oil-body protein in microspore-derived embryos of *Brassica napus*. *Plant Physiology* **97**, 1051–1058.

Huang, B. and Keller, W. A. (1989). Microspore culture technology. *Journal of Tissue Culture Methods* **12**, 171–178.

Ingram, D. S., Loh, C. S., MacDonald, M. V. and Newsholme, D. M. (1984). Secondary embryogenesis *in Brassica*: A tool for research and crop improvement. *Annual Proceeding of Phytochemistry Society of Europe* **23**, 219–242.

Ilić-Grubor, K., Attree, S. M. and Fowke, L. C. (1998). Induction of microspore-derived embryos of *Brassica napus* L. with polyethylene glycol (PEG) as osmoticum in a low sucrose medium. *Plant Cell Reports* **17**, 329–333.

Iqbal, M. C. M., Mollers, C. and Robbelen, G. (1994). Increased embryogenesis after colchicine treatment of microspore cultures of *Brassica napus* L. *Journal of Plant Physiology* **143**, 222–226.

Jain, R. K., Brune, N. and Friedt, W. (1989). Plant regeneration from *in vitro* cultures of cotyledon explants and anthers of *Sinapis alba* and its implications on breeding of crucifers. *Euphytica* **43**, 153–163.

Jain, S. M., Bagga, N., Bhalla-Sarin, N., Guha-Mukherjee, S. and Sopory, S. K. (1980). *In vitro* culture of anthers in *Petunia hybrida* and *Brassica oleracea*. *In* "Plant Tissue Culture, Genetic Manipulation and Somatic Hybridization of Plant Cells" (P. S. Rao, M. R. Heble and M. S. Chadha, eds.), pp. 85–92. Bhabha Atomic Research Centre, Bombay, India.

Johansson, L. (1983). Effects of activated charcoal in anther cultures. *Physiologia Plantarum* **59**, 397–403.

Johnson-Flanagan, A. M. and Singh, J. (1993). A method to study seed degreening using haploid embryos of *Brassica napus* cv. Topas. *Journal of Plant Physiology* **141**, 487–493.

Kameya, T. and Hinata, K. (1970). Induction of haploid plants from pollen grains of *Brassica*. *Japanese Journal of Breeding* **20**, 82–87.

Keller, W. A. and Armstrong, K. C. (1975). Stimulation of embryogenesis and haploid production in *Brassica campestris* anther cultures by elevated temperature treatments. *Theoretical and Applied Genetics* **55**, 65–67.

Keller, W. A. and Armstrong, K. C. (1978). High frequency production of microspore derived plants from *Brassica napus* anther cultures. *Zeitschriftfur Pflanzenzuchtung* **80**, 100–108.

Keller, W. A. and Armstrong, K. C. (1979). Stimulation and embryogenesis and haploid production in *Brassica campestris* anther cultures by elevated temperatue treatments. *Theoretical and Applied Genetics* **55**, 65–67.

Keller, W. A. and Armstrong, K. C. (1982). Production of haploids via anther culture in *Brassica oleracea* var. *italica*. *Euphytica* **32**, 151–159.

Klimaszewska, K. and Keller, W. A. (1983). The production of haploids production from *Brassica hirta* Moench (*Sinapis alba*) anther cultures. *Zeitschrift fuer Pflanzenphysiologie* **109**, 235–241.

Komatsu, Y. (1936). Investigations on the progeny of haploid plants and embryo sac formation by diploid and haploid plants of *Brassica napus*. *Proceeding of Crop Science Society Japan* **8**, 364–372.

Kott, L. S. (1996). Production of mutants using the rapeseed doubled haploid system. *In* "Induced Mutation and Molecular Techniques for Crop improvement. IAEA/FAO Proceedings of an International Symposium on the Use of Induced Mutations and Molecular Techniques for Crop Improvement," pp. 505–515. Vienne, Austria.

Kott, L. S. (1998). Application of doubled haploid technology in breeding of oilseed *Brassica napus*. *AgBiotech News and Information* **10**, 69–74.

Kott, L. S. and Beversdorf, W. D. (1990). Enhanced plant regeneration from microspore-derived embryos of *Brassica napus* by chilling, partial desiccation and age selection. *Plant Cell Tissue Organ Culture* **23**, 187–192.

Kott, L. S., Polsoni, L. and Beversdorf, W. D. (1988a). Cytological aspects of isolated microspore culture of *Brassica napus*. *Canadian Journal of Botany* **66**, 1658–1664.

Kott, L. S., Polsoni, L., Ellis, B. and Beversdorf, W. D. (1988b). Autotoxicity in isolated microspore cultures of *Brassica napus*. *Canadian Journal of Botany* **66**, 1665–1670.

Kuriyama, H. and Watanabe, Y. (1955). Studies on the haploid plant of *Brassica carinata*. *Journal of Breed* **5**, 1–5.

Kuginuki, Y., Nakamura, K., Hida, K. I. and Yosikawa, H. (1997). Varietal differences in embryogenic and regenerative ability in microspore culture of Chinese cabbage (*Brassica rapa* L. ssp. *pekinensis*). *Breeding Science* **47**, 341–346.

Leelavathi, S., Reddy, V. S. and Sen, S. K. (1984). Somatic cell genetics in *Brassica* species. I. High frequency production of haploid plants in *Brassica alba* (L.) H. F. + T. *Plant Cell Reports* **3**, 102–105.

Li, H. Z., Zhou, W. J., Zhang, Z. J., Gu, H. H., Takeuchi, Y. and Yoneyama, K. (2005). Effect of γ-radiation on development, yield and quality of microtubers *in vitro* in *Solanum tuberosum* L. *Biologia Plantarum* **49**, 625–628.

Li, L. C. and Hohlenbach, H. W. (1982). Somatic embryogenesis in quite a direct way in cultures of mesophyll protoplasts of *Brassica napus* L. *Plant Cell Reports* **1**, 209–211.

Lichter, R. (1981). Anther culture of *Brassica napus* in a liquid culture medium. *Zeitschrift fuer Pflanzenphysiologie* **103**, 229–237.

Lichter, R. (1982). Induction of haploid plants from isolated pollen of *Brassica napus*. *Zeitschrift fuer Pflanzenphysiologie* **103**, 229–237.

Lichter, R. (1989). Efficient yield of embryoids by culture of isolated microspores of different *Brassicaceae* species. *Plant Breeding* **103**, 119–123.

Lichter, R., De Groot, E., Fiebig, D., Schweiger, R. and Gland, A. (1988). Glucosinolates determined by HPLC in the seeds of microspore-derived homozygous lines rapeseed (*Brassica napus* L.). *Plant Breeding* **100**, 209–221.

Lionneton, E., Beuret, W., Delaitre, C., Ochatt, S. and Rancillac, M. (2001). Improved microspore culture and doubled-haploid plant regeneration in the brown condiment mustard (*Brassica juncea*). *Plant Cell Reports* **20**, 126–130.

Liu, C. H., Wang, Y. F., Lu, R. J., Sun, Y. F. and Huang, J. H. (2002). Estimation of plantlet ploidy level of *Brassica napus* by observing stoma guard cell circumferences. *Acta Agriculturae Shanghai* **18**(3), 35–38.

Liu, Y., Liu, H. Y. and Zheng, Z. Y. (1997). The application of microspore culture in screening disease resistance to *Sclerotinia sclerotiorum* in oilseed rape. Southwest China. *Journal of Agricultural Sciences* **10**(s), 108–112.

Loh, C. S. and Ingram, D. S. (1982). Production of haploid plants from anther cultures and secondary embryoids of winter oilseed rape, *Brassica napus* ssp. *oleifera*. *New Phytologist* **91**, 507–516.

Loh, C. S. and Lim, G. K. (1992). The influence of medium components on secondary embryogenesis of winter oilseed rape, *Brassica napus* L. ssp. Oleifera (Metzg.) Sinsk. *New phytologist* **121**, 425–430.

Loh, C. S., Ingram, D. S. and Hanke, D. E. (1983). Cytokinins and the regeneration of plantlets from secondary embryoids of winter oilseed rape, *Brassica napus* ssp. *oleifera*. *New Phytologist* **95**, 349–358.

Lott, C. S. and Haube, D. E. (1983). Cytokinins and the regeneration of plantlets from secondary embryoids of winter oilseed rape, *Brassica napus* ssp. *oleifera*. *New Phytologist* **95**, 349–350.

MacDonald, M. V., Newsholme, D. M. and Ingram, D. S. (1988). The biological effects of gamma irradiation on secondary embryoids of *Brassica napus* ssp. Oleifera Metzg. (Sinsk.) winter oil rape. *New phytologist* **110**, 255–259.

MacDonald, M. V., Hadwiger, M. A., Aslam, F. N. and Ingram, D. S. (1988a). The enhancement of anther culture efficiency in *Brassica napus* ssp. *oleifera* Metzg. (Sinsk) using low doses of gamma radiation. *New Phytologist* **110**, 101–107.

MacDonald, M. V., Newsholme, D. M. and Ingram, D. S. (1988b). The biological effects of gamma irradiation on secondary embryoids of *Brassica napus* ssp. *oleifera* Metzg. (Sinsk) winter oilseed rape. *New Phytologist* **110**, 255–259.

Maheshwari, S. C., Tyagi, A. K., Malhotra, K. and Sopory, S. K. (1980). Induction of haploidy from pollen grains in angiosperms- the current status. *Theoretical and Applied Genetics* **58**, 193–206.

Maluszynski, M, Hadwiger, B. S. and Sigurbjörnsson, B. (1995). Application of *in vitro* mutation techniques for crop improvement. *Euphytica* **85**, 303–315.

Mathias, R. (1998). An improved *in vitro* culture procedure for embryoids derived from isolated microspore of rape (*Brassica napus* L.). *Plant Breeding* **100**, 320–332.

Mathias, R. and Robbelen, G. (1991). Effective diploidization of microspore-derived haploids of rape (*Brassica napus* L.) by *in vitro* colchicine treatment. *Plant Breeding* **106**, 82–84.

Momoh, E. J. J., Zhou, W. J. and Kristiansson, B. (2002). Variation in the development of secondary dormancy in oilseed rape genotypes under conditions of stress. *Weed Research* **42**, 446–455.

Morinaga, T. and Fukushima, E. (1933). Karyological studies on a spontaneous haploid mutant of *Brassica napella*. *Cytologia* **4**, 457–460.

Möllers, C., Iqbal, M. C. M. and Robbelen, G. (1994). Efficient production of doubled haploid *Brassica napus* plants by colchicine treatment of microspores. *Euphytica* **75**, 95–104.

Murashige, T. and Skoog, F. (1962). A revised medium for rapid growth and bioassays with tobacco tissue cultures. *Physiologia Plantarum* **15**, 473–479.

Niu, Y. Z., Liu, Y. Z., Wang, L. Z., Yuan, Y. X., Li, S. C. and Fang, Q. J. (1999). A preliminary study on isolated microspore culture and plant regeneration of resynthesized *Brassica napus*. *Journal of Sichuan Agricultural University* **17**, 167–171.

Ockendon, D. J. (1985). Anther culture in Brussels sprouts (*Brassica oleracea* var. *gemmifera*). II. Effect of genotype on embryo yields. *Annals of Applied Biology* **107**, 101–104.

Osolnik, B., Bohanec, B. and Jelaska, S. (1993). Stimulation of androgenesis in white cabbage (*Brassica oleracea* var. *capita*) anthers by low temperature and anther dissection. *Plant Cell Tissue Organ Culture* **32**, 241–246.

Olsson, G. and Hagberg, A. (1955). Investigations on a haploid rape. *Hereditas* **41**, 227–237.

Palmer, C. E., Keller, W. A. and Arnison, P. G. (1996a). Experimental haploidy in *Brassica* species. *In* "*In Vitro* Haploid Production in Higher Plants" (S. M. Jain, S. K. Sopory and R. E. Veilleux, eds.), Vol. 3, pp. 143–172. Kluwer Academic Publishers, Dordrecht, The Netherlands.

Palmer, C. E., Keller, W. A. and Arnison, P. G. (1996b). Utilization of *Brassica* haploids. *In* "*In Vitro* Haploid Production in Higher Plants" (S. M. Jain, S. K. Sopory and R. E. Veilleux, eds.), Vol. 3, pp. 173–192. Kluwer Academic Publishers, Dordrecht, The Netherlands.

Pechan, P. M. and Keller, W. A. (1989). Induction of microspore embryogenesis in *Brassica napus* L. by gamma irradiation and ethanol stress. *In Vitro Cellular and Developmental Biology* **25**, 1073–1074.

Pechan, P. M. and Smykal, P. (2001). Androgenesis Affecting the fate of the male gametophyte. *Physiologia Plantarum* **111**, 1–8.

Phippen, C. and Ockendon, D. J. (1990). Genotype, plant, bud size and media factors affecting anther culture of cauliflower (*Brassica oleracea* var. *botrytis*). *Theoretical and Applied Genetics* **79**, 33–38.

Polsoni, L., Kott, L. S. and Beversdorf, W. D. (1988). Large-scale microspore culture technique for mutation-selection studies in *Brassica napus*. *Canadian Journal of Botony* **66**, 1681–1685.

Pomeroy, M. K., Kramer, J. K. G., Hunt, D. J. and Keller, W. A. (1991). Fatty acid changes during development of zygotic and microspore-derived embryos of *Brassica napus*. *Physiologia Plantarum* **81**, 447–454.

Prabhudesai, V. and Bhaskaran, S. (1991). A continuous culture system of direct somatic embryogenesis in microspore-derived embryos of *Brassica juncea*. *Plant Cell Reports* **12**, 289–292.

Pu, H. M. (2003). Review of genetically modified rapeseed with herbicide-resistance and ecologic safety. *Chinese Journal of Oil Crop Sciences* **25**, 89–93.

Raina, S. K. and Irfan, S. T. (1998). High-frequency embryogenesis and plantlet regeneration from isolated microspores of indica rice. *Plant Cell Reports* **17**, 957–962.

Ramanujam, S. (1941). A haploid plant in Toria (*Brassica campestris*). *Proceeding of Indian Academy Science Bulletin* **14**, 25–34.

Roberts-Oehlschlager, S. L. and Dunwell, J. D. (1990). Barley anther culture Pre-treatment on mannitol stimulates production of microspore-derived embryos. *Plant Cell Tissue Organ Culture* **20**, 235–240.

Roulund, N., Hansted, L., Anderson, S. B. and Farestveit, B. (1990). Effect of genotype, environment an carbohydrate on anther culture response in head cabbage (*Brassica oleracea* L. convar. *capitata* Alef). *Euphytica* **49**, 237–242.

Sato, S., Katoh, N., Iwai, S. and Hagimori, M. (2002). Effect of low temperature pretreatment of buds or inflorescence on isolated microspore culture in *Brassica rapa* (syn. *B. campestris*). *Breed Science* **52**, 23–26.

Sato, T., Nishio, T. and Hirai, M. (1989). Plant regeneration from isolated microspore cultures of Chinese cabbage (*Brassica campestris* spp. *pekinensis*). *Plant Cell Reports* **8**, 486–488.

Scarth, R., Sequin-Swarez, G. and Rakow, G. F. W. (1991). Application of doubled haploidy to *Brassica napus* breeding. *In* "GCIRC, July 9–11, 1991" (D. A. McGregor, ed.), pp. 1449–1453. Saskatoon, Saskatchewan.

Seguin-Swartz, G., Hutcheson, D. S. and Downey, R. K. (1983). Anther culture in *Brassica campestris*. *In* "Proceedings of the Sixth International Rapeseed Congress," pp. 246–251. Paris, France.

Sharma, K. K. and Bhojwani, S. S. (1985). Microspore embryogenesis in anther cultures of two Indian cultivars of *Brassica juncea* (L.) Czern. *Plant Cell Tissue Organ Culture* **4**, 235–239.

Shi, S. W., Wu, J. S. and Liu, H. L. (1995). *In vitro* selection of long-pod and dwarf mutants in *Brassica napus* L. *Acta Agriculturae Nucleatae Sinica* **9**(4), 252–253.

Shu, W. and Loh, C. S. (1987). Secondary embryogenesis in long term cultures of winter oilseed rape, *Brassica napus* ssp. oleifera. *New phytologist* **107**, 39–46.

Sopory, S. K. and Munshi, M. (1996). Anther culture. *In* "*In Vitro* Haploid Production in Higher Plants" (S. M. Jain, S. K. Sopory and R. E. Veilleux, eds.), Vol. 1, pp. 145–176. Kluwer Academic Publishers, Dordrecht, The Netherlands.

Stringam, G. R. and Downey, R. K. (1973). Haploid frequencies in *Brassica napus*. *Canadian Journal of Plant Science* **53**, 229–231.

Stringam, G. R. and Thiagarajah, M. R. (1991). Effectiveness of selection for early flowering in F-2 populations of *Brassica napus* L. A comparison of doubled haploid and single seed descent methods. *In* "GCIRC, July 9–11," (D. I. McGregor, ed.), pp. 70–75. Saskatoon, Saskatchewan.

Sunderland, N. (1971). Anther culture a progress report. *Science Progress Oxford* **59**, 527–549.

Swanson, E. B. (1990). Microspore culture in Brassica. *In* "Methods in Molecular Biology Vol. 6, Plant Cell and Tissue Culture" (J. W. Pollard and J. M. Walker, eds.), pp. 159–169. The Humana Press, Clifton, NJ.

Swanson, E. B., Coumans, M. P., Wu, S. C., Barsby, T. and Beversdorf, W. D. (1987). Efficient isolation of microspores and the production of microspore-derived embryos from *Brassica napus*. *Plant Cell Reports* **6**, 94–97.

Swanson, E. B., Coumans, M. P. and Brown, J. D. (1988). The characterization of herbicide tolerant plants in *Brassica napus* L. after *in vitro* selection of microspores and protoplasts. *Plant Cell Reports* **7**, 83–87.

Swanson, E. B., Herrgesell, M. J., Arnoldo, M., Sippell, D. and Wong, R. S. C. (1989). Microspore mutagenesis and selection Canola plants with field tolerance to the imidazolinones. *Theoretical and Applied Genetics* **78**, 525–530.

Takahata, Y., Brown, D. C. W. and Keller, W. A. (1991). Effect of donor plant age and inflorescence age on microspore culture of *Brassica napus* L. *Euphytica* **58**, 51–55.

Tang, G. X., Zhou, W. J., Li, H. Z., Mao, B. Z., He, Z. H. and Yoneyama, K. (2003). Medium, explant and genotype factors influencing shoot regeneration in oilseed *Brassica* spp. *Journal of Agronomy and Crop Science* **189**, 351–358.

Taylor, D. C., Weber, N., Underhill, E. W., Pomeroy, M. K., Keller, W. A., Scowcroft, R. W., Wilen, R. W., Moloney, M. M. and Holbrook, L. A. (1990). Storage protein regulation and lipid accumulation in microspore embryos of *Brassica napus* L. *Planta* **181**, 18–26.

Telmer, C. A., Newcomb, W. and Simmonds, D. H. (1993). Microspore development in *Brassica napus* and the effect of high temperature on division *in vivo* and *in vitro*. *Protoplasma* **172**, 154–165.

Thomas, E., Hoffmann, F., Potrykus, I. and Wenzel, G. (1976). Protoplast regeneration and stem embryogenesis of haploid androgenetic rape. *Molecular Genomes and Genetics* **145**, 245–247.

Thompson, K. F. (1956). Production of haploid plants of marrow-stem kale. *Nature (London)* **178**, 748.

Thompson, K. F. (1969). Frequencies of haploids in spring oilseed rape (*Brassica napus*). *Hereditas* **24**, 318–319.

Thompson, K. F. (1974). Homozygous diploid lines from naturally occurring haploids. In 4th Interul. Rapskongr Giessen pp. 119–124.

Thurling, N. and Chay, P. M. (1984). The influence of donor plant genotype and environment on production of multicellular microspores in cultured anthers of *Brassica napus* spp. *oleifera*. *Annals of Botany* **54**, 681–695.

Touraev, A., Vicente, O. and Heberle-Bors, E. (1997). Initiation of embryogenesis by stress. *Trends in Plant Science* **2**, 297–302.

Wang, H. Z., Liu, G. H., Zheng, Y. B., Wang, X. F. and Yang, Q. (2002). Breeding of *Brassica napus* cultivar Zhongshuang No. 9 with resistance to *Sclerotinia sclerotiorum*. *Chinese Journal of Oil Crop Sciences* **24**(4), 71–73.

Wang, M., Farnham, M. W. and Nannes, J. S. P. (1999). Ploidy of broccoli regenerated from microspore culture versus anther culture. *Plant Breeding* **118**, 249–252.

Wenzel, G. (1980). Recent progress in microspore culture of crop plants. *In* "The Plant Genome" (D. R. Davies and D. A. Opwood, eds.), pp. 185–213. The John Innes Charity, Norwich.

Wong, R. S. C., Zee, S. Y. and Swanson, E. B. (1996). Isolated microspore culture of Chinese flowering cabbage (*Brassica campestris* ssp. *para chinensis*). *Plant Cell Reports* **15**, 396–400.

Wu, J. S., Shi, S. W., Zhou, Y. M. and Liu, H. L. (1999). Breeding and studies of Hua Double 3 (*Brassica napus* L). *Journal of Huazhong Agricultural University* **18**(1), 1–4.

Xu, L., Zhang, G. Q., Gu, H. H. and Zhou, W. J. (2005). *In vitro* selection of herbicide resistance in microspore-derived embryos of oilseed rape (*Brassica napus* L.). *Journal of Agricultural Biotechnology* **13**, 411–415.

Zaki, M. and Dickinson, H. (1991). Microspore-derived embryos in *Brassica* the significance of division symmetry in pollen mitosis I to embryogenic development. *Sexual Plant Reproduction* **4**, 48–55.

Zhang, F. L. and Takahata, Y. (1999). Microspore mutagenesis and *in vitro* selection for resistance to soft rot disease in Chinese cabbage (*Brassica campestris* L. ssp. *pekinensis*). *Breeding Science* **49**, 161–166.

Zhang, F. L., Zhao, X. Y., Zhang, D. S., Liu, F. and Xu, J. B. (2001). High frequency production of doubled haploid plants of Chinese cabbage derived from microspore embryogenesis without colchicine treatment. *Cruciferae Newsletter* **23**, 31–32.

Zhang, G. Q., Zhou, W. J., Gu, H. H., Song, W. J. and Momoh, E. J. J. (2003). Plant regeneration from the hybridization of *Brassica juncea* and *B. napus* through embryo culture. *Journal of Agronomy and Crop Science* **189**, 347–350.

Zhang, G. Q., Tang, G. X., Song, W. J. and Zhou, W. J. (2004). Resynthesizing *Brassica napus* from interspecific hybridization between *Brassica rapa* and *B. oleracea* through ovary culture. *Euphytica* **140**, 181–187.

Zhang, Z. J., Zhou, W. J. and Li, H. Z. (2005). The role of GA, IAA and BAP in the regulation of *in vitro* shoot growth and microtuberization in potato. *Acta Physiologia Plantarum* **27**, 363–369.

Zhang, G. Q., Zhang, D. Q., Tang, G. X., He, Y. and Zhou, W. J. (2006). Plant development from microspore-derived embryos in oilseed rape as affected by chilling, desiccation and cotyledon excision. *Biologia Plantarum* **50**, 180–186.

Zhao, J. P., Simmonds, D. H. and Newcomb, W. (1996). High frequency production of doubled haploid plants of *Brassica napus* cv. Topas derived from colchicine-induced microspore embryogenesis without heat shock. *Plant Cell Reports* **15**, 668–671.

Zhou, W. J. and Hagberg, P. (2000). High frequency production of doubled haploid rapeseed plants by direct colchicine treatment of isolated microspores. *Journal of Zhejiang University (Agricultural and Life Science)* **26**, 125–126.

Zhou, W. J., Hagberg, P. and Tang, G. X. (2002a). Increasing embryogenesis and doubling efficiency by immediate colchicine treatment of isolated microspores in spring *Brassica napus*. *Euphytica* **128**, 27–34.

Zhou, W. J., Tang, G. X. and Hagberg, P. (2002b). Efficient production of doubled haploid plants by immediate colchicine treatment of isolated microspores in winter *Brassica napus*. *Plant Growth Regulation* **37**, 185–192.

Zhou, W. J., Mao, B. Z., Tang, GX. and Hagberg, P. (2002c). Studies on the determination of ploidy level of microspore-derived plants in *Brassica napus*. *Scientia Agricultura Sinica* **35**, 724–727.

Breeding for Apetalous Rape: Inheritance and Yield Physiology

LIXI JIANG

Institute of Crop Science, College of Agriculture and Biotechnology, Zhejiang University, 268 Kaixuan Road, Hangzhou 310029, China

ABSTRACT

Apetalous genotypes might be of interest in the breeding programs of Brassicas because of their more efficient photosynthesis and a lower probability of infection from the pathogens distributed by petals. In this chapter, the studies on apetalous mutants in rapeseed have been reviewed with special reference to the inheritance of the apetalous flower, the yield physiology, and the role of petals in the distribution of diseases.

Advances in Botanical Research, Vol. 45
Incorporating Advances in Plant Pathology

0065-2296/07 $35.00
DOI: 10.1016/S0065-2296(07)45008-X

I. INTRODUCTION

Rapeseed oil is a kind of desirable edible oils in view of human health (Downey and Rakow, 1987; Röbbelen, 1997), and it could substitute a part of fuel when petrolic oil becomes shortage (Peterson *et al.*, 1983; Withers and Noordam, 1996). The potential demand for rapeseed oil is, therefore, tremendous. The total production and the acreage of rapeseed increased continuously and remarkably in the recent years, in comparison to cottonseed and soybean (FAO production yearbook, 2000–2005).

A considerable proportion of the increment of the world production of rapeseed has been contributed by developing countries, in particular China and India. On the way toward industrialization, developing countries are confronted with increasing population and the corresponding demand for agricultural products, which is limited by the decline of arable land and enlarging recognition of the burden imposed on natural ecosystem aroused by intensive agricultural land uses. In this background, genetic improvement of crops is an effective way to increase production without additional external inputs by a more efficient transformation of solar energy and internal regulation of metabolic functions.

One of the effective approaches is to seek for the morphological ideotype, which appears with erect leaf posture, splitting leaf shape, and/or waxed leaf surface, and so on (Fray *et al.*, 1996; Virk *et al.*, 2004). In *Brassica* oilseeds, apart from the above features, apetalous flower is a trait of particular interest, as a *Brassica* plant has a mass of brightly colored flowers at the top layer for an extended period. Apetalous genotypes are thought to have advantages such as more efficient photosynthesis and less severe disease infection spread by petals (Jiang and Becker, 2003; Jiang *et al.*, 2006; Zhang and Fu, 2002; Zhang *et al.*, 2005).

II. ORIGIN OF APETALOUS RAPESEED

The morphological characters of major oilseed crops are documented in many literatures. As to the floral organs, the inflorescence of rapeseed looks like a head with the oldest buds at periphery and the youngest ones in the center. Normal flowers of Brassicas are yellow, widely spread when open. They are regular and cruciform, bisexual, complete, and hypogenous. The calyx has four unattached sepals in two whorls. The corolla consists of four free petals in one whorl, each with distinct limb and claw. Among the floral organs, petals take the greatest proportion of weight. Therefore, considerable photosynthesized assimilates are consumed for petal formation and respiration. Apetalous trait is rather a kind of floral abnormality, which was firstly reported in a naturally occurred mutant of turnip rape (*Brassica rapa*), where the apetalous character was regulated by single recessive gene subjecting to the influence of modifiers (Ramanujum, 1940).

Some years later, apetalous mutant was discovered in toria brown sarson (*B. campestris* var. brown sarson) (Singh, 1961b) and in *B. campestris* (Cours and Williams, 1977). According to their description, the segregation pattern in the F_2 progeny of a cross between the apetalous mutant and normal type exhibited a 3:1 (normal to apetalous) ratio, suggesting that the apetalous character in *B. campestris* was controlled by one recessive gene. Apart from the segregation ratio, it was observed that the apetalous condition was associated with a reduction in the number of stamens, yet the pod and seed were normal.

Liu (1985) produced apetalous plants by crossing *B. napus* with *B. campestris* (*B. napus* as mother), and had developed an apetalous line during 1976–1980 at Huazhong Agricultural University in Wuhan, China. He observed that rudimentary petals were more frequently on lower part of an inflorescence than on upper part in his apetalous plants. He crossed the apetalous line with a normally petaled variety. The F_1 plants had normal petals, yet the petalous degree (PDgr) segregated very complicatedly in F_2 population. No genetic hypothesis was given based on his observation (Liu, 1985).

Chen (1989) had bred an apetalous line, while he was studying the resynthesized *B. napus*. His original efforts stressed on breeding for yellow-seeded genetic lines, which had thinner seed coat associated with higher oil and protein, and lower crude fiber contents in comparison to black or brown seeds (Shirzadegan and Röbbelen, 1985). One of his resynthesized partially yellow-seeded plants, No. 7477, was crossed with another resynthesized *B. napus* line, No. 7406, which composed of genomes from yellow-seeded *B. campestris* var. yellow sarson K-151 and black-seeded *B. alboglabra*, No. 4003. Plants with apetalous flowers were found in the F_2 population. These plants, according to his description, were completely apetalous at early flowering stage but had one or two petals in some flowers at later stages. The genetic regulation of the apetalous character of his material was not studied (Chen, 1989).

Rana (1985) studied the inheritance of an apetalous mutant of Ethiopian mustard (*B. carinata* A. Braun) at University of California, Davis, USA. He made reciprocal crosses between apetalous and normal parents. The F_1 plants were found all similar to normal parent. His further studies indicated that two major genes governed the degree of petally. His explanation to the presence of variable numbers of partially petaled plants in segregating populations was the function of an unidentified number of modifying genes.

He did not find true apetalous type, but partially petaled type and suggested that the partially petalled types be possibly heterozygous but perhaps also be very sensitive to the internal and external environment. He hypothesized that the ancestral species of *B. carinata, B. nigra* and *B. oleracea* each contributed a gene that governed petally (Rana, 1985).

In canola, apetalous mutant was first reported by the Australian breeder Buzza (1983). The flowers of the Australian apetalous line were normal except for absence of petals. To determine the inheritance of the apetalous character, Buzza crossed his apetalous line reciprocally to a normal line "RU6" and observed the F_1 and F_2 generations. He concluded that the apetalous flowers were controlled by two recessive genes. A normal plant had four alleles for petals (*PPPP*) and the apetalous plant had four alleles for absence of petals (*pppp*). PDgr of the intermediate type depends on the additive effect of *P* allele for normal petal. However, the apetalous variant in *B. napus* whose apetalous trait was regulated by four recessive genes was also reported (Lü and Fu, 1990). The apetalous line, "APL 0256," was crossed with seven normal petaled genotypes, which had different petal sizes. It was revealed that segregations of apetalous trait in the F_2 and BC_1 were different for different crosses, ranging from 255: 1 to 16:1 in the F_2, or 16:1 to 4:1 in the BC_1, for normal petally to apetalous flowers. The deduction was that the petals of small size, for example in the genotype as "MI-CMS," were coded by $P_1P_1P_2P_2p_3p_3p_4p_4$. In such genotype, normal flowers were controlled by two pairs of dominant genes. The other two pairs of recessive genes remained homozygous in all other generations. However, the bigger petals, for example in some Chinese materials as "Rongshan" or "Zaxuan-841," were coded by $P_1P_1P_2P_2P_3P_3P_4P_4$. In such genotype, normal flowers were controlled by four pairs of dominant genes. Therefore, when apetalous genotype ($p_1p_1p_2p_2p_3p_3p_4p_4$) was crossed with the genotypes with small petals ($P_1P_1P_2P_2p_3p_3p_4p_4$), the segregation of plants with normal flowers to plants with apetalous flowers in F_2 and BC_1 was 15:1 and 3:1, whereas when it was crossed with genotypes with big petals ($P_1P_1P_2P_2P_3P_3P_4P_4$), the segregation in F_2 or BC_1 turned out to be 255:1 or 15:1, respectively.

Some other types of distinct apetalous variants (*B. napus*) were also identified (Kelly *et al.*, 1995). RFLP analysis was applied to map the genes controlling flower morphology. One of the apetalous phenotypes was determined by an epistatic interaction between recessive alleles at a pair of homoeologous loci and another was controlled by the interaction among the alleles at three loci. The apetalous line "N-o-11" with shriveled leaves demonstrated the cosegregation of genes controlling both shriveled leaves and apetalous morphologies. Two *STAP* loci controlling the production of flowers with stamenoid petals were mapped to homologous loci in the genome of *B. napus*

(Fray *et al.*, 1997). The *STAP* loci were thought to be duplicated genes because they exhibited an epistatic interaction such that only plants homozygous for recessive *stap* alleles at both loci expressed the variant phenotype. The *CURLY LEAF* (*CLF*) gene of *Arabidopsis thaliana* pleiotropically influences both flower and leaf morphologies. The cloned *CLF* gene of *A. thaliana* was homologous to a polymorphic *B. napus* coincident with one of the *B. napus STAP* loci. The possibility that *CLF* was a candidate gene for *STAP* suggested that the variant *stap* alleles of *B. napus* exert pleiotropic effects over both flower and leaf morphologies.

The apetalous character was also possibly governed by the interaction of cytoplasmic genes and nuclear genes in the case of "*ap-Tengbe*" mutant (Jiang and Becker, 2003). The plant with normal flowers had normal cytoplasm (C_N) and four nuclear alleles coding normal petals ($P_1P_1P_2P_2$), whereas the apetalous "*ap-Tangbe*" mutant had apetalous cytoplasm (C_{ap}) and four alleles coding the absence of petals ($p_1p_1p_2p_2$). Alleles for petals were dominant to those for apetalous flowers. The genetic interpretations for apetalous, partial apetalous, and normal petaled phenotypes have been summarized in Table I. The F_1 plants ("Falcon" × "*ap-Tengbe*"), which had C_N and the nuclear genes $P_1p_1P_2p_2$, showed normal petals. The RF_1 plants ("*ap-Tengbe*" × "Falcon"), which had C_{ap}, still showed normal petals due to dominant effect of both P_1 and P_2 alleles. In BC_{1-1} or F_2 plants with C_N, which influenced the expression of apetalous character from p_1 or p_2, only the genotypes with $p_1p_1p_2p_2$ were able to express a certain degree of apetalous character and the genotypes with three or less recessive alleles of p_1 and p_2 had normal petals. Segregations in BC_{1-1} and F_2 populations were 0:1:3 and 0:1:15, respectively for apetalous, intermediate, and normal petaled types. However, the RBC_{1-1} or RF_2 plants with C_{ap} cytoplasm showed a

TABLE I

Genetic Interpretations for Apetalous, Partial Apetalous, and Normal Petaled Phenotypes in the Materials with "Ap-Tengbe" Background

Type of cytoplasm	Nuclear genotype	Phenotype
C_N	$P_1\_\,\_\,\_$	Normal petaled
	$\_\,\_\,P_2\_$	Normal petaled
	$p_1p_1p_2p_2$	Partial apetalous
C_{ap}	$P_1\_\,P_2\_$	Normal petaled
	$P_1p_1\,p_2p_2$	Partial apetalous
	$p_1p_1P_2p_2$	Partial apetalous
	$p_1p_1p_2p_2$	Apetalous
	$P_1P_1p_2p_2$	Normal or partial petaled
	$p_1p_1P_2P_2$	Normal or partial petaled

segregation of 1:2:1 or 1:6:9 for apetalous, intermediate, and normal petaled types, respectively. Because the BC_{1-2} and the RBC_{1-2} plants had either C_N $(P_1\_\_\_)$ or C_N $(\_\_P_2\_)$, they exhibited normal petaled phenotype. No true apetalous type was observed in these two populations (Jiang and Becker, 2003).

More recently, a spontaneous mutant with apetalous flowers was obtained from a hybrid progeny in *B. napus* by Zhao and Wang (2004). The result of genetic analysis showed that the apetalous character was controlled only by one gene locus, normal flower exhibited incomplete dominance over apetalous flower and its expression was not affected by cytoplasmic factors.

The origin and the inheriting mode of different apetalous mutant in *Brassica* oilseeds reported in literature so far have been summarized in Table II. Taken together, the apetalous mutants either spontaneous occurred in nature, or were artificially created by interspecific/inner-specific crossing, or chemical mutagenesis. They were found in the major species of *Brassica* oilseeds, namely *B. napus*, *B. rapa*, *B. carinata*, and *B. juncea*. The inheriting mode of the apetalous character varies among different origins. It was possibly controlled by a single locus, two loci, three loci, or even four loci. The loci were independent as described in most reports, but epistatic interaction was reported too. Cytoplasmic effect on the inheritance of apetalous character was also reported. In some cases, the apetalous character was linked with male sterility at certain degree.

III. EFFECT OF ENVIRONMENT ON APETALOUS DEGREE IN RAPE AND EFFECT OF HORMONES ON DEVELOPMENT OF FLORAL ORGANS

The apetalous character in rapeseed was found unstable or easily affected by environmental conditions. More complete apetalous degree was observed at later flowering stage than at initial anthesis in different apetalous materials (Liu, 1985; Tengbe, 1990). In contrast to this phenomenon, some researchers observed a more complete apetalous degree in earlier flowering stage rather than in later stage (Chen, 1989; Rana, 1985). By scanning electron and light microscopy on developing buds of petalous and apetalous flowers, it was found that the sequence of floral development was sepals, androecium and gynoecium simultaneously, and finally the corolla. The development of apetalous and petalous types was similar until corolla initiation, and the development of petals appeared to be in delicate balance with the internal and external environment (Rana, 1985).

Balance of endogenous plant hormones may be important in the process of floral organ initiation and differentiation. Such hormones may likely be

TABLE II

Different Origins of Apetalous Mutant and the Respective Inheriting Modes in Brassica Oilseed Rape

Species	Origin	Inheriting mode	Researcher year reported
B. rapa	Naturally occurred	One recessive gene	Ramanujum, 1940
B. rapa	Naturally occurred	One recessive gene	Singh, 1961b
B. napus	Interspecies crossing	Unknown	Liu, 1985
B. rapa	Naturally occurred	One recessive gene	Cours and Williams, 1977
B. napus	Naturally occurred	Two recessive genes	Buzza, 1983
B. carinata	Unknown	Two major recessive genes + some modifying genes	Rana, 1985
B. napus	Interspecies crossing	Unknown	Chen, 1989
B. napus	Chemical mutagenesis	Cytoplasm + two recessive genes	Tengbe, 1990
B. napus	Interspecies crossing	Four recessive genes	Lü and Fu, 1990
B. napus	–	Varying from two to four recessive genes depends different line	Kelly *et al.*, 1995
B. napus	Interspecies crossing	Unknown	Pierre *et al.*, 1996
B. napus	Chemical mutagenesis	Two recessive genes	Fray *et al.*, 1997
B. napus	Innerspecies crossing	Two recessive genes	Rakow, 1999
B. juncea	Interspecies crossing	Cytoplasmic genes + nuclear genes (number of loci known)	Malik *et al.*, 1999
B. napus	Naturally occurred	One recessive gene	Zhao and Wang, 2004
B. napus	Innerspecies crossing	Apetalous character linked with male sterility	Zhang *et al.*, 2005

cytokinins and auxins (Davies, 1987; Metzger, 1987). The involvement of cytokinins in floral organ development was also studied and described in *Dianthus chinensis* L. (Zhou *et al.*, 1995), tomato (*Lycopersicon esculentum*) (Kaur *et al.*, 1996), *Hyacinthus orientalis* (Lu *et al.*, 1994), and *Chenopodium rubrum* (Vondrakova *et al.*, 1998). Auxins' significant activities in differentiation and the formation of floral meristem were discovered in the early growth stage in *Arabidopsis* (Oka *et al.*, 1999), *Bougainvillea glabra* L. (Awad *et al.*, 1987), and *Tanacetum cinerariaefolium* L. (Brown and Menary, 1994). Some other hormones or substances may also have influence on floral organ formation. These are lipids, gibberellins, and phosphorus. Venglat and Sawhney (1996) used *Arabidopsis* as a model crop to investigate the regulatory genes that control and coordinate the determination, differentiation, and morphogenesis of the floral meristem and floral organs. It was revealed that application of BAP, a cytokinin, to inflorescences at certain developmental stages resulted in: (1) increase in floral organ number, (2) formation of abnormal floral organs, and (3) induction of secondary floral buds in the axils of sepals. It was suggested that exogenous BAP suppresses the normal function of the genes for floral meristem identity and thereby affects flower development and the later stages of floral organ differentiation.

Application of exogenous hormones could disturb endogenous balance of plant hormones and impose influence on the process of floral organs development. Jiang and Becker (2001b) investigated the effects of flower developmental stage and the application of different phytohormones on the expression of apetalous character. It was observed that PDgr decreased from initial anthesis to end of flowering both in the partial apetalous and the normal petaled plants originating from the cross between "Falcon" and "ap-Tengbe"; and that the supply of auxins or cytokinins resulted in reduction of PDgr in the partial or the normal petaled DH plants. The results suggest that the unstable expression of PDgr was caused by endogenetic accumulation of auxins or/and cytokinins at later flowering stages.

IV. ADVANTAGES OF APETALOUS RAPE IN YIELD

Apetalous flower is an ideal morphological trait. The thick and brightly colored flowers at the top layer of the normal type prevent *Brassica* oilseeds from the efficient utilization of solar energy. Flowers are present over a period of ~4–6 weeks in typical cultivars of *Brassica* oilseeds. Flowering canopies reflect substantially more radiation and absorb less than vegetative canopies between photosynthesizing active radiation (PAR) of 500 and 700 nm but reflect less and absorb slightly more between PAR of 400 and 500 nm (Yates and Steven, 1987). Most leaves senesced soon after flowering commenced and green pods and stems became the major sources of photosynthesis. Breeding for genotypes with reduced petal size or without petals is of interest for a more efficient light distribution within the canopy for *Brassica* oilseeds.

Jiang and Becker (2001a) compared apetalous (AP), half apetalous (HAP), and normal petaled (NP) near isogenic lines with the parent "Falcon," a German cultivar of winter rapeseed, for leaves area index (LAI) under different combinations of nitrogen application and plant density. They found that petals had no negative effect until peak flowering stage, when, in general, the LAI of AP was about 28% higher than that of HAP, and the difference between HAP and NP was significant also. "Falcon" had the highest LAI before anthesis, but AP surpassed "Falcon" significantly at peak flowering by about 11.6%. Three weeks after the end of flowering, HAP surpassed Falcon for LAI too. AP and HAP had about 37.9% and 27.6% more remaining active green leaves than NP in average. However, petal's effect on LAI depends on nitrogen application and plant density. At low nitrogen level, the petals had no clear negative effect at peak flowering. However, with 200-kg N/ha, plus particularly low density (30 plants/m^2), AP obviously had a greater LAI than HAP and NP, and even greater than "Falcon." Two weeks after flowering, under high nitrogen application (200-kg N/ha), LAI of AP responded positively to increasing population density and reached the maximum with 120 plants/m^2. LAI of NP reached almost the maximum with 80 plants/m^2, and did not increase significantly when population density further increased.

Petal's negative effect on total biomass is also obvious (Jiang and Becker, 2001a). AP had a higher biomass than HAP, while HAP had a higher biomass than NP in most combinations of plant density and nitrogen application, in particular, with very high nitrogen application plus very high population density, where AP had ~35% more total biomass than HAP, and HAP had 26% more total biomass than NP.

The weight of petals is significantly and negatively correlated with the weight of pistil, pedicel, and calyx. This suggests that the elimination of petals benefits the development of the other floral organs (Jiang, 2001). NP had the largest weight of whole flower, which was 20.18% and 6.8% heavier than that of AP and HAP, respectively. The three lines differed significantly for the weight of all floral parts. AP had the heaviest floral organs except petal.

Without or with fewer petals, the plants maintain more photosynthesized assimilates in leaves or other floral organs, as the petal is not a photosynthesizing organ but the organ consuming photosynthesized assimilates. Under European field conditions, around 25,000 flowers/m^2 were produced by a typical commercial cultivar of winter rapeseed, which equivalents ~125- to 200-g dry matter/m^2, if we assume that petals of a normal flower weigh 6–8 mg.

The effect of apetalous flowers on grain yield differed for various genetic sources of apetalous flowers. By comparing the yield performance between apetalous lines and normal petaled cultivar "Marnoo," Mendham *et al.* (1991) reported that the yield of the apetalous lines remained high even without irrigation, in contrast to much lower yield of "Marnoo" under such condition. The apetalous line benefited not only from less evaporation, but also from higher root activity, which was promoted by longer active leaf life. Normal petaled crops generally showed little increase in root weight after flowering starts. Apetalous types not only maintained a consistently higher stomatal conductance than "Marnoo," but also a higher leaf turgor at lower osmotic potentials. All these led to better yield performance in a drought environment. However, no yield advantage was found in the apetalous line (AP) with "*ap-Tengbe*" background. The lower yield performance of AP may be due to the lower pollen production linked to the apetalous character (Jiang and Becker, 2001a,b).

A series of commercial varieties with apetalous flowers have been successfully released in China (Zhang and Fu, 2002). Breeders in China are now endeavoring to integrate the apetalous genes into systems of heterosis utilization (Zhang *et al.*, 2005).

V. ROLE OF PETALS IN SPREADING DISEASES

The apetalous type of *Brassica* oilseeds was thought to be less infected from diseases such as *Sclerotinia sclerotiorum*, which are distributed by petals that wilt and drop on leaf tissue. The infection can be identified by the symptom of white colored canker on stem during or after flowering, or by black *Sclerotinia* grains inside the stalks. The disease starts from ascospores, which develop in apothecia and are discharged in spring or early summer. Discharging of the ascospores is influenced by weather. During dry and slightly windy conditions, many spores are ejected. The spores may adhere to young flower petals and then germinate and penetrate the host with short germ tubes that resulted in collapse of the epidermal cells of petals (Jamaux *et al.*, 1994). Petals play an important role in infecting the plants further. Firstly, they favored the infection by serving as C-source that is needed by the invading fungus (Krüger, 1975). Secondly, petals that are covered with mycelium are the infection sites, where the mycelium invades leaf tissues. The ascospores that land directly on leaf surface do not germinate (Jamaux and Spire, 1999).

Genotypes that are immune or highly resistant to *Sclerotinia* disease were hardly found in the existing germplasm of *Brassica* oilseeds (Jiang *et al.*, 1995; Lu, 2003). However, crop husbandry measures, such as rotation, removal of stubble from the previous crop, stimulation of soil microorganisms, and prevention of plant injury, may be helpful to reduce the disease (Twengstrom *et al.*, 1998; Wahmhoff *et al.*, 1998), but are laborious. Genotypes with apetalous flowers may be an alternative to be applied in areas where the natural infestation is heavy. Not much information about the ability of apetalous genotypes in terms of the alleviation of disease infection is available in literature. Rakow and Séguin-Swartz (1999) reported that the incidence and severity of *Sclerotinia* disease was much lower in plots of apetalous *B. napus* (9% incidence, 6.6% severity) compared to plots of *B. napus* canola AC Excel (32.5% incidence, 27.5% severity). Zhao and Jiang (manuscript unpublished) found in field experiments that apetalous genotypes were lower in the incidence of *Sclerotinia* infection than the normal petaled, but there was no significant difference among the genotypes for the severity of infection. Zhao and Wang (2004) reported that apetalous lines were not significantly different from normal petaled lines when the disease pressure was low, but showed significant advantageous in the condition of heavy disease pressure. Henson and LeCare (2005) investigated the *Sclerotinia* infection of 26 canola cultivars, including the apetalous line Hylite 201, at a favorable environment for disease development by inoculation with ascospores (foliar sprayed) and misted. However, the apetalous line did not show any advantages (41% incidence, 3.9% severity) to the other 25 normal petaled cultivars (35.3% incidence, 3.9% severity in average). Therefore, so far the results were rather contradictory from the experiments investigating the relationship between petals and *Sclerotinia* infection to our knowledge.

Petal's role in distributing disease may be not only limited to *S. sclerotiorum*, but the spores of some other pathogens, for example, *Botrytis cinierea* and *Peronospora parasitica*, which cause symptoms during or shortly after flowering, were also found on petals. Lefol and Morrall (1996) even discovered more than 50 groups of ascospores existing on rape petals, among which some are aggressive. Genotypes with apetalous flowers might have the advantage to avoid these diseases too.

VI. FLORAL ABNORMALITIES AND INHERITANCE OF FLORAL CHARACTERS

In addition to apetalous character, some other flower abnormalities were reported in Brassicas and other genera. Singh (1961a) noted a plant in his inbred lines of brown sarson with flowers of four, five, and six petals. The abnormal numbers resulted from splitting of the petals. Palmer (1958) also reported split petals in some flowers of *Brassica napus, Brassica oleracea, Oriza sativa.*

Apart from petal number, the inheritances of other traits related to petal in oilseed Brassicas are reviewed as the following:

Petal color (*B. rapa*): Controlled by single pair of genes, yellow:white = 3:1 in F_2 (Richiharia, 1957)

Petal color (*B. carinata*): The yellow flower color was governed by two complementary dominant genes (Heyn, 1973); Controlled by single pair of genes, cream:yellow = 3:1 in F_2 (Getinet *et al.*, 1993)

Petal color (*B. juncea*): Controlled by two loci, yellow:pale yellow:white = 12:3:1 in F_2 (Brar *et al.*, 1991; Rawat and Anand, 1986; Singh *et al.*, 1964)

Petal color (*B. juncea*): Governed by two dominant genes, *Y* and *Cr*, which interacted epistatically, Y presents either single or doubly, resulted in normal yellow flowers and was epistatic to Cr that codes cream flowers, the double recessive genotype (yy crcr) bore white flowers (Rawat and Anand, 1986); Controlled by single pair of genes, yellow:white = 3:1 in F_2 (Brar *et al.*, 1991)

Petal color (*B. napus*): Controlled by three pairs of genes, AB for yellow, Ab for dark orange yellow; ab for pale orange yellow, c accounts for cream white. B locus is inactive in presence of aa and cc (Morice, 1960); Monogenically controlled by gene on C genome, yellow:termediate:white = 12:3:1 in F_2 (Chen and Haneen, 1990); Controlled by single pair of genes, lemon yellow:pale yellow = 3:1 in F_2 (Seguin-Swartz, 1988)
Petal shape (*B. rapa*): Controlled by single pair of genes, normal:tucked = 3:1 in F_2, normal:jagged-petal margin = 3:1 in F_2 (Hawk and Crowder, 1978)
Petal shape (*B. nigra* L.): The folded petal trait was thought to be controlled by two loci (Delwiche and Williams, 1977)
Anther condition (*B. rapa*): Controlled by single pair of genes, extrose:introse = 3:1 in F_2 (Ali and Sikka, 1937)

VII. CONCLUSION

Brassica oilseeds with apetalous flowers have many origins. Breeders in many countries endeavored to introgress apetalous genes into commercial cultivars for the ideotype of better solar energy utilization and less disease infection. It would be interesting to know if genetically manipulating/silencing of a single locus, for example a gene from the B cluster according to the ABC model describing the development of plant floral organs (Coen and Meyerowitz, 1991; Günter, 2001), would result in apetalous phenotype; how about the relationship among the genes that were identified in different apetalous origins? Do all these genes belong to the B cluster? How to break the linkage between apetalous character and some negative agronomic traits? Some undergoing studies may unveil part of the answers, which may favor breeders who are endeavoring to commercialize the apetalous varieties in *Brassica* oilseeds.

REFERENCES

Ali, M. and Sikka, S. M. (1937). Breeding investigations in some of the oleiferous *Brassicae* of Punjab. *Indian Journal of Agriculture of Science* **7**, 849–862.

Awad, A., Dawh, A. K. and Attya, M. (1987). Cutting thickness and auxin affecting the rooting and consequently the growth and flowering of *Bougainvillea glabra* L. *Acta Horticulturae* **2**, 445–454.

Brar, B. K., Dhillion, S. S., Singh, K. and Labana, K. S. (1991). Genetics of qualitative traits in Indian mustard. *Cruciferae Newsletter* **14/15**, 34–35.

Brown, P. H. and Menary, R. C. (1994). Flowering in pyrethrum (*Tanacetum cinerariaefolium* L.). II. Changes in plant growth regulator concentrations. *Journal of Horticultural Science* **69**, 985–992.

Buzza, G. C. (1983). The inheritance of an apetalous character in Canola (*B. napus* L.). *Cruciferae Newsletter* **8**, 11–12.

Chen, B. Y. (1989). Apetalous flower. Resynthesized *Brassica napus* L.: A potential in breeding and research. Dissertation, the Swedish University of Agricultural Sciences, Svalöv, pp. 3–29.

Chen, B. Y. and Haneen, W. K. (1990). Independent inheritance of flower color and male-fertility characters in *Brassica napus* L. *Plant Breeding* **104**, 81–84.

Coen, E. S. and Meyerowitz, E. M. (1991). The war of the whorls: Genetic interactions controlling flower development. *Nature* **353**, 31–37.

Cours, B. J. and Williams, P. H. (1977). Genetic studies in *Brassica campestris* L. *Plant Breeding Abstract* **51**, 1533.

Davies, P. J. (1987). The phytohormones: Their nature, occurrence, and functions. *In* "Phytohormones and Their Role in Plant Growth and Development" (P. J. Davies, ed.), p. 5. Martinus Nijhoff Publishers (Boston, USA).

Delwiche, P. A. and Williams, P. H. (1977). Genetic studies on *Brassic nigra* (L.) Koch. *Cruciferae* Newsletter **2**, 39.

Downey, R. K. and Rakow, G. (1987). "Principles of Cultivar Development—Crop Species" (W. R. Fehr, ed.), Vol. 2, pp. 437–486. Macmillan Publications Comp., New York.

FAO production yearbook (2000–2005). *FAO on line statistics http://www.fao.org/.*

Fray, M. J., Evan, E. J., Lydiate, D. J. and Arthur, A. (1996). Physiological assessment of apetalous flowers and erectophile pods in oilseed rape (*Brassica napus*). *Journal of Agriculture Science Cambridge* **127**, 193–200.

Fray, M. J., Puangsomlee, P., Goodrich., J., Coupland., G., Evans., E. J., Arthur, A. E. and Lydiate, D. J. (1997). The genetics of stamenoid petal production in oilseed rape (*Brassica napus*) and equivalent variation in *Arabidopsis thaliana*. *Theoretic and Applied Genetics* **94**, 731–736.

Getinet, A., Rakow, G. and Downey, R. K. (1993). Inheritance of a cream petal mutant in Ethiopian mustard. *Canadian Journal of Plant Science* **73**, 1075–1076.

Günter, T. (2001). Floral quartets. *Nature* **409**, 469–471.

Hawk, J. A. and Crowder, L. V. (1978). The inheritance of four mutants in early flowering. *Brassic campestris* L. *Journal of Heredity* **69**, 125–127.

Henson, B. and LeCare, D. (2005). Evaluation of Canola Cultivars for Resistance to *Sclerotinia.* Available at: http://www.ag.ndsu.nodak.edu/carringt/agronomy/Research/.

Heyn, F. W. (1973). Beiträge zum Auftreten unreduzierter Gameten und zur Genetik einiger Merkmale bei *Brassiceae*. Thesis. Georg-August-Universität Göttingen, 102 pp.

Jamaux, D. I. and Spire, D. (1999). Comparison of responses of ascospores and mycelium by ELISA with anti-mycelium and anti-ascospore antisera for the development of a method to detect *Sclerotinia sclerotiorum* on petals of oilseed rape. *Annals of Applied Biology* **134**, 171–179.

Jamaux, D. I., Gelie, B. and Lamarque, C. (1994). Early stages of infection of rapeseed petals and leaves by *Sclerotinia sclerotiorum* revealed by scanning electron microscopy. *Plant Pathology (Oxford)* **44**, 22–30.

Jiang, L. (2001). A mutant with apetalous flowers in oilseed rape (*Brassica napus*): Mode of inheritance and its effects on yield physiology and resistance to *Sclerotinia scleronitiorum*. Dissertation. (http://webdoc.sub.gwdg.de/diss/2001/jiang/).

Jiang, L. and Becker, H. C. (2001a). Effect of apetalous flowers on crop physiology in winter oilseed rape (*Brassica napus*). *Pflanzenbauwissenschaft* **5**, 58–63.

Jiang, L. and Becker, H. C. (2001b). Environment effects on expression of apetalous flowers in oilseed rape. *Journal of Applied Botany* **75**, 75–79.

Jiang, L. and Becker, H. C. (2003). Inheritance of apetalous flowers in a mutant of oilseed rape. *Crop Science* **43**, 508–510.

Jiang, L., Yu, J. B., Jin, M. S. and Jiang, W. L. (1995). Evaluation of resistance to *Sclerotinia sclerotiorum* and TuMV virus for 140 land races (*B. napus*) in Zhejiang. *Acta Agriculture Zhejiangensis* **7**, 202–205.

Jiang, L., Li, Y. and Meng, H. (2006). The apetalous mutants in *Brassica* oilseeds. *Crop Research* **32**, 462–468.

Kaur, S. R., Applewhite, P. B. and Galston, A. W. (1996). Formation *in vitro* of ripe tomato fruits from thin layer explants of flower pedicels. *Plant Growth Regulation* **18**, 191–199.

Kelly, A., Fray, M., Arthur, E. A. and Lydiate, D. J. (1995). The genetic control of petalless flowers and upright pods. Genetics and methods. *In* "Proceedings of the Ninth International Rapeseed Congress," pp. 732–733. Cambridge, UK.

Krüger, W. (1975). Die Beeinflussung der Apothezien- und Ascosporen- Entwicklung des Rapskrebserregers *Sclerotinia sclerotiorum* (Lib.) de Bary durch Umselt-faktoren. Pflanzen Krankheiten **2**, 101–108.

Lefol, C. and Morrall, R. A. A. (1996). Immunofluorescent staining of *sclerotinia* ascospores on canola petals. *Canadian Journal of Plant Pathology* **18**, 237–241.

Liu, H. L. (1985). Remote crossing. *In* "Genetics and Breeding of Rapeseed" (H. L. Liu, ed.), pp. 87–119. Shanghai Publishing House for Science and Technical Literatures, Shanghai.

Lu, G. (2003). Engineering *Sclerotinia sclerotiorum* resistance in oilseed crops. *African Journal of Biotechnology* **2**, 509–516.

Lu, W. L., Liang, B., Enomoto, K. and Fukunaga, Y. (1994). Role of exogenous hormones in inducing cells in different positions of perianth explants to regenerate flower buds in *Hyacinthus orientalis*. *Acta Botanica Sinica* **36**, 581–586.

Lü, Z. J. and Fu, S. Z. (1990). Inheritance of apetalous character in Rape (*Brassica napus* L.) and its implication in breeding. *Jiangsu Journal of Agriculture of Sciences* **6**, 30–36.

Malik, M., Vyas, P., Ranggaswamy, N. S. and Shivanna, K. R. (1999). Development of two new cytoplasmic male-sterile lines in *Brassica juncea* through wide hybridisation. *Plant Breeding* **118**, 75–78.

Mendham, N. J., Rao, M. S. S. and Buzza, G. C. (1991). The apetalous flower character as component of a high yielding ideotype. *In* "Proceedings of the GCIRC Eighth International Rapeseed Congress," pp. 596–600. Saskatoon, Canada.

Metzger, D. J. (1987). Hormones and reproductive development. *In* "Phytohormones and Their Role in Plant Growth and Development" (P. J. Davies, ed.), pp. 443–450. Martinus Nijhoff Publishers (Boston, USA).

Morice, J. (1960). Inheritance of flower color in rape, *B. napus* L. var. oleiferi Metzger. *Ann. Inst. Nat. Rech. Agron, Paris Ser. B.* **10**, 155–168.

Oka, M., Miyamoto, K., Okada, K. and Ueda, J. (1999). Auxin polar transport and flower formation in *Arabidopsis thaliana* transformed with indoleacetamide hydrolase (iaaH) gene. *Plant and Cell Physiology* **40**, 231–237.

Palmer, T. P. (1958). Split petals in Cruciferae. *Heredity* **12**, 417–428.

Peterson, C. L., Auld, D. L. and Korus, R. A. (1983). Winter Rape Oil Fuel for Diesel Engines: Recovery and Utilization. *Journal of American Oil Chemistry Society* **60**, 1579–1587.

Pierre, J., Marilleau, R., Pham-Delegue, M. H., Tanguy, X. and Renard, M (1996). Influence of the apetalous character in rape (*Brassica napus*) on foraging behaviour of honeybees (*Apis mellifera*). *Plant Breeding* **115**, 484–487.

Rakow, G. and Séguin-Swartz, G. (1999). Development of apetalous *Brassica napus*. *In* "Proceedings of the 10th International Rapeseed Congress," p. 69. Canberra, Australia.

Ramanujum, S. (1940). An apetalous mutant in turnip (*Brassica campestris* L.). *Nature* **145**, 552–553.

Rana, M. A. (1985). Development morphology of flower and inheritance of an apetalous mutant in *Brassica carinata*. Dissertation, University of California, Davis. Abstr. *International. B (Science and Engineering)* **45**, 3425B.

Rawat, D. S. and Anand, I. J. (1986). Inheritance of flower color in mustard mutant. *Indian Journal of Agriculture Sciences* **56**, 206–208.

Richiharia, R. H. (1957). Plant breeding with reference to oilseed crops. *Indian Journal of Genetics* **17**, 218–247.

Röbbelen, G. (1997). The evolution of rapeseed quality. *GCIRC Bulletin* **14**, 102–113.

Seguin-Swartz, G. (1988). Inheritance of pale yellow petal mutant of summer rape. *Cruciferae Newsletter* **13**, 24–25.

Shirzadegan, M. and Röbbelen, G. (1985). Influence of seed color and hull proportion on quality properties of seed in *Brassica napus* L. *Fetteund Seifen Anstrichmittel* **6**, 235–237.

Singh, D. (1961a). Heredity changes in the number of petals in Brown Sarson. *Indian Oilseed Journal* **5**, 190–193.

Singh, D. (1961b). An apetalous mutant in toria brown sarson (*Brassica campestris* var. brown sarson) and its inheritance. *Current Science* **30**, 62–63.

Singh, D. P., Hauhan, S. C. and Magnath, K. S. (1964). Inheritance of characters in *B. juncea*. *Indian Journal of Genetics* **24**, 288–290.

Tengbe, M. A. (1990). Züchterische Nutzung von induzierten Mutanten bei Raps (*Brassica napus* L.) - veränderte Polyenfettsäure-Zusammensetzung und Blütenblattlosigkeit. Diss. Goerg-August-Universität Göttingen, pp. 34–45.

Twengstrom, E., Kopmans, E., Sigvald, R. and Svensson, C. (1998). Influence of different irrigation regimes on carpogenic germination of sclerotia of *Sclerotinia sclerotiorum*. *Journal of Phytopathology* **146**(10), 487–493.

Venglat, S. P. and Sawhney, V. K. (1996). Benzylaminopurine induces phenocopies of floral meristem and organ identity mutants in wild-type *Arabidopsis* plants. *Planta (Heidelberg)* **198**, 480–487.

Virk, P. S., Khush, G. S. and Peng, S. (2004). Breeding to enhance yield potential in rice at IRRI. *Institute of Restorative and Rehabilitation Neuroscience* **29**, 5–9.

Vondrakova, Z., Krekule, J. and Machackova, I. (1998). Is the root effect on flowering of *Chenopodium rubrum* mediated by cytokinins? *Journal of Plant Growth Regulation* **17**, 115–119.

Wahmhoff, W., Hedke, K. V., Tiedemann, A., Nitzsche, O. and Ulber, B. (1998). Impact of crop rotation and soil cultivation on the development of pests and diseases of rapeseed. *Zeitschrift für Pflanzenkrankheiten und Pflanzenschutz* **106**, 57–73.

Withers, R. V. and Noordam, M. (1996). Biodiesel from canola: An economic feasibility analysis. Liquid fuels and industrial products from renewable resources. Proceedings of the Third Liquid Fuel Conference, September 15–17, Nashville, Tennessee 211–231.

Yates, D. J. and Steven, M. D. (1987). Reflection and absorption of solar radiation by flowering canopies of oil-seed rape (*Brassica napus* L.). *Journal of Agriculture of Science Cambridge* **109**, 495–502.

Zhang, J. and Fu, S. Z. (2002). Advances of Genetics and Breeding in apetalous rapeseed. *Jiangsu Journal of Agricultural Sciences* **18**, 181–187.

Zhang, W. X., Li, D. R., Tian, J. H. and Yang, C. L. (2005). The discovery and study of apetalous sterility mutant in *Brassica napus* L. *Journal of Oil Crops in China* **27**, 13–15.

Zhao, Y. and Wang, M. L. (2004). Inheritance and agronomic performance of an apetalous flower mutant in *Brassica napus* L. *Euphytica* **137**, 381–386.

Zhou, Y. Y., Wu, Z. B., Liu, Y. Q. and Yang, Y. (1995). Mechanism of direct redifferentiation of flower bud from explant of *Dianthus chinensis* L. by cytokinin inducement. *Acta Phytophysiologica Sinica* **21**, 209–214.

Breeding Herbicide-Tolerant Oilseed Rape Cultivars

PETER B. E. MCVETTY AND CARLA D. ZELMER

Department of Plant Science, University of Manitoba, Winnipeg, Canada

Advances in Botanical Research, Vol. 45
Incorporating Advances in Plant Pathology

0065-2296/07 $35.00
DOI: 10.1016/S0065-2296(07)45009-1

ABSTRACT

Numerous novel herbicide-tolerant oilseed rape cultivar types have been developed in Canada in recent years. These include triazine-tolerant, glyphosate-tolerant, imidazolinone-tolerant, glufosinate-tolerant, and bromoxynil-tolerant oilseed rape cultivars. The registration of Canada's first novel herbicide-tolerant oilseed rape cultivars and the regulatory approvals required to permit herbicide-tolerant oilseed rape production in Canada are discussed. A review of the adoption of novel herbicide-tolerant oilseed rape cultivars by producers and of Canadian consumer acceptance of the oil derived from these cultivars is included. This chapter concludes with comments about the future of herbicide-tolerant oilseed rape cultivar breeding in Canada.

I. INTRODUCTION

Oilseed rape, *Brassica napus* L. is a small-seeded dicotyledonous crop used as a major source of vegetable oil in the temperate climates of the world (Kimber and McGregor, 1995). The seedlings are very small at emergence and they can take up to several weeks to grow sufficiently large to provide strong competition with weeds. As a consequence, weeds are the most important limiting factor for spring oilseed rape production and an important factor for winter oilseed rape production in most parts of the world (Orson, 1995).

Common weeds in oilseed rape production include volunteer cereals, wild oats (*Avena fatua*), green foxtail (*Setaria viridis*), wild mustard (*Sinapis arvensis*), common chickweed (*Stellaria media*), annual meadow grass (*Poa annua*), wild buckwheat (*Fallopia convolvulus*), annual stinkweed (*Thlaspi arvense*), lamb's quarters (*Chenopodium album*), and red root pigweed (*Amaranthus retroflexus*) (Orson, 1995; Thomas, 1992).

Traditionally, a range of nonselective herbicides, including glufosinate ammonium (Liberty®), glyphosate (Roundup®), or viologen (Paraquat®), are used, either just prior to seeding or just after seeding (and before crop emergence) to control weeds that have emerged. In conventional oilseed rape production, a primary cultivation precedes the use of herbicides, while in minimum or zero tillage oilseed rape production herbicides only are used for weed control.

A wide range of selective herbicides have also been developed for use in weed control in oilseed rape. These herbicides are designed to suppress weed growth or kill weeds in the crop with minimal negative effects on the crop. There are the preseeding, soil incorporated herbicides such as trifuralin (Treflan®) and ethylfuralin (Edge®). There are also postemergence herbicides used in oilseed rape production, including grass killers such as ethoxydim (Poast®) or clethodim (Select®), thistle control herbicides such as clopyralid (Lontrel®), and wild mustard killers such as ethametsulfuron-methyl (Muster®) (Orson, 1995).

In spite of the availability of the above herbicides for use in oilseed rape production, there is still a shortage of broadleaved herbicides to be used in oilseed rape since there are many broadleaved weeds not controlled by the currently available selective herbicides. From the 1940s until the 1990s, this shortage of selective herbicides was addressed through the development of new herbicides to which the crop (such as oilseed rape) was naturally tolerant (resistant) (Duke, 1996).

The phrase "herbicide tolerance" will be consistently used in this chapter instead of "herbicide resistance." Herbicide tolerance is becoming the universally accepted standard descriptor to describe crop cultivars that safely tolerate commercial rate applications of (usually nonselective) herbicides.

Completely different approaches to weed control in oilseed rape began in the 1970s (intra- and interspecific transfer of naturally occurring herbicide tolerance) and the 1980s (mutagenic treatments to create herbicide-tolerant oilseed rape and interspecies herbicide tolerance gene transfers via transformation). All of these approaches are examples of breeding a crop such as oilseed rape to tolerate a novel herbicide. The idea in general was to develop oilseed rape, which would safely tolerate agronomically relevant applications of novel non-selective herbicides. This was a revolution in the use of herbicides in oilseed rape. The use of herbicide-tolerant oilseed rape is steadily increasing throughout the world currently, with near total conversion to exclusively herbicide-tolerant oilseed rape production already achieved in Canada. Canada has taken a leadership position in the development, registration, and commercial production of herbicide-tolerant oilseed rape and will be the focus of the remaining chapter.

This chapter will review (A) sources of herbicide tolerance genes to selected herbicides, (B) the mechanism of action for selected herbicides, (C) genetics of herbicide tolerance for several novel herbicide tolerances, (D) development of oilseed rape lines possessing herbicide tolerance genes, (E) the regulatory approvals for herbicide-tolerant oilseed rape lines in Canada, (F) breeding herbicide-tolerant oilseed rape cultivars in Canada, (G) the registration of herbicide-tolerant oilseed rape cultivars in Canada, (H) the adoption of herbicide-tolerant oilseed rape cultivars by producers in Canada, (I) consumer acceptance of oilseed rape oil derived from Canadian-produced herbicide-tolerant oilseed rape cultivars, and finally (J) the future of herbicide-tolerant oilseed rape cultivar breeding in Canada.

Five different herbicide-tolerant oilseed rape cultivar types have been developed in Canada over the last 20–25 years. These include oilseed rape cultivars with tolerance to one of triazine, glufosinate ammonium, glyphosate, imidazolinone, or bromoxynil herbicides.

Breeding herbicide-tolerant oilseed rape cultivars is a logical and straightforward extension of breeding nonherbicide-tolerant oilseed rape cultivars. All previous oilseed rape cultivar-breeding objectives including improving seed yield, phenological traits, lodging resistance, seed quality, and disease and insect resistance (Downey and Röbbelen, 1989) are still relevant. Similarly, all previous breeding approaches including crossing and backcrossing combined with pedigree selection for open pollinated population oilseed rape cultivar development, or doubled haploid line, synthetic and hybrid oilseed rape cultivar development (Downey and Röbbelen, 1989) are still relevant. Hybrid oilseed rape cultivar development can make use of a number of pollination control systems, including cytoplasmic male sterility (CMS) (McVetty, 1997), genetic male sterility (GMS) (Sawhney, 1997), self-incompatibility (SI) (McCubbin and Dickinson, 1997), and nuclear male sterility (NMS) (Williams *et al.*, 1997). Herbicide tolerance is compatible with all of these pollination control systems and an integral part of the NMS pollination control system (Williams *et al.*, 1997).

Herbicide tolerance can be introduced into crops using three basic approaches: (1) overexpression of the herbicide target enzyme, (2) detoxification of the herbicide, and (3) expression of an herbicide-insensitive target enzyme. Combinations of these three basic approaches can also be used (Dill, 2005). Breeding herbicide-tolerant oilseed rape cultivars can begin when one or more herbicide tolerance genes (either cytoplasmic or nuclear) have been discovered in, created in, or transferred into, via transformation technology, oilseed rape (Dyer, 1996). Specific herbicide-tolerant oilseed rape-breeding procedures depend on the nature of the inheritance of the herbicide tolerance being used (with cytoplasmic, semidominant nuclear, or dominant nuclear gene-based herbicide tolerance all possible). Additional herbicide tolerance screening procedures are added to the breeding program procedures list, with indoor single plant screening and outdoor row and plot screening for herbicide tolerance routinely done. In addition, multiple station-year herbicide tolerance tests of all advanced herbicide-tolerant oilseed rape lines prior to their registration in Canada are also required (Stringam *et al.*, 2003). Extensive herbicide tolerance transgenic construct purity testing is also required for some herbicide tolerance systems, both during and after the herbicide-tolerant oilseed rape-breeding phase.

Plants with novel traits including most of the herbicide-tolerant oilseed rape types developed and registered in Canada require regulatory approvals from the Canadian Federal Government prior to their commercial production in this country. Plants with novel traits are defined in Canada as "new plant varieties developed either through traditional breeding techniques or modern biotechnology. Plants with novel traits have genetic characteristics, or traits, different from those of similar crops grown in Canada" (Canadian Food Inspection Agency, 1994). In addition, many countries purchasing Canadian-produced oilseed rape must approve the use of herbicide-tolerant transgenic, genetically modified, oilseed rape seed, oil, and meal in their countries before importation can occur.

II. TRIAZINE-TOLERANT OILSEED RAPE

The first herbicide-tolerant oilseed rape cultivars to be developed and registered in Canada were triazine tolerant.

A. SOURCES OF TRIAZINE TOLERANCE GENES

Triazine-tolerant bird's rape (*B. rapa*) plants were discovered in two corn fields near Sherbrooke, Quebec in the mid-1970s (Maltais and Bouchard, 1978). These mutant bird's rape plants were used as the source of triazine tolerance for all triazine-tolerant oilseed rape cultivars developed in Canada and elsewhere.

B. TRIAZINE MECHANISM OF ACTION

The site of action of triazine herbicides is in the chloroplasts where they inhibit photosynthesis. Triazine herbicides block electron transport in photosystem II by irreversibly binding to protein D1 of photosystem II, thereby blocking plastoquinone Q_B from its binding site on protein D1 (Fuerst and Norman, 1991). This inhibits electron transport through photosystem II, which in turn reduces the amount of ATP and NADPH produced in the chloroplast, resulting in physiologically detrimental consequences to the plant (Dodge, 1983). In addition, when triazine herbicides block the normal flow of electrons in photosystem II, high levels of singlet oxygen are produced in the chloroplast, which results in lipid peroxidation, loss of chloroplast membrane structure, chlorosis, cell death, and finally plant necrosis (Fuerst and Norman, 1991).

C. GENETICS OF TRIAZINE TOLERANCE

The triazine tolerance found in bird's rape displayed uniparental inheritance through the maternal parent; therefore, it was conditioned by one or more cytoplasmic genes (Souza Machado *et al.*, 1978). The tolerance to triazine herbicides first found in bird's rape is due to an alteration of the biding site for triazine herbicides in the D1 protein (Fuerst and Norman, 1991). A mutation occurring on the psbA gene results in the substitution of a serine amino acid by a glycine amino acid at amino acid position 264 on the D1 protein. This single amino acid substitution reduces the binding of triazines to the D1 protein by 100–1000 times (Van Oorschot, 1991). The psbA gene mutation results in impaired electron transport in photosystem II, which accounts for some, but not all, of the seed yield reduction seen in triazine-tolerant oilseed rape (Hall *et al.*, 1996).

D. DEVELOPMENT OF NEW TRIAZINE-TOLERANT OILSEED RAPE LINES

A triazine-tolerant oilseed rape-breeding program was initiated in 1976 at the University of Guelph, Guelph Canada to take advantage of the naturally occurring triazine tolerance found in wild bird's rape (Beversdorf *et al.*, 1980a). Since *B. rapa* is one of the parent species of oilseed rape (*B. napus*), crosses and backcrosses of these two species can be done to transfer traits between *B. rapa* and *B. napus*. Initially, a double-low (low erucic acid, low glucosinolate)-quality triazine-tolerant *B. rapa* population, referred to as ATR-5Ca, was developed by crossing and backcrossing the triazine-tolerant bird's rape biotype as the female to the canola quality *B. rapa* cv. Candle (Beversdorf *et al.*, 1980b). ATR-5Ca was widely distributed, without encumbrances, to oilseed rape-breeding organizations ca 1978. Subsequently, a double-low-quality oilseed rape (*B. napus*) line, ATR-5Tw, was developed by crossing and backcrossing the triazine-tolerant bird's rape biotype as the female to the double-low-quality oilseed rape cv. Tower (Beversdorf *et al.*, 1980c). ATR-5Tw was widely distributed, without encumbrances, to oilseed rape-breeding organizations ca. 1980.

E. CANADIAN REGULATORY APPROVALS

Triazine-tolerant oilseed rape lines were developed in the 1980s in Canada. There were no environmental safety, feed safety, or food safety approvals required for triazine-tolerant oilseed rape lines developed in Canada since the Canadian Federal Government regulations governing plants with novel traits were not brought into effect until 1994 (Canadian Food Inspection Agency, 1994). Similarly, there were no export destination country approvals required prior to the importation of triazine-tolerant oilseed rape seed, oil, or meal from Canada. In retrospect, these were the simple and easy days for herbicide-tolerant oilseed rape cultivar development, registration, and commercialization in Canada!

F. BREEDING TRIAZINE-TOLERANT OILSEED RAPE CULTIVARS

Breeding of triazine-tolerant open pollinated population oilseed rape cultivars used a backcross program approach with the bird's rape source of triazine tolerance as the female parent since the trait was conferred by a chloroplast gene. As little as four backcrosses were used to generate new triazine-tolerant open pollinated population oilseed rape cultivars with most of the performance characteristics of the recurrent parent (Beversdorf and Hume, 1984). There was at least a threefold safety margin for commercial rate triazine application conferred by the bird's rape cytoplasmic mutation, so the herbicide tolerance was suitable to develop triazine-tolerant oilseed rape cultivars. To breed for the herbicide tolerance trait, it is presumed that all plants in each backcross generation were sprayed with triazine herbicide at approximately the normal commercial application rate and triazine tolerance tests of all advanced triazine-tolerant lines sprayed at a minimum double the commercial application rate were conducted. This ensured that the new triazine-tolerant open pollinated population oilseed rape cultivars were pure breeding for herbicide tolerance and that the crop had adequate levels of tolerance to triazines.

The triazine tolerance conferred on oilseed rape plants by the mutation in the psbA gene, resulting in an alteration of the D1 protein of photosystem II, carried a substantial biological cost with it. On average over many trials grown in Canada from 1980 to 1988, triazine-tolerant open pollinated population oilseed rape cultivars yielded 20–25% less than their triazine susceptible open pollinated population sister line checks, displayed 2 day delays to first flowering and 1 day delays to maturity (Beversdorf *et al.*, 1988). The triazine-tolerant open pollinated population oilseed rape cultivars also had on average 2% lower oil concentration than comparable triazine susceptible oilseed rape checks and much reduced seedling vigor (Thomas, 1992). These biological costs were consistent with the reduced photosynthetic rates observed in all triazine-tolerant open pollinated population oilseed rape cultivars (Hobbs, 1987).

Considerable efforts were made at the University of Guelph to overcome the biological costs associated with triazine tolerance in oilseed rape using leaf fluorescence measurements to try and identify triazine-tolerant oilseed rape lines, which did not have impaired photosynthesis. No triazine-tolerant oilseed rape lines with improved photosynthetic efficiency were found (Hall *et al.*, 1996). It therefore became clear that triazine tolerance conferred using this source of herbicide tolerance and the biological costs associated with it were inseparable. All triazine-tolerant open pollinated population oilseed rape-breeding programs in Canada were terminated in the early 1990s, well before any triazine-tolerant hybrid oilseed rape cultivars were developed.

G. REGISTRATION OF TRIAZINE-TOLERANT OILSEED RAPE CULTIVARS

Backcrosses and selection for low erucic acid and low glucosinolates in the ATR-5Tw × Tower cross were used at the University of Guelph to produce the world's first triazine-tolerant open pollinated population oilseed rape cultivar, OAC Triton. OAC Triton was registered in 1984 (Beversdorf and Hume, 1984). Commercial production of OAC Triton began in 1985. Subsequently, the triazine-tolerant open pollinated population oilseed rape cultivar, Tribute, developed by the Saskatoon Research Centre of Agriculture and Agri-Food Canada was registered in 1985 (Rakow and Downey, 1993). This was followed by OAC Triumph developed by the University of Guelph and registered in 1987 (Beversdorf *et al.*, 1990). Two additional triazine-tolerant open pollinated population oilseed rape cultivars, AC Tristar, developed by the Saskatoon Research Centre of Agriculture and Agri-Food Canada and registered in 1990 (Rakow, 1992), and Stallion, developed by Svalof Weibull and registered in 1990 (Canadian Food Inspection Agency Variety Registration Office, 2006), were grown in Canada.

H. PRODUCER ADOPTION

Many Canadian producers, being rapid adopters of new technology, tried triazine-tolerant open pollinated population oilseed rape cultivars when they first became available. However, the production area of triazine-tolerant open pollinated population oilseed rape cultivars in Canada remained small, perhaps because of the substantial biological costs associated with the triazine tolerance provided by the bird's rape mutant source. All commercial production of triazine-tolerant open pollinated population oilseed rape cultivars in Canada stopped in 1993 (Hall *et al.*, 1996).

I. CONSUMER ACCEPTANCE

Consumer acceptance in Canada and throughout the world of the oil produced by triazine-tolerant oilseed rape was not an issue since these new triazine-tolerant open pollinated population oilseed rape cultivars were considered to produce oil substantially equivalent to double-low-quality oilseed rape cultivars of the day. A single amino acid change in a single nonnovel protein conferred triazine tolerance in oilseed rape. In addition, protein does not end up in the oil, the only direct use human food derived from oilseed rape. The seed from the triazine-tolerant oilseed rape cultivars produced in Canada was blended into the oilseed rape seed stream and a composite oil and meal produced. This is still the case in countries, such as Australia, where triazine-tolerant open pollinated population oilseed rape cultivars are still being developed, registered, and grown and the seed, oil, and meal either used domestically or exported (Norton, 2003).

J. FUTURE OF TRIAZINE-TOLERANT OILSEED RAPE BREEDING

There will not be any new triazine-tolerant open pollinated population or hybrid oilseed rape cultivars developed and registered in Canada in the foreseeable future since Canadian oilseed rape producers now have several better herbicide-tolerant open pollinated population and hybrid oilseed rape types to grow.

III. GLUFOSINATE-TOLERANT OILSEED RAPE

The next herbicide-tolerant oilseed rape type to be developed in Canada was glufosinate (ammonium) tolerance.

A. SOURCES OF GLUFOSINATE-TOLERANT GENES

Numerous attempts were made to discover oilseed rape or other related species genotypes that were naturally tolerant to glufosinate but none were found, so tolerance to glufosinate was introduced into oilseed rape using a detoxification of the herbicide approach (Tan *et al.*, 2006; Vasil, 1996). This approach to the development of glufosinate-tolerant oilseed rape required plant transformation using one or more transgenes and generated transgenic, genetically modified plants.

B. GLUFOSINATE MECHANISM OF ACTION

The site of action of glufosinate is in the cytoplasm and in the chloroplasts where glufosinate is an amino acid biosynthesis inhibitor. Glufosinate binds to the L-glutamate-binding site of the glutamine synthetase enzyme, a nuclear gene coded enzyme (Hess, 2000). This blocks the conversion of L-glutamate to L-glutamine through the assimilation of ammonia in the cytoplasm and the plastids (Vaughn and Duke, 1991). Inhibition of glutamine synthetase by glufosinate results in a glutamine deficiency, inhibiting amino acid biosynthesis and producing a toxic accumulation of ammonia in the plant. In addition, glufosinate indirectly inhibits electron flow in photosynthetic light reactions and causes chloroplast membrane disruption (Hess, 2000).

C. GENETICS OF GLUFOSINATE TOLERANCE

Tolerance to glufosinate in oilseed rape was created by Agrevo (now Bayer CropScience, Calgary, Canada) by using a gene from a soil actinomycete, *Streptomyces viridochromogenes* which coded for a glufosinate-detoxifying enzyme, phosphinothricin *N*-acetyltransferase. Phosphinothricin *N*-acetyltransferase readily converts L-glufosinate to a nonphytotoxic metabolite, *N*-acetyl-L-glufosinate, through acetylation of glufosinate, greatly reducing the possibility that glufosinate will inhibit glutamine synthetase (Droge *et al.*, 1992). Two different transgenic constructs using the phosphinothricin *N*-acetyltransferase gene from *S. viridochromogenes* were created. The first was an antibiotic resistant version, containing both the phosphinothricin *N*-acetyltransferase gene and a bacterial npt II gene for kanamycin resistance as a selectable marker. A constitutive promoter was used for the phosphinothricin *N*-acetyltransferase gene and a different, weak constitutive promoter was used for the kanamycin resistance gene (Health Canada, 1995b). This transgenic construct was used to transform oilseed rape using a disarmed, nonpathogenic *Agrobacterium tumefaciens* vector. A second transgenic construct, a nonantibiotic resistant version was also developed by Agrevo which used the phosphinothricin *N*-acetyltransferase gene from *S. viridochromogenes* alone linked to a constitutive promoter.

This construct was used to transform oilseed rape using a disarmed, nonpathogenic *A. tumefaciens* vector (Health Canada, 1997b).

Several different additional glufosinate tolerance transgenic constructs were also developed by Plant Genetic Systems, Inc. (now Bayer CropScience). These glufosinate tolerance transgenic constructs, identified as RF1, RF2, and RF3, included a male fertility restorer gene for the NMS system (Williams *et al.*, 1997) as well as a glufosinate tolerance gene. The RF1, RF2, and RF3 transgenic constructs in oilseed rape each consists of two genes, the barstar gene, from *Bacillus amyloquefaciens* coding for the barstar enzyme, and the bar gene, from a soil actinomycete, *Streptomyces hygroscopicus*, coding for a phosphinothricin *N*-acetyltransferase enzyme (Williams *et al.*, 1997). The barstar gene is linked to the PTA29 promoter, an anther-specific promoter from *Nicotiana tabacum*. The bar gene is linked to the PSsuAra promoter from *Arabidopsis thaliana*. Posttranslational targeting of the bar gene product was accomplished by fusion of the 5′-terminal-coding sequence with the chloroplast transit peptide DNA sequence from *A. thaliana* (Health Canada, 1997a). The phosphinothricin *N*-acetyltransferase genes from *S. hygroscopicus* and *S. viridochromogenes* are highly homologous, even though they are from different species (Vasil, 1996).

D. DEVELOPMENT OF NEW GLUFOSINATE-TOLERANT OILSEED RAPE LINES

Agrevo transformed the oilseed rape cv. Topas with the antibiotic resistant version of the glufosinate tolerance construct and then crossed this line to two other oilseed rape lines to generate a glufosinate-tolerant oilseed rape line, HCN92 (Canadian Food Inspection Agency, 1995a). Molecular analysis showed that HCN92 had two copies of the transferred DNA at a single locus in the plant genome (Health Canada, 1995b). Agrevo also used the nonantibiotic resistant version glufosinate tolerance construct to transform the oilseed rape cv. AC Excel using a disarmed, nonpathogenic *A. tumefaciens* vector. This line was backcrossed to AC Excel and then two generations of single-seed descent were used to generate another glufosinate-tolerant oilseed rape line, HCN28 (with the T45 transgenic construct). Molecular analysis showed that HCN28 (T45) had a single copy of the transferred DNA in the plant genome (Health Canada, 1997b).

Independently, Plant Genetic Systems developed several glufosinate-tolerant oilseed rape lines. The oilseed rape cv. Drakkar was transformed with three different versions of the RF transgenes. The RF genes are a combination of the barstar gene from a soil bacterium *B. amyloliquefaciens*, coding for an inhibitor of the barnase ribonuclease enzyme, and the bar gene from *S. hygroscopicus*, coding for the phosphinothricin *N*-acetyltransferase enzyme, linked to two different promoters using a disarmed nonpathogenic

A. tumefaciens vector (Canadian Food Inspection Agency, 1996d). The most recent glufosinate tolerance transgene RF version to be used in glufosinate-tolerant oilseed rape-breeding programs is RF3. Molecular analysis showed that the Drakkar RF3 line had a single copy of the RF3 transgene (Health Canada, 1997a). Success in transferring the RF3 transgene to *B. oleracea* indicated that RF3 is on the C genome of *B. napus* (Health Canada, 1997a). It appears that the RF1 and RF2 versions of the glufosinate tolerance transgenic construct in oilseed rape will be phased out so that RF3 will become the sole glufosinate tolerance transgene used in oilseed rape in coming years.

Agrevo and Plant Genetic Systems patented these glufosinate tolerance transgenes in most or all oilseed rape-producing countries in the world.

Access to the Agrevo and Plant Genetic Systems-patented glufosinate tolerance transgenes is controlled by research and/or commercial licenses between Bayer CropScience and oilseed rape-breeding organizations.

Bayer CropScience also trademarked the term Liberty Link™ in 2000 as the name of the herbicide tolerance system that uses glufosinate-tolerant oilseed rape crops developed using Bayer CropScience-patented glufosinate tolerance transgenes combined with glufosinate ammonium herbicide (Canadian Intellectual Property Office, 2000a,c).

E. CANADIAN REGULATORY APPROVALS

Environmental safety, feed safety, and food safety approvals were required for glufosinate tolerance constructs in oilseed rape lines since the Canadian Federal Government regulations governing plants with novel traits and transgenic, genetically modified plants came into effect in 1994 (Canadian Food Inspection Agency, 1994). Glufosinate-tolerant oilseed rape line HCN92 was grown in confined trials in Canada from 1990 to 1994 to generate the data required by Canadian Federal Government regulators. The glufosinate tolerance construct present in HCN92 received environmental safety approval from the Canadian Federal Government in 1995, feed safety approval in 1995 (Canadian Food Inspection Agency, 1995a), and food safety approval in 1995 (Health Canada, 1995b). Glufosinate-tolerant oilseed rape line HCN28 was grown in confined trials in Canada from 1993 to 1995 to generate the data required by Canadian Federal Government regulators. The glufosinate tolerance construct present in HCN28 received environmental safety approval from the Canadian Federal Government in 1996 (Canadian Food Inspection Agency, 1996b), feed safety approval in 1996 (Canadian Food Inspection Agency, 1996c), and food safety approval in 1997 (Health Canada, 1997b).

Glufosinate-tolerant oilseed rape lines with the RF3 transgene were grown in Canada from 1994 to 1996 to generate data required by Canadian Federal Government regulators. The RF3 glufosinate tolerance constructs in Drakkar (Drakkar RF3) received environmental safety approval from the Canadian Federal Government in 1996, feed safety approval in 1996 (Canadian Food Inspection Agency, 1996d), and food safety approval in 1997 (Health Canada, 1997a). These Canadian Federal Government approvals were combined with the Canadian Federal Government approvals for the MS8 transgene which produces male sterility in oilseed rape lines. The MS8 transgenic construct also includes a glufosinate tolerance gene as well as male sterility-inducing gene (Canadian Food Inspection Agency, 1996d). The MS8/RF3 transgenes create the NMS pollination control system (Williams *et al.*, 1997) that is used by Bayer CropScience to develop glufosinate-tolerant hybrid oilseed rape cultivars (Canadian Food Inspection Agency, 1996d).

F. BREEDING GLUFOSINATE-TOLERANT OILSEED RAPE CULTIVARS

The breeding of glufosinate-tolerant open pollinated population oilseed rape cultivars began in Canada in the mid-1980s at Agrevo. Glufosinate-tolerant open pollinated population oilseed rape cultivars were developed using two different glufosinate tolerance transgenic constructs (HCN92 and HCN28). An efficient approach to the breeding of glufosinate-tolerant open pollinated population oilseed rape cultivars would use a crossing program combined with pedigree selection (Downey and Röbbelen, 1989). There was at least a threefold safety margin for glufosinate application at commercial rate conferred by the HCN92 and HCN28 tolerance transgenes. To efficiently breed for the herbicide tolerance trait, it is presumed that all plants in each cross generation, all rows, and all plots were sprayed with glufosinate herbicide at approximately the normal commercial application rate and that glufosinate tolerance tests of all advanced glufosinate-tolerant lines at a minimum double the commercial application rate were conducted. This ensured that the new glufosinate-tolerant open pollinated population oilseed rape cultivars were pure breeding for herbicide tolerance and that the crop had adequate levels of tolerance to glufosinate. Breeding of glufosinate-tolerant hybrid oilseed rape cultivars used the NMS system developed and patented by Plant Genetics Systems-generated glufosinate-tolerant hybrids automatically since the glufosinate tolerance transgenes are deployed in both the female and male parent lines (Mariani *et al.*, 1992; Williams *et al.*, 1997). The MS8 and RF3 glufosinate tolerance transgenes each confers at least a threefold safety margin to commercial application rate of Liberty. An efficient breeding approach to glufosinate-tolerant hybrid oilseed rape cultivar development would result in the development of glufosinate-tolerant doubled haploid lines for both female and male parents to be used in the hybrid (Palmer and Keller, 1997). To efficiently breed for the herbicide tolerance trait, it is presumed that all parent plants and hybrids were sprayed with glufosinate herbicide at approximately the normal commercial application rate, and glufosinate tolerance tests of all hybrids at a minimum double the commercial application rate were conducted. This ensured that the new glufosinate-tolerant hybrid oilseed rape cultivars were pure breeding for herbicide tolerance and that the crop had adequate levels of tolerance to glufosinate. Only Plant Genetic Systems has developed and registered glufosinate-tolerant hybrid oilseed rape cultivars in Canada to date, all of these developed using the NMS system.

Alternate glufosinate-tolerant hybrid-breeding approaches may also be possible since the glufosinate tolerance transgene acts like a Mendelian-dominant gene. Therefore, the RF3 glufosinate tolerance transgene may be deployed in just one parent used to make glufosinate-tolerant hybrids. This could speed the development of new glufosinate-tolerant hybrids and also make it possible to monitor hybridity of glufosinate-tolerant hybrid seed lots using herbicide tolerance.

G. REGISTRATION OF GLUFOSINATE-TOLERANT OILSEED RAPE CULTIVARS

Agrevo used the (glufosinate-tolerant Topas/ACSN3//AC Excel) cross to develop the first glufosinate-tolerant open pollinated population oilseed rape cultivar in Canada, Innovator, registered in 1995 (J. Lindo, Canadian Food Inspection Agency Variety Registration Office, personal communication, 2006). Innovator had the antibiotic-resistance gene version of the glufosinate tolerance transgene present in HCN92. Agrevo also developed and registered Canada's second glufosinate-tolerant open pollinated population oilseed rape cultivar, Independence in 1996 (Canadian Food Inspection Agency Variety Registration Office, 2006).

Additional glufosinate-tolerant open pollinated population oilseed rape cultivars were developed by other breeding organizations in Canada. DSV Canada developed the glufosinate-tolerant open pollinated population oilseed rape cv. LBD2393 LL, registered in 1997 (Canadian Food Inspection Agency Variety Registration Office, 2006). Svalof Weibull developed three glufosinate-tolerant open pollinated population oilseed rape cultivars, 2631 LL, registered in 1998, SW Legion LL, registered in 1999, and SW Flare LL, registered in 2000 (Canadian Food Inspection Agency Variety Registration Office, 2006).

The first glufosinate-tolerant hybrid oilseed rape cultivar 2153, developed by Plant Genetic Systems, was registered in 1997 (J. Lindo, personal communication, 2006). Fifteen additional new glufosinate-tolerant hybrid oilseed rape cultivars have been developed by Bayer CropScience and registered since 1999 (Canadian Food Inspection Agency Variety Registration Office, 2006).

H. PRODUCER ADOPTION

Many Canadian producers, being rapid adopters of new technology, tried glufosinate-tolerant open pollinated population oilseed rape cultivars when they first became available in the 1990s. The production area of glufosinate-tolerant open pollinated population oilseed rape cultivars has grown steadily since 1996. The switch to glufosinate-tolerant hybrid oilseed rape cultivars in recent years has resulted in steady increases in the production area of glufosinate-tolerant oilseed rape cultivars. Glufosinate-tolerant hybrid oilseed rape cultivars currently command a significant and growing market share (20–25% of planted oilseed rape area) in Canada (Canola Council of Canada, 2005; Crop Life Canada, 2002; Stringam *et al.*, 2003).

I. CONSUMER ACCEPTANCE

Consumer acceptance of glufosinate-tolerant oilseed rape-derived oil was, and continues to be, good in Canada and in several export destinations for Canadian-produced glufosinate-tolerant oilseed rape (Canola Council of Canada, 2005). The RF3 glufosinate tolerance transgene codes for two novel proteins. Proteins do not end up in the oil, the only direct use human food derived from oilseed rape. Glufosinate-tolerant oilseed rape was one of the first transgenic, genetically modified, crops grown on substantial areas in Canada and one of the first transgenic, genetically modified oilseed rape types grown in the world. The seed from the glufosinate-tolerant oilseed rape cultivars produced in Canada is blended into the oilseed rape seed stream and a composite oil and meal produced. The Canadian Federal Government approval process for new transgenic, genetically modified crops, based on substantial equivalency appears adequate to assure most consumers of the safety of the glufosinate-tolerant transgenic, genetically modified oilseed rape crop grown in this country (Canadian Food Inspection Agency, 1994; Health Canada, 1994a).

J. FUTURE OF GLUFOSINATE-TOLERANT OILSEED RAPE BREEDING

Glufosinate-tolerant oilseed rape cultivars, especially glufosinate-tolerant hybrid ones, are popular with Canadian producers. The glufosinate-tolerant hybrid oilseed rape cultivars command significant and steadily growing production area in Canada. More glufosinate-tolerant hybrid oilseed rape cultivars will be developed, registered, and grown in Canada in coming years. As other oilseed rape-producing areas of the world approve local production and consumption of transgenic, genetically modified oilseed rape, glufosinate-tolerant oilseed rape cultivars will be developed, registered, and grown commercially more widely throughout the world. Currently, it is difficult to predict when transgenic, genetically modified crops will be accepted elsewhere in the world and therefore difficult to predict when glufosinate-tolerant oilseed rape-breeding programs will begin in other oilseed rape-producing areas of the world.

IV. GLYPHOSATE-TOLERANT OILSEED RAPE

Another herbicide-tolerant oilseed rape type to be developed for use in Canada was glyphosate tolerance.

A. SOURCES OF GLYPHOSATE TOLERANCE GENES

Since no crops have naturally occurring tolerance to glyphosate, tolerance to glyphosate was introduced into oilseed rape and other crops using a combined approach of expression of a herbicide-insensitive target enzyme and detoxification of the herbicide approach (Dill, 2005). This required plant transformation using one or more transgenes and generated transgenic, genetically modified plants.

B. GLYPHOSATE MECHANISM OF ACTION

The site of action of glyphosate herbicide is in the chloroplasts where glyphosate is an amino acid biosynthesis inhibitor. Glyphosate binds nearly irreversibly to the phosphoenolpyruvate-binding site of the 5-enolpyruvylshikimate-3-phosphate synthase enzyme (Amrhein *et al.*, 1980), blocking the transfer of the enolpyruvyl moiety of phosphoenolpyruvate to shikimate-3-phosphate. The transfer of enolpyruvyl moiety from phosphoenolpyruvate to shikimate-3-phosphate is a critical step in the biosynthesis of aromatic amino acids, plant hormones, and other plant metabolites (Amrhein *et al.*, 1980).

C. GENETICS OF GLYPHOSATE TOLERANCE

Tolerance to glyphosate in oilseed rape is conferred by several genes, one of which is a gene from *A. tumefaciens* strain CP4 that possessed a gene coding for a glyphosate-insensitive enzyme, CP4 5-enolpyruvylshikimate-3-phosphate synthase (Dill, 2005). The CP4 strain enzyme has a high affinity for phosphoenolpyruvate coupled with a tolerance to glyphosate. Since the binding site for phosphoenolpyruvate on CP4 5-enolpyruvylshikimate-3-phosphate synthase is identical to the binding site for phosphoenolpyruvate in glyphosate-sensitive

5-enolpyruvylshikimate-3-phosphate synthase enzymes, it appears that changes in amino acid sequence outside the phosphoenolpyruvate-binding site create conformational changes in the enzyme, which are responsible for the lack of glyphosate binding at the phosphoenolpyruvate-binding site (Dill, 2005). While the CP4 5-enolpyruvylshikimate-3-phosphate synthase glyphosate enzyme is used in nearly all glyphosate-tolerant crops currently grown, use of this transgene-coded mechanism of glyphosate tolerance alone is insufficient to confer adequate levels of tolerance to glyphosate in oilseed rape (Dill, 2005). A second gene coding for an enzyme that detoxifies glyphosate to enhance glyphosate tolerance is used in oilseed rape. A gene from the soil bacterium *Ochrobactrum anthropi* (formerly *Achromobacter* sp.) strain LBAA coding for an enzyme, glyphosate oxidoreductase capable of degrading glyphosate to the nontoxic products glyoxylate and aminomethylphosphonate, is also used (Barry *et al.*, 1992; Cajacob *et al.*, 2004; Hallas *et al.*, 1988). The original glyphosate oxidoreductase transgene confers some tolerance to glyphosate, but its expression level was insufficient to confer adequate levels of glyphosate tolerance in oilseed rape (Dill, 2005). Modification of the coding region of the glyphosate oxidoreductase gene resulted in substantially increased glyphosate oxidoreductase expression and enhanced glyphosate degradation (Dill, 2005). However, even this enhanced level of glyphosate oxidoreductase expression was still insufficient to confer adequate glyphosate tolerance in oilseed rape. Since the target of action of glyphosate *in planta* is 5-enolpyruvylshikimate-3-phosphate synthase in the chloroplast, the glyphosate oxidoreductase gene was fused with a chloroplast transit peptide from the SSU1 gene of *A. thaliana* (Timko *et al.*, 1988). This resulted in a much enhanced glyphosate oxidoreductase enzyme activity in transformed plants (but still not enough to confer adequate tolerance to glyphosate in oilseed rape). Ultimately, adequate levels of glyphosate tolerance in oilseed rape required the combination of a constitutive promoter, the chloroplast transit peptide, the CP4 5-enolpyruvylshikimate-3-phosphate synthase gene, and the modified glyphosate oxidoreductase gene, all in one transgenic construct (Dill, 2005). This multigene glyphosate tolerance transgenic construct confers at least threefold commercial application rate tolerance to glyphosate. This construct was identified as the GT73 transgenic construct by Monsanto (Dill, 2005). A second glyphosate tolerance transgenic construct, similar to GT73, was also created and called the GT200 transgenic construct by Monsanto.

D. DEVELOPMENT OF NEW GLYPHOSATE-TOLERANT OILSEED RAPE LINES

The breeding glyphosate-tolerant transgenic oilseed rape lines began at Monsanto in the mid-1980s. Monsanto developed glyphosate-tolerant oilseed rape lines in the oilseed rape cv. Westar using *A. tumefaciens* mediated plant transformation using an antibiotic free version of the above multigene transgenic constructs (GT73 and GT200) ca. 1990. The GT73 and GT200 transgenic constructs, even though they contain three different transgenes, act like a single Mendelian-dominant transgenes. Molecular analysis showed that both the GT73 line and the GT200 line contained single copies of their respective glyphosate tolerance transgenes (Canadian Food Inspection Agency, 1995b, 1996a).

The GT73 and GT200 glyphosate tolerance transgenic constructs were patented by Monsanto in most or all oilseed rape-producing countries throughout the world. Since Monsanto was successful in attempts to transfer this transgene to *B. rapa* by interspecific crosses and backcrosses, it was concluded that GT73 was on the A genome of *B. napus* (Canadian Food Inspection Agency, 1995b).

Monsanto also trademarked the term Roundup Ready™ in 1997 to use as the name of the glyphosate herbicide tolerance system for oilseed rape crops developed using Monsanto-patented glyphosate tolerance transgenes combined with Roundup herbicide (Canadian Intellectual Property Office, 1997).

Access to the Monsanto-patented GT73 and GT200 glyphosate tolerance transgenes construct is controlled by research and/or commercial licenses between Monsanto and oilseed rape-breeding organizations.

E. CANADIAN REGULATORY APPROVALS

Environmental safety, feed safety, and food safety approvals were required for glyphosate tolerance constructs in oilseed rape lines since the Canadian Federal Government regulations governing plants with novel traits and transgenic, genetically modified plants came into effect in 1994 (Canadian Food Inspection Agency, 1994). Glyphosate-tolerant oilseed rape line Westar GT73 transgenic constructs were grown in confined trials in Canada from 1992 to 1994 to generate the data required by Canadian Federal Government regulators. The glyphosate-tolerant construct transgene present in line Westar GT73 received environmental safety approval from the Canadian Federal Government in 1995, feed safety approval in 1995 (Canadian Food Inspection Agency, 1995b), and food safety approval in 1994 (Health Canada, 1994b). Glyphosate-tolerant oilseed rape line Westar GT200 transgenic construct was grown in confined trials in Canada from 1992 to 1994 to generate the data required by Canadian Federal Government regulators. The glyphosate tolerance transgenic construct present in line Westar GT200 received environmental safety approval from the Canadian Federal Government in 1996, feed safety approval in 1996 (Canadian Food Inspection Agency, 1996a), and food safety approval in 1997 (Health Canada, 1997c).

F. BREEDING GLYPHOSATE-TOLERANT OILSEED RAPE CULTIVARS

The breeding of glyphosate-tolerant open pollinated population oilseed rape cultivars began in Canada in the late 1980s to early 1990s at Monsanto and several other breeding organizations in Canada. Many oilseed rape-breeding organizations throughout the world initially began their glyphosate-tolerant oilseed rape-breeding programs using both the GT73 line and transgenic construct and the GT200 line and transgenic construct. Ultimately, use of the GT200 line and transgenic construct for glyphosate-tolerant oilseed rape cultivar development was discontinued in the 1990s, leaving all further glyphosate-tolerant oilseed rape breeding to be done with the GT73 line and transgenic construct only. Breeding of glyphosate-tolerant open pollinated population oilseed rape cultivars used a crossing program combined with pedigree selection. There was at least a threefold safety margin for glyphosate application at commercial rate conferred by the GT73 glyphosate tolerance transgene. To efficiently breed for the herbicide tolerance trait, it is presumed that all plants in each cross generation, rows, and plots were sprayed with glyphosate herbicide at approximately the normal commercial application rate, and glyphosate tolerance tests of all advanced glyphosate-tolerant lines at a minimum double the commercial application rate were conducted. This ensured that the new glyphosate-tolerant open pollinated population oilseed rape cultivars were pure breeding for herbicide tolerance and that the crop had adequate levels of tolerance to glyphosate. Breeding of glyphosate-tolerant hybrid oilseed rape cultivars used a CMS system with the GT73 glyphosate tolerance transgene deployed in both the female and male parent lines. An efficient approach to glyphosate-tolerant hybrid oilseed rape cultivar development would result in the development of glyphosate-tolerant doubled haploid lines for both female and male parents to be used in the hybrid (Palmer and Keller, 1997). To efficiently breed for the herbicide tolerance trait, it is presumed that all parent plants and hybrids were sprayed with glyphosate herbicide at approximately the normal commercial application rate and glyphosate tolerance tests of all hybrids at a minimum double the commercial application rate were conducted. This ensured that the new glyphosate-tolerant hybrid oilseed rape cultivars were pure breeding for herbicide tolerance and that the crop had adequate levels of tolerance to glyphosate.

Alternate glyphosate-tolerant hybrid-breeding approaches may also be possible since the glyphosate tolerance transgenic construct acts like a Mendelian-dominant gene. Therefore, the GT73 glyphosate tolerance transgene may be deployed in just one parent used to make glyphosate-tolerant hybrids. This could speed the development of new glyphosate-tolerant hybrids and also make it possible to monitor hybridity of hybrid seed lots using herbicide tolerance.

To ensure the complete elimination of the GT200 glyphosate tolerance transgene from oilseed rape-breeding programs, monitoring for the presence of GT73 and the absence of GT200 transgenes in all glyphosate-tolerant oilseed rape plants and the absence of the G200 transgene in all non-glyphosate-tolerant oilseed rape plants used in the development of glyphosate-tolerant oilseed rape cultivars continues. All glyphosate-tolerant oilseed rape pedigree seed lots are also monitored for the presence of GT73 and the absence of GT200 transgenes. Glyphosate-tolerant oilseed rape cultivars are developed using the GT73 glyphosate tolerance transgenic construct.

G. REGISTRATION OF NEW GLYPHOSATE-TOLERANT OILSEED RAPE CULTIVARS

The GT73 glyphosate tolerance transgene has been used by numerous oilseed rape-breeding organizations to develop glyphosate-tolerant open pollinated population cultivars initially and more recently glyphosate-tolerant hybrid cultivars. The glyphosate-tolerant hybrid oilseed rape cultivars are being developed using both CMS and GMS pollination control systems.

The first glyphosate-tolerant open pollinated population oilseed rape cultivar registered in Canada was LG 3295, developed by Limagrain (now Monsanto) and registered in 1996 (J. Lindo, personal communication, 2006). The first glyphosate-tolerant hybrid oilseed rape cultivar was 357 RR, developed by Advanta (now Monsanto) and registered in 1997 (J. Lindo, personal communication, 2006). Nearly 100 additional glyphosate-tolerant open pollinated population and hybrid oilseed rape cultivars have been developed by numerous oilseed rape-breeding organizations and registered in Canada since 1996 (Canadian Food Inspection Agency Variety Registration Office, 2006).

H. PRODUCER ADOPTION

Canada was the first country in the world to grow glyphosate-tolerant oilseed rape commercially. Beginning with a small production area in 1996, glyphosate-tolerant oilseed rape cultivars are now grown on a very large area in Canada and are the dominant herbicide-tolerant oilseed rape cultivar type grown in Canada currently.

Glyphosate-tolerant oilseed rape has ~50% market share based on planted area in Canada currently (Canola Council of Canada, 2005 Crop Life Canada, 2002; Stringam *et al.*, 2003).

I. CONSUMER ACCEPTANCE

Consumer acceptance of glyphosate-tolerant oilseed rape oil was, and continues to be, good in Canada or several export destinations for Canadian-produced oilseed rape (Canola Council of Canada, 2005). The GT73 glyphosate tolerance transgene codes for two novel proteins. Proteins do not end up in the oil, the only direct use human food derived from oilseed rape. The seed from the glyphosate-tolerant oilseed rape cultivars produced in Canada is blended into the oilseed rape seed stream and a composite oil and meal produced. Glyphosate tolerant oilseed rape was one of the first transgenic, genetically modified crops grown on substantial areas in Canada and one of the first grown in the world. The Canadian Federal Government approval process for new transgenic, genetically modified crops based on substantial equivalency appears adequate to assure most consumers of the safety of the glyphosate-tolerant transgenic, genetically modified oilseed rape crop grown in this country.

J. FUTURE OF GLYPHOSATE-TOLERANT OILSEED RAPE BREEDING

Glyphosate-tolerant oilseed rape cultivars, especially glyphosate-tolerant hybrid ones, are popular with Canadian producers. The glyphosate-tolerant hybrid oilseed rape cultivars command significant and steadily growing production area in Canada. More glyphosate-tolerant hybrid oilseed rape cultivars will be developed, registered, and grown in Canada in coming years. As other oilseed rape-producing areas of the world approve local production and consumption of transgenic, genetically modified oilseed rape, glyphosate-tolerant oilseed rape cultivars will be developed, registered, and grown commercially more widely throughout the world. It is difficult to predict when transgenic, genetically modified crops will be accepted elsewhere in the world and therefore difficult to predict when glyphosate-tolerant oilseed rape-breeding programs will begin in other oilseed rape-producing areas of the world.

V. IMIDAZOLINONE-TOLERANT OILSEED RAPE

The second nontransgenic, nongenetically modified herbicide-tolerant oilseed rape type to be developed in Canada and the world was imidazolinone tolerance.

A. SOURCE OF IMIDAZOLINONE TOLERANCE GENES

Imidazolinone herbicide tolerance was developed using microspore mutagenesis. It is the first and the only herbicide-tolerant oilseed rape type in Canada developed by mutagenic techniques to date.

B. IMIDAZOLINONE MECHANISM OF ACTION

The site of action of imidazolinone herbicide is the plastid where imidazolinone is an amino acid biosynthesis inhibitor. Imidazolinone binding to acetohydroxyacid synthase greatly reduces or blocks acetohydroxyacid synthase catalytic activity and results in the deficiency of branched-chain amino acids. The acetohydroxyacid synthase enzyme catalyzes two reactions in the branched-chain amino acid biosynthesis pathway: the condensation of two pyruvate molecules to acetolactate and the biosynthesis of 2-acetohydroxybutyrate from pyruvate and 2-ketobutyrate (Singh and Shaner, 1995). Acetolactate leads to the biosynthesis of leucine and valine, while 2-acetohydroxybutyrate is a precursor of isoleucine. Acetohydroxyacid synthase is coded by a nuclear gene but the enzyme is transported into plastids where the biosynthesis of branched-chain amino acids occurs (Singh and Shaner, 1995). The imidazolinones bind to the entry site for the acetohydroxyacid synthase substrate or the substrate access channel (Ott *et al.*, 1996). Plants eventually die from the lack of leucine, valine, and isoleucine amino acids (Shaner and Singh, 1992).

C. GENETICS OF IMIDAZOLINONE TOLERANCE

Tolerance to imidazolinone was created by mutagenic treatment of the genes coding for acetohydroxyacid synthase, also known as acetolactate synthase. There are five acetohydroxyacid synthase nuclear genes in oilseed rape, with acetohydroxyacid synthase 2, acetohydroxyacid synthase 3, and acetohydroxyacid synthase 4 on the A genome of *B. napus* and acetohydroxyacid synthase 1 and acetohydroxyacid synthase 5 on the C genome (Tan *et al.*, 2005). Two mutant lines PM1 and PM2 exist. Line PM1 has a mutation in the acetohydroxyacid synthase 1 gene, while line PM2 has a mutation in the acetohydroxyacid synthase 3 gene (Tan *et al.*, 2005). These mutations occur in acetohydroxyacid synthase 1 codon position 653 and acetohydroxyacid synthase 3 codon position 574 in reference to the *A. thaliana* acetohydroxyacid synthase gene sequence (Tranel and Wright, 2002). The amino acid substitution at codon position 653 is asparagine substituted for serine, while the amino acid substitution at codon position 574 is leucine substituted for tryptophan (Tan *et al.*, 2005). These two mutations in the acetohydroxyacid synthase gene are semidominant. As a result, crop tolerance to imidazolinones increases with gene (allele) dosage.

The PM1 and PM2 mutant genes are apparently not patented and are therefore publicly available throughout the world.

A commercial license with American Cyanamid (now BASF) is required prior to the commercialization of imidazolinone herbicide-tolerant *Brassica* cultivars developed using the PM1 and PM2 mutant genes.

D. DEVELOPMENT OF NEW IMIDAZOLINONE-TOLERANT OILSEED RAPE LINES

The development of imidazolinone-tolerant oilseed rape lines was a collaborative research effort between American Cyanamid and Allelix Crop Technologies (Swanson *et al.*, 1989). Microspore mutagenesis of microspores derived from the oilseed rape cv. Topas was done using 20-M ethyl nitrosourea as the mutagen and 40-g/liter imazethapyr as the imidazolinone herbicide-selection agent. The microspores were chromosome doubled with colchicine and the doubled haploid lines grown and selfed. The selfed progeny of two doubled haploid lines showed levels of tolerance to imidazolinone higher than seen in the susceptible oilseed rape check cv. Topas. Line PM1 tolerated 50-g/ha imazethapyr (one type of imidazolinone herbicide), line PM2 tolerated 500-g/ha imazethapyr, while Topas was killed with only 20-g/ha imazethapyr. Crossing line PM1 with line PM2 produced progeny that were more tolerant of imazethapyr than either parent, showing an additive effect of the two tolerance sources (Swanson *et al.*, 1988, 1989). The Topas PM1 + PM2 mutant gene line was crossed to Topas or Regent and the progeny selfed for several generations to develop three imidazolinone-tolerant lines, NS738, NS1471, and NS1473 (Canadian Food Inspection Agency, 1995c).

E. CANADIAN REGULATORY APPROVALS

Three imidazolinone-tolerant lines, NC738, NS1471, and NS1473, containing the Topas PM1 + PM2 mutant genes in homozygous condition, were grown in trials in Canada from 1989 to 1994 (NS738) or from 1991 to 1994 (NS1471 and NS1473) to generate the data required by Canadian Federal Government regulators. Since imidazolinone-tolerant oilseed rape cultivars were developed using mutagenesis (i.e., using traditional breeding methods), they are considered to be novel but not transgenic, genetically modified plants. In Canada where plants with novel traits, not just transgenic, genetically modified plants, are reviewed by regulatory agencies (Canadian Food Inspection Agency, 1994), the imidazolinone-tolerant oilseed rape line underwent Canadian Federal Government regulatory review. The imidazolinone-tolerant PM1 + PM2 mutant gene lines NS738, NS1471, and NS1473 received environmental safety approval from the Canadian Federal Government in 1995, feed safety approval in 1995 (Canadian Food Inspection Agency, 1995c), and food safety approval in 1995 (Health Canada, 1995a).

F. BREEDING IMIDAZOLINONE-TOLERANT OILSEED RAPE CULTIVARS

The breeding of imidazolinone-tolerant open pollinated population or doubled haploid line oilseed rape cultivars began in Canada in the late 1980s at Pioneer Hi-Bred. Imidazolinone-tolerant oilseed rape cultivars were developed using the (Topas PM1 + PM2) line as the original source of imidazolinone tolerance (Swanson *et al.*, 1989). Breeding of imidazolinone-tolerant open pollinated population oilseed rape cultivars using a crossing program combined with pedigree selection was possible. However, since two independently segregating, semidominant genes are required for adequate imidazolinone tolerance, a more efficient approach to the development of imidazolinone-tolerant conventional cultivars was the development of imidazolinone-tolerant doubled haploid lines (Palmer and Keller, 1997). There was at least a threefold safety margin for imidazolinone application at commercial rate conferred by the PM1 + PM2 mutant gene combination. In initial attempts to efficiently breed for the herbicide tolerance trait, all plants in each cross generation or all doubled haploid lines were sprayed with imidazolinone herbicide at application rates sufficient to identify the homozygous PM1 + PM2 homozygous lines (presumed to be approximately one to two times commercial application rate) and imidazolinone tolerance tests of all advanced imidazolinone-tolerant lines at a minimum double the commercial application rate were conducted. This procedure was later supplemented with molecular marker assisted section of lines containing both the PM1 and PM2 mutant genes. This ensured that the new imidazolinone-tolerant conventional oilseed rape cultivars were pure breeding for herbicide tolerance and that the crop had adequate levels of tolerance to imidazolinone.

Breeding of imidazolinone-tolerant hybrid oilseed rape cultivars used a CMS system. Since the imidazolinone tolerance genes are semidominant, they must be deployed in homozygous condition in both the female and male parent lines. An efficient approach to imidazolinone-tolerant hybrid oilseed rape cultivar development would result in the development of imidazolinone-tolerant doubled haploid lines for both female and male parents to be used in the hybrid. To efficiently breed for the herbicide tolerance trait, it is presumed that all doubled haploid line parent plants and hybrids were sprayed with imidazolinone herbicide at approximately one to two times commercial application rate and imidazolinone tolerance tests of all hybrids at a minimum double the commercial application rate were conducted. This ensured that the new imidazolinone-tolerant hybrid oilseed rape cultivars were pure breeding for herbicide tolerance and that the crop had adequate levels of tolerance to imidazolinone.

G. REGISTRATION OF IMIDAZOLINONE-TOLERANT OILSEED RAPE CULTIVARS

Pioneer Hi-Bred developed Canada's first imidazolinone-tolerant conventional oilseed rape cultivar, 45A71, registered in 1997 (Canadian Food Inspection Agency Variety Registration Office, 2006). This variety was first grown commercially in 1998 under the Smart^{TM} canola name. Seventeen additional imidazolinone-tolerant open pollinated population and hybrid oilseed rape cultivars, including the first glufosinate-tolerant hybrid oilseed rape cultivar, SP 442 CL developed by Saskatchewan Pool and registered in 2004, were developed by several breeding organizations and registered in Canada since 1997 (Canadian Food Inspection Agency Variety Registration Office, 2006).

Since 2000, imidazolinone-tolerant oilseed rape cultivars have been marketed under the Clearfield^{TM} name (Canadian Intellectual Property Office, 2000d).

H. PRODUCER ADOPTION

Imidazolinone-tolerant oilseed rape cultivars have grown from a small production area in 1998 to a large production area in Canada in recent years. They currently have a 20–25% market share based on planted area for herbicide-tolerant oilseed rape types grown in Canada (Canola Council of Canada, 2005; Crop Life Canada, 2002; Stringam *et al.*, 2003).

Imidazolinone-tolerant oilseed rape cultivars are also grown in many other oilseed rape-producing areas throughout the world. Imidazolinone-tolerant oilseed rape cultivars were developed using mutagenesis (i.e., traditional breeding methods), so they are considered to be novel but not transgenic, genetically modified plants. Since the imidazolinone-tolerant oilseed rape cultivars were developed using traditional breeding methods, there have been fewer regulatory hurdles in the path to registration and commercialization of imidazolinone-tolerant oilseed rape cultivars in many countries of the world (Tan *et al.*, 2005, 2006) with the exception of Canada where plants with novel traits, not just transgenic or genetically modified plants, are reviewed by Federal Government regulatory agencies (Canadian Food Inspection Agency, 1994).

I. CONSUMER ACCEPTANCE

Consumers in Canada and throughout the world have readily accepted imidazolinone-tolerant oilseed rape-derived oil since the imidazolinone-tolerant oilseed rape cultivars are considered by consumers to be traditionally developed cultivars. Single-amino acid changes in two nonnovel proteins confer imidazolinone tolerance in oilseed rape. In addition, protein does not end up in the oil, the only direct use human food derived from oilseed rape. The seed from the imidazolinone-tolerant oilseed rape cultivars produced in Canada is blended into the oilseed rape seed stream and a composite oil and meal produced. In Canada, imidazolinone-tolerant oilseed rape was subjected to the same Federal Government mandated regulatory reviews as transgenic herbicide-tolerant oilseed rape. Extensive Canadian Government regulation based, in part, on the principle of substantial equivalency provided added assurance to most consumers of the safety of imidazolinone-tolerant oilseed rape oil.

J. FUTURE OF IMIDAZOLINONE-TOLERANT OILSEED RAPE BREEDING

New imidazolinone-tolerant open pollinated population and hybrid oilseed rape cultivars are being developed for Canada and many other oilseed rape-growing areas of the world. The development, registration, and commercialization of imidazolinone-tolerant oilseed rape cultivars is easier than for other herbicide-tolerant oilseed rape types because they are novel but not transgenic, genetically modified plants. The development of imidazolinone-tolerant hybrid oilseed rape cultivars in Canada and in other countries will likely increase the production area and market share for imidazolinone-tolerant hybrid oilseed rape in oilseed rape-producing areas of the world in coming years.

VI. BROMOXYNIL-TOLERANT OILSEED RAPE

The latest herbicide-tolerant oilseed rape type to be developed in Canada was bromoxynil tolerance.

A. SOURCES OF BROMOXYNIL TOLERANCE GENES

Since neither oilseed rape nor any related crops displayed naturally occurring tolerance to bromoxynil, tolerance to bromoxynil was introduced into crops using detoxification of the herbicide approach (Freyssinet *et al.*, 1996). This approach to the development of bromoxynil-tolerant crops required the use of transgenes and plant transformation and generated transgenic, genetically modified plants.

B. BROMOXYNIL MECHANISM OF ACTION

The site of action of bromoxynil herbicide is the chloroplast where bromoxynil is a photosynthesis inhibitor. Bromoxynil binds to the D1 protein of photosystem II and blocks electron transport. Bromoxynil binds nearly irreversibly to the plastoquinone Q_B-binding site on protein D1 of photosystem II blocking electron transport (Ashton and Crafts, 1981; Duke, 1990; Oettmeier *et al.*, 1982; Sanders and Pallett, 1985). This inhibits electron transport through photosystem II, which in turn reduces the amount of ATP and NADPH produced in the chloroplast, resulting in physiologically detrimental consequences to the plant (Dodge, 1983). In addition, when bromoxynil herbicides block the normal flow of electrons in photosystem II, high levels of singlet oxygen are produced in the chloroplast, which results in lipid peroxidation, loss of chloroplast membrane structure, chlorosis, cell death, and finally plant necrosis (Duke, 1990).

C. GENETICS OF BROMOXYNIL TOLERANCE

Tolerance to bromoxynil in oilseed rape is conferred by a gene from the soil bacterium *Klebsiella pneumoniae* subsp. *ozaenae*, which possessed a gene, the OXY gene, coding for a nitrilase enzyme capable of detoxifying bromoxynil (McBride *et al.*, 1986; Stalker and McBride, 1987; Stalker *et al.*, 1988). Nitrilase converts bromoxynil to the nonphytotoxic product, 5-dobromo-4-hydroxy benzoic acid, by hydrolysis removal of its cyano group (Freyssinet *et al.*, 1989; McBride *et al.*, 1986). Two different transgenic constructs using the OXY gene were created at Rhône-Poulenc Agrochimie (now Bayer CropScience). The first transgenic construct type had the OXY gene linked to a constitutive promoter, 35S CaMV (Odell *et al.*, 1985), while the second transgenic construct had the OXY gene linked to a light inducible promoter, the sunflower RuBisCo small subunit promoter (Waksman *et al.*, 1987). The bromoxynil-tolerant transgenic construct with the 35S CaMV promoter was identified as the 235 construct, while the bromoxynil-tolerant transgenic construct with the sunflower RuBisCo promoter was identified as the 237 construct (G. Freyssinate, personal communication). There was no antibiotic resistance gene in either the 235 or 237 transgenic constructs. The bromoxynil tolerance gene was used as the selectable marker.

D. DEVELOPMENT OF NEW BROMOXYNIL-TOLERANT OILSEED RAPE LINES

Rhône-Poulenc Agrochimie developed bromoxynil-tolerant oilseed rape in the oilseed rape cv. Westar using *A. tumefaciens*-mediated plant transformation using an antibiotic-free version of the above two OXY transgenic constructs ca 1988 (Freyssinet *et al.*, 1996). Several different inserts for each OXY gene construct were created. Several different transgene insert lines for both 235 and 237 construct types were produced in the cv. Westar using disarmed *A. tumefaciens* as a vector. Molecular analysis showed that the Westar OXY 235 line contained a single copy of the bromoxynil tolerance transgene (Canadian Food Inspection Agency, 1998). Molecular analysis showed that the Westar OXY 237 line contained a single copy of the bromoxynil tolerance transgene (Cuthbert and McVetty, 2001).

Since the University of Manitoba was successful in attempts to transfer the OXY 235 transgene to *B. rapa* by interspecific crosses and backcrosses, it was concluded that OXY 235 was on the A genome of *B. napus*.

The OXY 235 and OXY 237 bromoxynil tolerance transgenic constructs were patented by Rhône-Poulenc Agrochimie in most or all oilseed rape-producing countries throughout the world.

Access to the Rhône-Poulenc Agrochimie-patented OXY 235 and OXY 237 transgenic constructs was controlled by research and/or commercial licenses between Rhône-Poulenc Agrochimie and oilseed rape-breeding organizations.

Rhône-Poulenc Agrochimie also trademarked the term Navigator™ Compas™ in 2000 to use as the brand name of bromoxynil-tolerant oilseed rape cultivars developed using Rhône-Poulenc Agrochimie-patented bromoxynil tolerance transgenes used with a bromoxynil plus clethodim herbicide mix (Canadian Intellectual Property Office, 2000b).

E. CANADIAN REGULATORY APPROVALS

Environmental safety, feed safety, and food safety approvals were required for bromoxynil tolerance constructs in oilseed rape lines since the Canadian Federal Government regulations governing plants with novel traits and transgenic, genetically modified plants came into effect in 1994 (Canadian Food Inspection Agency, 1994). Bromoxynil-tolerant Westar transgenic lines containing the OXY 235 and OXY 237 transgenic constructs were tested in confined field trials in Canada from 1992 to 1996 to generate the data required for the regulatory approval of bromoxynil-tolerant oilseed rape in Canada. The bromoxynil tolerance transgenic construct present in line Westar OXY 235 received environmental safety approval from the Canadian Federal Government in 1997, feed safety approval in 1997 (Canadian Food Inspection Agency, 1998), and food safety approval in 1999 (Health Canada, 1999).

F. BREEDING BROMOXYNIL-TOLERANT OILSEED RAPE CULTIVARS

Several different inserts of both the OXY 235 transgenic construct and the OXY 237 transgenic construct in bromoxynil-tolerant Westar transgenic lines were tested by the University of Manitoba in confined field trails from 1992 to 1993 to identify the best OXY 235 and OXY 237 insert lines among several different insert lines. OXY 235–2 and OXY 237–1 were identified as the best insert lines for these two different bromoxynil tolerance transgenic constructs. These bromoxynil-tolerant lines were used in all further bromoxynil-tolerant oilseed rape-breeding work. Both the OXY 235 and

OXY 237 transgenic constructs conferred at least a tenfold safety margin to herbicide application at commercial rates.

Breeding of bromoxynil-tolerant oilseed rape cultivars began at the University of Manitoba in 1994. Initially, the University of Manitoba used both OXY 235 and OXY 237 transgenic constructs in its bromoxynil-tolerant oilseed-breeding program. Crosses of the bromoxynil-tolerant Westar OXY 235 and Westar OXY 237 transgenic lines to blackleg resistant canola cultivars combined with pedigree selection was used to develop new bromoxynil-tolerant open pollinated population oilseed rape cultivars. Bromoxynil-tolerant oilseed rape cultivar development with the OXY 237 construct was discontinued in early 1996, leaving the OXY 235 transgenic construct as the sole source of bromoxynil tolerance in oilseed rape. The University of Manitoba and Svalof Weibull began a collaborative bromoxynil oilseed rape cultivar development program in 1996, using only the OXY 235 bromoxynil tolerance transgene.

Bromoxynil-tolerant hybrid oilseed rape cultivars were under development at the University of Manitoba using CMS pollination control systems. The OXY 235 transgene was incorporated in both the female and male parents for selected hybrid oilseed rape cultivars, ensuring complete purity for bromoxynil tolerance in the hybrid seed lots produced, regardless of the hybridity of these seed lots. Alternate glyphosate-tolerant hybrid-breeding approaches were also possible since the glyphosate tolerance transgenic construct acted like a Mendelian-dominant gene (Cuthbert and McVetty, 2001). Therefore, the OXY 235 bromoxynil tolerance transgene could be deployed in just one parent used to make bromoxynil-tolerant hybrids. This would have sped the development of new bromoxynil-tolerant hybrids and also made it possible to monitor hybridity of hybrid seed lots using herbicide tolerance.

To ensure the complete elimination of the OXY 237 bromoxynil tolerance transgene from oilseed rape-breeding programs, monitoring for the presence of the OXY 235 transgene and the absence of the OXY 237 transgene in all bromoxynil-tolerant oilseed pedigree seed lots before they left the University of Manitoba greenhouses was done using PCR-based techniques.

After a thorough review of all bromoxynil-tolerant oilseed rape related business matters, Bayer CropScience, the owner of the OXY 235 transgene patents and trademarks in oilseed rape and the manufacturer of Compas™ herbicide, decided to withdraw bromoxynil-tolerant oilseed rape from the Canadian market in early 2002. Bromoxynil-tolerant oilseed rape-breeding programs ceased in 2002 and no bromoxynil-tolerant oilseed rape-breeding materials were planted in the field after 2001. As a consequence, the frequency of the OXY 235 transgene in the field in Canada is declining steadily, and at this time 5 years after the last commercial production of bromoxynil-tolerant oilseed rape in Canada, the frequency of the OXY 235 transgene should already be very low. Monitoring of oilseed rape lines bred at former bromoxynil-tolerant oilseed rape-breeding organizations including the University of Manitoba, using PCR-based screening for the presence of the OXY 235 transgene, continues to ensure that no further inadvertent introduction of the OXY 235 transgene into the fields of Canada occurs. It is, however, difficult to predict when the OXY 235 transgene will be completely eliminated from Canadian fields. Oilseed rape breeders may find it is much easier to introduce a new transgene into oilseed rape than it is to completely remove it. The introduction of bromoxynil-tolerant open pollinated population oilseed rape cultivars up to the commercial production level in Canada followed by the complete removal of the OXY 235 transgene from all oilseed rape grown in Canada is providing a good learning experience in this regard.

G. REGISTRATION OF BROMOXYNIL-TOLERANT OILSEED RAPE CULTIVARS

The OXY 235 transgene was used by the University of Manitoba, in collaboration with Svalof Weibull, to develop several open pollinated population bromoxynil-tolerant cultivars. Four bromoxynil-tolerant oilseed rape cultivars were registered in Canada in 2000: Armor BX (McVetty *et al.*, 2001a), Cartier BX (McVetty *et al.*, 2001b), 295 BX (McVetty *et al.*, 2001c), and Zodiac BX (McVetty *et al.*, 2001d). One additional bromoxynil-tolerant oilseed rape cultivar, Renegade BX, was registered in 2001 (McVetty *et al.*, 2002). The registrations of all bromoxynil-tolerant open pollinated population oilseed rape cultivars in Canada were canceled in 2003 since Bayer CropScience decided to withdraw bromoxynil-tolerant oilseed rape from the Canadian market in 2002.

H. PRODUCER ADOPTION

Canada was the first country in the world to grow bromoxynil-tolerant oilseed rape commercially. Bromoxynil-tolerant open pollinated population oilseed rape production began on a small production area in 1999, and then on moderately increasing production areas in 2000 and 2001. Bromoxynil-tolerant oilseed rape cultivars had ~1% market share of planted oilseed rape area by 2001, the last year for bromoxynil-tolerant oilseed rape commercial production in Canada (Stringam *et al.*, 2003). Bayer CropScience decided to withdraw bromoxynil-tolerant oilseed rape from the Canadian market in early 2002. There was, therefore, no commercial production of bromoxynil-tolerant open pollinated population oilseed rape in Canada after 2001.

I. CONSUMER ACCEPTANCE

Consumer acceptance of bromoxynil-tolerant oilseed rape cultivar-derived oil was good in Canada or several export destinations for Canadian-produced oilseed rape. The OXY 235 bromoxynil tolerance transgene codes for a single novel protein. Proteins do not end up in the oil, the only direct use human food derived from oilseed rape. The seed from the bromoxynil-tolerant oilseed rape cultivars produced in Canada was blended into the oilseed rape seed stream and a composite oil and meal produced. Bromoxynil-tolerant oilseed rape was one of the first genetically modified crops grown in Canada, and Canada was the first and only area in the world where bromoxynil-tolerant oilseed rape was grown.

J. FUTURE OF BROMOXYNIL-TOLERANT OILSEED RAPE BREEDING

It appears that new bromoxynil-tolerant oilseed rape cultivars will not be developed in Canada or in any other oilseed rape-producing areas of the world in the foreseeable future.

VII. CONCLUSION

Several different herbicide-tolerant oilseed rape types have been developed, registered, and grown commercially in Canada in the last 20 years. The first herbicide-tolerant oilseed rape cultivars developed, registered, and grown commercially in Canada were triazine tolerant. These cultivars were developed using a backcrossing program since the herbicide tolerance was a cytoplasmically inherited trait. However, triazine-tolerant oilseed rape cultivars had several agronomic deficiencies that limited their success in Canada. Triazine-tolerant oilseed rape cultivars were first grown in Canada in 1987 and were last grown in 1993. Triazine-tolerant oilseed rape cultivars are still being developed, registered, and grown commercially in Australia (Norton, 2003).

The next herbicide-tolerant oilseed rape cultivars developed, registered, and grown commercially in Canada were glufosinate-tolerant ones, with the first cultivar, Innovator, grown commercially in 1995. New glufosinate-tolerant oilseed rape cultivars continue to be developed and released with hybrid glufosinate-tolerant cultivars predominating in recent years. These cultivars were developed using pedigree selection since the glufosinate tolerance transgene acts like a Mendelian-dominant gene that can be easily selected. However, because glufosinate-tolerant oilseed rape materials are transgenic, genetically modified plants, there were additional breeding challenges initially. Glufosinate-tolerant hybrid oilseed rape cultivars command a significant share of the oilseed rape market in Canada currently. Glufosinate-tolerant oilseed rape cultivars are transgenic, and they are grown in few other oilseed rape production areas of the world outside of Canada because of their transgenic, genetically modified status.

The next herbicide-tolerant oilseed rape cultivars developed, registered, and grown commercially in Canada were glyphosate-tolerant ones, with LG 3295 being the first glyphosate-tolerant cultivar grown commercially in 1996. New glyphosate-tolerant oilseed rape cultivars continue to be developed and released with glyphosate-tolerant hybrid cultivars predominating in recent years. These cultivars were developed using pedigree selection since the glyphosate tolerance transgene acts like a Mendelian-dominant gene that can be easily selected. However, because glyphosate-tolerant oilseed rape materials are transgenic, genetically modified plants, there were additional breeding challenges initially. Glyphosate-tolerant open pollinated population and hybrid oilseed rape cultivars command a dominant share of the oilseed rape market in Canada currently. Glyphosate-tolerant oilseed rape cultivars are transgenic, and they are grown in few other oilseed rape-producing areas of the world outside of Canada because of their transgenic, genetically modified plant status.

The next herbicide-tolerant oilseed rape cultivars developed, registered, and grown commercially in Canada were imidazolinone tolerant. These cultivars were developed using traditional breeding techniques, including double haploid line development and pedigree selection because this herbicide tolerance is conferred by two independently segregating, semidominant genes. These genes were created by mutagenesis so this herbicide tolerance is nontransgenic, nongenetically modified. Imidazolinone-tolerant oilseed rape cultivars were first grown commercially in Canada in 1997. New imidazolinone-tolerant oilseed rape cultivars continue to be developed and released with more hybrid imidazolinone-tolerant cultivars being released in recent years. Imidazolinone-tolerant open pollinated population and hybrid oilseed rape cultivars command a significant share of the oilseed rape market in Canada currently. Imidazolinone-tolerant oilseed rape cultivars are nontransgenic, nongenetically modified plants that are grown in many other oilseed rape-producing areas of the world outside of Canada already.

The most recent herbicide-tolerant oilseed rape cultivars developed, registered, and grown commercially in Canada were bromoxynil-tolerant. These cultivars were developed using pedigree selection because the herbicide tolerance transgene acts like a Mendelian-dominant gene that can be easily selected. However, because bromoxynil-tolerant oilseed rape materials are transgenic, genetically modified, there were additional breeding challenges initially. Bromoxynil-tolerant oilseed rape cultivars were first grown commercially in Canada in 1999 and last grown commercially in 2001. Bromoxynil-tolerant oilseed rape cultivars were only grown commercially in Canada. Efforts continue to completely remove the OXY 235 transgene from all oilseed rape grown in Canada.

There are three successful herbicide-tolerant oilseed rape types currently grown in Canada, one nontransgenic, nongenetically modified (imidazolinone) tolerance and two transgenic, genetically modified (glufosinate and glyphosate) tolerance. It appears that these may be all of the herbicide-tolerant oilseed rape types that will be developed, registered, and grown commercially in Canada for quite sometime since it is becoming increasingly difficult and expensive to develop, register, and then commercialize any new herbicide tolerance types, especially transgenic types. It has been estimated that it will take a minimum of 8–10 years from the discovery of a new herbicide tolerance gene to the commercialization of a new herbicide-tolerant crop and cost between \$50 and \$100 million (Dill, 2005).

It is likely that commercial production of herbicide-tolerant oilseed rape throughout the world will increase as more oilseed rape-growing areas adopt this technology. It is, however, difficult, at the present time, to put a timetable on these developments.

Herbicide-tolerant oilseed rape breeding has been very successful in Canada in recent years. Over 90% of all oilseed rape hectares grown in Canada are planted to herbicide-tolerant oilseed rape cultivars now (Canola Council of Canada, 2005). Herbicide-tolerant oilseed rape appears here to stay in Canada!

REFERENCES

Amrhein, N., Deus, B., Gehrke, P. and Steinrucken, H. C. (1980). The site of the inhibition of the shikimate pathway by glyphosate. 2. Interference of glyphosate with chorismate formation *in vivo* and *in vitro*. *Plant Physiology* **66**, 830–834.

Ashton, F. M. and Crafts, A. S. (1981). Nitriles. *In* "Mode of Action of Herbicides," 2nd Edn, Chapter 16, pp. 254–271. John Wiley & Sons, New York.

Barry, G., Kishore, G., Taylor, M., Padgette, S. M., Kolacz, K., Weldon, M., Re, D., Eichholtz, D., Fincher, K. and Hallas, K. (1992). Inhibitors of amino acid biosynthesis: strategies for imparting glyphosate tolerance to crop plants. *In* "Biosynthesis and Molecular Regulation of Amino Acids in Plants" (B. K. Singh, H. E. Flores and J. C. Shannon, eds.), pp. 139–145. American Society of Plant Physiologists, Rockville, MD.

Beversdorf, W. D. and Hume, D. J. (1984). OAC Triton spring rapeseed. *Canadian Journal of Plant Science* **64**, 1007–1009.

Beversdorf, W. D., Weiss-Lerman, J., Erickson, L. R. and Souza Machado, V. (1980a). Transfer of cytoplasmically-inherited triazine resistance from birds rape to cultivated oilseed rape (*Brassica campestris* and *Brassica napus*). *Canadian Journal of Genetics and Cytology* **22**, 167–172.

Beversdorf, W. D., Weiss-Lerman, J. and Erickson, L. R. (1980b). Registration of triazine resistant *Brassica rapa* germplasm. *Crop Science* **20**, 289.

Beversdorf, W. D., Weiss-Lerman, J. and Erickson, L. R. (1980c). Registration of triazine resistant *Brassica napus* germplasm. *Crop Science* **20**, 289.

Beversdorf, W. D., Hume, D. J. and Donnelly-Vanderloo, M. J. (1988). Agronomic performance of triazine-resistant and susceptible spring canola hybrids. *Crop Science* **28**, 932–934.

Beversdorf, W. D., Hume, D. J., Gostovic, P., Chu, G., Montminy, W. and Muscat, J. (1990). OAC Triumph summer rape. *Canadian Journal of Plant Science* **70**, 857–859.

Cajacob, C. A., Feng, P. C., Heck, G. R., Murtaza, F. A., Sammons, R. D. and Padgette, S. R. (2004). Engineering resistance to herbicides. *In* "Handbook of Biotechnology" (P. Christou and H. Klee, eds.), Chapter 1, pp. 353–372. John Wiley and Sons, Chichester, UK.

Canadian Food Inspection Agency (1994).(CFIA) Directive 94–08 Dir94–08 (1994). Assessment criteria for determining environmental safety of plants with novel traits. (http://www.inspection.gc.ca/english/plaveg/bio/dir/dir9408e.shtml).

Canadian Food Inspection Agency (1995a).(CFIA) Decision Document DD95–01. Determination of environmental safety (and feed safety) of Agrevo Canada Inc.'s glufosinate ammonium-tolerant canola. (http://www.inspection.gc.ca/english/plaveg/bio/dd/dd9501e.shtml).

Canadian Food Inspection Agency (1995b).(CFIA) Decision Document DD95–02. Determination of environmental safety (and feed safety) of Monsanto Canada's Roundup® herbicide –tolerant Brassica napus canola line GT73. (http://www.inspection.gc.ca/english/plaveg/bio/dd/dd9502e.shtml).

Canadian Food Inspection Agency (1995c).(CFIA), Decision Document DD 95–03. Determination of environmental safety (and feed safety) of Pioneer Hi-Bred International Inc.'s imidazolinone-tolerant canola. (http://www.inspection.gc.ca/english/plaveg/bio/dd/dd9503e.shtml).

Canadian Food Inspection Agency (1996a).(CFIA), Decision Document DD96–07. Determination of environmental safety of Monsanto Canada's Roundup® herbicide—tolerant *Brassica napus* canola line GT200. (http://www.inspection.gc.ca/english/plaveg/bio/dd/dd9607e.shtml).

Canadian Food Inspection Agency (1996b).(CFIA), Decision Document DD96–11. Determination of environmental safety of Agrevo Canada Inc.'s glufosinate ammonium-tolerant canola line HCN28. (http://www.inspection.gc.ca/english/plaveg/bio/dd/dd9611e.shtml).

Canadian Food Inspection Agency (1996c).(CFIA) Decision Document DD96–11. Supplement to decisions document DD96–11: Determination of Livestock feed safety of Agrevo Canada's Inc.'s glufosinate ammonium tolerant canola line HCN28. (http://www.inspection.gc.ca/english/plaveg/bio/dd/dd9611supe.shtml).

Canadian Food Inspection Agency (1996d).(CFIA) Decision Document DD96–17. Determination of environmental safety (and feed safety) of Plant Genetic Systems Inc.'s (PGS) novel hybridization system for rapeseed (*Brassica napus* L.). (http://www.inspection.gc.ca/english/plaveg/bio/dd/dd9617e.shtml).

Canadian Food Inspection Agency (1998).(CFIA) Decision Document DD98–25. Determination of environmental (and feed safety) safety of Rhone-Poulenc's oxynil herbicide-tolerant *Brassica napus* canola line Westar Oxy-235. (http://www.inspection.gc.ca/english/plaveg/bio/dd/dd9825e.shtml).

Canadian Food Inspection Agency Variety Registration Office (2006).Variety Registration Office for oilseed rape cultivars grown in Canada. (http://www.inspection.gc.ca/english/plaveg/variet/vartoce.shtml).

Canadian Intellectual Property Office (1997). Canadian Intellectual Properties Office. Trademark registration for Roundup Ready. Reg. # TMA470473.

Canadian Intellectual Property Office (2000a). Canadian Intellectual Properties Office. Trademark registration for Navigator. Reg. # TMA522502.

Canadian Intellectual Property Office (2000b). Canadian Intellectual Properties Office. Trademark registration for Compass. Reg. # TMA525556.

Canadian Intellectual Property Office (2000c). Canadian Intellectual Properties Office. Trademark registration for Liberty Link. Reg. # TMA526007.

Canadian Intellectual Property Office (2000d). Canadian Intellectual Properties Office. Trademark registration for Clearfield. Reg. # TMAS531173.

Canola Council of Canada (2005).Overview of Canada's Canola Industry. (http://www.canola-council.org/ind-overview.html).

Crop Life Canada (2002). Trends here and abroad: Farmers are voting with their seeders. *Biocrop News* **1**(1), 2 (http://www.croplife.ca/english/pdf/Biocrop_2002v1.pdf).

Cuthbert, J. L. and McVetty, P. B. E. (2001). Comparison of the performance of bromoxynil-resistant and susceptible near-isogenic populations of oilseed rape. *Canadian Journal of Plant Science* **81**, 367–372.

Dill, G. M. (2005). Glyphosate-resistant crops: History, status and future. *Pest Management Science* **61**, 219–224.

Dodge, A. D. (1983). Toxic oxygen species and herbicide action. *In* "Pesticide Chemistry: Human Welfare and the Environment" (J. Miyamoto and P. C. Kearney, eds.), pp. 55–59. Proceedings of the fifth International Congress Pestic. Chem., Pergamon Press, Oxford.

Downey, R. K. and Röbbelen, G. (1989). Brassica species. *In* "Oil Crops of the World" (G. Röbbelen, R. K. Downey and A. Ashri, eds.), Chapter 16, pp. 339–362. McGraw-Hill, New York.

Droge, W., Broer, I. and Puhler, A. (1992). Transgenic plants containing the phosphinothricin-N-acetyltransferase gene metabolize the herbicide L-phosphinothricin (glufosinate) differently from untransformed plants. *Planta* **187**, 142–151.

Duke, S. O. (1990). Overview of herbicide mechanisms of action. *Environmental Health Perspectives* **87**, 263–271.

Duke, S. O. (1996). Herbicide-resistant crops—background and perspectives. *In* "Herbicide Resistant Crops" (S. O. Duke, ed.), Chapter 1, pp. 1–10. CRC Press Inc., Boca Raton, FL.

Dyer, W. E. (1996). Techniques for producing herbicide resistant cultivars. *In* "Herbicide Resistant Crops" (S. O. Duke, ed.), Chapter 3, pp. 38–51. CRC Press Inc., Boca Raton, FL.

Freyssinet, G., Leroux, B., Lebrun, M., Pelissier, B., Sailland, A. and Pallet, K. E. (1989). Transfer of bromoxynil resistance into two crops. Brighton Crop Protection Conference – Weeds, pp. 1225–1234.

Freyssinet, M., Creange, P., Renard, M., McVetty, P. B. E., Derose, P. and Freyssinet, G. (1996). Development of transgenic oilseed rape resistant to oxynil herbicides. *In* "Crop Protection: Forecasting and Chemical Control. pp. 974–976." International Rapeseed Congress.

Fuerst, E. P. and Norman, M. A. (1991). Interactions of herbicides with photosynthetic electron transport. *Weed Science* **39**, 458–464.

Hall, J. C., Donnelly-Vanderloo, M. J. and Hume, D. J. (1996). Triazine-resistant crops: The agronomic impact and physiological consequences of chloroplast mutation. *In* "Herbicide-Resistant Crops: Agricultural, Environmental, Economic, Regulatory and Technical Aspects" (S. O. Duke, ed.), Chapter 7, pp. 107–126, 420 pp. CRC Press, Inc., Boca Raton, FL.

Hallas, L. E., Hahn, E. M. and Korndorfer, C. (1988). Characterization of microbial traits associated with glyphosate biodegradation in industrial activated sludge. *Journal of Industrial Microbiology* **3**, 377–385.

Health Canada (1994a). Guidelines for the safety assessment of novel foods—Volume 1. (http://hc-sc.gc.ca/fn-an/legislation/guide-ld/nvvli_e/html).

Health Canada (1994b). Glyphosate tolerant canola, GT73. FD/OFB-094–325-A. (http://hc-sc.gc.ca/fn-an/gmf-agm/appro/ofb-094–325-a_e.html).

Health Canada (1995a). Imidazolinone herbicide tolerant canola lines NS738, NS1471, NS1473. (http://hc-sc.gc.ca/fn-an/gmf-agm/appro/ofb-095–115-a_e.html).

Health Canada (1995b). Glufosinate ammonium tolerant canola (HCN92). (http://hc-sc.gc.ca/fn-an/gmf-agm/appro/ofb-094–325-a_e.html).

Health Canada (1997a). Novel hybridization system for canola (MS8/RF3). (http://hc-sc.gc.ca/fn-an/gmf-agm/appro/28bg_pgs_e.html).

Health Canada (1997b). Glufosinate ammonium tolerant canola (T45). (http://hc-sc.gc.ca/fn-an/gmf-agm/appro/31bg_agrevo-ct_agrevo_e.html).

Health Canada (1997c). Glyphosate tolerant canola, GT200. FD/OFB-097–255-A. (http://hc-sc.gc.ca/fn-an/gmf-agm/appro/ofb-097–255-a_e.html).

Health Canada (1999). Bromoxynil tolerant canola (Westar-Oxy-235). (http://hc-sc.gc.ca/fn-an/gmf-agm/appro/36sum_e.html).

Hess, F. D. (2000). Light-dependent herbicides: an overview. *Weed Science* **48**, 160–170.

Hobbs, S. L. A. (1987). Comparison of photosynthesis in normal and triazine-resistant *Brassica*. *Canadian Journal of Plant Science* **67**, 457–466.

Kimber, D. S. and McGregor, D. I. (1995). Brassica oilseeds. The species and their origin, cultivation and world production. *In* "Brassica Oilseeds Production and Utilization" (D. S. Kimber and D. I. McGregor, eds.), Chapter 1, pp. 1–7. CAB International, Wallingford, UK.

Maltais, B. and Bouchard, C. J. (1978). Une moutarde des oiseaux (*Brassica rapa* L.) resistante a l'atrazine. *Phytoprotection* **59**, 117–119.

Mariani, C., Gossele, V., De Beuckeleer, M., De Block, M., Goldberg, R. B., De Greef, W. and Leemans, J. (1992). A chimaeric ribonuclease-inhibitor gene restores fertility to male sterile plants. *Nature* **357**, 384–387.

McBride, K. E., Kenny, J. W. and Stalker, D. M. (1986). Metabolism of the herbicide bromoxynil by *Klebsiella pneumoniae* subsp. *ozaenae*. *Applied and Environmental Microbiology* **52**, 325–330.

McCubbin, A. and Dickinson, H. G. (1997). Self-incompatibility. *In* "Pollen Biotechnology for Crop Production and Improvement" (K. R. Shivanna and V. K. Sawhney, eds.), Chapter 9, pp. 199–217. Cambridge University Press, Cambridge, UK.

McVetty, P. B. E. (1997). Cytoplasmic male sterility. *In* "Pollen Biotechnology for Crop Production and Improvement" (K. R. Shivanna and V. K. Sawhney, eds.), Chapter 7, pp. 155–182. Cambridge University Press, Cambridge, UK.

McVetty, P. B. E., Rimmer, S. R. and Scarth, R. (2001a). Armor BX summer oilseed rape. *Canadian Journal of Plant Science* **81**, 93–95.

McVetty, P. B. E., Rimmer, S. R. and Scarth, R. (2001b). Cartier BX summer oilseed rape. *Canadian Journal of Plant Science* **81**, 105–106.

McVetty, P. B. E., Scarth, R. and Rimmer, S. R. (2001c). 295 BX summer oilseed rape. *Canadian Journal of Plant Science* **81**(1), 97–99.

McVetty, P. B. E., Rimmer, S. R. and Scarth, R. (2001d). Zodiak BX summer oilseed rape. *Canadian Journal of Plant Science* **81**, 101–103.

McVetty, P. B. E., Rimmer, S. R. and Scarth, R. (2002). Renegade BX summer oilseed rape. *Canadian Journal of Plant Science* **82**, 435–436.

Norton, R. (2003). Conservation farming systems and canola. Report of the University of Melbourne, Melbourne, 3010 Australia, 26pp.

Odell, J. T., Nagy, F. and Chua, N. H. (1985). Identification of sequences required for activity of the cauliflower mosaic virus 35S promoter. *Nature* **313**, 810–812.

Oettmeier, W., Masson, K. and Johanningmeier, V. (1982). Evidence for two different herbicide binding proteins at the reducing site of photosystem II. *Biochimica et Biophysica Acta* **679**, 376–383.

Orson, J. (1995). Weeds and their control. *In* "*Brassica* Oilseeds Production and Utilization" (D. S. Kimber and D. I. McGregor, eds.), Chapter 4, pp. 93–109. CAB international, UK.

Ott, K., Kwagh, J., Stockton, G. W., Sidorov, V. and Kakefuda, G. (1996). Rational molecular design and genetic engineering of herbicide resistant crops by structure modeling and site-directed mutagenesis of acetohydroxyacid synthase. *Journal of Molecular Biology* **263**, 359–368.

Palmer, C. E. and Keller, W. F. (1997). Pollen embryos. *In* "Pollen Biotechnology for Crop Production and Improvement" (K. R. Shivanna and V. K. Sawhney, eds.), Chapter 20, pp. 392–422. Cambridge University Press, Cambridge, UK.

Rakow, G. (1992). AC Tristar summer rape. *Canadian Journal of Plant Science* **72**, 1239–1240.

Rakow, G. and Downey, R. K. (1993). Tribute summer rape. *Canadian Journal of Plant Science* **73**, 189–191.

Sanders, G. E. and Pallett, K. E. (1985). *In vitro* activity and binding characteristics of the hydroxybenzonitriles in chloroplasts isolated from *Matricaria inodora* and *Viola arvensis*. *Pesticide Biochemistry and Physiology* **24**, 317–325.

Sawhney, V. K. (1997). Genic male sterility. *In* "Pollen Biotechnology for Crop Production and Improvement" (K. R. Shivanna and V. K. Sawhney, eds.), Cambridge University Press, Cambridge, UK.

Shaner, D. L. and Singh, B. K. (1992). How does inhibition of amino acid biosynthesis kill plants? *In* "Biosynthesis and Molecular Regulation of Amino Acids in Plants" (B. K. Singh, H. E. Flores and J. C. Shannon, eds.), pp. 174–183. American Society of Plant Physiologists, Rockville Maryland.

Singh, B. K. and Shaner, D. L. (1995). Biosynthesis of branched chain amino acids: from test tube to field. *Plant Cell* **7**, 935–944.

Souza Machado, V., Bandeen, J. D., Sephenson, G. R. and Lavigne, P. (1978). Uniparental inheritance of chloroplast atrazine tolerance in *Brassica campestris*. *Canadian Journal of Plant Science* **58**, 977–981.

Stalker, D. M. and McBride, K. E. (1987). Cloning and expression in *E. coli* of a *Klebsiella ozaenae* plasmid-borne gene encoding a nitrilase specific for the herbicide bromoxynil. *Journal of Bacteriology* **169**, 955–960.

Stalker, D. M., Malyj, L. D. and McBride, K. E. (1988). Purification and properties of a nitrilase specific for the herbicide bromoxynil and corresponding nucleotide sequence analysis for the *bxn* gene. *Journal of Biological Chemistry* **263**, 6310–6314.

Stringam, G. R., Ripley, V. L., Love, H. K. and Mitchell, A. (2003). Transgenic herbicide tolerant canola-the Canadian experience. *Crop Science* **43**, 1590–1593.

Swanson, E. B., Coumans, M. P., Brown, G. L., Patel, J. D. and Beversdorf, W. D. (1988). The characterization of herbicide tolerant plants in *Brassica napus* L.

after *in vitro* selection of microspores and protoplasts. *Plant Cell Reports* **7**, 83–87.

Swanson, E. B., Herrgesell, M. J., Arnaldo, M., Sipell, D. W. and Wong, R. S. C. (1989). Microspore mutagenesis and selection-canola plants with field tolerance to the imidazolinones. *Theoretical and Applied Genetics* **78**, 525–530.

Tan, S., Evans, R. R., Dahmer, M. L., Singh, B. K. and Shaner, D. L. (2005). Imidazolinone-tolerant crops: history, current status and future. *Pest Management Science* **61**, 246–257.

Tan, S, Evans, R. and Singh, B. (2006). Herbicidal inhibitors of amino acid biosynthesis and herbicide-tolerant crops. *Amino Acids* **30**, 195–204.

Thomas, P. (1992). "Canola Growers' Manual," 200 pp. Canola Council of Canada, Winnipeg, Canada.

Timko, M. P., Herdies, L., De Alameida, E., Cashmore, A. R., Leemans, J. and Krebbers, E. (1988). Genetic engineering of nuclear-encoded components of the photosynthetic apparatus in *Arabidopsis*. *In* "The Impact of Chemistry on Biotechnology—a Multidisciplinary Discussion" (M. Phillips, S. P. Shoemaker, R. D. Middlekauff and R. M. Ottenbrite, eds.), pp. 279–295. ACS Books, Washington, DC.

Tranel, P. J. and Wright, T. R. (2002). Resistance of weeds to ALS-inhibiting herbicides: What have we learned? *Weed Science* **50**, 700–712.

Van Oorschot, J. L. P. (1991). Chloroplastic resistance of weeds to triazines in Europe. *In* "Herbicide Resistance in Weeds and Crops" (J. C. Caseley, G. W. Cussans and R. K. Atkin, eds.), pp. 87–99. Long Ashton Int. Symp. Butterworth-Heinemann, Boston.

Vasil, I. K. (1996). Phosphinothricin-resistant crops. *In* "Herbicide Resistant Crops" (S. O. Duke, ed.), pp. 85–91. CRC Press, Boca Raton, FL.

Vaughn, K. C. and Duke, S. O. (1991). Biochemical basis of herbicide resistance. *Chemical Plant Protection* **7**, 141–169.

Waksman, G., Lubrun, M. and Freyssinet, G. (1987). Nucleotide sequence of a gene encoding sunflower ribulose-1,5-biphosphate carboxylase/oxygenase small subunit (rbcs). *Nucleic Acids Research* **15**, 7181.

Williams, M. E., Leemans, J. and Michiels, F. (1997). Male sterility through recombinant DNA technology. *In* "Pollen Biotechnology for Crop Production and Improvement" (K. R. Shivanna and V. K. Sawhney, eds.), Chapter 11, pp. 237–258. Cambridge University Press, Cambridge, UK.

Breeding for Blackleg Resistance: The Biology and Epidemiology

W. G. DILANTHA FERNANDO, YU CHEN AND KAVEH GHANBARNIA

Department of Plant Science, University of Manitoba, Winnipeg, Canada

Advances in Botanical Research, Vol. 45
Incorporating Advances in Plant Pathology

0065-2296/07 $35.00
DOI: 10.1016/S0065-2296(07)45010-8

ABSTRACT

Canola/rapeseed is the third largest global vegetable oil source, after soybean and palm oil. Blackleg caused by the fungus *Leptosphaeria maculans* (Desm.) Ces. & de Not. (anamorph *Phoma lingam* Tode ex Fr.), is a devastating disease present all over rapeseed/canola-growing areas except China. The pathogen can infect all parts of the plant, but stem canker is the most serious symptom as it causes plant lodging and yield loss. Populations of blackleg pathogen are a complex of at least two genetically distinct groups, now referred to as two different species, *L. maculans* (highly virulent, A-group) and *L. biglobosa* (weakly virulent, B-group). According to the interaction phenotype (IP), isolates can be classified into at least five pathogenicity groups (PGs): PG-1, PG-2, PG-3, PG-4, and PGT. Population genetic diversity of *L. maculans* that can break down novel resistance sources is mainly derived from three factors: sexual recombination, large population size and a high gene flow through large-scale dissemination of ascospores. Human activity too may contribute to this. The gene-for-gene hypothesis has been proven in *Brassica*–blackleg pathosystem. To date, nine avirulence (*Avr*) genes *AvrLm1–9* and their corresponding resistance genes have been characterized following genetic studies on both the plant and the pathogen. Two types of the induced resistance, systemic acquired resistance (SAR) and induced systemic resistance (ISR), have been suggested in *Brassica*–blackleg pathosystem. The disease control methods such as crop rotation, resistant cultivars and seed treatment with fungicide have proven to be effective. However, the most economical and environmentally friendly method of disease control to the farmer is resistance (R-gene resistance) in the host. This durable resistance against the Blackleg pathogen is sort after by all breeding groups. Although single gene resistance has been effective, with the advent of pathogen strains belonging to different pathotypes and carrying different *Avr* genes, it has become necessary for most breeding programs to concentrate their efforts on gene pyramiding for multiple R genes in the host. This is effectively achieved by marker-assisted selection.

I. INTRODUCTION

Leptosphaeria maculans (Desmaz.) Ces. & De Not. (anamorph *Phoma lingam* Tode ex Fr.), an ascomycetic fungus, is a pathogen causing seedling death, stem canker (also termed blackleg), and early senescence on rapeseed/canola plants. This pathogen is a hemibiotrophic parasite, that is, a facultative necrotroph for short periods at the early and end of the season to produce leaf spot and stem canker by toxins and degradation enzymes, and a biotroph during endophytic latent phase from leaf lesion to stem lesion (cryptic intercellular growth in leaf and stem). Both the sexual and asexual spores of *L. maculans* are pathogenic.

The pathogen was first described by Tode in 1791 on dead stems of red cabbage (*Brassica oleracea*) and it was named *Sphaeria lingam*. In 1849, Desmazieres found the same fungus on living plants and transferred it to the genus *Phoma* (*P. lingam*). In 1956, H. C. Smith first discovered the perfect (sexual) stage of *P. lingam* and named it *L. napi*. In 1964, H. C. Smith and B. C. Sutton changed the name to *L. maculans* (Smith and Sutton, 1964). The complete taxonomy of *L. maculans* is kingdom Fungi, phylum Ascomycota, subphylum Pezizomycotina, class Dothideomycetes (Loculoascomycetes), order Pleosporales, family Leptosphaeriaceae, genus *Leptosphaeria*, and species *maculans* (Rouxel and Balesdent, 2005). In 2001, *L. biglobosa* was assigned for weakly virulent (WV) isolates from *L. maculans* (Shoemaker and Brun, 2001).

II. HOST RANGE

L. maculans has been found to be strictly a pathogen of crucifers, mainly of *Brassica* crops such as *B. napus*, *B. rapa*, *B. juncea*, *B. oleracea*, and so on. Numerous wild crucifer weed species like *Thlaspi* spp., *Sisymbrium* spp., *Descurainia* spp., *Capsella* spp., *Lepidium* spp., *Raphanus* spp., and *Erysimum* spp. can also be infected by the pathogen (Johnson and Lewis, 1994; William and Fitt, 1999).

III. GENOME OF THE SPECIES

L. maculans is a strictly heterothallic, haploid, and outcrossing fungus (Cozijnsen and Howlett, 2003). Heterothallic here means that there are two mating types with alternate forms of mating type (*MAT*) locus required for sexual reproduction (Somda *et al.*, 1997). Haploid here refers to a single set of 15–16 unpaired chromosomes with a size range from 0.7 to 3.5 million base pairs (Mb) each, indicating that the genome size of *L. maculans* is smaller (about 34 Mb, 30 times smaller than that of *Brassica* species) and predicted to encode ca. 10,000–12,000 genes (Howlett, 2004; Kuhn *et al.*, 2006). Some of the chromosomes are of a dispensable "B" type (Leclair *et al.*, 1996). The genome contains two linear plasmids, 9 and 10 kb, respectively, but no function of the plasmids has been revealed (Lim and Howlett, 1994). A high level of dispersed repetitive DNA sequences in *L. maculans* genome was observed and chromosomal length polymorphisms were found using molecular markers bound to every chromosome and pulsed field gel electrophoresis (Plummer and Howlett, 1995). The linkage groups in the *L. maculans* genome have been investigated with clamped homogeneous electrical field (CHEF) electrophoresis and amplified fragments length polymorphism (AFLP) molecular markers (Kuhn *et al.*, 2006). According to the analysis of 126 expressed sequence tags (ESTs) isolated from hyphae, 11 protein-coding genes such as *znr*1 and *maa*1, have been identified so far in the *L. maculans* genome but only 6 genes are deposited in the GenBank database (Idnurm and Howlett, 2001; Idnurm *et al.*, 2003). Five of them may have possible roles in infection such as ABC transporters, cellulases, and endopolygalacturonases (Howlett *et al.*, 2001).

Genetic and physical maps of *L. maculans* will be useful for genetic and functional genomic studies. Because of the high degree of dispersed repetitive DNA (Cozijnsen *et al.*, 2000), chromosome walking and AFLP markers may be difficult in this fungus because many markers will map to every chromosome (Howlett *et al.*, 2001).

IV. IMPACT OF BLACKLEG DISEASE ON RAPESEED/CANOLA INDUSTRY

Blackleg disease caused by *L. maculans* is of economic importance in the main rapeseed/canola-growing areas except in China. China has the largest rapeseed/canola production area (7.2 million ha in 2003) in the world. In the early 1970s, the most severe epidemics of blackleg limited the expansion of the emerging rapeseed industry in Australia until the development of resistant cultivars (West *et al.*, 2001). In Canada, a virulent strain of the fungus was discovered on spring rapeseed in Saskatchewan in 1975, and it spread to most canola-growing provinces very quickly, often causing significant economic losses. Several factors helped blackleg to become epidemic across the country during the 1980s. For instance, large-scale monoculture of highly susceptible cultivars such as "Westar" for high yield lasted at least 8 years (up to 95% of the rapeseed acreage) since 1984 when it was commercially released. In 1987, the cancellation of import quotas of rapeseed into Japan enhanced the growth of canola in Canada. Decline of wheat prices on the world market led to a rapid conversion to rapeseed cultivation. A lack of rapeseed cultivation and experience with its pathogens encouraged farmers to shorten rotations for increased income. All those combined elements made the nationwide outbreak and epidemic of blackleg in canola in Canada (Juska *et al.*, 1997a).

In Ontario, 60–100% of winter rape fields and 27–31% of spring rape fields were found affected by stem cankers in the late 1980s (Hall *et al.*, 1993), whereas in Western Canada, blackleg has been responsible for yield losses estimated to be as much as 50% in individual fields (Gugel and Petrie, 1992).

In the United Kingdom, blackleg was first observed on susceptible winter oilseed rape cultivars in 1977 and has become a damaging disease with yield loss in excess of €56 million per season (Fitt *et al.*, 2006). In France, blackleg was reported in the 1950s and has been responsible for mean yield losses of 5–20% (Aubertot *et al.*, 2004). Losses as high as 50% have occurred in Germany (Gugel and Petrie, 1992). Blackleg was first detected in North Dakota (United States) in 1991 and 75–90% losses have been reported (Lamey and Hershman, 1993). In Brazil the disease was first reported in 2003 (Fernando and Parks, 2003) and in Hungary in 2006 (Magyar *et al.*, 2006; Szlávik *et al.*, 2006).

A. DISEASE SYMPTOMS

L. maculans can infect all parts of the plant at any stage of crop development. When the disease is seed borne, seedlings of the susceptible cultivars may be killed at emergence or shortly thereafter. However, leaf and stem symptoms are observed more often. Foliar lesions called "phoma leaf spot" are usually the first visible symptoms. The leaf spots are a tan or buff colored and round to irregular in shape. The center of leaf spots normally bears tiny black spots (pycnidia) (Guo and Fernando, 2005).

Secondary symptoms are seen either on the upper stem or on basal stem. "phoma stem lesion" is a term normally used to refer to upper stem lesions, which are white to gray in color and shallow with little or no girdling of the stem. Phoma stem lesions usually do not appear until mid-season. But if they appear at a relatively early stage, yield losses have been observed in Australia, Canada, Europe, and especially in Poland (Hall *et al.*, 1993; Jedryczka *et al.*, 1999).

Lesions at stem base appear at the end of growing season and are called either blackleg, crown canker, stem (basal) canker, or collar rot (West *et al.*, 2001). Stem cankers are dark gray to black in color and have strict brown margins. They usually become sunken and may girdle the stem. Pycnidia can be seen on stem lesions originating from leaf lesions early in the season. Stem cankers cause premature dying of the plant and may also result in lodging. Pods and seeds may also be infected. Infected seeds may be shriveled and gray in color, resulting in seed loss.

B. DISEASE CYCLE

L. maculans overwinters as a saprophyte either on rapeseed/canola stubble in the form of pycnidia and pseudothecia (Williams, 1992), or on seed coat as mycelia or pycnidia (Hall, 1992; Keri, 1999). Blackleg-infested stubble is a major source of inoculum for subsequent spread from one season to next and from one area to another until it is decomposed. Large pieces of stubble are found to take up to 4 years or longer to break down during a series of dry seasons (Guo *et al.*, 2005).

Two opposite mating types of *L. maculans* isolates colonizing debris can cross to produce sexual fruiting bodies—pseudothecia—at any time (Cozijnsen and Howlett, 2003). Mature pseudothecia produce and release ascospores over an extended period of time under favorable conditions. In Western Canada, pseudothecia can be produced on stubble and ascospores released as early as May, 9 months after harvest (Petrie, 1995). Released ascospores can be airborne for long-distance dispersal (1–2 km, maximum up to 10 km) (Gladders and Musa, 1980; Keri, 1999). After landing on cotyledons or leaves, ascospores germinate and penetrate the plants via stomata or wounds without the aid of specialized infection structures such as appressoria. The fungus then grows into intercellular spaces between mesophyll and causes leaf lesions in a few days. The fungus colonizing young leaves develops an endophytic symptomless growth from the leaf spots down the petiole and finally invades and kills cells of the stem cortex, resulting in basal stem canker (Rouxel and Balesdent, 2005). The earlier the infection of the seedling stage occurs, the greater the likelihood of basal stem canker development and more severe the yield loss (West *et al.*, 1999).

In Western Canada, asexual spores called pycnidiospores from pycnidia (Sing: pycnidium) were found to be primary inoculum as well (Guo, 2003). During the growing season, pycnidiospores released from leaf lesions can spread to short distances within a field, mainly by rain splash within 1 m, causing secondary infections on nearby plants.

Regional dispersal of the disease occurs through the movement of infected seed. When blackleg-infested seed is sown, the seedlings may develop cotyledon, leaf, or stem infections. In Canada, the incidence of infected oilseed rape at sowing was found to be correlated with the incidence of infected plants with stem canker and the incidence of infected seed at harvest (Hall *et al.*, 1996). Seedling infection arises either from infected seed or from airborne spores. Seedling infection derived from infected seed can be prevented by using a fungicide seed treatment, but this treatment will not protect the seedling from infection by airborne spores. Once leaves are infected, the fungus can move systemically in the plant's vascular tissue to the stem base.

Infections on cotyledons and leaves early in the season produce the most damaging stem cankers. Stem cankers girdle the stem base, preventing the flow of water up the stem and often result in lodging of the crop. Infections occurring beyond the six-leaf stage cause less damage. Pod infection can result in infected seed, which serves as another source of inoculum for future infestation (Hall, 1992).

C. DISEASE EPIDEMIOLOGY

In Australia, Canada, and Europe, blackleg disease in rapeseed/canola is generally initiated by airborne ascospores and further transmitted by rain-splashed pycnidiospores. Blackleg disease is usually considered monocyclic because secondary cycles by pycnidiospores produced on infected plants do not affect yield significantly (Hall, 1992; Salam *et al.*, 2006; West *et al.*, 2001). Epidemics of this disease result from the interaction among three elements: virulence and amount of *L. maculans* inoculum, cultivar susceptibility and density, and favorable environmental conditions including temperature, humidity, and soil type.

In Western Canada, blackleg disease varies with growth stage and plant vigor but most yield loss results from stem cankers, which develop principally by infection of young plants from the cotyledon to sixth-leaf stages, rather than secondary infection caused by pycnidiospores during the late season (Hall, 1992). The first six leaves are found to be more susceptible to infections than those produced later, but symptom expression on the first six leaves is less severe than on aging host tissue (Badawy *et al.*, 1991). On the aging leaves, even nonaggressive isolates can cause severe symptoms with heavy sporulation (Badawy *et al.*, 1992). These differences in symptom expression could explain why correlations between incidence of leaf lesions and severity of stem cankers are not always apparent (Schramm and Hoffmann, 1991).

Temperature influences the germination of spores, pathogen infection, length of incubation period from penetration to appearance of lesions, and the rate of growth of the pathogen down the leaf petioles into stems. Under controlled environmental conditions, the production of leaf lesions derived from ascospores was found to require 5 days at 20°C and 2 weeks at 8°C (Biddulph *et al.*, 1999), and the time between the appearances from leaf lesions to stem lesions is 77 days at 18°C and 175 days at 3°C (Hammond *et al.*, 1985). With high temperatures, severity of symptom on cotyledons, leaves, and stems appears to be enhanced. Badawy *et al.* (1992) noticed that incompatible reactions (small necrotic lesions) on the cotyledons inoculated with avirulent isolates at 18°C changed to completely compatible reactions (enlarged lesions) at 27°C. This supports a theory that resistance genes in young plants may be temperature sensitive, producing more disease at 24°C than at 14°C (West *et al.*, 2001). The symptoms of phoma stem canker were found much less severe at 12°C than at 18°C. In Western Canada, high summer temperatures may lead to severe epidemics of blackleg.

Pseudothecia maturation depends on both temperature and wetness, with an optimum at 14–15°C (Petrie, 1994). In Western Canada, pseudothecia usually form 9 to 10 months after harvest on stand stubble because subzero winter temperatures delay the pseudothecia maturation (Kharbanda, 1993). The period of ascospore release varies between regions but usually coincides with the presence of young, susceptible plants and high moisture from rainfall or heavy dews (West *et al.*, 2001). The ascospore release period can last 3–4 months, depending on the environmental conditions, or longer with a peak usually occurring 1–2 months after its onset (Guo and Fernando, 2005). After ascospore release, a minimum wetness period of 8 h at 4–28 °C is required for ascospores to germinate and produce hyphae to penetrate plant cells, with optimum of 15–20 °C and 48 h of wetness (Biddulph *et al.*, 1999).

Survival of *L. maculans* on rapeseed/canola stubble plays an important role in stem canker epidemics as pycnidiospores and ascospores liberated from infected residues can both serve as primary inoculum for initiating infection of new canola crop in spring (Guo, 2003). Weather conditions and cultural practices affect the persistence of *L. maculans* on infected residues. The rate of degradation of residues depends on wetness and soil temperature, with pathogen survival favored by dry summers and cold winters. In Western Canada, infected canola stubble on the soil surface continues to discharge ascospores up to 3–5 years because winters are very cold and summers are dry and hot (Petrie, 1995). Survival of the pathogen on canola stubble for 3–5 years greatly exceeds the mean 3-year length of rotations away from canola followed by farmers. In western Australia, infected canola debris do not degrade in the hot, dry summers, and *L. maculans* can survive on the debris for up to 4 years (West *et al.*, 2001). In the United Kingdom, the climate is mild and wet, and rapeseed residues generally decompose within 2 years (West *et al.*, 1999).

Physical and wind damage could be other factors increasing blackleg severity. Hammond *et al.* (1985) noticed that pycnidiospores were unlikely to infect intact leaves in greenhouse unless leaves, petioles, and stems were wounded. Therefore, wind and hail which are commonly observed in Western Canada could enhance disease infection in the field.

D. DISEASE CONTROL

Successful disease control requires a thorough knowledge of the causal agent and the disease cycle, host–pathogen interactions in relation to environmental factors, and cost. Management practices of blackleg in canola include: cultivar resistance, cultural, physical, chemical, and biological control.

1. Cultivar resistance

Breeding of resistant cultivars must be given a high priority to cope with the great variability of *L. maculans* populations as it is the most environmentally friendly and reliable method of disease control. It is also the most inexpensive method of disease control for the farmer. The genetic basis of resistance in adult and seedling plants may differ (Rimmer and vandenBerg, 1992), even though in many cases it is identical (Li and Cowling, 2003). Adult resistance, normally conferred by quantitative trait loci (QTL), stops the development of stem canker or phoma stem lesions, and seedling resistance, normally conferred by dominant loci, retards the growth of pathogen in the leaf lamina or spread from the leaves to the petioles, the hypocotyls or to stems. In other words, dominant resistance aims at preventing the fungus from entering the plant. The evaluation of genetic resistance should be done not only by inoculating the wounded cotyledons in greenhouse but by challenging adult plants in the field with high disease pressure. In Canada, *B. napus* cultivars were found to be susceptible to *L. maculans* in the 1980s, but currently grown cultivars are resistant or moderately resistant to *L. maculans.* As this chapter deals primarily with breeding for disease resistance, we will address this in detail later.

2. Cultural control

a. Crop rotation. Crop rotation is one of the oldest and most effective cultural control strategies. It is a management tool to reduce certain pathogen populations to a level where they will have little or no economic impact when the crop is grown again. The principle of crop rotation is to plan the order of specific crops planted on the same field so that decomposition of infected stubble and a reduction in the viability of pathogen survival structure will take place. The pathogen's ability to produce inoculum as a potential source of a disease will be reduced or eliminated (Bockus and Shroyer, 1998). With blackleg, 3–5 years of crop rotation are usually recommended to decrease the risk of infection by ascospores released from colonized stubble (Guo *et al.*, 2005). However, Morrall *et al.* (1999) found in their survey that rotation length did not affect blackleg incidence because other factors, such as the presence of volunteer canola, might contribute to disease incidence.

b. Tillage. Tillage raises the soil temperature and increases soil aeration, leading to faster decomposition of stubble. Tillage systems, on the basis of the amount of surface residues, can broadly be classified into three categories: conventional tillage, reduced tillage, and conservation tillage (Workneh and Yang, 2000).

Infested residue is the usual source of primary inoculum (i.e., pycnidiospores and ascospores). Appropriate tillage and crop rotation could effectively reduce stem canker of blackleg (Guo *et al.*, 2005, 2006). Appropriate residue management such as burying infected debris by tillage before autumn has proven to be a good control for blackleg (West *et al.*, 2001). The greatest problem with infected residues in Western Canada is zero tillage following canola, which increases the amount of infected stubble remaining on the soil surface.

c. Other cultural methods. Late sowing dates in Australia and Europe, and early sowing dates in England and western Australia have been reported to be associated with lower levels of stem canker development. However, sowing date was found to have no overall effect on stem canker development on spring canola in Western Canada because ascospores are released throughout the growing season (Kharbanda and Tewari, 1996).

High nitrogen concentrations are often found to be capable of increasing the susceptibility of plants to disease. However, several Canadian studies reported that the amount of nitrogen fertilizer did not affect the development of stem canker (Aubertot *et al.*, 2004; Hall *et al.*, 1993).

3. Chemical control

Application of fungicides depends on high-yield potential of crop, high levels of inoculum, or low resistance of cultivars. Seed treatments or soil or foliar sprays with fungicides are used to control blackleg in different regions. In Canada, seed treatment commenced in 1978 and was recommended even if blackleg was not found in seed lots (Guo, 2003). The fungicides Cloak (mix of carbathiin:thiram:lindane), Rovral ST (mix of iprodione:lindane), and Vitavax RS powder (mix of carbathiin:thiram:lindane) are often used in Canada for seed dressing. Tilt 250 (propiconazole), registered in 1994 for blackleg control in Canada, has been recommended for application during the canola rosette stage, but it does not completely control the disease. The timing of fungicide sprays in the field is crucial as fungicides are effective for only limited periods of time, generally 2–3 weeks. In addition, once *L. maculans* infects the stem it can no longer be controlled by chemicals. In Western Canada, there is a need to prevent the pathogen from reaching the stem by fungicide application in early spring when possible. In England, control of phoma stem canker depends heavily on the use of fungicide since resistance in current cultivars is not sufficient for complete control (West *et al.*, 2002).

4. Biological control

Biological control is an environmentally friendly method of plant disease management. In past decades, attempt to develop a biological agent active against *L. maculans* has been conducted in several labs. But no effective agent is available for use at present (Beatty and Jensen, 2002). However, interest in the development of biological control against *L. maculans* is continuously increasing because of public concern for fungicide use and environmental safety. Biological agents under investigation for the control of blackleg include many genera of bacteria and filamentous fungi. Two fungal species, *Cyathus striatus* and *C. olla*, have been found to be capable of biological control because they decrease the stubble food base on which *L. maculans* survives (West *et al.*, 2001). Pseudophomins B, a cyclic lipodepsipeptide isolated from *Pseudomonas fluorescens* strain BRG100, showed high antifungal activity against *L. maculans* (Pedras *et al.*, 2003). A bacterium *Paenibacillus polymyxa* strain PKB1 was investigated as antifungal agent because it produces two fusaricidin peptides that inhibit the growth of *L. maculans* in culture, on leaves, on stems, and on stubble (West *et al.*, 2001). The function of this bacterium was not affected by fungicides and herbicides approved for use on canola in Canada, so it has a great potential to be directly used as a biological agent. A few *Pseudomonas* and *Bacillus* bacteria strains isolated from canola phyllosphere and rhizosphere were found to be strong suppressers of blackleg due to the production of fungitoxic antibiotics like pyrrolnitrin, phenazine, and zwittermicin A (Ramarathnam and Fernando, 2006). *Bacillus* species are well known for their ability to control plant pathogens through the production of secondary metabolites. Gram-positive bacteria, isolated as endophytes from canola leaves, which exhibited agar-diffusible antifungal activity in plates and blackleg disease suppression in plants, carried biosynthetic genes of peptide synthetases of iturin A, bacillomycin D surfactin, and zwittermicin A, along with the zwittermicin A self-resistance protein (Ramarathnam, 2006). Identification of the different antibiotics produced by antagonistic bacteria would help us understand the mechanism of action in biocontrol. These bacteria, which synthesize these antibiotics, could act as natural genetic sources for genetic engineering of the peptide synthetases for production of novel antifungal antibiotics (Ramarathnam and Fernando, 2006).

Suppression of virulent strains of *L. maculans* was found after inoculating with WV strains (Mahuku *et al.*, 1996; Petrie, 1982). Li *et al.* (2006) reported that the collapse of resistance in an Australian cv. "Surpass 400" could be prevented by coinoculation of pycnidiospores of an avirulent strain with highly virulent (HV) strains. The mechanism involved was due to a hypersensitive response rather than systemic acquired resistance (SAR) found by Mahuku *et al.* (1996). Chen and Fernando (2006a) demonstrated induced resistance in canola plants when inoculated with the WV strain of *L. biglobosa* under field conditions. On six-leaf stage plants of Westar, percent lesions/leaf area declined significantly compared to the control when the lower leaves were treated with either the WV *L. biglobosa* strain or salicylic acid (SA) and then challenged with a virulent *L. maculans* isolate 24 h later. In addition, the activity of four plant enzymes (chitinase, β-1,3-glucanase, peroxidase, and phenylalanine ammonia lyase) was greatly enhanced at 48 and 72 h when canola cotyledons were inoculated first with *L. biglobosa* followed by

L. maculans 24 h later, compared to a water control treatment. Field experiments showed decreased blackleg severity in plants inoculated with *L. biglobosa* alone or prior to *L. maculans* compared to plants inoculated with *L. maculans* alone or prior to *L. biglobosa* inoculation (Chen and Fernando, 2006a). If used at the right time, this method may be useful in managing blackleg disease.

V. SPECIES COMPLEX: TWO SPECIES

Populations of *L. maculans* are a complex of at least two genetically distinct groups, described as HV and WV (Taylor *et al.*, 1991), virulent and avirulent (McGee and Petrie, 1978), aggressive and nonaggressive (Hammond and Lewis, 1987), Tox^{+} and Tox^{0} (Balesdent *et al.*, 1992), or A- and B-group (Johnson and Lewis, 1994). These two groups are now referred to as two different species, *L. maculans* (HV, A-group) and *L. biglobosa* (WV, B-group) on the basis of differences in pseudothecial morphology and sexual incompatibility (Bonman *et al.*, 1981; Shoemaker and Brun, 2001), culture, symptoms, biochemical factors, and molecular genetics. More recently, seven distinct intraspecific groups of the species complex have been elucidated based on the sequence of the internal transcribed spacer (ITS) region of the ribosomal DNA (rDNA) repeat (Mendes-Pereira *et al.*, 2003).

A. DIFFERENTIATION OF PATHOGEN ISOLATES

1. Symptom identification

The characteristics of leaf or stem lesions caused by different groups can be easily differentiated both in the field and in the greenhouse (Ansan-Melayah *et al.*, 1997; Brun *et al.*, 1997; Johnson and Lewis, 1994). In the field, lesions caused by B-group ascospores on the leaves and stems have few pycnidia and are darker, smaller, and less noticeable than the larger, pale gray lesions with many pycnidia produced by A-group ascospores. In addition, B-group isolates cause only either leaf spots (phoma leaf spot lesions) or superficial upper stem lesions (phoma stem lesions), neither of which causes yield loss; A-group isolates cause severe basal stem cankers, which leads to plant lodging and yield loss. In the greenhouse, many B-group infections remain as pinpoint lesions, suggesting an incompatible reaction (Toscano-Underwood *et al.*, 2001).

In Western Canada, 95% of ascospores released from pseudothecia in early May are A-group isolates. B-group isolates appear late in the season during pod maturation and the phoma stem lesions caused by these isolates are superficial and cause little damage (Kharbanda, 1993; West *et al.*, 2001).

2. Morphological markers

The differences between A- and B-group isolates have been found in rate of mycelial growth, pycnidial production, and spore size (Petrie, 1988). A-group (Tox^+) isolates are characterized by slow growth *in vitro* and mass production of pycnidia, whereas B-group (Tox^0) isolates grow more rapidly *in vitro* with little production of pycnidia. Morphological markers are often limited and may not be accurate in determining genetic relationships in pathogen populations since their heritability may be low, and they may be subject to pleiotropic effects due to the influence of the environment (Keri, 1999).

3. Physiological and biochemical markers

In liquid culture, A-group isolates did not produce water-soluble pigments but B-group isolates accumulated a yellow-brown pigment (McGee and Petrie, 1978). However, Sippell and Hall (1995) reported that A- and B-group isolates may produce pigments in culture. The amount of pigment produced may vary considerably in single-spore isolates of a culture or between isolates, which makes pigmentation an unreliable marker.

Profiles of phytotoxins produced in liquid cultures have been used to differentiate A- and B-group isolates. A-group isolates have ability to produce nonhost-specific phytotoxins named sirodesmin PL, which B-group isolates are unable to produce (Pedras *et al.*, 1995). Phytotoxins are secondary metabolites. At least nine phytotoxins have been found to form A-group isolates, including sirodesmin PL, deacylsirodesmin PL, sirodesmin H, sirodesmin J, sirodesmin K, phomalide, phomalizarine, and so on (Pedras and SeguinSwartz, 1992). High performance liquid chromatography (HPLC) analyses allowed the differentiation of the blackleg isolates into three groups (Soledade *et al.*, 2000): (1) All A-group (Tox^+) isolates produce sirodesmin PL, deacetylsirodesmin PL, and phomamide; (2) Polish B-group (Tox^0) isolates do not produce either sirodesmin PL or phomamide but produce indolyl dioxopiperazines; and (3) Canadian (Tox^0) isolates produce very low amounts of metabolites such as polyketides (low phytotoxicity), but no indolyl dioxopiperazines, which differ from the profile of Polish isolates. The chemicals produced in culture can be used as biochemical markers to differentiate or group blackleg isolates.

4. Genetic markers

Isozyme differences have been used to classify isolates of A- and B-group (Gall *et al.*, 1995). Isozymes share a common substrate but exhibit different electrophoretic mobility. When different electromorphs are identified by genetic analysis as allelic, they are called allozymes. Polymorphisms of isozymes as genetic markers have been used to estimate heterozygosity originating from dominant, recessive, or epistatic alleles or loci. The esterase band profiles generated in native polyacrylamide gel electrophoresis, and soluble protein band patterns obtained by isoelectric focusing showed differences between A- and B-groups (Balesdent *et al.*, 1992). Two monoclonal antibodies were found to be capable of distinguishing A-group isolates from B-group isolates (Stacesmith *et al.*, 1993).

The quantitative measurement of extracellular enzymes showed elevated activity levels for cellulose, α- and β-glucanases, and polygalacturonase in A-group isolates compared to B-group isolates. These are believed to be important in pathogen infection (Hassan *et al.*, 1991).

Different isolates may vary in the number and size of chromosomes. The separation of whole chromosomes from *L. maculans* by transverse alternating field electrophoresis showed the different karyotypes between the two groups (Taylor *et al.*, 1991). A-group isolates had 6–8 distinct chromosomal bands whereas B-group isolates had 12–14 bands. Chromosome size polymorphisms between two type isolates were confirmed by hybridization of DNA fragments to chromosomes (Morales *et al.*, 1993b).

5. Molecular markers

DNA markers have been used to differentiate or fingerprint A- and B-group isolates. According to random amplified polymorphic DNA (RAPD) analysis of isolates collected across Canada (Goodwin and Annis, 1991), virulent pathotypes including PG-2 have recently been introduced in Canada and diverged relatively little; but PG-1 isolates have probably been present in Canada for a longer time and diverged with geographic isolation, that is, PG-1 isolates from Western Canada or Eastern Canada were genetically different. Informative bands for distinguishing A- and B-group isolates could be obtained using four PCR primers (Schafer and Wostemeyer, 1992). After investigating differential RAPD bands, A- and B-group-specific PCR primer pairs named RAPD-PCR were designed (Voigt *et al.*, 1998).

DNA polymorphism of two distinct genetic groups was first described using restriction fragment length polymorphism (RFLP) by Johnson and Lewis (1990). A- and B-group isolates can clearly be characterized by RFLP using molecular probes for actin, ubiquitin, ribosomal RNA, and the gene for elongation factor EF1 (TEF) (Hassan *et al.*, 1991). A phylogenetic tree for 28 isolates was constructed using RFLP with 2 restriction enzymes and 42 probes of cloned nuclear DNA sequences from *L. maculans* (Koch *et al.*, 1991), in which A- and B-group isolates are phylogenetically distant. Genomic DNA of 16 isolates were digested with 6 restriction enzymes and probed with labeled synthetic oligonucleotides, consistently grouping A- and B-group isolates together (Meyer *et al.*, 1992). Phylogenetic analyses of

RFLP patterns indicated that A- and B-isolates are not monophyletic (Taylor and Borgmann, 1994). After heat shock, responsive genes and cloned genomic DNA sequences from a virulent isolate were used to probe restriction fragments of genomic DNA of *L. maculans* isolates, the RFLP resulted in distinct banding patterns, clearly separating isolates into two groups (Patterson and Kapoor, 1995).

AFLPs are PCR-based markers for the rapid screening of genetic diversity. Pongam *et al.* (1999) found that 49 virulent isolates representing PG-2, PG-3, and PG-4 collected from around the world could be clustered into two big groups: AFLP group I comprised the isolates from Western Canada, United States, and one from United Kingdom, and they were mostly of the same PG group, whereas AFLP group II consisted of the isolates from Ontario, Australia, and Europe, which could be further divided into two subgroups and included three PG groups. Similarity between North Dakota isolates (25 PG-2 isolates) collected in 1995–1996 and Western Canadian isolates (4 PG-2 isolates) collected from 1980s was noticed, suggesting that *L. maculans* was introduced into North Dakota from Western Canada and the populations have remained unchanged over the 10 years. The genetic diversity and geographic differentiation of 100 Australian, European, and North American isolates of *L. maculans* were analyzed with AFLP and they could be classified into 5 types (Purwantara *et al.*, 2000). Their results indicated that Australian and European populations were separate adjacent clusters while the North American populations partially overlapped both of these. North American virulent isolates were less diverse than those from Australia and Europe.

Phylogenetic relationships and PG identification of *L. maculans* populations by means of rDNA molecular assay have been studied since the early 1990s. On the basis of the DNA sequencing of 17S and 5.8S genes in the ITS1 region of *L. maculans*, the specific PCR primer pairs HV17S/5.8S for A-group isolates and WV17S/5.8S for B-group isolates were developed (Mahuku *et al.*, 1996; Xue *et al.*, 1992). The major differences in both ITS1 and ITS2 sequences that correlated with the PG groups of *L. maculans* were confirmed by Morales *et al.* (1993a). In practice, the amplification and restriction of ITS regions by using just *L. maculans* pycnidia as a PCR template have been achieved (Balesdent *et al.*, 1998b).

Besides sequence divergence in ITS sequences, a 5238-bp repetitive sequence present only in A-group isolates was utilized to develop a set of PCR primers for detection of seed contaminated with A-group isolates (Taylor, 1993).

It should be noted that utility of all investigations mentioned above is restricted to the differentiation of A- from B-group. It seems that only the

AFLP method is suitable to characterize phylogenetic relationship within A-group isolates.

VI. PATHOGENICITY GROUP, ITS DISTRIBUTION, AND APPLICATION

Determination of pathogenicity groups (PGs) is based on pathogenicity of isolates with different avirulence (*Avr*) genes when inoculated onto cotyledons of three *B. napus* differential cv. Westar (susceptible), Glacier (*Rlm2* and *Rlm3* resistance genes), and Quinta (*Rlm1*- and *Rlm4*-linked resistance genes) (Koch *et al.*, 1991; Mengistu *et al.*, 1991). According to the interaction phenotype (IP), isolates can be classified into at least four PGs: PG-1 (avirulent on Westar, Glacier, and Quinta), PG-2 (virulent on Westar and avirulent on Glacier and Quinta), PG-3 (virulent on Westar and Glacier and avirulent on Quinta), and PG-4 (virulent on Westar, Glacier, and Quinta). In addition PGT (virulent on Westar and Quinta, but avirulent on Glacier) was added into PGs (Rimmer, 2006). PG identification is based on gene-for-gene relationship, where the IP (resistance or susceptibility) depends on the presence of one major gene for resistance in the plant (*Rlm*) and one corresponding *Avr* gene in the pathogen. For instance, the dominant single resistant allele *Rlm2* in Glacier is in response to avirulent gene *AvrLm2* in PG-2 isolates of *L. maculans*, resulting in "Glacier-PG-2" incompatible interaction. Pair matching between two dominantly linked *Rlm1* and *Rlm4* genes in Quinta and two avirulent unlinked genes *AvrLm1* and *AvrLm4* in *L. maculans* isolates controls the "Quinta-PG-2" interaction. Gene match between *Rlm1* in Quinta and its complementary *AvrLm1* in PG-3 isolates determines the "Quinta-PG-3" incompatible interaction (Ansan-Melayah *et al.*, 1998).

Many additional IPs between *B. napus* and *L. maculans* were described in order to combine the genetic correspondence between *Rlm*/*Avr* interactions. Isolates of *L. maculans*, according to the differential reactions on cv. Erfurter Zwerg, Glacier, Runde, Quinta, and Jet Neuf, were classified as pathotype group A (aggressive) or pathotype group NA (nonaggressive) in which five subgroups A0–A4 under pathotype group A were determined (Badawy *et al.*, 1991). The variation in pathogenicity of *L. maculans* populations collected from Manitoba, Saskatchewan, and western Australia was investigated on a range of *B. napus* cv. Glacier, Global, Karat, Marnoo, Quinta, Regent, UM3132, and two DH lines R83-14.DH47 and R83-14.DH26 (Kutcher *et al.*, 1993). The research suggested that isolates from western Australia were more pathogenic than those from Manitoba and Saskatchewan; isolates from Manitoba and Saskatchewan had a similar pathogenicity spectrum.

However, isolates of *L. maculans* obtained from a single field varied in pathogenicity (Mahuku *et al.*, 1997). An extended differential set (cv. Lirabon, Glacier, Quinta, and Jet Neuf) was tested to distinguish six PGs (A1–A6) (Kuswinanti *et al.*, 1995). In 1999, on the basis of six *B. napus* cv. Westar, Glacier, Quinta, Quantum, Sentry, Sprint, and two lines Val-1 and Dac-1, *L. maculans* PG-2 isolates were subdivided into 15 distinct PGs, and PG-3 and PG-4 were subdivided into 8 PGs (Keri, 1999). It is evident that the various results obtained in these studies may not be consistent or comparable because different isolates of *L. maculans* and different host cultivars were used.

PG-1 isolates (*L. biglobosa*) are present in all blackleg devastating regions, but specifically prevalent in Poland. In Canada, PG-1 isolates were first detected in the 1960s (Vanterpool, 1961). PG-2 has been reported in Western Canada since 1975 (Kutcher *et al.*, 1993; Mahuku *et al.*, 1997; Mengistu *et al.*, 1991). As *L. maculans* has now spread into Eastern Europe where only *L. biglobosa* was previously identified (Magyar *et al.*, 2006; Szlávik *et al.*, 2006), it poses a potential threat to Asian countries such as China where rapeseed production is in large scale (Fitt *et al.*, 2006). Isolates of PG-3 and PG-4 are widely distributed in Australia and Western European countries like United Kingdom and France. However, a high proportion of PG-4 (80%) and PG-3 (11%) isolates was found in Ontario, Canada (Mahuku *et al.*, 1997).

Assessment of PG can provide valuable information about the dynamics of *L. maculans* populations and take into account the potential risk of new virulent isolates in specific locations where they might be absent. It also helps with choice of cultivars because *Avr* genes conditioning pathogenicity of different PGs can be used to predict the corresponding resistance presented in cultivars. Care should be taken in the determination of PGs since variability within batches of differential seeds, and the effect of temperature and other environmental conditions can alter the phenotype interaction. A more precise method is required.

VII. POPULATION GENETIC DIVERSITY AND HOST RESISTANCE

A. SEXUAL RECOMBINATION

Population genetic diversity of *L. manculans* that can break down novel resistance sources is mainly derived from three factors: sexual recombination, large population size, and a high gene flow through large-scale dissemination of ascospores (Rouxel and Balesdent, 2005). *L. maculans* belongs to the fungal class Dothideomycetes. It has a single *MAT* locus with two alternative forms (idiomorphs), which means that two isolates with alternate forms at *MAT* locus are required for sexual reproduction. Sexual reproduction of *L. maculans* is extremely important in the disease cycle for generating genetic variation. Any pathogen population with a high evolutionary/variable potential and a high potential for gene flow will pose the greatest risk of breaking down resistance genes (McDonald and Linde, 2002). Molecular studies using RAPD (Mahuku *et al.*, 1997), AFLP (Barrins *et al.*, 2004), and microsatellite markers (Hayden *et al.*, 2004) have shown that the populations of *L. maculans* in Australia and Canada are composed of many different genotypes. A multiplex PCR assay is available to identity the two mating types, *MAT1-1* and *MAT1-2*, in the *L. maculans* population (Cozijnsen and Howlett, 2003). The coexistence of two or more genotypes including two mating types in one lesion (Linde *et al.*, 2002; Mahuku *et al.*, 1996) indicates the genetic diversity in *L. maculans*.

B. CONCEPT OF GENE-FOR-GENE HYPOTHESIS

Flor's genetic analysis of both the host plant and the pathogen allowed him to summarize his findings in the gene-for-gene hypothesis to explain interaction between resistance gene in pathogen and resistance gene in plant (Flor, 1942, 1955, 1971). The main concept of this hypothesis is that "for each gene conditioning resistance in the host there is a specific gene conditioning avirulence in the parasite." Since the time the gene-for-gene hypothesis was proposed, a lot of effort has been concentrated toward understanding the molecular basis for resistance in two areas, the identification and cloning of the *Avr* and resistance (*R*) genes and the characterization of the cellular defensive responses that are induced after recognition (Leach and White, 1996). Induced plant defensive response is characterized by rapid calcium and ion fluxes, an extracellular oxidative burst, transcriptional reprogramming within and around the infection sites, and, in most cases, a localized programmed cell death, which is termed the hypersensitive response (Greenberg and Vinatzer, 2003; Nimchuk *et al.*, 2003). There are two hypotheses for the interaction of *R* genes and *Avr* genes. In the simplest one, the *Avr* genes encode or generate "specific ligands" which interact physically with a "receptor" that is encoded by the corresponding plant resistance gene (Belhadir *et al.*, 2004; Jia *et al.*, 2000). In another hypothesis, the product (s) of the resistance gene recognizes the action of avirulence factors as they modify or perturb host cellular targets (Belhadir *et al.*, 2004; Danl and Jones, 2001; Van der Beizen and Jones, 1998). This model of indirect recognition has been termed "the guard hypothesis," that is *R* genes have a surveillance role in cell homeostasis (Danl and Jones, 2001; Van der Beizen and Jones, 1998). So far nine fungal avirulence isolates have been cloned and deposited in the data base (Aubertot *et al.*, 2006). However, in most cases there is no functional homology among cloned *Avr* genes (Attard *et al.*, 2002).

C. TYPES OF RESISTANCE IN OILSEED RAPE

Genetic studies have allowed two main types of resistance to *L. maculans* to be distinguished: qualitative resistance at the seedling stage (Ansan-Melayah *et al.*, 1998; Dion *et al.*, 1995; Ferreria *et al.*, 1995; Mendes-Pereira *et al.*, 2003) and quantitative adult-plant resistance (Ferreria *et al.*, 1995; Pang and Halloran, 1996; Pilet *et al.*, 1998). The first type of resistance is race specific, which is usually effective as early as cotyledon stage. This type is controlled by one *R* gene, which enables the plant to recognize corresponding pathogen *Avr* gene products in a gene-for-gene manner and causes a resistance response (Rimmer, 2006). The second type of resistance is quantitative and is usually expressed at the adult-plant stage and controlled by a group of resistance genes (Rimmer, 2006). In contrast to qualitative resistance, the adult-plant resistance is race nonspecific. There have been different studies which try to understand the relationship of these two types of resistance with each other. Some researchers suggested that adult-plant resistance in the field is monogenic (Sawatsky, 1989; Stringam *et al.*, 1994) but some other studies mentioned that it is a polygenic resistance (Ferreria *et al.*, 1995; Pang and Halloran, 1996). Some authors mention a significant relationship between seedling and adult-plant resistance (Bansal *et al.*, 1994; MacNabb *et al.*, 1993) but the most recent studies indicate that they are under different genetic controls (Ballinger and Salisbury, 1996; Ferreria *et al.*, 1995; Pang and Halloran, 1996). The summary of race-specific resistance genes in *Brassica* species to *L. maculans* and their map locations in *B. napus* have been summarized in Table I.

There are some studies on identification of QTL related to field resistance. Three QTLs associated with seedling and field resistances have been identified (Ferreria *et al.*, 1995). One of them associated with seedling resistance is located in the seedling resistance gene (*Rlm1*) on linkage group N7 and the other two are associated with field resistance on linkage groups LG12 and LG21. These QTLs were found in cv. "Major." In another study, 4 out of 10 QTLs associated with reduced disease severity in a cross of winter-type *B. napus* cv. "Darmor-bzh × Yudal" were identified (36). Pilet *et al.* (2001) identified four and six genomic regions in the cross "Darmor × Samourai" for field blackleg resistance which were associated with DH and $F_{2:3}$ families,

TABLE I
Known Resistance Genes in Brassica Species and Their Location on a Genetic Map

Resistance gene	Cultivar/line	Linkage group	Source
LEM1	Major	N6	*B. napus*
LmFl	Cresor	N7	*B. napus*
LmR1	Shiralee	N7	*B. napus*
Rlm1	Quinta, Columbus	LG10	*B. napus*
Rlm2	Bristol, Glacier	LG16	*B. napus*
Rlm3	Glacier, 22-1-1	LG10	*B. napus*
Rlm4	Quinta, Falcon, Jet Neuf	LG10	*B. napus*
Rlm5	Aurea, Picra	–	*B. juncea*
Rlm6	Aurea, Picra	LG17	*B. juncea*
Rlm7	149-2-1	LG10	*B. napus*
Rlm8	French accession	–	*B. rapa*
Rlm9	Darmor-bzh×Yudal DH progeny	LG10	*B. napus*
LepR1	Breeding line	N2	*B. rapa* subsp. *sylvestris*
LepR2	Breeding line	N10	*B. rapa* subsp. *sylvestris*
LepR3	Surpass 400	N10	*B. rapa* subsp. *sylvestris*

respectively. They collectively explained 36–42% of the variation within each population. Four out of 16 QTLs identified in the 2 crosses Darmor-bzh × Yudal and Darmor × Samourai were consistent over the 2 crosses. The authors believe that the inconsistencies observed among populations could be the result of different genetic background and disease infestation level. Pang and Halloran (1996) studied the F_2 and first backcross populations of the cross" Maluka × Niklas." They found that disease severity scores in these populations are discontinuous based on external lesion length, indicating major gene control for these measures of resistance; but those for the percentage of crown girdling were continuous, indicating quantitative genetic control for this measure.

D. GENETIC STUDIES AND MOLECULAR MAPPING OF AVIRULENCE GENES

To date nine avirulence genes *AvrLm1–9* (Table II) have been characterized following genetic studies on both the plant and the pathogen (Ansan-Melayah *et al.*, 1995, 1998; Balesdent *et al.*, 1998a, 2002; Pongam *et al.*, 1998). *AvrLm1* was the first avirulence gene that was characterized genetically through crossings of PG-3 and PG-4 isolates (Ansan-Melayah *et al.*, 1995). The genetic analysis of the interaction between the pathogen and cv. Quinta showed that a

TABLE II
Known Avirulence in L. maculans and Its Corresponding Resistance Genes

Avirulence gene	Corresponding resistance gene	Source of resistance
AvrLm1	*Rlm1*	*B. napus*
AvrLm2	*Rlm2*	*B. napus*
AvrLm3	*Rlm3*	*B. napus*
AvrLm4(alm1)[a]	*Rlm4*	*B. napus*
AvrLm7	*Rlm7*	*B. napus*
AvrLm9	*Rlm9*	*B. napus*
AvrLm5	*Rlm5*	*B. juncea*
AvrLm6	*Rlm6*	*B. juncea*
AvrLm8	*Rlm8*	*B. rapa*

[a]Delourme *et al.* (2004).

single genetic locus (*AvrLm1*) in a PG-3 isolate controlled the IPs with this cultivar (Ansan-Melayah *et al.*, 1995). However, the authors doubted the dominance of this gene due to the fact that the infective stage of this ascomycete is haploid. Ansan-Melayah *et al.* (1998) performed a tetrad analysis following PG-3×PG-4 and PG-2×PG-4 crosses. *AvrLm2* was introduced following the analysis of F_1, F_2, and test cross population indicating that race-specific resistance of Glacier to PG-2 isolate was conditioned by the new corresponding *Avr* gene. Two isolates IBCN78 and IBCN79 also showed a new specific interaction. They were avirulent on cv. Glacier (*Rlm2, Rlm3*) but virulent on cv. Bristol (*Rlm3*). Induction of incompatible interaction on line 22-1-1 (*Rlm3*) was also seen. The authors concluded that this new interaction was due to a new avirulence gene in both isolates termed *AvrLm3* (Balesdent *et al.*, 2002). Balesdent *et al.* (1998a) showed incompatibility of isolate IBCN5 toward Jet Neuf (*Rlm4*). Incompatibility of this isolate toward cv. Jet Neuf segregated as a single trait in cross with a virulent isolate on this cultivar. Half of the progeny displayed an interaction like the isolate IBCN5. This isolate also induced resistance on Quinta (*Rlm1, Rlm4*). Therefore, the authors suggested that this new interaction is due to new avirulence gene called *AvrLm4*. Screening of European PG-4 isolates which possess virulent alleles at *AvrLm1, AvrLm2*, and *AvrLm4* loci, and selecting lines which were resistant to PG-4 but susceptible to IBCN78 and IBCN79 (*Rlm3*), suggesting that all these lines were governed by a new resistance gene. The progeny of the cross of these isolates and virulence on each of these lines showed the 50:50 ratio expected for a single gene control of avirulence. Therefore, the authors postulated that one single locus termed *AvrLm7* governed avirulence on these new resistant lines possessing *Rlm7* (Balesdent *et al.*, 2005). A new *Rlm9* was termed due to interaction of cv. Darmor or Darmor-bzh with the isolate

IBCN56 which was found to be virulent on these two lines but avirulent on Yudal. The progeny of cross Darmor × Yudal indicated two QTLs, one with major effect close to LG10 and another with minor effect on LG16. The major one termed *Rlm9* and its corresponding avirulence gene on IBCN56 was termed *AvrLm9* (Leach *et al.*, 2001). Classical genetic analysis indicated that avirulence on *B. juncea* cv. Aurrea and Picra is governed by two pairs of genes termed *ArLm5*/*Rlm5* and *AvrLm6*/*Rlm6*, respectively (Balesdent *et al.*, 2002). *AvrLm8* was introduced following a 2:2 A/V segregation ratio on cross IBCN18 (avirulence) and 20.3.04 (virulent), suggesting that there is a new single gene controlling incompatibility of the interaction on *B. rapa* line 156-2-1. This new *Avr* gene segregated independently from other *Avr* genes (Balesdent *et al.*, 2002). Linkage analysis displayed that two avirulence genes *AvrLm1* and *AvrLm2* are linked together with a genetic distance of 1.5–8.7 cM.This analysis showed that *AvrLm6* is also tightly linked to *AvrLm1* and *AvrLm2* and located between them. These three avirulence genes (*AvrLm1*–*AvrLm2*–*AvrLm6*) are considered as a cluster of *Avr* genes in a genetic distance of less than 10 cM (Balesdent *et al.*, 2002). They indicated that three other avirulence genes, *AvrLm4, AvrLm3*, and *AvrLm7*, are either three different alleles of a single locus or they are truly distinct, but clustered, *Avr* genes. Also their study demonstrated that four unlinked regions of *L. maculans, AvrLm1*–*AvrLm2*–*AvrLm6* cluster, *AvrLm3*–*AvrLm4*–*AvrLm7* region, *AvrLm5*, and *AvrLm8* are associated with host specificity toward *B. napus, B. rapa*, and *B. juncea*, respectively (Balesdent *et al.*, 2002).

Among all avirulence genes, the location of *AvrLm1, AvrLm4*, and *AvrLm7* have been displayed on the genetic linkage map of *L. maculans* which was developed by different researchers (Cozijnsen *et al.*, 2000; Kuhn *et al.*, 2006; Pongam *et al.*, 1998). Pongam *et al.* (1998) constructed a genetic map with 9 linkage groups based on analysis of segregating F_1 progenies and 14 test crosses introduced a new avirulence gene termed *alm1*(*AvrLm4*). This *Avr* gene is located at linkage group 1 and two flanking AFLP markers are linked to alm1 with 3.1 and 8.1 cM. Kuhn *et al.* (2006) developed a more detailed map with 15 major groups and overall coverage of 2129 cM. Also they considered another 14 additional groups in terms of number of markers in a very small genetic distance. On the basis of this map, *AvrLm1* is located on linkage group 2 and *AvrLm4* and *AvrLm7* are located on LG33. There is no report assigning the location of other *Avr* genes on the above linkage maps.

E. RACE STRUCTURE OF THE *L. MACULANS* ISOLATES

L. maculans race structure was analyzed based on isolates from different parts of the world (Balesdent *et al.*, 2005). The isolates were classified based on PGs and genotyping of isolates at all AVR loci. Genotyping of isolates from different countries in terms of AVR frequency showed that there are differences among isolates for *AvrLm1, AvrLm2, AvrLm3, AvrLm7, AvrLm8*, and *AvrLm9* in terms of race structure, diversity, and complexity (Balesdent *et al.*, 2005). Twenty-six distinct races were identified, most of them from Europe and Australia. The authors also mentioned that mean number of virulence alleles per isolate in Australia was higher than in Europe and Canada (Balesdent *et al.*, 2005). This study indicated that European isolates possess eight distinct races with two major ones, Av5-6-7-8 and Av1-5-6- 7-8. All of these isolates contain *AvrLm6* and *AvrLm7* and shared the absence of *AvrLm3* and *AvrLm9*. The occurrence of winter-type oilseed rape cultivars which possess *Rlm9* implies the absence of *AvrLm9* in European isolates (Balesdent *et al.*, 2005). They found that three virulence alleles (*AvrLm2, AvrLm3*, and *AvrLm9)* were present in all isolates and none of them has virulence allele in *AvrLm6*. Polymorphism has been observed for *AvrLm1, AvrLm4, AvrLm5*, and *AvrLm7* alleles in which virulence alleles at *AvrLm1* and *AvrLm4* and avirulence alleles at *AvrLm7* and *AvrLm5* loci were predominant in the population (Stachowiak *et al.*, 2006). They also indicated that more than 90% of isolates belonged to racesAv5-6-7 and Av6-7. They concluded that *Rlm6* and *Rlm7* still can be considered as a main source of resistance in Europe (Stringam *et al.*, 1994). Race structure of *L. maculans* also has been surveyed in France (Balesdent *et al.*, 2006). The results of this study are the same as the above survey with a few differences. The authors mentioned that the frequency of *AvrLm6* and *AvrLm7* was higher than 99%, whereas *AvrLm2* and *AvrLm9* were fixed in population. *AvrLm1, AvrLm4, AvrLm5*, and *AvrLm8* were polymorphic and the frequency of *AvrLm3* was very low. They also showed that the predominant race in France was Av5-6-7 which represents about 65% of the population (Balesdent *et al.*, 2006). It has also been shown that *Avr* allele combinations in Australian isolates are more diverse and this high level of diversity may explain the ability of the races to overcome the host resistance such as breaking down of single dominant resistance (*LepR3)* of *B. rapa* subsp. *sylvestris*-derived resistance (Balesdent *et al.*, 2005) and of resistance in *B. juncea, B. nigra*, and *B. carinata* (Ballinger and Salisbury, 1996; Li *et al.*, 2005; Purwantara *et al.*, 1998). Australian isolates possess many virulence genes. It has been postulated that alternate hosts, including wild mustards and mustards used as green manure, were the main reason for the selection pressure on the AVR loci, leading to the complex race pattern (Balesdent *et al.*, 2005).

Balesdent *et al.* (2005) also mentioned that general conclusions about Canadian isolates were very difficult to make due to small number of isolates used in their study. It seems that most isolates from Ontario are clustered with European isolates shown on a neighbor-joining denderogram but isolates from Saskatchewan are different from them. They also mentioned that most Canadian isolates shared the absence of *AvrLm7* and *AvrLm8* but the presence of *AvrLm1, AvrLm2, AvrLm3*, and *AvrLm9*. Therefore, it can be concluded that *Rlm7* can be a source of resistance in Europe but not in Canada and Australia. In turn *Rlm1, Rlm2, Rlm3*, and *Rlm9* can be sources of resistance in Canada. However, it is necessary to define AVR alleles of the new PG-3 isolates identified in Western Canada (Fernando and Chen, 2003). These data suggested that the structure of pathogen can be determined by the local cropping system (Balesdent *et al.*, 2006). However, these studies showed that the frequency of *AvrLm5* and *AvrLm8* is high while commercial crops possessing *Rlm5* and *Rlm8* have never been used (Balesdent *et al.*, 2006). The authors believed that this is due to resistance in *B. juncea, B. rapa*, or other mustard cultivars used as green manure or vegetable crops. They also argue that the ancient agricultural use of *B. rapa* and *B. juncea* could be the reason for the presence of these virulence alleles in the population (Balesdent *et al.*, 2006).

F. "COST OF FITNESS" IN *L. MACULANS*

Fitness is the combined ability of an organism to survive and reproduce (Leonard and Czochor, 1980). Van der Plank (1968) considered the pathogen fitness as a driving force in stability of a plant–pathogen ecosystem. He believed that in a freely mutating ecosystem, unnecessary virulence genes in a pathogen population constitute a genetic load such that selection will favor that genotype with minimal number of these genes (Leonard and Czochor, 1980; Van der Plank, 1968). He suggested that a mutation from avirulence to virulence would only occur if it is necessary to overcome an *R* gene. Therefore, a mutation to virulence is associated with an increase in fitness if the *R* gene is present. Leonard and Czochor (1980) included a cost of virulence, *k*, to his model which developed based on Van der Plank's hypothesis to better understanding the coevolution of avirulent and virulent phenotypes in gene-for-gene systems. On the basis of this model, each additional gene for virulence is assumed to reduce the pathogen's intrinsic rate of reproduction where the virulence is unnecessary. Thus, if loss of avirulence function is associated with a cost to fitness (*k*), then the greater the cost of fitness, the more durable the resistance is likely to be (MacNabb *et al.*, 1993; Stachowiak *et al.*, 2006). Flor (1971) mentioned that races with a wider virulence range, which likely possess more mutations in *Avr* genes, competed poorly with races that had a narrow virulence range. It has also been observed that the *Avr* genes with low mutation rates correspond to more durable resistance genes (Van der Plank, 1968). The concept of "cost of fitness" has been studied in *L. maculans* which confirmed the Van der Plank hypothesis

(Li *et al.*, 2005). Huang *et al.* (2006) used near isogenic lines possessing *AvrLm4* and *avrLm4*, indicating that *AvrLm4* and *avrLm4 L. maculans* isolates differ for some components of phoma leaf spot development (leaf lesion, incubation period, growth rate), suggesting that there is a fitness cost of virulence. Plants inoculated with ascospores of *AvrLm4* have more and larger lesions than those inoculated with ascospores of *avrLm4* isolates at all different temperatures tested. However, there were no differences between these two isolates in terms of incubation period and growth rate. There was also an increase in the frequency of *AvrLm4* isolates. They concluded that there was a fitness cost of virulence and *AvrLm4* isolates are fit than *avrLm4* isolates on susceptible cultivars. Loss of *AvrLm4* function may also affect the ability of the virulent isolates to compete with avirulent isolates during growth, down to the petiole to reach the stem base and in colonizing the basal plant tissues in natural field conditions (Huang *et al.*, 2006).

G. DURABLE RESISTANCE IN *BRASSICA* SPECIES

Durable resistance has been defined as resistance that remains effective when used in a large growing area over a long period of time under environments favorable to disease development (Leach and White, 1996). On the basis of this definition, a durable resistance source can only be identified after its long-term deployment and, in most cases, after its breakdown. Durable resistance can be governed by either qualitative or quantitative traits. Most examples of durable resistance against bacterial and fungal pathogens are quantitative. However, there are several cases demonstrating that single *R* genes can confer highly durable resistance (Leonard and Czochor, 1980).

It has been observed that major gene resistance is overcome by changes in virulence of *L. maculans* populations in France and Australia. In Australia, a *B. napus* cultivar (Surpass 400) with resistance derived from *B. napus* spp. *sylvestris* broke down 3 years after its commercial release. Disease severity was higher in cultivars with sylvestris-derived resistance than cultivars with polygenic resistance. *Rlm1*-derived resistance, in France, has been broken down after 5 years under commercial cropping practices (Sprague *et al.*, 2006). The authors believed that widespread use of sylvestris-derived resistance in Australia and cultivars possessing *Rlm1* in France is the main reason for breakdown of the resistance. They also mentioned that other factors probably contributed to the changing of the frequency of alleles in France. These factors include increased areas of oilseed rape in France leading to the larger population size, high inoculum concentrations, promoting colonization of plants by *avrLm1*-bearing isolates, and subsequent sexual recombination and changes in cultural practices, including shorter crop rotations and development of minimum tillage practices (Sprague *et al.*, 2006). All these factors including high sexual recombination of the pathogen may have played a role in the breakdown of resistance in a blackleg-resistant cv. Q2 in Western Canada (Chen, 2006; Chen and Fernando, 2006b) and the occurrence and detection of new PGs in Western Canada (Chen and Fernando, 2005; Fernando and Chen, 2003) and North Dakota (Bradley *et al.*, 2005). The presence of all known PGs has been isolated from a single field in Manitoba, Canada (Chen, 2006).

H. DURABLE RESISTANCE IN THE MANAGEMENT OF BLACKLEG DISEASE

Different methods have been introduced for management of blackleg, but in most cases they are inefficient to control the disease completely. Different studies showed that the best method to control the blackleg is to use resistant cultivar/lines. However, due to strong selection pressure in most cases, using resistant lines/cultivars are not durable. Different factors are involved in the stability of resistance such as type of resistance (monogenic/polygenic), the type of reproduction of the pathogen, the capacity of the pathogen for dispersal, different resistant deployment (pyramiding of specific resistance, mixture of cultivars, spatiotemporal alteration), and the size of pathogen population (Aubertot *et al.*, 2006). Durability of resistance can be gained from different ways. Aubertot *et al.* (2006) mentioned that, in some cases, durability of resistance occurs at the level of the gene and there must be some mutations (in different *Avr* genes) till a pathogen is able to be virulent. Since one mutation in avirulence is enough to convert it to a virulent one and also virulent isolates are present before the introduction of corresponding resistance gene in the field, reducing the size of the population at the local level may decrease the occurrence of new mutants. In this case, virulence alleles may be lost through genetic drift. Another way is combination or pyramiding of resistance genes in one individual. It is difficult for a pathogen to acquire several mutations to become a virulent isolate. Introducing resistant gene(s) in lines with polygenic resistance in the adult-plant stage can increase the durability of resistance in the presence of a resistant genetic background (Rimmer, 2006). Another approach that has been mentioned is use of non-durable genes in a manner that prevents adaptation of the pathogen population, despite its ability to overcome individual genes or gene combinations (Aubertot *et al.*, 2006). The main goal of this approach is maintaining low, stable frequencies of virulent alleles which increase durability of resistance. To reach this goal the pathogen population in both susceptible cultivars (as source of virulent mutants) and resistance cultivars (on which virulence alleles will multiply) must be controlled in such a manner that composition of virulence alleles in the population will not change over time.

As a matter of fact, the management of pathogen populations must be focused on to prevent the resistant cultivar to contribute progeny of virulence alleles to the next generation. Cultural practices can be considered as a main tool for enhancing the durability of resistance if used on both resistance and susceptible cultivars. Decreasing the amount of inoculum (soil tillage, burning, use of pesticide) and preventing inoculum dispersal on resistant cultivars must be taken into account. Decreasing the population size (sexual stage) on resistant cultivars prevents the individuals from finding an opposite mating types. Decreasing the population size in susceptible cultivars can help reduce the pathogen population in resistance cultivars. Reducing the number of virulent mutants, lowering the probability that such mutants will occur in a background with low fitness costs, enhancing the loss of virulence alleles through genetic drift, and lowering allelic diversity so that fewer allelic combinations are generated by genetic drift can be outcome of the reduction of pathogen population size on susceptible cultivars on resistant ones (Aubertot *et al.*, 2006).

VIII. INTEGRATED AVIRULENCE MANAGEMENT

Aubertot *et al.* (2006) proposed the concept of integrated avirulence management (IAM) to enhance durability of specific resistance. Limiting the selection pressure exerted by specific resistance on pathogen population using gene deployment strategy and also limiting the inoculum size of the pathogen populations using appropriate combinations of cultural, physical, biological, or chemical methods are the main goal of this method. Pyramiding at the plant level, cultivar mixtures and species mixtures at the field level, and interfield diversification can be considered as different gene deployment strategies. Sequentially releasing resistance lines (releasing a new resistant line after breakdown occurs) or using different resistance lines in different seasons is another aspect of gene deployment strategy in regard to time. To handle the complexity of the system, the authors proposed a model in which all cultural practices such as changes over time in the regional spatial distribution of oilseed cultivars with different specific and quantitative resistances, stubble management (tillage or burning), application of chemicals to leaves or stubble, sowing date and nitrogen fertilizer, and environmental factors (temperature, rainfall, and wind) are taken into account (Aubertot *et al.*, 2006).

A. INDUCED RESISTANCE AND TYPES

Induced resistance to disease in plants is a state of enhanced plant defense dependent on the host plant's physical and chemical barriers, activated by biotic or abiotic elicitors (Baker, 2000). Induced resistance is the generation of resistance where there is none, by the activation of latent defense mechanisms (Durrant and Dong, 2004). Induced resistance can be subdivided into two broad categories. The first type is SAR, which develops locally such as the hypersensitive reaction (HR) or systemically in response to a pathogen that causes a necrotic lesions (Metraux *et al.*, 2002). SAR is associated with the production of pathogenesis-related (PR) proteins and normally mediated via an SA signal pathway.

The second type of induced resistance is induced systemic resistance (ISR), which is developed systemically in response to different biotic or abiotic stress such as wounding, plant growth-promoting rhizobacteria (PGPR), or insect feeding. ISR is mediated via a jasmonate/ethylene signal pathway but might have crosstalk with the SA signal pathway (Bostock, 2005).

B. AVIRULENT ISOLATES: TRIGGERING SAR

SAR has been described in over 30 plant species (Baker, 2000). Elicitors of SAR comprise biotic or abiotic agents, and biotic agents can be avirulent pathogens or other organisms. Avirulent isolates have been found to induce resistance in the host plant against virulent isolates in many pathosystems: two avirulent races triggered systemic resistance in watermelon against virulent races of *Fusarium oxysporum* f. sp. *niveum* (Martyn *et al.*, 1991), WV strains of *Phytophthora infestans* ISR in potato to late blight (Stromberg, 1995), avirulent isolates activated systemic resistance in broccoli against downy mildew (Monot *et al.*, 2002), induced resistance against Fusarium wilt by nonpathogenic races of *F. oxysporum* f. sp. *ciceris*, and *F. oxysporum* f. sp. *lycopersici* was detected in chickpea and tomato (Fuchs *et al.*, 1999; Hervas *et al.*, 1995), respectively, and resistance in susceptible mustard against the HV *Alternaria brassicae* isolate A (AbA) and moderately virulent isolate C (AbC) was induced using an avirulent isolate D (AbD) of *A. brassicae* (Vishwanath *et al.*, 1999).

C. SAR IN *BRASSICA*–BLACKLEG SYSTEM

In the *Brassica*–blackleg system, the HR was found on *Brassica* plants pretreated with avirulent (WV) isolates. For instance, rapid necrosis of guard cells in the leaves of Indian mustard (*B. juncea*) to arrest hypha growth was observed after the inoculation of an avirulent isolate of *L. maculans* (Chen and Howlett, 1996). The accumulation of phytoalexins in *Brassica* spp. in relation to an HR to *L. maculans* was detected (Pedras *et al.*, 2002). An elicitin (cryptogein) and an avirulent isolate of *L. maculans* were found to be capable of triggering HR on cotyledons of *B. napus* (Roussel *et al.*, 1999).

Accumulation of pectin in the lumen of xylem vessels and the production of callose and lignin were confirmed to play a role in inhibiting systemic infection caused by *L. maculans* (Hammond *et al.*, 1985).

All the studies mentioned above were focused on the HR in *Brassica* plants with preapplication of an avirulent isolate of *L. maculans*. However, HR is a localized defense reaction. It will be more useful if avirulent isolates of *L. maculans* can trigger SAR against the infection caused by virulent isolates for following reason: SAR rather than HR can provide the plant with longer-lasting protection, and against a broad spectrum of pathogens.

Mahuku *et al.* (1996) reported SAR in the canola–blackleg system. In this experiment, the size of blackleg lesions on cotyledons of Westar was significantly reduced when a WV isolate was introduced 64 h prior to the infection by an HV isolate. After WV inoculation, the size of lesions caused by the HV pathotype on the same leaf, on neighboring leaves (above and below the inoculated leaf position), and in the stem was reduced. However, Chen and Fernando (2006a) while demonstrating SAR in a similar study in the greenhouse also showed the induction of chemicals (plant defense enzymes: chitinase, β-1,3-glucanase, peroxidase, and phenylalanine ammonia lyase) and blackleg disease reduction through induced resistance under field conditions.

In addition to the avirulent isolate, a nonhost pathogen, that is an isolate of tomato pathogen *Cladosporium fulvum* with *Avr9* gene was found to trigger a temporary induced resistance to *L. maculans* (Hennin *et al.*, 2001). Chemicals other than SA, such as acibenzolar-*S*-methyl (BTH) and menadione sodium bisulphite (MSB), were proven to be effective in activating locally or systemically induced resistance to *L. maculans* (Borges *et al.*, 2003).

D. THE MECHANISMS OF INDUCED RESISTANCE IN THE *BRASSICA*–BLACKLEG SYSTEM

A few papers have documented the mechanisms of induced resistance involved in *Brassica*–blackleg system. For instance, induction of PR-proteins such as β-1,3-glucanase and chitinase in leaves of resistant and susceptible *Brassica* genotypes has been observed (Chen and Fernando, 2006a; Dixelius, 1994; Hennin *et al.*, 2001; Rasmussen *et al.*, 1992). A systemic enhancement of ascorbate peroxidase (APX) in MSB-pretreated first true leaves of cv. Bristol was detected by real-time PCR (Borges *et al.*, 2003). The accumulation of five indole phytoalexins, that is brassilexin, brassinin, cyclobrassinin, cyclobrassinin sulphoxide, and methoxybrassinin, was found in the cotyledons pretreated with CuC_{12} (Rouxel *et al.*, 1991). At the proteome level, antioxidant enzymes, including superoxide dismutase, nitrate reductase, and carbonic anhydrase, were elevated in the incompatible reaction between *Brassicas* and *L. maculans* using two-dimensional electrophoresis followed by tandem mass spectrometry (MS) (Subramanian *et al.*, 2005).

The complexity of defense response at the molecular level was confirmed by sequencing 277 cDNA (ESTs) derived from mRNA of *B. napus* cv. Glacier inoculated with *L. maculans* PG-2, where about 31% of ESTs were sequences homologous to previously identified defense genes (56 ESTs), disease resistant genes (6 ESTs), and stress-related genes (25 ESTs) (Fristensky *et al.*, 1999).

SAR is normally dependent on SA signaling, and ISR is dependent on jasmonic acid/ethylene signaling. But resistance in *Arabidopsis thaliana* against *L. maculans* was found to be partially dependent on phytoalexin camalexin biosynthesis, which is independent of the SA, jasmonic acid/ethylene pathway (Bohman, 2001), suggesting that the defense response to *L. maculans* might involve complex, multiple, and previously uncharacterized pathways.

WV-induced resistance is not observed under natural field conditions for two reasons: WV isolates appear late in the season during pod maturation (West *et al.*, 2001) and WV isolates may cause compatible interactions with aging host tissue under high temperature (Badawy *et al.*, 1992). However, three independent studies carried out in Canada, United Kingdom, and Australia showed successful induced resistance triggered by WV isolates (*L. biglobosa*) when an early inoculation was done in the field with *L. biglobosa* (Chen and Fernando, 2006a; Li *et al.*, 2006; Liu *et al.*, 2006) giving hope to a new method of biological disease control.

IX. CONCLUSION

Blackleg, caused by the fungus *L. maculans* (Desm.) Ces. & de Not. (anamorph *P. lingam* Tode ex Fr.), is a devastating disease present all over rapeseed/canola-growing areas (West *et al.*, 2001) except China. The pathogen can infect all parts of the plant, but stem canker is the most serious symptom as it causes plant lodging and yield loss. The disease control methods, such as crop rotation, resistant cultivars, and seed treatment with fungicide, have proven to be effective.

REFERENCES

Ansan-Melayah, D., Balesdent, M. H., Buee, M. and Rouxel, T. (1995). Genetic characterization of *AvrLm1*, the first avirulence gene of *Leptospaeria maculans*. *Phytopathology* **85**, 1525–1529.

Ansan-Melayah, D., Rouxel, T., Bertrandy, J., Letarnace, B., Mendes-Pererira, E. and Balesdent, M. H. (1997). Field efficiency of Brassica napus specific resistance correlates with Leptosphaeria maculans population structure. *European Journal of Plant Pathology* **103**, 835–841.

Ansan-Melayah, D., Balesdent, M. H., Delourme, R., Pilet, M. L., Tanguy, X., Renard, M. and Rouxel, T. (1998). Genes for race-specific resistance against blackleg disease in *Brassica napus* L. *Plant Breeding* **117**, 373–378.

Attard, A., Got, L. and Rouxel, T. (2002). Analysis of molecular markers genetically linked to the *Leptosphaeria maculans* avirulence gene *AvrLm1* in field populations indicates a highly conserved event leading to virulence on *Rlm1* genotypes. *Phytopathology* **15**, 672–682.

Aubertot, J. N., Pinochet, X. and Dore, T. (2004). The effects of sowing date and nitrogen availability during vegetative stages on *Leptosphaeria maculans* development on winter oilseed rape. *Crop Protection* **23**, 635–645.

Aubertot, J. N., West, J. S., Bousset-Vaslin, L., Salam, M. U., Barbetti, M. J. and Diggle, A. J. (2006). Improved resistance management for durable disease control: A case study of phoma stem canker of oilseed rape. *European Journal of Plant Pathology* **114**, 91–106.

Badawy, H. M. A., Hoppe, H. H. and Koch, E. (1991). Differential reactions between the genus *Brassica* and aggressive single spore isolates of *Leptosphaeria maculans*. *Journal of Phytopathology* **131**, 109–119.

Badawy, H. M. A., Kakau, J. and Hoppe, H. H. (1992). Temperature and aging of host tissue affect the interactions between different oilseed rape cultivars and pathotype groups of *Leptosphaeria maculans*. *Journal of Phytopathology* **134**, 255–263.

Baker, C. (2000). Systemic acquired resistance. *In* "Molecular Plant Pathology. Annual Plant Reviews" (M. Dickinson and J. Beynon, eds.), Vol. 4, pp. 198–217. Sheffield Academic Press, Sheffield, UK.

Balesdent, M. H., Gall, C., Robin, P. and Rouxel, T. (1992). Intraspecific variation in soluble mycelial protein and esterase patterns of *Leptosphaeria maculans* French isolates. *Mycological Research* **96**, 677–684.

Balesdent, M. H., Attard, A., Ansan-Melayah, D. and Rouxel, T. (1998a). Genetic control and host range of avirulence towards *Brassica napus* cvs. Quinta and Jet Neuf in *Leptosphaeria maculans*. *Phytopathology* **91**, 70–76.

Balesdent, M. H., Jedryczka, M., Jain, L., Mendes-Pereira, E., Bertrandy, J. and Rouxel, T. (1998b). Conidia as a substrate for internal transcribed spacer-based PCR identification of members of the *Leptosphaeria maculans* species complex. *Phytopathology* **88**, 1210–1217.

Balesdent, M. H., Attard, A., Kuhn, M. L. and Rouxel, T. (2002). New avirulence genes in the phytopathogenic fungus *Leptosphaeria maculans*. *Phytopathology* **92**, 1122–1133.

Balesdent, M. H., Barbetti, M. J., Li, Hua, Sivasithamparam, K., Gout, L. and Rouxel, T. (2005). Analysis of *Leptosphaeria maculans* race structure in a worldwide collection of isolates. *Phytopathology* **95**, 1061–1071.

Balesdent, M. H., Louvard, K., Pinochet, X. and Rouxel, T. (2006). A large-scale survey of races of *Leptosphaeria maculans* occurring on oilseed rape in France. *European Journal of Plant Pathology* **114**(1), 53–65.

Ballinger, D. J. and Salisbury, P. A. (1996). Seedling and adult plant evaluation of race variability in *Leptosphaeria maculans* on *Brassica* species in Australia. *Australian Journal of Experimental Agriculture* **36**, 485–488.

Bansal, V. K., Kharbanda, P. D., Stringam, G. R., Thiagarajah, M. R. and Tewari, J. P. (1994). A comparison of greenhouse and field screening methods for blackleg resistance in doubled haploid of *Brassica napus*. *Plant Disease* **78**, 276–281.

Barrins, J. M., Ades, P. K., Salisbury, P. A. and Howlett, B. J. (2004). Genetic diversity of Australian isolates of *Leptosphaeria maculans*, the fungus that causes blackleg of canola (*Brassica napus*). *Australasian Plant Pathology* **33**, 529–536.

Beatty, P. H. and Jensen, S. E. (2002). *Paenibacillus polymyxa* produces fusaricidin-type antifungal antibiotics active against *Leptosphaeria maculans*, the causative agent of blackleg disease of canola. *Canadian Journal of Plant Pathology* **48**, 159–169.

Belhadir, Y., Subramaniam, R. and Dang, J. L. (2004). Plant disease resistance protein signaling: NBS-LRR proteins and their partners. *Current Opinion in Plant Biology* **7**, 391–399.

Biddulph, J. E., Fitt, B. D. L., Leech, P. K., Welham, S. J. and Gladders, P. (1999). Effects of temperature and wetness duration on infection of oilseed rape leaves by ascospores of *Leptosphaeria maculans* (stem canker). *European Journal of Plant Pathology* **105**, 769–781.

Bockus, W. W. and Shroyer, J. P. (1998). The impact of reduced tillage on soilborne plant pathogens. *Annual Review of Phytopathology* **36**, 485–500.

Bohman, S. (2001). Molecular studies of *Arabidopsis* and *Brassica* with focus on resistance to *Leptosphaeria maculans*. Doctoral Thesis. Swedish University of Agricultural Sciences, Uppsala, Sweden.

Bonman, J. M., Gabrielson, R. L., Williams, P. H. and Delwiche, P. A. (1981). The role of virulence of *Phoma lingam* strains in cabbage blackleg epidemics. *Phytopathology* **71**, 103.

Borges, A. A., Cools, H. J. and Lucas, J. A. (2003). Menadione sodium bisulphite: A novel plant defence activator which enhances local and systemic resistance to infection by *Leptosphaeria maculans* in oilseed rape. *Plant Pathology* **52**, 429–436.

Bostock, R. M. (2005). Signal crosstalk and induced resistance: Straddling the line between cost and benefit. *Annual Review of Phytopathology* **43**, 545–580.

Bradley, C., Parks, P. S., Chen, Y. and Fernando, W. G. D. (2005). First report of pathogenicity groups 3 and 4 of *Leptosphaeria maculans* on canola in North Dakota. *Plant Disease* **89**, 776.

Brun, H., Levivier, S., Eber, F., Renard, M. and Chevre, A. M. (1997). Electrophoretic analysis of natural populations of *Leptosphaeria maculans* directly from leaf lesions. *Plant Pathology* **46**, 147–154.

Chen, C. Y. and Howlett, B. J. (1996). Rapid necrosis of guard cells is associated with the arrest of fungal growth in leaves of Indian mustard (*Brassica juncea*) inoculated with avirulent isolates of *Leptosphaeria maculans*. *Physiological and Molecular Plant Pathology* **48**, 73–81.

Chen, Y. (2006). Molecular and phenotypic characterization of populations of *Leptosphaeria maculans* and induced resistance to blackleg disease of canola by *Leptosphaeria biglobosa*. Ph.D. Thesis. University of Manitoba, Winnipeg, MB, Canada.

Chen, Y. and Fernando, W. G. D. (2005). First report of canola blackleg caused by pathogenicity group 4 of *Leptosphaeria maculans* in Manitoba. *Plant Disease* **89**, 339.

Chen, Y. and Fernando, W. G. D. (2006a). Induced resistance in blackleg (*Leptosphaeria maculans*) disease in canola (*Brassica napus*) caused by a weakly virulent isolate of *Leptosphaeria biglobosa*. *Plant Disease* **90**, 1059–1064.

Chen, Y. and Fernando, W. G. D. (2006b). Prevalence and pathogenicity groups of *Leptosphaeria maculans* in Western Canada and North Dakota. *Canadian Journal of Plant Pathology* **28**, 533–539.

Cozijnsen, A. J. and Howlett, B. J. (2003). Characterisation of the mating-type locus of the plant pathogenic ascomycete *Leptosphaeria maculans*. *Current Genetics* **43**, 351–357.

Cozijnsen, A. J., Popa, K. M., Purwantara, A., Rolls, B. D. and Hawllet, B. J. (2000). Genome analysis of the plant pathogenic ascomycetes *Leptosphaeria maculans*; mapping mating type and host specificity loci. *Molecular Plant Pathology* **1**, 293–302.

Danl, J. L. and Jones, J. D. G. (2001). Plant pathogens and integrated defense responses to infection. *Nature* **411**, 826–833.

Delourme, R., Pilet-Nayel, M. L., Archipiano, M., Horvais, R., Tanguy, X., Rouxel, T., Brun, H., Renard, M. and Balesdent, M. H. (2004). A cluster of major specific resistance genes to *Leptosphaeria maculans* in *Brassica napus*. *Phytopathology* **94**, 578–583.

Dion, Y., Gugel, R. K., Rakow, G. F. W., SeguinSwartz, G. and Landry, B. S. (1995). RFLP mapping of resistance to the blackleg disease [causal agent, *Leptosphaeria maculans* (Desm) Ces et de Not] in canola (*Brassica napus* L.). *Theoretical and Applied Genetics* **91**, 1190–1194.

Dixelius, C. (1994). Presence of the pathogenesis-related protein-2, protein-Q and protein-S in stressed *Brassica napus* and *Brassica nigra* plantlets. *Physiological and Molecular Plant Pathology* **44**, 1–8.

Durrant, W. E. and Dong, X. (2004). Systemic acquired resistance. *Annual Review of Phytopathology* **42**, 185–209.

Fernando, W. G. D. and Chen, Y. (2003). First report on the presence of *Leptosphaeria maculans* pathogenicity group-3, causal agent of blackleg of canola in Manitoba. *Plant Disease* **87**, 1268.

Fernando, W. G. D. and Parks, P. S. (2003). First report of blackleg disease caused by *Leptosphaeria maculans* on canola in Brazil. *Plant Disease* **87**, 314.

Ferreria, M., Rimmer, S. R., Williams, P. and Osborn, T. (1995). Mapping loci controlling *Brassica napus* resistance to *Leptosphaeria maculans* under different screening conditions. *Phytopathology* **85**, 213–217.

Fitt, B. D. L., Brun, H., Barbetti, M. J. and Rimmer, S. R. (2006). World-wide importance of phoma stem canker (*Leptosphaeria maculans* and *L. biglobosa*) on oilseed rape (*Brassica napus*). *European Journal of Plant Pathology* **114**(1), 3–15.

Flor, H. H. (1942). Inheritance of pathogenicity in a cross between physiologic races 22 and 24 of *Melampsora lini*. *Phytopathology* **32**, 5 (Abstract).

Flor, H. H. (1955). Host-parasite interaction in flax rust-its genetics and other implications. *Phytopathology* **45**, 680–685.

Flor, H. H. (1971). Current status of the gene-for-gene concept. *Annual Review of Phytopathology* **9**, 275–296.

Fristensky, B., Balcerzak, M., He, D. F. and Zhang, P. (1999). Expressed sequence tags from the defense response of *Brassica napus* to *Leptosphaeria maculans*. *Molecular Plant Pathology*. Available from http://www.bspp.org.uk/mppol/1999–0301FRISTENSKY.

Fuchs, J. G., Moenne-Loccoz, Y. and Defago, G. (1999). Ability of nonpathogenic *Fusarium oxysporum* Fo47 to protect tomato against *Fusarium* wilt. *Biological Control* **14**, 105–110.

Gall, C., Balesdent, M. H., Desthieux, I., Robin, P. and Rouxel, T. (1995). Polymorphism of Tox^{o}*Leptosphaeria maculans* isolates as revealed by soluble protein and isozyme electrophoresis. *Mycological Research* **99**, 221–229.

Gladders, P. and Musa, T. M. (1980). Observations on the epidemiology of *Leptosphaeria maculans* stem canker in winter oilseed rape. *Plant Pathology* **29**, 28–37.

Goodwin, P. H. and Annis, S. L. (1991). Rapid identification of genetic variation and pathotype of *Leptosphaeria maculans* by random amplified polymorphic DNA assay. *Applied and Environmental Microbiology* **57**, 2482–2486.

Greenberg, J. T. and Vinatzer, B. A. (2003). Identifying type III effectors of plant pathogens and analyzing their interaction with plant cells. *Current Opinion in Microbiology* **6**, 20–28.

Gugel, R. K. and Petrie, G. A. (1992). History, occurrence, impact, and control of blackleg of rapeseed. *Canadian Journal of Plant Pathology* **14**, 36–45.

Guo, X. W. (2003). Effects of crop rotation and tillage on *Leptosphaeria maculans* of *Brassica napus*. Master's Thesis. University of Manitoba, Winnipeg, Canada.

Guo, X. W. and Fernando, W. G. D. (2005). Seasonal and diurnal patterns of spore dispersal by *Leptosphaeria maculans* from canola stubble in relation to environmental conditions. *Plant Disease* **89**, 97–104.

Guo, X. W., Fernando, W. G. D. and Entz, M. (2005). Effects of crop rotation and tillage on blackleg disease of canola. *Canadian Journal of Plant Pathology* **27**, 53–57.

Guo, X. W., Fernando, W. G. D. and Entz, M. H. (2006). Dynamics of infection by *Leptosphaeria maculans* on canola (*Brassica napus*) as influenced by crop rotation and tillage archives of phytopath. *Plant Protection* (*in press*). Published-in-on-line: DOI: 10.1080/03235400600655388.

Hall, R. (1992). Epidemiology of blackleg of oilseed rape. *Canadian Journal of Plant Pathology* **14**, 46–55.

Hall, R., Peters, R. D. and Assabgui, R. A. (1993). Occurrence and impact of blackleg on oilseed rape in Ontario. *Canadian Journal of Plant Pathology* **15**, 305–313.

Hall, R., Chigogora, J. L. and Phillips, L. G. (1996). Role of seedborne inoculum of *Leptosphaeria maculans* in development of blackleg on oilseed rape. *Canadian Journal of Plant Pathology* **18**, 35–42.

Hammond, K. E. and Lewis, B. G. (1987). Variation in stem infections caused by aggressive and nonaggressive isolates of *Leptosphaeria maculans* on *Brassica napus* var *oleifera*. *Plant Pathology* **36**, 53–65.

Hammond, K. E., Lewis, B. G. and Musa, T. M. (1985). A systemic pathway in the infection of oilseed rape plants by *Leptosphaeria maculans*. *Plant Pathology* **34**, 557–565.

Hassan, A. K., Schulz, C., Sacristan, M. D. and Wostemeyer, J. (1991). Biochemical and molecular tools for the differentiation of aggressive and nonaggressive isolates of the oilseed rape pathogen, *Phoma lingam*. *Journal of Phytopathology* **131**, 120–136.

Hayden, H. L., Wilson, L. M., Cozijnsen, A. J. and Howlett, B. J. (2004). Characterization and cross-species amplification of microsatellite loci in the plant pathogenic fungus *Leptosphaeria maculans*. *Molecular Ecology Notes* **4**, 480–481.

Hennin, C., Diederichsen, E. and Hofte, M. (2001). Local and systemic resistance to fungal pathogens triggered by an AVR9-mediated hypersensitive response in tomato and oilseed rape carrying the Cf-9 resistance gene. *Physiological and Molecular Plant Pathology* **59**, 287–295.

Hervas, A., Traperocasas, J. L. and Jimenezdiaz, R. M. (1995). Induced resistance against *Fusarium* wilt of chickpea by nonpathogenic races of *Fusarium oxysporum* f. sp *ciceris* and nonpathogenic isolates of *Fusarium oxysporum*. *Plant Disease* **79**, 1110–1116.

Howlett, B. J. (2004). Current knowledge of the interaction between *Brassica napus* and *Leptosphaeria maculans*. *Canadian Journal of Plant Pathology* **26**, 245–252.

Howlett, B. J., Idnurm, A. and Pedras, M. S. C. (2001). *Leptosphaeria maculans*, the causal agent of blackleg disease of *Brassicas*. *Fungal Genetics and Biology* **33**, 1–14.

Huang, Y. J., Li, Z. Q. E., vans, N., Rouxel, T., Fitt, B. D. and Balesdent, M. H. (2006). Fitness cost associated with loss of the *AvrLm4* avirulence function in *Leptosphaeria maculans* (phoma stem canker of oilseed rape). *European Journal of Plant Pathology* **114**, 77–89.

Idnurm, A. and Howlett, B. J. (2001). Characterization of an opsin gene from the ascomycete *Leptosphaeria maculans*. *Genome* **44**, 167–171.

Idnurm, A., Taylor, J. L., Pedras, M. S. C. and Howlett, B. J. (2003). Small scale functional genomics of the blackleg fungus, *Leptosphaeria maculans*: Analysis of a 38 kb region. *Australasian Plant Pathology* **32**, 511–519.

Jedryczka, M., Fitt, B. D. L., Kachlicki, P., Lewartowska, E., Balesdent, M. H. and Rouxel, T. (1999). Comparison between Polish and United Kingdom populations of *Leptosphaeria maculans*, cause of stem canker of winter oilseed rape. *Journal of Plant Diseases and Protection* **106**, 608–617.

Jia, Y., McAdams, S. A., Bryan, G. T., Hershey, H. P. and Valent, B. (2000). Direct interaction of resistance gene and avirulence gene products confers rice blast resistance. *The EMBO Journal* **19**, 4004–4014.

Johnson, R. D. and Lewis, B. G. (1990). DNA polymorphism in *Leptosphaeria maculans*. *Physiological and Molecular Plant Pathology* **37**, 417–424.

Johnson, R. D. and Lewis, B. G. (1994). Variation in host-range, systemic infection and epidemiology of *Leptosphaeria maculans*. *Plant Pathology* **43**, 269–277.

Juska, A., Busch, L. and Tanaka, K. (1997a). The blackleg epidemic in Canadian rapeseed as a 'normal agricultural accident'. *Ecological Applications* **7**, 1350–1356.

Keri, M. (1999). Genetic studies of host-pathogen interaction between *Brassica napus* and *Leptosphaeria maculans*. Ph.D. Thesis. University of Manitoba, Winnipeg, Canada.

Kharbanda, P. D. (1993). Blackleg of canola in Alberta: Investigations on biology, epidemiology and management. Vegreville, AB, Canada: AECV-R5.

Kharbanda, P. D. and Tewari, J. P. (1996). Integrated management of canola diseases using cultural methods. *Canadian Journal of Plant Pathology* **18**, 168–175.

Koch, E., Song, K., Osborn, T. C. and Williams, P. H. (1991). Relationship between pathogenicity and phylogeny based on restriction fragment length polymorphism in *Leptosphaeria maculans*. *Molecular Plant-Microbe Interactions* **4**, 341–349.

Kuhn, M. L., Gout, L., Howlett, B. J., Melayah, D., Meyer, M., Balesdent, M. H. and Rouxel, T. (2006). Genetic linkage maps and genomic organization in *Leptosphaeria maculans*. *European Journal of Plant Pathology* **114**(1), 17–31.

Kuswinanti, T., Sock, J. and Hoppe, H. H. (1995). Variation in virulence of aggressive isolates of *Leptosphaeria maculans* based on cotyledon reactions on an extended differential set. *In* "Proceedings of the Ninth International Rapeseed Congress," pp. 1248–1250. Cambridge, UK.

Kutcher, H. R., vandenBerg, C. G. J. and Rimmer, S. R. (1993). Variation in pathogenicity of *Leptosphaeria maculans* on *Brassica* spp. based on cotyledon and stem reactions. *Canadian Journal of Plant Pathology* **15**, 253–258.

Lamey, H. A. and Hershman, D. E. (1993). Blackleg canola (*Brassica napus*) caused by *Leptosphaeria maculans* in North Dakota. *Plant Disease* **77**, 1263.

Leach, J. E. and White, F. F. (1996). Bacterial avirulence genes. *Annual Review of Phytopathology* **34**, 153–179.

Leach, J. E., Vera Cruz, C. M., Bai, J. and Leung, H. (2001). Pathogen fitness penalty as a predictor of durability of disease resistance genes. *Annual Review of Phytopathology* **39**, 187–224.

Leclair, S., AnsanMelayah, D., Rouxel, T. and Balesdent, M. H. (1996). Meiotic behaviour of the minichromosome in the phytopathogenic ascomycete *Leptosphaeria maculans*. *Current Genetics* **30**, 541–548.

Leonard, K. J. and Czochor, R. J. (1980). Theory of genetic interactions among populations of plants and their pathogens. *Annual Review of Phytopathology* **18**, 237–258.

Li, C. X. and Cowling, W. A. (2003). Identification of a single dominant allele for resistance to blackleg in *Brassica napus* 'Surpass 400.' *Plant Breeding* **122**, 485–488.

Li, H., Barbetti, M. J. and Sivasithamparam, K. (2005). Hazard from reliance on cruciferous hosts as sources of major gene-based resistance for managing blackleg (*Leptosphaeria maculans*) disease. *Field Crops Research* **91**, 185–198.

Li, H., Barbetti, M. J. and Sivasithamparam, K. (2006). Concomitant inoculation of an avirulent strain of *Leptosphaeria maculans* prevents break-down of a single dominant gene-based resistance in *Brassica napus* cv. Surpass 400 by a virulent strain. *Field Crops Research* **95**, 206–211.

Lim, L. and Howlett, B. L. (1994). Linear plasmids, plm9 and plm10, can be isolated from the phytopathogenic ascomycete *Leptosphaeria maculans* by pulsed-field gel electrophoresis. *Current Genetics* **26**, 276–280.

Linde, C. C., Zhan, J. and McDonald, B. A. (2002). Population structure of *Mycosphaerella graminicola*: From lesions to continents. *Phytopathology* **92**, 946–955.

Liu, S. Y., Liu, Z., Fitt, B. D. L., Evans, N., Foster, S. J., Huang, Y. J., Latunde-Dada, A. O. and Lucas, J. A. (2006). Resistance to *Leptosphaeria maculans* (phoma stem canker) in *Brassica napus* (oilseed rape) induced by *Leptosphaeria biglobosa* and chemical defence activators in field and controlled environments. *Plant Pathology* **55**(3), 401–412.

MacNabb, W. M., Van Den Berg, C. G. J. and Rimmer, S. R. (1993). Comparison of inoculation methods for selection of plant resistant to *Leptosphaeria maculans* in *Brassica napus*. *Canadian Journal of Plant Pathology* **73**, 1199–1207.

Magyar, D., Barasits, T., Fischl, G. and Fernando, W. G. D. (2006). First record of the natural occurrence of the teleomorph of *Leptosphaeria maculans* on oilseed rape and airborne dispersal of ascospores in Hungary. *Journal of Phytopathology* **154**, 428–431.

Mahuku, G. S., Hall, R. and Goodwin, P. H. (1996). Co-infection and induction of systemic acquired resistance by weakly and highly virulent isolates of *Leptosphaeria maculans* in oilseed rape. *Physiological and Molecular Plant Pathology* **49**, 61–72.

Mahuku, G. S., Goodwin, P. H., Hall, R. and Hsiang, T. (1997). Variability in the highly virulent type of *Leptosphaeria maculans* within and between oilseed rape fields. *Canadian Journal of Botany* **75**, 1485–1492.

Martyn, R. D., Biles, C. L. and Dillard, E. A. (1991). Induced resistance to *Fusarium* wilt of watermelon under simulated field conditions. *Plant Disease* **75**, 874–877.

McDonald, B. A. and Linde, C. (2002). Pathogen population genetics, evolutionary potential, and durable resistance. *Annual Review of Phytopathology* **40**, 349–379.

McGee, D. C. and Petrie, G. A. (1978). Variability of *Leptosphaeria maculans* in relation to blackleg of oilseed rape. *Phytopathology* **68**, 625–630.

Mendes-Pereira, E., Balesdent, M. H., Brun, H. and Rouxel, T. (2003). Molecular phylogeny of the *Leptosphaeria maculans–L. biglobosa* species complex. *Mycological Research* **107**, 1287–1304.

Mengistu, A., Rimmer, S. R., Koch, E. and Williams, P. H. (1991). Pathogenicity grouping of isolates of *Leptosphaeria maculans* on *Brassica napus* cultivars and their disease reaction profiles on rapid-cycling *Brassicas*. *Plant Disease* **75**, 1279–1282.

Metraux, J. P., Nawrath, C. and Genoud, T. (2002). Systemic acquired resistance. *Euphytica* **124**(2), 237–243.

Meyer, W., Lieckfeldt, E., Wostemeyer, J. and Borner, T. (1992). DNA fingerprinting for differentiating aggressiveness groups of the rape seed pathogen *Leptosphaeria maculans*. *Mycological Research* **96**, 651–657.

Monot, C., Pajot, E., Le Corre, D. and Silue, D. (2002). Induction of systemic resistance in broccoli (*Brassica oleracea* var. *botrytis*) against downy mildew (*Peronospora parasitica*) by avirulent isolates. *Biological Control* **24**, 75–81.

Morales, V. M., Pelcher, L. E. and Taylor, J. L. (1993a). Comparison of the 5.8s rDNA and internal transcribed spacer sequences of isolates of *Leptosphaeria maculans* from different pathogenicity groups. *Current Genetics* **23**, 490–495.

Morales, V. M., SeguinSwartz, G. and Taylor, J. L. (1993b). Chromosome size polymorphism in *Leptosphaeria maculans*. *Phytopathology* **83**, 503–509.

Morrall, R. A. A., Kaminski, D. A. and Kaminski, L. A. (1999). The 1996–1998 western Canada disease survey; Do agronomic practices affect diseases or *vice versa? In* "Proceedings of the 10th International Rapeseed Congress." Canberra, Australia. Available from http://www.regional.org.au/au/gcirc/3/164.htm?PHPSESSID=312ea62463c546f2f6cf72af93a4f2d5.

Nimchuk, Z., Eulgem, T., Holt, B. F. and Dangl, J. L. (2003). Recognition and response in the plant immune system. *Annual Reviews in Genetics* **37**, 579–609.

Pang, E. and Halloran, G. (1996). The genetics of blackleg disease [casual agent, *Leptosphaeria maculans* (Desm.) ces. Et. De Not.] resistance in rapeseed (*Brassica napus*). II. Seedling and adult-plant resistance as quantitative traits. *Theoretical and Applied Genetics* **93**, 941–949.

Patterson, N. A. and Kapoor, M. (1995). Developmentally regulated expression of heat-shock genes in *Leptosphaeria-Maculans*. *Canadian Journal of Microbiology* **41**, 499–507.

Pedras, M. S. C. and SeguinSwartz, G. (1992). The blackleg fungus—phytotoxins and phytoalexins. *Canadian Journal of Plant Pathology* **14**, 67–75.

Pedras, M. S. C., Taylor, J. L. and Morales, V. M. (1995). Phomaligin A and other yellow pigments in *Phoma lingam* and *P. wasabiae*. *Phytochemistry* **38**, 1215–1222.

Pedras, M. S. C., Nycholat, C. M., Montaut, S., Xu, Y. M. and Khan, A. Q. (2002). Chemical defenses of crucifers: Elicitation and metabolism of phytoalexins and indole-3-acetonitrile in brown mustard and turnip. *Phytochemistry* **59**, 611–625.

Pedras, M. S. C., Ismail, N., Quail, J. W. and Boytchko, S. M. (2003). Structure, chemistry, and biological activity of pseudophomins A and B, new cyclic lipodepsipeptides isolated from the biocontrol bacterium *Pseudomonas fluorescens*. *Phytochemistry* **62**, 1105–1114.

Petrie, G. A. (1982). Blackleg of rapeseed (canola) caused by *Leptosphaeria maculans*: Interaction of virulent and weakly virulent strains and implications for biological control. *Canadian Journal of Plant Pathology* **4**, 309 (Abstract).

Petrie, G. A. (1988). The rapid differentiation of virulent and weakly virulent strains of *Leptosphaeria maculans* (blackleg or stem canker) and related pycnidial fungi from *Brassica* seeds and stems. *Canadian Journal of Plant Pathology* **10**, 188–190.

Petrie, G. A. (1994). Effects of temperature and moisture on the number, size and septation of ascospores produced by *Leptosphaeria maculans* (blackleg) on rapeseed stubble. *Canadian Plant Disease Survey* **74**, 141–151.

Petrie, G. A. (1995). Long-term survival and sporulation of *Leptosphaeria maculans* (blackleg) on naturally-infected rapeseed/canola stubble in Saskatchewan. *Canadian Plant Disease Survey* **75**, 23–34.

Pilet, M., Delourme, R., Foisset, N. and Renard, M. (1998). Identification of QTL involved in field resistance to light leaf spot (*Pyrenopeziza brassicae*) and blackleg resistance (*Leptosphaeria maculans*) in winter rapeseed (*Brassica napus* L.). *Theoretical and Applied Genetics* **97**, 398–406.

Pilet, M. L., Dulpan, G., Archipiano, M., Barret, P., Baron, C., Horvais, R., Tanguy, X., Lucas, M. O., Renard, M. and Delourme, R. (2001). Stability of QTL for field resistance to blackleg across two genetic backgrounds in oilseed rape. *Crop Science* **41**, 197–205.

Plummer, K. M. and Howlett, B. J. (1995). Inheritance of chromosomal length polymorphisms in the ascomycete *Leptosphaeria maculans*. *Molecular and General Genetics* **247**, 416–422.

Pongam, P., Osborn, T. C. and Williams, P. H. (1998). Genetic analysis and identification of amplified fragment length polymorphism markers linked to the *alm1* avirulence gene of *Leptosphaeria maculans*. *Phytopathology* **88**, 1068–1072.

Pongam, P., Osborn, T. C. and Williams, P. H. (1999). Assessment of genetic variation among *Leptosphaeria maculans* isolates using pathogenicity data and AFLP analysis. *Plant Disease* **83**, 149–154.

Purwantara, A., Salisbury, P. A., Burton, W. A. and Howlett, B. J. (1998). Reaction of *Brassica juncea* (Indian mustard) lines to Australian isolates of *Leptosphaeria maculans* under glasshouse and field conditions. *European Journal of Plant Pathology* **104**, 895–902.

Purwantara, A., Barrins, J. M., Cozijnsen, A. J., Ades, P. K. and Howle, B. J. (2000). Genetic diversity of isolates of the *Leptosphaeria maculans* species complex from Australia, Europe and North America using amplified fragment length polymorphism analysis. *Mycological Research* **104**, 772–781.

Ramarathnam, R. (2006). Phyllosphere bacterial biological control of *Leptosphaeria maculans*, the blackleg pathogen of canola (*Brassica napus* L.): Screening for potential antibiotic producers, investigation of the mechanism of control, biochemical detection of the antifungal compounds, and establishment of the role of antibiosis, p. 198. Ph.D. Thesis, University of Manitoba, Winnipeg, MB, Canada.

Ramarathnam, R. and Fernando, W. G. D. (2006). Preliminary phenotypic and molecular screening for potential bacterial biocontrol agents of *Leptosphaeria maculans,* the blackleg pathogen of canola. *Biocontrol Science and Technology* **16**(6), 567–582.

Rasmussen, U., Bojsen, K. and Collinge, D. B. (1992). Cloning and characterization of a pathogen-induced chitinase in *Brassica napus*. *Plant Molecular Biology* **20**, 277–287.

Rimmer, S. R. (2006). Resistance genes to *Leptosphaeria maculans* in *Brassica napus*. *Canadian Journal of Plant Pathology* **28**, S288–S297.

Rimmer, S. R. and vandenBerg, C. G. J. (1992). Resistance of oilseed *Brassica* spp. to blackleg caused by *Leptosphaeria maculans*. *Canadian Journal of Plant Pathology* **14**, 56–66.

Roussel, S., Nicole, M., Lopez, F., Ricci, P., Geiger, J. P., Renard, M. and Brun, H. (1999). *Leptosphaeria maculans* and cryptogein induce similar vascular responses in tissues undergoing the hypersensitive reaction in *Brassica napus*. *Plant Science* **144**, 17–28.

Rouxel, T. and Balesdent, M. H. (2005). The stem canker (blackleg) fungus, *Leptosphaeria maculans*, enters the genomic era. *Molecular Plant Pathology* **6**, 225–241.

Rouxel, T., Kollmann, A., Boulidard, L. and Mithen, R. (1991). Abiotic elicitation of indole phytoalexins and resistance to *Leptosphaeria maculans* within *Brassicae*. *Planta* **184**, 271–278.

Salam, M. U., Fitt, B. D. L., Aubertot, J.-N., Diggle, A. J., Huang, Y. J., Barbetti, M. J., Gladders, P., Jedryczka, M., Khangura, R. K., Wratten, N., Fernando, W. G.D., Pinochet, X., *et al.* (2006). Modeling first release of Leptosphaeria ascospores: Two weather-based models for predicting onset of seasonal release of ascospores of *Leptosphaeria maculan*. *Plant Pathology* (in press).

Sawatsky, W. M. (1989). Evaluation of screening techniques for resistance to *Leptosphaeria maculans* and genetic studies of resistance to the disease in *Brassica napus*. M.Sc. thesis, University of Manitoba, Winnipeg, Manitoba.

Schafer, C. and Wostemeyer, J. (1992). Random primer dependent PCR differentiates aggressive from nonaggressive isolates of the oilseed rape pathogen *Phoma lingam* (*Leptosphaeria maculans*). *Journal of Phytopathology* **136**, 124–136.

Schramm, H. and Hoffmann, G. M. (1991). Biological bases for integrated control of *Phoma lingam* (teleomorph, *Leptosphaeria maculans* (Desm) Ces and Denot) the cause of crown and stem canker in winter oilseed rape. *Journal of Plant Diseases and Protection* **98**, 581–596.

Shoemaker, R. A. and Brun, H. (2001). The teleomorph of the weakly aggressive segregate of *Leptosphaeria maculans*. *Canadian Journal of Botany* **79**, 412–419.

Sippell, D. W. and Hall, R. (1995). Glucose phosphate isomerase polymorphisms distinguish weakly virulent from highly virulent strains of *Leptosphaeria maculans*. *Canadian Journal of Plant Pathology* **17**, 1–6.

Smith, H. C. and Sutton, B. C. (1964). *Leptosphaeria maculans*, the ascogenous state of *Phoma lingam*. *Transactions of the British Mycological Society* **47**, 159–165.

Soledade, M., Pedras, C. and Biesenthal, C. J. (2000). HPLC analyses of cultures of *Phoma* spp.: Differentiation among groups and species through secondary metabolite profiles. *Canadian Journal of Microbiology* **46**, 685–691.

Somda, I., Harkous, S. and Brun, H. (1997). Bipolar heterothallism in B-group isolates of *Leptosphaeria maculans*. *Plant Pathology* **46**, 890–896.

Sprague, S. J., Balesdent, M., Brun, H., Hayden, H. L., Marcroft, S. J., Pinochet, X., Rouxel, T. and Howlet, B. J. (2006). Major gene resistance in *Brassica napus* (oilseed rape) is overcome by changes in virulence of populations of *Leptosphaeria maculans* in France and Australia. *European Journal of Plant Pathology* **114**, 33–40.

Stacesmith, R., Bowler, G., Mackenzie, D. J. and Ellis, P. (1993). Monoclonal antibodies differentiate the weakly virulent from the highly virulent strain of *Leptosphaeria maculans*, the organism causing blackleg of canola. *Canadian Journal of Plant Pathology* **15**, 127–133.

Stachowiak, A., Olechnowicz, J., Jedryczka, M., Rouxel, T., Balesdent, M. H., Happstadius, I., Gladders, P., Latunde-Dada, A. and Evans, N. (2006). Frequency of avirulence alleles in field populations of *Leptosphaeria maculans* in Europe. *European Journal of Plant Pathology* **114**(1), 67–75.

Stringam, G. R., Bansal, V. K., Thiagarajah, M. R., Degenhardt, D. F. and Tewari, J. P. (1994). Development of agronomically superior resistant canola cultivar in *Brassica napus* L. using doubled haploidy. *Canadian Journal of Plant Science* **75**, 437–439.

Stromberg, A. (1995). Systemically induced resistance in potato cultivars with different degree of resistance to late blight caused by *Phytophthora infestans* (Mont) de Bary. *Journal of Phytopathology* **143**, 27–31.

Subramanian, B., Bansal, V. K. and Kav, N. N. V. (2005). Proteome-level investigation of *Brassica carinata*-derived resistance to *Leptosphaeria maculans*. *Journal of Agricultural and Food Chemistry* **53**, 313–324.

Szlávik, S., Barasits, T. and Fernando, W. G. D. (2006). First report of pathogenicity group-3 of Leptosphaeria maculans on winter rape in Hungary. *Plant Disease* **90**(5), 684.

Taylor, J. L. (1993). A simple, sensitive, and rapid method for detecting seed contaminated with highly virulent *Leptosphaeria maculans*. *Applied and Environmental Microbiology* **59**, 3681–3685.

Taylor, J. L. and Borgmann, I. E. (1994). An unusual repetitive element from highly virulent isolates of *Leptosphaeria maculans* and evidence of its transfer to a weakly virulent isolate. *Molecular Plant-Microbe Interactions* **7**, 181–188.

Taylor, J. L., Borgmann, I. and SeguinSwartz, G. (1991). Electrophoretic karyotyping of *Leptosphaeria maculans* differentiates highly virulent from weakly virulent isolates. *Current Genetics* **19**, 273–277.

Toscano-Underwood, C., West, J. S., Fitt, B. D. L., Todd, A. D. and Jedryczka, M. (2001). Development of phoma lesions on oilseed rape leaves inoculated with ascospores of A-group or B-group *Leptosphaeria maculans* (stem canker) at different temperatures and wetness durations. *Plant Pathology* **50**, 28–41.

Van der Beizen, E. A. and Jones, J. D. G. (1998). Plant disease-resistance proteins and the gene-for-gene concept. *Trends in Biochemical Sciences* **2**, 181–184.

Van der Plank, J. E. (1968). "Disease Resistance in Plants," p. 206. Academic Press, London, New York.

Vanterpool, T. C. (1961). Rape diseases in Saskatchewan in 1961. *Canadian Plant Disease Survey* **44**(5), 372–373.

Vishwanath, Kolte, S. J., Singh, M. P. and Awasthi, R. P. (1999). Induction of resistance in mustard (*Brassica juncea*) against *Alternaria* black spot with an avirulent *Alternaria brassicae* isolate-D. *European Journal of Plant Pathology* **105**, 217–220.

Voigt, K., Schleier, S. and Wostemeyer, J. (1998). RAPD-based molecular probes for the blackleg fungus *Leptosphaeria maculans* (*Phoma lingam*): Evidence for pathogenicity group-specific sequences in the fungal genomes. *Journal of Phytopathology* **146**, 567–576.

West, J. S., Biddulph, J. E., Fitt, B. D. L. and Gladders, P. (1999). Epidemiology of *Leptosphaeria maculans* in relation to forecasting stem canker severity on winter oilseed rape in the UK. *The Annals of Applied Biology* **135**, 535–546.

West, J. S., Kharbanda, P. D., Barbetti, M. J. and Fitt, B. D. L. (2001). Epidemiology and management of *Leptosphaeria maculans* (phoma stem canker) on oilseed rape in Australia, Canada and Europe. *Plant Pathology* **50**, 10–27.

West, J. S., Fitt, B. D. L., Leech, P. K., Biddulph, J. E., Huang, Y. J. and Balesdent, M. H. (2002). Effects of timing of *Leptosphaeria maculans* ascospore release and fungicide regime on phoma leaf spot and phoma stem canker development on winter oilseed rape (*Brassica napus*) in southern England. *Plant Pathology* **51**, 454–463.

Williams, P. H. (1992). Biology of *Leptosphaeria maculans*. *Canadian Journal of Plant Pathology* **14**, 30–35.

William, R. H. and Fitt, B. D. L. (1999). Differentiating A and B groups of *Leptosphaeria maculans*, causal agent of stem canker (blackleg) of oilseed rape. *Plant Pathology* **48**(2), 161–175.

Workneh, F. and Yang, X. B. (2000). Prevalence of *Sclerotinia* stem rot of soybeans in the north-central United States in relation to tillage, climate, and latitudinal positions. *Phytopathology* **90**, 1375–1382.

Xue, B., Goodwin, P. H. and Annis, S. L. (1992). Pathotype identification of *Leptosphaeria maculans* with PCR and oligonucleotide primers from ribosomal internal transcribed spacer sequences. *Physiological and Molecular Plant Pathology* **41**, 179–188.

Development of Alloplasmic Rape

MICHAL STARZYCKI, ELIGIA STARZYCKI AND JAN PSZCZOLA

Plant Breeding and Acclimatization Institute, Poland

ABSTRACT

In rapeseed-breeding program, research on resistance to biotic and abiotic stresses plays an important role in biological improvement of new materials. In this chapter, two aspects of rapeseed breeding have been discussed: development of alloplasmic interspecific hybrids and their resistance to the important fungal pathogens. The new methods have allowed to receive new hybrids with cytoplasms of brussel sprout, curly leaf kale, and pasture cabbage. New forms of winter rapeseed have been obtained,

Advances in Botanical Research, Vol. 45
Incorporating Advances in Plant Pathology

0065-2296/07 $35.00
DOI: 10.1016/S0065-2296(07)45011-X

which are more resistant to frost and fungal pathogens like *Leptosphaeria maculans* (Desm.) Ces. et de Not., conidial stage *Phoma lingam* (Tode ex Fr.) Desm. (lately a new subspecies was determined by sequentioning of DNA; *L. biglobosa* nov.), *Sclerotinia sclerotiorum* (Lib.) de Bary., and *Alternaria* spp. In order to identify the newly received rapeseed hybrid forms, DNA molecular markers have been searched that unambiguously give information about origin of the new rapeseed hybrids. This approach fully responses to recent expectation of worldwide rapeseed breeders. Some other issues such as generating interspecific rapeseed hybrids by isolated embryo *in vitro* technique, cytological analysis of DNA of species used in remote crossing as well as their progeny, and advantages of using the new hybrids in resistance-breeding program have also been discussed.

I. INTRODUCTION

In Poland as well as in other European countries of moderate climate, rapeseed is the major oilseed crop. Rapid increase of importance of this plant was a consequence of improving its quality by breeding methods. Rapeseed oil has become economically important and wanted edible fat, and rapeseed cake is a rich source of protein used in animal feeding. Deficiency of forage protein is the main factor stimulating increase of rapeseed cultivation in Europe. In the EU countries as well as in Poland, only 25% of nutrition protein demands are covered by local production. The rest is imported from North and South America. Apart from high requirements for nutrition purposes, the enhancement of rapeseed production is also due to the increase of rapeseed oil consumption, not only for edible purposes but also in chemical industry and for biofuel production.

Rapeseed belongs to Cruciferae family. In oil, it contents erucic acid that is harmful for human health and antinutritional sulfur-containing compounds called glucosinolates are present in its meal. Both the unwanted components had been eliminated by breeding, when Professor Jan Krzymański found spring rapeseed var. Bronowski in Poland. This cultivar has no erucic acid and very low (2–4 μM/g) glucosinolate content. The breeding progress was possible due to genetic investigations of inheritance of the traits and modern laboratory techniques like high performance liquid chromatography (HPLC), gas chromatography (GC), nuclear magnetic resonance (NMR), infrared and ultraviolet spectrometry, as well as DNA-PCR analyses. Rapeseed breeding methods are still improved by achievements of population genetics, genome analysis, and biotechnology methods like *in vitro* culture, androgenesis, mutagenesis, and interspecific crossbreeding methods. Resistance breeding to biotic and abiotic factors is equally important element of a breeding progress.

This chapter reports results from investigation on interspecific and alloplasmatic hybrids and their resistance to the most dangerous fungal diseases.

II. INTERSPECIFIC HYBRIDS BY *IN VITRO* TECHNIQUE OF ISOLATED EMBRYO

The main goal of interspecific crossings is to receive alloplasmatic genotypes and rapeseed genotypes with *Brassica oleracea* L. cytoplasm to enhance winterhardiness and fungal disease resistance.

It is relatively easy to learn the *in vitro* technique of receiving hybrids. The interspecific hybrids have new combinations of traits that are valuable from a breeding point of view. Investigations on rapeseed (*B. napus* L.) resynthesis from *B. campestris* and *B. oleracea* in order to broaden its genetic base had been conducted in Resistance Breeding Methods Laboratory of Plant Breeding and Acclimatization Institute in Poznan for many years. Later, to receive fungal disease-resistant rapeseed, specific forms of *B. oleracea* were crossed with *B. napus*. As a result, after many backcrosses, rapeseed with cytoplasm of cabbage was obtained.

III. PRIMARY SPECIES OF BRASSICAS AND ALLOPLOIDS: SECONDARY POLYPLOIDS–PALEOPOLYPLOIDS

Brassica – ancestral genome x = 6 chromosomes a b c d e f

B. campestris L., genome – A; 10 chromosomes number — a a b c d d e f f f

B. nigra L., genome – B; 8 chromosomes number — a b c d d e f f

B. oleracea L., genome – C; 9 chromosomes number — a b b c c d e e f

A. ALLOPOLYPLOIDS IN *BRASSICA*

- **Rapeseed - *B. napus* L. 2*n* = 4x = 38 chromosomes number (AACC) in somatic cells**

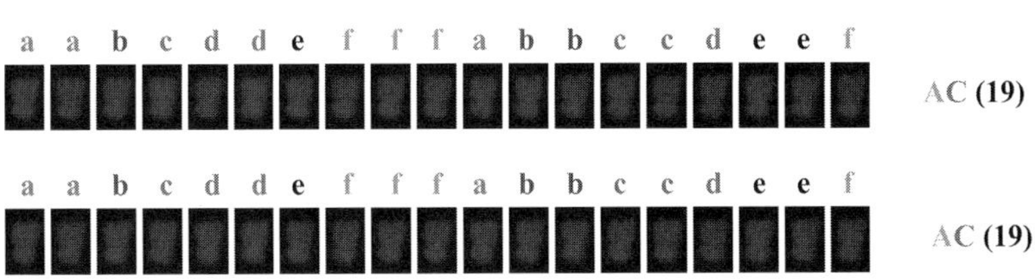

- ***B. juncea* L. 2*n* = 4x = 36 chromosomes number (AABB) in somatic cells**

- ***B. carinata* L. 2*n* = 4x = 34 chromosomes number (BBCC) in somatic cells**

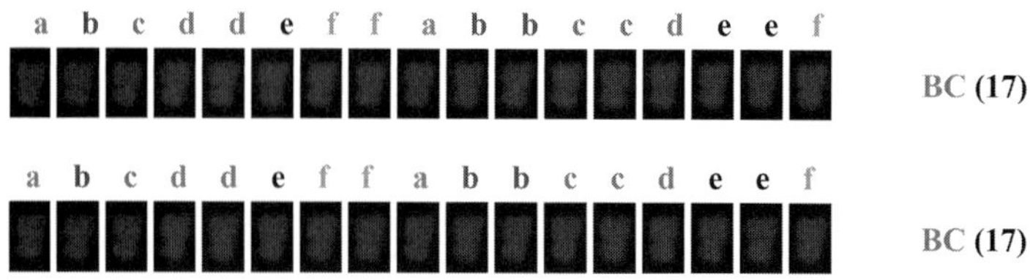

As mentioned earlier, *B. napus* L. is the most important among alloploids; it came into being spontaneously in natural conditions by crossing *B. campestris* L.(A) and *B. oleracea* L.(C).

In artificial conditions, hybridization between the species with A and C genomes has different obstacles. The most important of them is incompatibility that results in suppression of pollen emergence or pollen tube growth. In genus *Brassica*, sporophytic incompatibility system occurs, which is determined by maternal plants. In most cases, researchers evade this difficulty by pollination in bud stage. Many researchers received higher efficacy when buds were pollinated instead of open flowers. Apart from sporophytic incompatibility, the next obstacle is incompatibility between endosperm and embryo. This problem can be resolved by *in vitro* growing of isolated embryos. Chemical endosperm composition and its ploidy strongly affect the embryo development, but it is difficult to determine how much the nutrition and hormonal elements of endosperm are specific for particular species.

IV. METHODS OF DEVELOPMENT OF ALLOPLASMATIC HYBRIDS

In this part of the chapter, the more important stages of *in vitro* techniques used for receiving of interspecific hybrids are described. Very important for good results is a careful performance of crossings between parental plants chosen for embryo preparation (Fig. 1). Emasculated maternal plants were isolated by bags made of two layers of cellophane and greaseproof paper or made of perforated polypropylene foil. It allowed observing ovaries—the future pods (Fig. 2).

In isolation bags, there are visible pods from which immature seeds were taken. Hybrid embryos were obtained from these parts. Three weeks after pollination, the whole rapeseed branch was cut out. Marked branch was placed in water and then stored in a refrigerator at temperature of 1–4°C. This was followed by the preparation of B_5 Gamborg agar medium with 10% sucrose content, followed by sterilization, for embryo incubation and with 2% sucrose content for roots.

Fig. 1. Flowering maternal *B. oleracea* plant before emasculation prepared for pollination (A) and flowering *B. napus* "OO" plant as pollen donor. (See color Plate 6.)

Fig. 2. Maternal *B. oleracea* plants 15 days after pollination with pollen of *B. napus*. (See color Plate 7.)

V. COMPOSITION OF B_5 GAMBORG AGAR MEDIUM (SOME INGREDIENTS MODIFIED)

Macroelements	
$NaH_2PO_4 \cdot H_2O$	150 mg/liter
KNO_3	2500 mg/liter
$(NH_4)_2SO_4$	134 mg/liter
$MgSO_4 \cdot 7H_2O$	250 mg/liter
$CaCl_2$	113 mg/liter
Na_2EDTA	37.3 mg/liter
$FeSO_4 \cdot 7H_2O$	27.8 mg/liter
Microelements	
KJ	0.75 mg/liter
H_3BO_3	3.0 mg/liter
$MnSO_4$	8.93 mg/liter
$ZnSO_4$	2.0 mg/liter
$Na_2MoO_4 \cdot 2H_2O$	0.25 mg/liter
$CuSO_4$	0.025 mg/liter
$CoCl_2$	0.013 mg/liter
Organic compounds	
Nicotinic acid	0.1 mg/liter
Thiamine HCl	0.1 mg/liter
Pyridoxine HCl	0.1 mg/liter
Myoinositol	50.0 mg/liter
L-Serine	50.0 mg/liter
L-Glutamine	400.0 mg/liter
$C_{12}H_{22}O_{11}$	10% or 2%
Others	
H_2O—deionized	1000 ml

Every micro ingredient was prepared separately in a 100-ml flask and 1 ml of ingredient solution was added to 1-liter medium. Nicotinic acid and pyridoxine were prepared in the same way as micro ingredients. No hormones were used.

After 30-min sterilization in autoclave in temperature 120°C, the medium was poured in thin (2 mm) layer into Petri dishes and then the dishes were protected with a parafilm.

Before preparation, ovaries were immersed in 70% ethanol and then sterilized with atomic oxygen (3% peroxidated water). The preparations were then washed with sterile water followed by the preparation of ovules. The seeds were separated with scalpel from young pod tissues, the ovaries (Fig. 3). The ovules so prepared were put into 20% sucrose solution (Fig. 4). Preparation was performed with stereoscopic microscope at 10× magnification.

The embryos after removal of integument and endosperm were transferred into B_5 medium that contained 10% of $C_{12}H_{22}O_{11}$. The dishes were placed on a glass table at 20°C with 12-h photoperiod. It was observed that during next

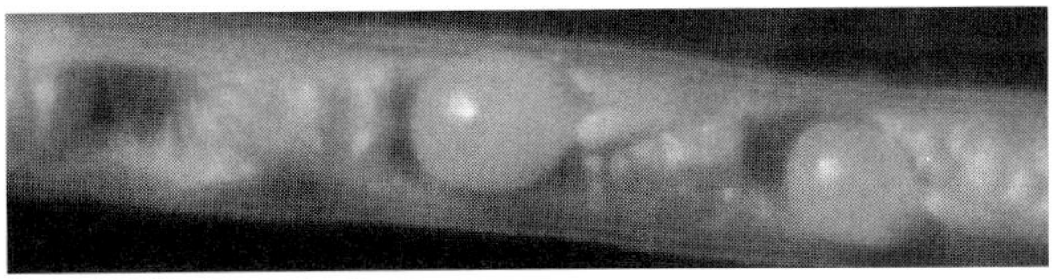

Fig. 3. Two-week-old pod after preparation with fertilized embryos covered with integument and liquid endosperm. (See color Plate 8.)

Fig. 4. Fifteen-day-old seeds of *B. oleracea* × *B. napus* hybrids in sucrose solution (20%—$C_{12}H_{22}O_{11}$), ready for embryo preparation. (See color Plate 9.)

10 days of growing, the embryos underwent all characteristic stages similar to normal development (Fig. 5). They were then transferred on 2% B_5 agar medium.

VI. CHARACTERISTICS OF EMBRYO PLACED ON AGAR MEDIUM WITH 10% $C_{12}H_{22}O_{11}$

A. Globular white: very little percentage would still grow, the embryos remained albino with no growth.
B. Globular green: this type of embryos went *in vitro* through the following growing phases:

Little heart stage	1 day of growth
Big heart stage	1 day
Little torpedo stage	2 days
Big torpedo stage	1 day
Walking-stick stage	3 days
"U" stage	2 days
Total	10 days of growth

C. The hybrid embryos obtained on agar medium had the same characteristic development stages as mentioned above.

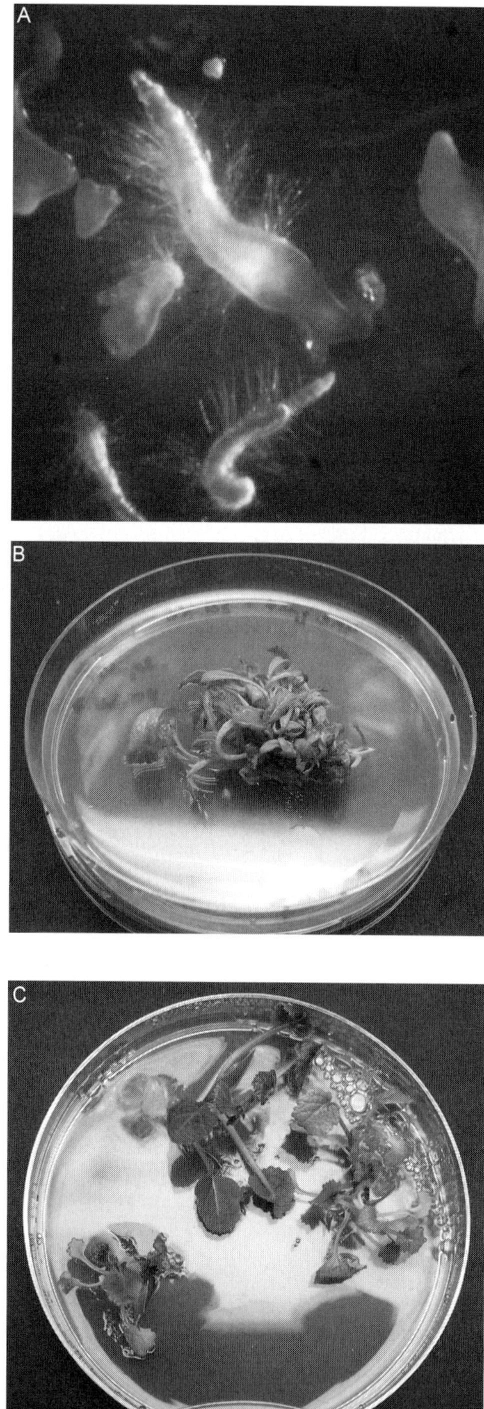

Fig. 5. Slightly developed 5-day-old hybrid embryos grown on B_5 agar medium (A), and 4-week-old plants in seedling stage (B) and (C). (See color Plate 10.)

The agar medium with 2% sucrose content promoted rooting of all embryos beginning from big torpedo stage. A very small number of older embryos had possibilities to develop roots on agar medium with 10% of $C_{12}H_{22}O_{11}$ content. After 3–4 days, rooted seedlings were transferred separately into polypropylene cups with steam sterilized soil (Fig. 6). The biggest plant losses were observed during repotting into nonsterile, natural conditions; that was the reason that in later investigation, hydroponic cultures were used.

The seedlings were transferred into mineral substrate in the stage of well-developed cotyledons and roots with many hair roots. Pots with seedlings were maintained at 25–28 °C, 80% humidity, and 12-h photoperiod. After the development of 3–4 leaves, young plants survived and were in good condition during the winters. Plot in which hybrids were grown was covered with polyethylene foil tunnel. The minimal temperature in the tunnel was about 4°C and hybrid plants of *B. oleracea* × *B. napus* F_1 were vernalized in a climatic chamber (Fig. 7). In the spring, plants that grew in natural condition, as well as plants grown in controlled (chamber) conditions, flowered at the same time.

In order to prove the effectiveness of *in vitro* isolated embryo cultures and to develop interspecific hybrids, all crossings were repeated. Two equivalent crossings were done on the same plant. One crossing was used for embryo culture; the second one was used to obtain seeds in traditional way. In any case, seeds of F_1 hybrids were not received apart form embryo culture technique (Tables I and II).

After the first cycle of interspecific hybrid preparation within *Brassica* genus, the second cycle (*B. oleracea* × *B. napus*) was performed for improving

Fig. 6. Five-week-old plants of F_1 generation of interspecific *B. oleracea* × *B. napus* hybrids, rooted in hydroponic culture to obtain noninfected seedlings after repotting into soil. (See color Plate 11.)

Fig. 7. Interspecific hybrids type *B. oleracea* × *B. napus* F_1 generation and vernalized in phytotron conditions (temperature from 4 to 6 °C, photoperiod—8-h light, 16-h darkness). After flowering, plants were pollinated with *B. napus* pollen. During the next 6 years, the same backcrosses were made. (See color Plate 12.)

TABLE I
Subspecies and Botanical Varieties of Genus Brassica *Used for Interspecific Crossings*

Brassica genome	Number of chromosomes	Latin names	English names
A = 10	AA = 20	*B. campestris* L.	Field cabbage
		B. campestris L. forma I *biennis*	Yellow-seeded winter
		B. campestris L. forma II *biennis*	turnip forma I and forma II
C = 9	CC = 18	*B. oleracea* L.	Cabbage
		var. gemmifera (D.C)	Brussel sprouts
		var. acephala	Fodder cabbage
		var. acephala subvar. Lacinista	Curly leaf kale, Scotch kale
		B. oleracea from Göttingen	Yellow-seeded cabbage
CB = 17	CCBB = 34	*B. carinata* L. f. *annua*	Ethiopian mustard
AC = 18	AACC = 36	*B. napus* L. f. *biennis*	Winter rapeseed

effectiveness of developing new forms of double-improved rapeseed that is more attractive for breeders (Table III).

In a crossing program, the next species were used as maternal plants: brussel sprouts *B. oleracea* var. *gemmifera* $2n = 18$, fodder cabbage *B. oleracea* var. *acephala* $2n = 18$, and curly leaf kale *B. oleracea* var. *acephala* subvar. *Lacinista* $2n = 18$. Each time winter rapeseed $2n = 38$ was used as paternal form. All crosses were made in such a way that the progeny had cabbage

TABLE II
Interspecific Hybrids Developed Through In Vitro *Isolated Embryos (First Cycle, I, of Crossings)*

S. No.	Successful combinations of interspecific crossings: Maternal form (♀) × Paternal form (♂)	Number of prepared ovules: Total	Number of prepared ovules: Globular stage	Embryos developed into seedling stage (no.)	Number of flowering hybrids after vernalization
1	*B. campestris* II × *B. oleracea* Göttingen	29	7	18	15
2	*B. campestris* II × *B. oleracea* Brussel	3	–	3	3
3	*B. campestris* II × *B. oleracea* Brussel	1	–	1	1
4	*B. campestris* II × *B. oleracea.* Göttingen	6	–	6	4
5	*B. oleracea* (pastewna) × *B. campestris* II	33	16	6	4
6	*B. oleracea* (curly leaf kale) × *B. campestris* II	4	2	–	–
7	*B. oleracea* (curly leaf kale) × *B. campestris*II	12	7	9	2
8	*B. oleracea* (brussel sprouts) × *B. campestris* II	26	14	13	3
9	*B. oleracea* (curly leaf kale) × *B. campestris* II	17	12	9	5
10	*B. oleracea* (curly leaf kale) × *B. campestris* II	6	3	2	1
11	*B. oleracea* Fodder × *B. campestris* II	3	1	2	1
12	*B. oleracea* (curly leaf kale) × *B. campestris* II	8	3	6	1
13	*B. oleracea* (curly leaf kale) × *B. campestris* I	24	14	8	3
14	*B. oleracea* (pastewna) × *B. campestris* I	5	3	3	1
15	*B. oleracea* (curly leaf kale) × *B. campestris* II	5	2	1	1
16	*B. oleracea* (curly leaf kale) × *B. campestris* II	2	2	–	–
17	*B. oleracea* (brussel sprouts) × *B. campestris* II	1	–	–	–
18	*B. oleracea* (curly leaf kale) × *B. campestris* I	10	6	5	2
19	*B. oleracea* (pastewna) × *B. campestris* II	4	2	1	1
20	*B. carinata* × *B. campestris* II	4	4	–	–
Total number after the first cycle		204	98	96	48

cytoplasm. Hybrid embryos were obtained 3 weeks after pollination. Green globular forms were placed on B_5 medium enriched with 10% sucrose.

Bigger embryos were also placed on B_5 medium with 2% sucrose content. In order to receive higher number of hybrid plants, all seedlings were rooted

TABLE III

Comparison of the Number of In Vitro *Embryos and Hybrid Plants Obtained in the Second Cycle (II)*

Parental forms	Number of isolated ovules	Number of embryos obtained	Number of plants obtained
B. oleracea var. *gemmifera* × *B. napus*	1500	12	65
B. oleracea var. *acephala* subvar. *Lacinista* × *B. napus*	700	53	20
B. oleracea var. *acephala* × *B. napus*	621	15	53
Total number after the second cycle (II)	2821	80	138

in hydroponics and replanted into pots filled with soil. Every year beginning from HF_1 up to BC_4 generation, all received plants were backcrossed with winter rapeseed according to the following scheme:

B. oleracea × *B. napus* Year 1
(n = 9) ↓ (n = 19)
HF_1 × *B. napus* Year 2
(hybrid 28) ↓
BC_1 × *B. napus* Year 3
↓
BC_2 × *B. napus* Year 4
↓
BC_3 × *B. napus* Year 5
↓
BC_4 × *B. napus* Year 6

The hybrid character of sterile F_1 generation plants was visible in each cross combination. The progeny plants contained traits of both parents.

VII. CYTOLOGICAL ANALYSIS OF HYBRID FORMS AND DNA ANALYSIS OF SPECIES USED FOR CROSSINGS AND ANALYSIS OF THEIR PROGENY

During vegetation period from each hybrid plants (HF_1), leaves were taken for cytological investigation due to determination of chromosome number in metaphase stage of mitotic division. The cytological analysis was performed by orcein method. Mixture of $C_2H_5OH + CH_3COOH$ in proportion 2:1 was used as fixative. Orcein was dissolved in vinegar acid. Two crystals of $FeSO_4$ were added to the 1 liter of a fixative.

On the basis of phenotype observation, hybrid character of all plants was confirmed, although in some plants there were difficulties to determine the number of chromosomes (Table IV).

TABLE IV
Number of Chromosomes Obtained After the First Cycle (I) of Investigation in Individual Hybrid Plants Received In Vitro

S. No.	Crossing combination	Number of chromosomes in somatic cells
1	*B. campestris* II × *B. oleracea* Göttingen	19
2	*B. campestris* II × *B. oleracea* Göttingen	19
3	*B. campestris* II × *B. oleracea* Göttingen	19
4	*B. campestris* II × *B. oleracea* Göttingen	19
5	*B. campestris* II × *B. oleracea* Göttingen	19
6	*B. campestris* II × *B. oleracea* Göttingen	19
7	*B. campestris* II × *B. oleracea* Göttingen	19
8	*B. campestris* II × *B. oleracea* Göttingen	19
9	*B. campestris* II × *B. oleracea* Göttingen	19
10	*B. campestris* II × *B. oleracea* Göttingen	?
11	*B. campestris* II × *B. oleracea* Göttingen	19
12	*B. campestris* II × *B. oleracea* Göttingen	?
13	*B. campestris* II × *B. oleracea* Göttingen	19
14	*B. campestris* II × *B. oleracea* Göttingen	19
15	*B. campestris* II × *B. oleracea* Göttingen	19
16	*B. campestris* II × *B. oleracea* (brussel sprouts)	19
17	*B. campestris* II × *B. oleracea* (brussel sprouts)	19
18	*B. campestris* II × *B. oleracea* Göttingen	19
19	*B. campestris* II × *B. oleracea* Göttingen	19
20	*B. campestris* II × *B. oleracea* Göttingen	19
21	*B. campestris* II × *B. oleracea* Göttingen	19
22	*B. oleracea* (pastewna) × *B. campestris* II	19
23	*B. oleracea* (pastewna) × *B. campestris* II	19
24	*B. oleracea* (forage cabbage) × *B. campestris* II	?
25	*B. oleracea* (forage cabbage) × *B. campestris* II	19
26	*B. oleracea* (curly leaf kale) × *B. campestris* II	19
27	*B. oleracea* (curly leaf kale) × *B. campestris* II	19
28	*B. oleracea* (brussel sprouts) × *B. campestris* II	19
29	*B. oleracea* (brussel sprouts) × *B. campestris* II	19
30	*B. oleracea* (curly leaf kale) × *B. campestris* II	19
31	*B. oleracea* (curly leaf kale) × *B. campestris* II	19
32	*B. oleracea* (curly leaf kale) × *B. campestris* II	19
33	*B. oleracea* (curly leaf kale) × *B. campestris* II	19
34	*B. oleracea* (curly leaf kale) × *B. campestris* II	19
35	*B. oleracea* (curly leaf kale) × *B. campestris* II	19
36	*B. oleracea* (pastewna) × *B. campestris* II	19
37	*B. oleracea* (curly leaf kale) × *B. campestris* II	19
38	*B. oleracea* (curly leaf kale) × *B. campestris* II	19
39	*B. oleracea* (curly leaf kale) × *B. campestris* II	19
40	*B. oleracea* (forage cabbage) × *B. campestris* I	19
41	*B. oleracea* (curly leaf kale) × *B. campestris* II	19
42	*B. oleracea* (curly leaf kale) × *B. campestris* I	?

In the second crossing cycle (II), the number of chromosomes in parental plants hybrid HF_1 and BC_4 plants were determined by orcein method. Hybrid character of fertile plants was visible in each cross combination, and progeny plants contained traits of both parents. In generation HF_1 and BC_4, all hybrid plants had in their somatic cells 28 (ACC = 28) and 38 chromosomes, respectively. After six backcrossings (elimination of chromosome as effect of backcrosses), new forms of rapeseed were received. To confirm presence of cabbage cytoplasm in obtained plants, mainly in BC_4 generation, DNA-PCR analysis was used.

Isolation of the whole DNA was done by phenol method according to Pastuglia *et al.* (1997) procedure. PCR analysis was carried in which a pair of coxIII starters and two RAPD starters OPY10 and OPW05 were used. Reactive mixture contained the next ingredients: dATP, dGTP, dCTP, dTTP (Sigma), each of concentration 100 μM, 2.5 pmol of primer, 0–5 units of Taq polymerase (Eurobio), $MgCl_2$ at concentration 1.5 mM, 25 ng of genomic DNA. DNA amplification was made in thermocycler GeneAmp PCR System 2400 (Perkin Elmer) with the following reaction parameters: preliminary denaturation 94°C for 5 min, one cycle; DNA replication 45 cycles, denaturation 94°C for 1 min, hybrydization 53°C for 2 min, elongation 72°C for 3 min; final elongation 72°C for 5 min, one cycle. Products of reaction were analyzed with electrophoresis on 1.2% agarose gel. After separation of DNA fragments in electric field, gels were stained with ethidium bromide and the DNA fragments were observed under UV light. As a marker of amplified DNA, DNA of phage lambda digested with *Hin*dIII i *Eco*RI enzymes (Eurobio) was used.

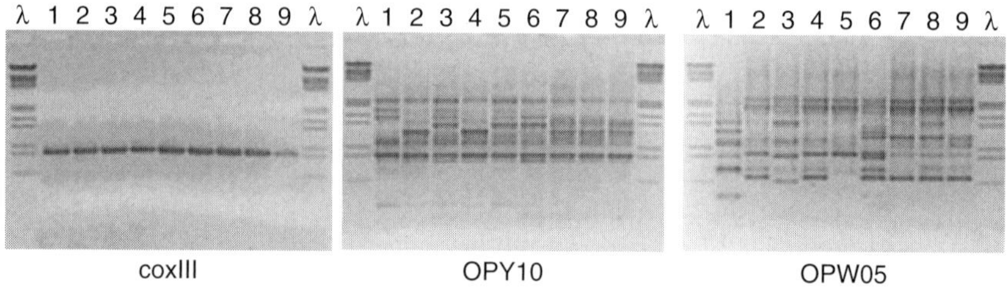

1. *B. oleracea* var. *gemmifera* – brussel sprouts;
2. *B. oleracea* var. *acephala* subvar. *Lacinista* – curly leaf kale;
3. *B .oleracea* var. *acephala* – fodder cabbage;
4. *B. napus* L. – winter rapeseed;
5. HF_1 *B. oleracea* var. *acephala* × *B. napus* L.;
6. HF_1 *B. oleracea* var. *gemmifera* × *B. napus* L.;
7. BC_4 *B. oleracea* var. *gemmifera* × *B. napus* L.;
8. BC_4 *B. oleracea* var. *acephala* subvar. *Lacinista* × *B. napus* L.;

9. BC_4 *B. oleracea* var. *acephala* × *B. napus* L.
λ. DNA lenght standard

The pair of mitochondrial gene coxIII starters amplified DNA in all variants. RAPD patterns observed in generation BC_4 were identical for plants containing cytoplasm of all cabbages. The biggest differences were observed in generation F_1. The banding pattern for curly leaf kale was similar to rapeseed pattern, and starters used for analyses determined three botanical cabbage variants. This work is essential for "intellectual property rights protection" of newly created varieties. In future, every breeder will know profiles of amplified DNA (patented starters), which will allow identifying his own original genotypes.

VIII. PRACTICAL USAGE OF INITIAL MATERIALS OBTAINED FROM INTERSPECIFIC CROSSINGS WITHIN *BRASSICA* GENUS IN BREEDING FOR RESISTANCE

Usually the material obtained from interspecific hybrids undergoes testing for biotic and abiotic stresses (frost resistance) in field condition (Fig. 8). After harvest, such antinutritional compounds as erucic acid (desired level—0%) and glucosinolates (desired level—less than 7 μM/g of seed) are measured.

Resistance of individual strains to fungal diseases that are the most dangerous in Poland, *L. maculans* (Desm.) Ces. et de Not., conidial stage *P. lingam* (Tode ex Fr.) Desm. (recently a new subsp. *L. biglobosa* nov. was determined) (Fig. 9), *S. sclerotiorum* (Lib.) de Bary. (Fig. 10), and *Alternaria* spp., was tested during the vegetation period. A mixture of medium aggressive pathotypes was used in the resistance tests (Table V).

After 12 years of resistance studies using three forms of *B. oleracea*, fodder cabbage, curly leaf kale, and brussel sprouts, it was observed that only curly leaf kale was more strongly infected with powdery mildew *Erysiphe cruciferarum*. The most dangerous rapeseed pathogens, that is blackleg and sclerotinia stem blight, did not attack any of the investigated cabbages.

Later on, investigation of fungal disease resistance of two species *B. taurica* L. and *Eruca sativa* L. received from IHAR Plant Gene Collection, Radzików was carried out. *B. taurica* was found to be more strongly infected with *Alternaria* species (Table VI).

Interspecific hybrids of *Brassicaceae* family are areas of interest of many researchers (Batra *et al.*, 1990; Brown *et al.*, 1997; Earle *et al.*, 1992; Gundimeda *et al.*, 1992; Landgren and Glimelius, 1994; Liu *et al.*, 1995; Nanda Kumar and Shivanna, 1993; Nothnagel *et al.*, 1997; Sundberg and

Fig. 8. Frost resistant *B. napus* genotype with forage cabbage *B. oleracea* var. *acephala* cytoplasm (A) and normal *B. napus* as control (B). Field condition, Plant Breeding Experiment Station in Malyszyn (year 2005).

Glimelius, 1991; Wojciechowski, 1985; Wojciechowski *et al.*, 1997). The main goal of the studies is creation of new forms, which combine traits of both parents. In the Plant Breeding and Acclimatization Institute in Poznan, research works in this subject have been conducted for many years with an objective to increase winter rapeseed resistance to fungal diseases. After first interspecific crossings, the cabbage cytoplasm to the best rapeseed-breeding material was performed. Four backcrosses were enough to establish rapeseed lines that were morphologically similar to pollinator strains (Figs. 11 and 12). To confirm the hybrid character of received plants, mtDNA, ptDNA probes, and isozyme markers were used by some authors (Brown *et al.*, 1997; Earle *et al.*, 1992; Landgren and Glimelius, 1994; Nothnagel *et al.*, 1997).

In the first crossing cycle (I) of the work, half of total number of maternal plants was kept in natural conditions and no seeds were observed. This effect may be explained due to sporophytic (in the case of lack of pods) or with

Fig. 9. Comparison of field rapeseed resistance to *Leptosphaeria maculans* (Desm.) Ces. et de Not., conidial stage *P. lingam* (Tode ex Fr.) Desm. (A) of alloplasmatic forms of rapeseed and (B) control form of *B. napus* Plant Experiment Station in Zielecin (year 2005). (See color Plate 13.)

embryo–endosperm incompatibility (in case when pods without seeds were developed). Therefore, *in vitro* techniques are necessary to obtain good efficiency in interspecies hybrid production.

After overpollination, the following abnormalities in the pods (ovaries) of maternal plants were observed:

- Pod developed (ovary), lack of embryos
- Developed pod (ovary), germs red in color, lack of embryos or very low number of globular white embryos observed
- Germs are transparent and big, only integument cover is filled with liquid similar to endosperm, lack of embryos
- Lack of endosperm in germs, but small embryo (no bigger than globular stage) is present
- In germ, all typical tissues apart from embryo are present

Fig. 10. Comparison of resistance to *S. sclerotiorum* (Lib.) de Bary of a rapeseed *B. napus* with *B. oleracea* cytoplasm (A) and control form (B). For plant inoculation, cereal grain with pathogen mycelium were placed under aluminium foil Plant Breeding Experiment Station in Malyszyn (year 2005). (See color Plate 14.)

TABLE V

Effectiveness of Interspecific Hybrids Resistant to P. lingam, S. sclerotiorum, Alternaria *spp. Genotype of* B. oleracea *Göttingen* × B. Napus *"00" was Used as Parental Component of Tested Strains in Generation HF_4*[a]

Plot number	Yield [kg/10m^2]	*P. lingam*	*S. sclerotiorum*	*Alternaria* spp.
4916/02	4.80	0	0	1
4910/02	4.60	1	0	0
4918/02	4.60	0	0	1
4921/02	4.60	0	0	1
4971/02	4.60	0	0	0
4980/02	4.60	0	0	0
4915/02	4.50	1	0	0
4920/02	4.50	1	0	1
4932/02	4.50	2	0	1
4972/02	4.50	0	0	0
4973/02	4.40	0	0	0
B. napus var. Lisek	4.40	1	2	1

[a]Evaluation scale: 0—lack of pathogen (no infection); 1—average infection; 2—strong infection.

TABLE VI
Efficiency of Interspecific Hybrid Production Using Isolated Embryos In Vitro *in 2002*

S. No.	Crossing combination	Number of plants received	Seed color
1	*B. napus* × *E. sativa*	21	Yellow
2	*B. napus* × *B. taurica*	14	Black
3	*B. oleracea* (curly leaf kale) × *B. taurica*	20	Black
4	*B. oleracea* (pasture cabbage) × *B. taurica*	19	Black
5	*B. taurica* × *B. oleracea* (pasture cabbage)	15	Black
6	*B. taurica* × *B napus* (KANA)	4	Black
$\sum = 93$			

Fig. 11. Backcrosses of rapeseed with *B. oleracea* cytoplasm (generation F_3 in field condition, Plant Breeding Experiment Station in Malyszyn (year 2006)). (See color Plate 15.)

In many cases, anatomical structure of ovary that was pollinated with alien pollen differs in shape from "germ"—maternal plant pod shape. Endosperm, embryo, seed cover, and pericarp develop harmoniously in the way typical for the exact species or variety.

In interspecific crossings this harmony is upset. The most often the first symptoms of disturbance were present in endosperm as untypical mitosis, excessive tumorlike growth what finishes with total degeneration.

Fig. 12. Alloplasmatic forms of *B. napus* (before harvest, Plant Breeding Station in Zielecin (year 2005)). (See color Plate 16.)

Endosperm abnormality results in weakness and death of embryo. Embryos early enough separated from defective endosperm are able to develop correctly on specific *in vitro* culture. It means that the embryo has no congenital defects. It was stated that the reason of embryo death is determinated by endosperm. The endosperm lost its development possibilities due to unbalanced chromosomal arrangement in hybrid genome or because of lack of endosperm adaptation to conditions that are in maternal organism.

Attempts to rear small globular embryos generally are not successful, even on media with addition of natural endosperm. It is not known, if embryo degeneration arises due to lack of proper nutrition, partial physiological autonomy, or because of physical reasons like damages during isolation, improper osmotic conditions, and so on. Embryos in heart-like stage and bigger ones are generally able to synthesize all necessary storage substances from monosaccharides present in the medium. In this stage, embryo is often called as autotrophic or prototrophic because despite of developed chloroplasts, the embryo still needs sugar from the medium. This is the reason that two different concentrations of $C_{12}H_{22}O_{11}$ were used—richer (10% of sucrose) for smaller embryo stages and poorer (2% of sucrose) for larger embryo stages. Losses of embryos in globular stages were often observed. As the reasons of hybrid embryo development disorder, only endosperm and embryo underdevelopment is indicated. In many cases, death of hybrid embryo can be caused by lethal genes that are active in embryogenesis. The most losses of young hybrid plants were observed during transfer from medium into aseptic soil. Hydroponics culture allows eliminating this kind of risk. The better developed seedlings with leaves and roots have more chances to survive after the transfer.

During growth of seedlings and young plants on medium, fungal infection was observed, which reduced and efficacy of hybrid plants production. The weaker seedlings were strongly infected and died. Specific morphological shape and sterility (lack of pollen) facilitate interspecies hybrids determination.

Most publications on interspecific hybrids among *Brassica* family concern on crossings of taxonomically distant species like *Sinapis alba* × *B. napus* or *S. alba* × *B. oleracea* (Brown *et al.*, 1997; Nothnagel *et al.*, 1997). In the first combination, to receive high efficiency of the process, *in vitro* rearing of isolate embryo and protoplast fusion was used. In second cycle (II) of our research, only two species *B. oleracea* and *B. napus* as well as *in vitro* isolated embryo technique were applied.

In *Brassica* family, sporophytic system of self-incompatibility determined by maternal plant was recognized (Nettancourt, 1969). It occurs when pods do not develop after cross-pollination. When deformed embryos can be observed in pods, it shows on endosperm type of incompatibility. In interspecific crossings, normal relationships among endosperm, embryo, seed cover, and pericarp are upset and first symptoms are seen in endosperm as untypical mitoses and tumorlike growth.

Abnormal endosperm causes weakness and death of embryo, but early developed embryos are able to grow on agar medium *in vitro* condition. In our investigation, no hybrid embryo development disturbance was observed. Cytological analysis conducted on HF_1 plants confirmed their hybrid character because all plants have 28 chromosomes (ACC = 28). Plants in BC_4 were found to be similar in shape to typical rapeseed.

Cytoplasmic molecular markers were used to identify hybrid plants. Marker coxIII amplified DNA in all analyzed samples, which indicates right choice of the method used for cytoplasmaic DNA isolation. Because this marker could not determinate cytoplasms from cabbage and rapeseed, markers taken from catalogue were used. RAPD stripe pattern for BC_4 generation was not satisfactory enough for good determination of new rapeseed forms with cabbage cytoplasm. For this reason, we made attempts to look for new markers.

IX. CONCLUSION

Works on utilization of interspecific hybrids in rapeseed are focused on enhancing rapeseed variability in terms of utility traits and also on production of genotypes with *B. oleracea* cytoplasm. By using modified *in vitro* isolated embryo techniques, we received new hybrids containing cytoplasm from brussel sprouts, curly leaf kales and fodder cabbage. After four to five backcrosses, new forms morphologically similar to rapeseed pollinator line were obtained. These lines were more resistant to frost and to the most important fungal pathogens. To identify the origin of our new hybrids, molecular DNA markers are still being searched.

ACKNOWLEDGMENTS

This work is funded in part by Ministry of Science and Higher Education, Poland, grant (nr. 2 PO6A 018 26).

REFERENCES

Batra, V., Prakash, S. and Shivanna, K. R. (1990). Intergeneric hybridization between *Diplotaxis siifolia*, a wild species and crop *Brassicas*. *Theoretical and Applied Genetics* **80**, 537–541.

Brown, J., Brown, A. P., Davis, J. B. and Erickson, D. (1997). Intergeneric hybridization between *Sinapis alba* and *Brassica napus*. *Euphytica* **93**, 163–168.

Earle, E. D., Temple, M. and Walters, T. W. (1992). Organelle assortment and mitochondrial DNA rearrangements in *Brassica* somatic hybrids and cybrids. *Physiologia Plantarum* **85**, 325–333.

Gundimeda, H. R., Prakash, S. and Shivanna, K. R. (1992). Intergeneric hybridization between *Enarthrocarpus lyratus*, a wild species and crop *brassicas*. *Theoretical and Applied Genetics* **83**, 655–662.

Landgren, M. and Glimelius, K. (1994). A high frequency of intergenomic mitochondrial recombination and an overall biased segregation of *B. campestris* or recombined *B. campestris* mitochondria were found in somatic hybrids made within *Brassicaceae*. *Theoretical and Applied Genetics* **87**, 854–862.

Liu, J. H., Dixelius, C., Eriksson, I. and Glimelius, K. (1995). *Brassica napus* (+) *B. tournefortii*, a somatic hybrid containing traits of agronomic importance for rapeseed breeding. *Plant Science* **109**, 75–86.

Nanda Kumar, P. B. A. and Shivanna, K. R. (1993). Intergeneric hybridization between *Diplotaxis siettiana* and crop *brassicas* for the production of alloplasmic lines. *Theoretical and Applied Genetics* **85**, 770–776.

Nothnagel, T., Budahn, H., Straka, P. and Schrader, O. (1997). Successful backcrosses of somatic hybrids between *Sinapis alba* and *Brassica oleracea* with the *Brassica oleracea* parent. *Plant Breeding* **116**, 89–97.

Nettancourt de, D. (1969). Badation effects on one Locus-gametophytic system on self-incompatibility in higher plants. *Theoretical and Applied Genetics* **39**, 187–196.

Pastuglia, M., Ruffio-Chable, V., Delorme, V., Gaude, T., Dumas, C. and Cock, M. (1997). A functional s locus anther gene is not required for the self-incompatibility response in *Brassica oleracea*. *The Plant Cell* **9**, 2065–2076.

Sundberg, E. and Glimelius, K. (1991). Production of cybrid plants within *Brassicaceae* by fusing protoplasts and plasmolytically induced cytoplasts. *Plant Science* **79**, 205–216.

Wojciechowski, A. (1985). Interspecific hybrids between *Brassica campestris* L. And *B. oleracea* L. I. Effectiveness of crossing, pollen tube growth, embriogenesis. *Genetica Polonica* **26/4**, 423–436.

Wojciechowski, A., Nowak, K. and Olejniczak, J. (1997). Wstepne wyniki krzyzowan miedzygatunkowych pomiedzy *Brassica napus, B. oleracea, B. campestris* i *B. fruticulosa. Rosliny Oleiste* **XVIII**, 91–98.

Honeybees and Rapeseed: A Pollinator–Plant Interaction

D. P. ABROL

Division of Entomology, Sher-e-Kashmir, University of Agricultural Sciences and Technology, Faculty of Agriculture, Udheywalla, Jammu 180002, Jammu and Kashmir, India

ABSTRACT

Bees and *Brassica* plants have mutualistic relationship and coevolved during the long course of their evolutionary history. Since most of the oilseed crops are cross-pollinated, adequate pollination is vital for any significant seed production. Among the various pollinating agents, honeybees play a predominant role in increasing the yield of rapeseed. Almost one-third of total cropped area under oilseeds has been reported to be entomophilous. The yield of rapeseed and mustard can be doubled through pollination by insects. Pollinators not only enhance the yield of the crop but also contribute to uniform and early pod setting. Therefore, planned honeybee

Advances in Botanical Research, Vol. 45
Incorporating Advances in Plant Pathology

0065-2296/07 $35.00
DOI: 10.1016/S0065-2296(07)45012-1

pollination could result in increased productivity and improvement in other parameters through the process of heterosis. Both protective applications of pesticides against pests and use of bees should be integrated in a manner to boost oilseed production and honey production. The relevant information on the mutualistic interactions between honeybees and rapeseed is discussed in this chapter to stimulate further research efforts to enhance productivity of oilseed crops.

I. POLLINATION

Bees and *Brassica* plants have coevolved during the course of their evolutionary history each dependent on one another to live, the bees for nectar and pollen and the Brassicas for pollination (Palmer, 1959). Most of the oilseed crops are self-incompatible and depend on pollinating insects for seed set. Even those self-fertile improve qualitatively and quantitatively if pollinated by insects. Interdependence of bees and *Brassica* is manifested from the fact that pollen of *Brassica* is very sticky and needs insect pollen vectors for its transfer. For *Brassica* plants, bees are marvelously coevolved pollen transferring devices.

The bee's body is covered with feathery hairs and has another anatomical modification suited for carrying and transferring pollen from one flower to another (Figs. 1–7). As the bee dips her head into the flower to sip the sweet nectar deep inside the blossom, its hairy body rubs against the anthers

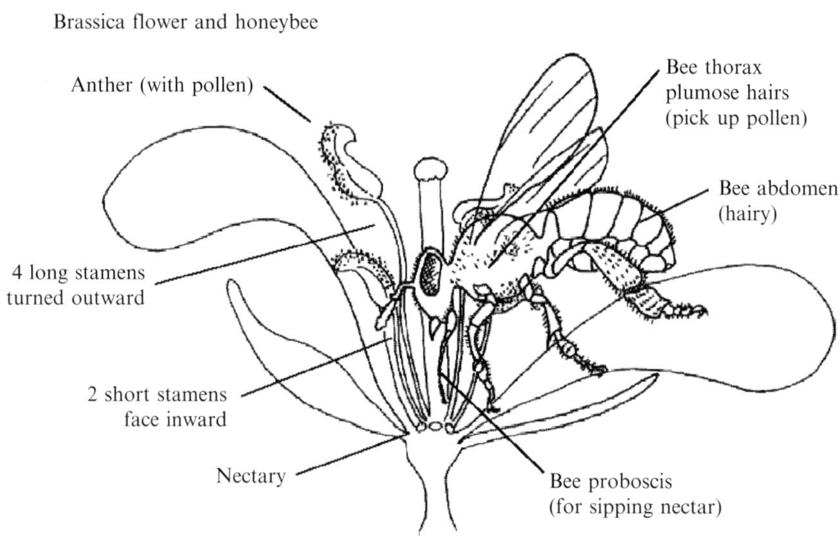

Fig. 1. Showing bee flower interaction in relation to pollination of *Brassica* species.

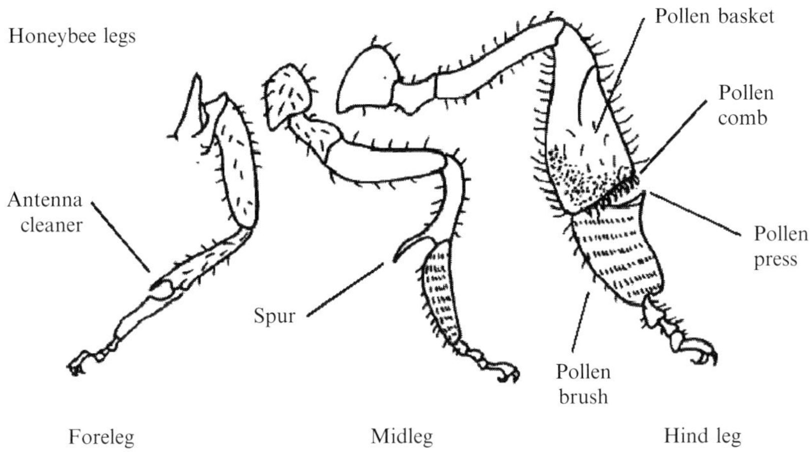

Fig. 2. Showing various anatomical adaptations of honeybee in relation to pollination of *Brassica* species.

holding the pollen. When she flies to next flower, some of the pollen drop from her body and fall on stigma, thereby affecting pollination (Fig. 8).

The oilseed production in India (1022 kg/ha) is still much less than the world average (1451 kg/ha). The low productivity of oilseeds can be attributed for various factors such as heavy infestation of pests and diseases, false siliqua formation, and growing in the marginal and submarginal areas of the country. Out of the various factors, one of the most important factors for low production of oilseed crop is due to the failure of seed setting attributable to the lack of proper pollination (Rao *et al.*, 1980; Panda and Nanda, 1989; Abrol and Kapil, 1996; Ahmad, 1999). Pollination is one of the most important natural factors enhancing crop production. Among the various pollinating agents, insect pollinators play a predominant role in increasing the yield of oilseed crop. Almost one-third of total cropped area under oilseeds has been reported to be entomophilous (Mishra *et al.*, 1988). The yield of rapeseed and mustard can be doubled through pollination by insects. Pollinators not only enhance the yield of the crop but also contribute to uniform and early pod setting. Being a cross-pollinated crop rapeseed attracts a large number of insect pollinators.

A. POLLINATION NEEDS OF RAPESEED

Most of the *Brassica* species are self-incompatible to some extent but this varies with the species and cultivar concerned and even with the age of the plant (Sampson, 1957; Olsson, 1958; Hasler and Maurizio, 1949, 1950). With increased interest in the production of hybrid seed, perfectly self-incompatible

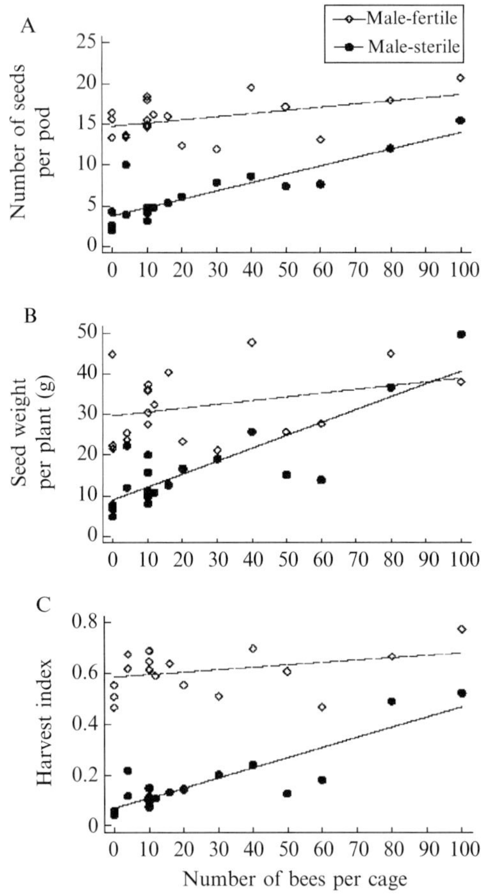

Fig. 3. Relationships between pollinator density of *O. rufa* (females and males) and yield components of a male-sterile (MS) and a male fertile (MF) oilseed rape line in 19 cages. (A) Number of seeds per pod, (B) seed weight per plant, and (C) harvest index (seed weight per plant/plant dry weight). Source: Steffan-Dewenter (2003).

cultivars have an important place in hybrid seed production of several species. However, considering the importance of these crops very little is known about their pollination requirements. Oilseed rape plants have been reported to benefit greatly from bee pollination.

Since most of the crops are cross-pollinated, adequate pollination is vital for any significant seed production. Among the oilseed crops, rapeseed-mustard occupies an important position in Indian economy as producers of edible oil. There are at least four distinct groups of mustard which range from self-sterile to self-fertile. The varieties toria (*Brassica campestris* L. var. toria), brown sarson (*B. campestris* L. var. dichotoma), black mustard (*B. nigra*

Fig. 4. Showing different honeybees pollinating *Brassica* flowers (Top Left) *Apis mellifera*, (Top Right) *Apis dorsata*, (Bottom Left) *Apis florea*, (Bottom Right) *Apis cerana*. (See color Plate 17.)

Fig. 5. Showing impact of open pollination on seed yield in *Brassica*. (See color Plate 18.)

Koch.), and taramira (*Eruca sativa* Mill) are self-incompatible, whereas Rai or Indian mustard (*B. juncea* Czern & Coss.) and yellow sarson (*B. campestris* L. var. sarson) are self-fertile.

Fig. 6. Showing apiary of *Apis mellifera*. (See color Plate 19.)

B. FLORAL BIOLOGY AND POLLINATION REQUIREMENTS OF RAPESEED CROPS

The flowers of cruciferae are hermaphrodite widely spread and attractive to insects when open. They are regular and cruciform, bisexual, complete, and hypogenous. The calyx has four unattached sepals in two whorls. The corolla consists of four free petals in one whorl, each with distinct limb and claw. The androecium consists of six stamens in two whorls, the outer two of which are shorter than the inner (tetradynamous). The ovary is superior, at first one celled but subsequently two celled because of the development of a false septum. Among the floral organs, petals take the greatest proportion of weight. Rapeseed flowers appear very bright to eye during flowering, indicating a high reflection of the visible spectrum.

Muller (1883) stated that the position of the anthers in relation to the nectaries and stigma makes cross-fertilization likely but by no means inevitable on the visit of pollinating insects. The flower is so constructed that pollen from another flower is likely to be transported to it before its own pollen comes in contact with the stigma. Some of the self-pollen may contact the stigma without the aid of insects, but this contact can be abetted by the bees' visit to the flower.

Flowering begins with the opening of the lowest bud on the main stem and continues upward with three to five or more flowers opening per day. Flowering at the base of the first secondary branch begins 2–3 days after the first flower opens on the main stem. Under reasonable growing conditions, flowering of the

Fig. 7. (A) Showing apiary of *Apis mellifera* in *Brassica* field for pollination. (B) Showing *Apis dorsata* working on *Brassica* flowers. (See color Plate 20.)

main stem will continue for 2–3 weeks in both *B. napus* and *B. rapa*. Flowers begin opening early in the morning and as the petals completely unfold pollen is shed and dispersed by both wind and insects. Flowers remain receptive to pollen for up to 3 days after opening.

If favorable, warm and dry weather will result in nearly all the pollen being shed the first day the flower opens. In the evening, the flower partially closes and opens again the following morning. Fertilization occurs within 24 h of pollination. After fertilization, the flower remains closed and the petals wilt and drop. A young pod becomes visible in the center of the flower a day after the petals drop. During flowering, the branches continue to grow longer as buds open into flowers and as flowers develop into pods. In this way, the first buds to open become the pods lowest on the main stem or secondary

Fig. 8. Showing (A) *Apis cerana* (B) *A. mellifera* working on *Brassica* flowers. (See color Plate 21.)

branches. Above them are the open flowers and above them the buds which are yet to open. All of the buds that will develop into open flowers on the main stem will likely be visible in *B. napus* within 3 days of the start of flowering and within 10 days in *B. rapa.*

Maisymora (1990) reported that rate of flowering in rapeseed was dependent on metrological conditions. Chand and Hameed (1994a,b) reported that under Pusa conditions flowering of rapeseed and mustard occurred during first fortnight of January. Mahanta and Rahman (1997) observed that flowering duration of rapeseed var. M-37 was 35 days having 3 days for single flower. Anthesis and dehiscence of flower took place at 5.30 AM and 12 noon hours, respectively.

C. POLLINATION BEHAVIOR

Both bees and flowers of rapeseed are mutually adapted to each other. Bees depend on the flower for nectar and pollen for their survival (Masierowska, 2003). Sugars in the nectar provide carbohydrates to power flight and life activities and pollen is the primary source of proteins, fats, vitamins, and minerals to build muscular, glandular, and skeletal tissues. A worker bee foraging for pollen will hover momentarily over the flower as she uses her highly adapted legs for pollen collection. As the bees move from plant to plant, cross-pollination occurs and genetic information is widely transferred. The honeybees have evolved branched hairs on which they can carry 5 million pollen grains, their intricate pollen baskets, specialized behavior for communication, efficient distribution over available territory; all related to their association with angiosperm plants. Among all the pollinating insects, honeybees make an important pollinator of rapeseed-mustard crops (Free, 1993; Langridge and Goodman, 1982; Mayerhoff, 1958; McGregor, 1976; Pierre, 2000; Sharma *et al.*, 2000). The foreleg is equipped with the *antenna cleaner*, a deep semicircular notch with a row of small spines. This is quickly passed over the antenna. Using the large flat *pollen brushes* on the midlegs, the bee quickly brushes the sticky pollen from her head, thorax, and forelegs. The pollen is transferred to *pollen baskets* by special adaptive features of the hind legs. First the pollen captured on the midleg brushes is raked off by the *pollen combs* onto the *pollen press*. This press is a deep notch located in the joint just below the pollen basket. Flexing the leg, the bee packs the pollen into the baskets which are enclosed spaces on the upper hind leg formed by a concave outer surface fringed with long curved hairs. When the baskets are filled, the worker bee returns to the hive with her supplies to feed the colony—nectar in her honey stomach, pollen in her baskets. In the process, the continuation of a new generation of Brassicas is ensured through her pollination activities.

D. IMPACT OF BEE POLLINATION ON OILSEED CROPS

Most of the oilseed crops are cross-pollinated and adequate pollination is vital for any significant seed production. The pollinator behavior of different crops is given in Table I. An increase in seed yield as a result of insect pollination has been reported in mustard (Latif *et al.*, 1960; Mohammad, 1935). Meyerhoff (1954) concluded that in "Lembke's" winter rape honeybees increased the number of pods per plant by 53.2%, pod length by 6.1%, and seeds per pod by 12.6%. Olsson (1952) obtained a set of 64.7% of the flowers, with 2.46 seeds per pod and 1.75 g per pod with bees excluded, but with bees present these values were increased to 95.3, 4.08, and 2.69, respectively, more than doubling total production. Koutensky (1959) also showed that the seed yield of white mustard was increased 66% by honeybee pollination. Olsson (1955) also showed that the presence of bees in cages of white mustard doubled the number of seeds per pod and increased the pod set by 50%.

Latif *et al.* (1960) showed that rapeseed production in fields with bees was more than double compared to the fields where bees (*Apis cerana* F.) were absent. They further found that *A. cerana* colonies kept near sarson and toria fields increased the seed yield by 60%. Free and Spencer-Booth (1963) found that seed production was double in *B. alba* fields where bees were provided. In *B. juncea*, production was increased only by 14%, an amount that was not

TABLE I

Oilseed Brassica Grown in India with Their Pollination Behavior

Crop/variety	Pollination behavior	Pollination requirement
B. juncea ($2n = 3$)	Self-pollinated	Self-compatible
B. campestris ($2n = 20$)		
Yellow sarson	Self-pollinated	Self-compatible
Brown sarson		
Lotani brown sarson	Cross-pollinated	Self-incompatible
Tora brown sarson	Self-pollinated	Self-compatible
Toria	Cross-pollinated	Self-incompatible
B. napus	Self-pollinated	Self-compatible
B. carinata	Self-pollinated	Self-compatible
B. toumefortii	Self-pollinated	Self-compatible
B. nigra	Cross-pollinated	Self-incompatible
E. sativa	Cross-pollinated	Self-incompatible

Source: Saran and Prasad (2002).

statistically significant in their test, but could be of great significance to the grower. Pritsch (1965) also obtained significantly greater yields of white mustard in cages with bees than in cages where bees were excluded.

There seems little doubt that *B. campestris* requires insects for cross-pollination and seed production. Koutensky (1958) showed that production in fields with apiaries beside them was 2095 kg/ha, but with apiaries 2.4 km (1.4 miles) distant the production was only 1275 kg/ha. Pritsch (1965) studied the pollination of *B. campestris* in cages with bees compared to cages with bees excluded, although smaller insects had access to both cages, and obtained significantly greater production with bee pollination. Downey and Bolton (1961) reported that the yield of seed in fields stocked with bees was at least 30% higher than fields not supplied with bees.

Koutensky (1959) reported that bee pollination increased seed yields of *B. napus* var. *arvensis* by 64%. Vesely (1962) reported that bee activity increased *B. napus* var. *oleifera* seed production by 25% and that cross-pollinated plants set the crop of seed and ceased flowering earlier than plants not visited by bees. Pritsch (1965) also reported significantly greater yield of seed in cages with bees than with bees excluded. Free and Nuttall (1968) reported a 13% increase in seed yield from cages with bees compared to cages without bees—an amount they did not consider of significance. Downey *et al.* (1970) stated that *B. campestris* is almost completely self-sterile and bees must be provided.

According to Free and Nuttall (1968), Fujita (1939) reported that *B. napus* plants caged with bees produced 25% more seed than plants caged without bees. Von Rhein (1952) reported that bees caused 17.4% more seed per pod and 9.7% heavier seeds than were produced on plants not visited by bees.

Louveaux and Verge (1952) reported a 50% increase in seeds per pod on plants growing near a large apiary as compared to plants caged to exclude bees. Jenkinson and Jones (1953) reported that although the relationship of anthers to the stigma in individual flowers favors self-pollination, the presence of bees resulted in increased yields, for example 8.8 seeds per pod with bees present versus 5.3 seeds per pod with bees excluded.

Downey *et al.* (1970) found that in mustard (*B. hirta*) seed production provided with bees was double than that where no bees were provided. This indicates that the provision of bee colonies to yellow mustard fields in adequate numbers should be encouraged. *B. napus* which is about 85% self-pollinating. But when supplemented with honeybee colonies, the seed production ranged from 0.67 to 1.05 tons per hectare. Cross-pollination increased yields only if carried out within 36 h after the flowers open. Most flowers are fully open by 9 AM and remain open for 2–3 days. Seed production (Canada) for *B. campestris* was 2674 kg/ha and for *B. napus* was 3050 kg/ha (Langridge and Goodman, 1975). In general, two to three hives per hectare are required for optimum crop production.

Downey *et al.* (1970) stated that *B. napus* is largely self-pollinated, and thus a good uniform set of seed can be obtained without bees. Turnip rape, however, is almost completely self-sterile and requires cross-pollination to set seed. Wind can carry pollen from one plant to another, but insects, particularly bees, are important. Experiments show that when fields of turnip rape (*B. campestris*) are stocked with bees, earlier and more uniform maturity results. Downey (1964) indicated that when bees were excluded from "Arlo" cv. of rape only two-third as much seed set as when bees were present at flowering. The ratio of seeds per flower with bees present was 6.7 compared to 2.8 without bee visitation.

Deodikar and Suryanaryana (1972) studied the impact of bee pollination on increase in yield of several oilseed crops, viz., *B. napus, B. campestris* var. toria, *B. campestris* var. dichotoma, *B. juncea*, and *B. alba* and found that bee pollination significantly increased the seed yield over self-pollinated and the percentage increase ranged from 12.8 to 222 in different crops (Table II).

Langridge and Goodman (1975) found that in var. *Midas*, honeybees increased yield by 2.5% over control plots. They further found that plants *B. napus* kept in the relatively still air of greenhouses only produced a third to half of normal production. This suggests that either wind or insect pollination was needed to obtain a high seed set. Langridge and Goodman (1975) found that in a trial on an oilseed rape (*B. campestris*) crop grown on dry land in northeastern Victoria, plots to which bees and larger insects had access produced 58% more seeds per plant and 46% greater weight of seed per plant than plots from which these insects had been excluded. The weight of

TABLE II
Yield in Self Pollinated and Bee Pollinated Crops

Crop (botanical name)	Reported range of seed increase from bee pollinated over self-pollinated: Percent more	Times more
Brassica napus L. (rape)	12.8–139.3	1.128–2.39
Brassica campestris L. var. toria (Toria)	66.0–220.9	1.66–3.20
Brassica campestris L. var. dichotoma (sarson)	222	3.22
Brassica juncea Czern & Coss. (Rai, Indian mustard)	18.4	1.184
Brassica alba Boiss. (white mustard)	128.1–151.8	2.28–2.51

Source: Deodikar and Suryanaryana (1972).

1000 seeds was significantly greater in the enclosed plots than in the open ones, but oil content of seed and germination did not differ significantly.

In case of *B. campestris* pollination benefits according to Langridge and Goodman (1975) in Australia were as described below: 775 kg/ha increase over control plot and 600 kg/ha increase over the control plot. Honeybees increased yield by 61.2% over control plots of var. *Arlo* and oil production by 10–20%. Wind helps in pollination. Hover flies are also important. For var. *Arlo*, 58% more seeds per plant and 46% greater weight of seed were achieved compared to plots excluded from insects. Ninety-seven percent were self-sterile in glasshouse studies. For var. *Golden*, 85% were self-sterile in glasshouse studies. The density of honeybees in the field was significantly related to the distance from the apiary (Sweden). Results indicated that higher densities of honeybees increased both seed yield and oil content of seed (Sweden).

Similar increase in yield has been reported in oilseed rape (Kisselhagen, 1977; Langridge and Goodman, 1975) and *B. campestris* var. Jambuck (Kubisova *et al.*, 1980).

Bisht *et al.* (1980) found that flowers of rapeseed visited by *Apis* species had higher pod set, more number of seeds per pod, and weight of seed was also higher than those deprived of pollinator visit. Mishra *et al.* (1988) found that in *B. campestris* percent pod setting, number of pods per plant, and proportion of healthy seeds were significantly higher in open pollinated flowers than in net caged and muslin bagged ones. Similarly, average weight of seeds and oil content was higher in open pollinated flowers (Table III). *A. cerana* was the most common pollinating species. The other pollinators observed included *A. mellifera*, syrphid flies, and so on. In a similar study, Prasad *et al.* (1989)

TABLE III
Yield and Oil Potential of Brassica campestris *var. sarson Under Three Conditions of Pollination*

Parameter	Open pollination (X ± S.E.)	Net caged (X ± S.E.)	Muslin bagged (X ± S.E.)	Critical difference	Transformation
Pod setting (%)	95.21 ± 2.46	70.27 ± 4.38	58.81 ± 6.57	(10.09)	Angular
Seeds/pod	12.72 ± 0.82	5.55 ± 1.18	3.95 ± 0.85	(0.73)	$\sqrt{n} \pm 1$
Healthy seeds (%)	84.78 ± 2.76	41.45 ± 8.25	48.37 ± 12.28	18.41	Angular
Seed weight (mg per 100 seeds)	478.50 ± 21.83	459.40 ± 23.41	542.20 ± 11.00	(1.31)	$\sqrt{n}$
Oil content (%)	39.02 ± 0.29	40.76 ± 0.90	41.64 ± 0.27	1.81	–
Oil yield (mg) pod	19.03 ± 0.24	3.02 ± 0.62	1.95 ± 0.12		
Increase in oil yield over muslin bagged	9.76	1.55	1.0		

Source: Mishra *et al.* (1988).

found that pollination of *B. juncea* by *A. cerana* resulted in more siliqua setting, increased length of siliqua, seed weight increased yield and had a pronounced effect on oil contents and germination.

Kevan and Eisikowitch (1990) reported that the presence of pollinators, for example honeybees, on canola (cv. O.A.C Triton) increases the germinability of resulting seeds from 83 to 96%. Khan and Chaudhary (1991) revealed that self-pollinated sarson crop in Pakistan produced 34.34 g seeds per plant while honeybee (*A. cerana*) and other insect-pollinated crop yielded 90.02 g seeds and 50 g seeds per plant, respectively. Mishra and Kaushik (1992) reported that yield and oil content was higher in honeybee-pollinated crops in comparison to self-pollinated. The results showed that higher percentage of pod setting was obtained in open pollinated crops of *B. nigra, B. campestris, B. carinata, B. napus*, and *E. sativa* having 80.00, 84.42, 71.90, and 70.30%, respectively. Average weight of 1000 healthy seeds varied between 3.0 and 5.1 g under open pollination, whereas it was 1.8–3.7 g under self-pollination. Mahanta and Rahman (1997) reported that pollination of rapeseed crop by *A. cerana* caused a maximum yield of 7.25 quintal per hectare which was 437.51% higher than self-pollinated treatment. Langridge and Goodman (2003), in a trial on oilseed rape (*B. napus*) cv. Midas in northern Victoria, found significant increase in yield parameters from plots where bees and larger insects had access compared with plots enclosed in cages of 2.5-mm mesh to exclude these insects. The weight of 1000 seeds was slightly greater in the enclosed plots (Table IV). Bees and a syrphid, *Melangyna viridiceps*, were the predominant insects visiting the flowers.

Singh *et al.* (2000) found that siliqua setting in plots caged with bees resulted in an increase of 239.77% over plots pollinated without insects and 2.65% over open pollinated flowers. Similarly number of seeds, seed weight, and seed germination were improved. Moreover, seeds from bee-pollinated plants showed 21.3% higher oil content than plants without bees and 4.1% than open pollinated plants (Table V). In a similar study, Prasad *et al.* (1989) found that pollination of *B. juncea* by *A. cerana* resulted in more siliqua setting, increased length of siliqua, seed weight increased yield, and had a pronounced effect on oil contents and germination.

TABLE IV
Rapeseed Crop with and Without Pollinators

Attribute	Open plots	Closed plots	Significance of difference
Plant density	459	406	
Seed yield	725	453	$p < 0.01$
Seed/plant	1.61 g	1.10 g	$p < 0.01$
Weight of 1000 seeds	1.78 g	1.88 g	$p < 0.01$
Number seeds/plant	927	588	$p < 0.01$
germination (%)	95.3	97.1	ns
Oil content (%)	37.9	36.1	ns

Source: Langridge and Goodman (1975).

TABLE V
Quantitative and Qualitative Effects of Apis cerana *Himalaya on Rapeseed*

Yield parameter	Control	Open pollinated	Bee pollinated	Critical difference	Percent increase over PWI	Percent increase over OP
Siliqua set (%)	20.67	68.42	70.23	5.00	239.77	2.65
Siliqua length (cm)	3.94	4.16	4.52	0.07	14.72	8.65
Number of seeds per Siliqua	9.12	11.65	13.73	0.24	50.55	17.85
Weight of 1000 seeds (g)	2.01	2.31	2.36	ns	17.41	2.16
Seeds yield (q/ha)	1.96	12.37	13.45	2.90	586.22	8.73
Seeds germination (%)	55.00	81.00	83.00		50.90	2.30
Oil content (%)	33.25	38.75	40.33		21.30	4.10

Source: Singh *et al.* (2000).

Manning and Boland (2000) found that the number of pods/plant decreased as distance from the apiary increased with a predicted pod loss of 15.3 pods/plant over a distance of 1000 m from an apiary. This was equivalent to a 16% loss based on an average of 59 plants/m^2 and average pod production of 5666 pods/m^2. For a 2-ton/ha crop this would be equivalent to about 320 kg/ha, thereby indicating the importance of honeybees in rapeseed pollination.

Sabbahi *et al.* (2005) studied the impact of pollination on seed production of canola, *B. napus* L. with three treatments: 0, 1.5, and 3 hives per hectare and also the plants caged to exclude pollinators and bee visitations. After the harvest, yield analyses were done in relation to the bee density gradient created, by using pod set, number of seeds per plant, and weight of 1000 seeds and found that results showed an improvement in seed yield of 46% in the presence of three honeybee hives per hectare, compared with the absence of hives. The introduction of honeybees contributed to production and consequently, these pollinators represented a beneficial and important pollen vector for the optimal yield of canola.

Steffan-Dewenter (2003) studied the effects of pollinator density on the seed set of a male-sterile (MS) and a male-fertile (MF) line of winter oilseed rape (*B. napus*) and found that increasing densities of *O. rufa* had a significant effect on almost all yield components of the MS line, but only marginal effects on the MF line (Table VI, Fig. 3).

II. INSECT POLLINATORS

Rahman (1940) studied the pollinators of *B. napus* in India. He concluded that the dwarf honeybee of India (*A. florea* F.), the wild bees (*Andrena ilerda* Cam. and *Halictus* sp.), and the fly (*Eristalis tenax* L.) were the most important pollinators. Honeybees are the primary pollinators of rape (Belozerova, 1960; Nikitina, 1950; Radchenko, 1964; Vesely, 1962). The plant is highly attractive to honeybees, providing both nectar and pollen, and the honeybee is of appropriate size for effective transfer of pollen from anthers to stigma. Hammer (1952) reported as many as 20,000 bees per hectare of rape in fields 31/2 to 4 km from the apiary. Each bee was returning to the hive with 30- to 60-mg loads of nectar, roughly half the weight of a worker bee. Belozerova (1960) noted that *B. napus* had 2.326-mg nectar per flower at the beginning of bloom, 1.950 mg during the peak, and 1.350 mg per flower toward the end of blooming. He noted that 96.3, 95.3, and 72.9% of the floral visitors at the three different periods were honeybees. Other pollinators in India include *A. florea, A. dorsata, A. cerana*, and *A. ilerda*

TABLE VI

Mean Yield Components of a Male-Sterile and Male-Fertile Oilseed Rape Line: Comparison of Controls (No Pollinator = Three Cages) Versus High Pollinator Densities (Honeybee Colony and Five Male Mason Bees, n = Five Cages)

Variable	Male-sterile line			Male-fertile line		
	Control (no pollinator)	High pollinator density	*p*-value of difference	Control (no pollinator)	High pollinator density	*p*-value of difference
Number of seeds per pod	2.9	15.9	0.0001	15.1	18.4	0.009
Number of seeds per plant	1018	10740	<0.0001	6709	7515	0.603
Seed weight per plant	6.5	56.8	<0.0001	29.5	29.1	0.95
Number of pods per plant	356.9	677.0	0.0004	440.6	409.3	0.735
Plant dry weight (g)	125.4	93.1	0.0059	57.4	49.1	0.432
1000 seeds weight (g)	6.49	5.34	0.0336	4.43	3.87	0.0078
Number of seeds per plant dry weight (g)	8.03	117.3	0.002	115.5	152.9	0.0091
Harvest index (seeds weight/plant dry weight	0.052	0.62	<0.0001	0.51	0.59	0.050

The significant probability of one-way Anova is given.
Source: Steffan-Dewenter (2003).

(Kapil *et al.*, 1969). Free and Nuttall (1968) studied the activity of honeybees on *B. napus*. They reported that all bees that visited the flowers collected nectar, although some collected pollen also. All became covered with pollen, but some removed and discarded it. Those that collected the pollen did so primarily during the morning hours. Kapil *et al.* (1971) recorded 13 species of bees belonging to 5 families visiting rapeseed-mustard flowers in Punjab and Haryana. The bee species identified included *A. florea*, *A. dorsata*, *A. mellifera*, *A. ilerda*, *A. leaena*, *Megachile lanata*, *M. flavipes*, *M. cephalotes*, *Xylocopa fenestrate*, *X. pubescens*, *Nomia* spp., *Halictus* spp., and *Pithitis smargdula*. Out of the various bee pollinators, *A. florea* and *A. dorsata* were the most important and efficient pollinators of all the crops. Bhalla *et al.* (1983) recorded seven species of insects visiting rapeseed crop under mid-hill conditions of Himachal Pradesh. Mishra *et al.* (1988) recoded *A. cerana* as the most important pollinator of *B. campestris* var. sarson accounting 69.47% of the total flower visitors while *A. mellifera* was the least with 2.95%. The other insect visitors included dipteran flies and syrphids, and so on. Prasad *et al.* (1989a,b) fount that among all the insect pollinators *A. cerana* was the predominant pollinator of brown mustard. In a similar study, Kumar *et al.* (1994) observed 18 species of insects foraging on toria. Among all the foragers, *A. mellifera* was found to be the dominant one accounting 57.45% but *A. ilerda* was the most efficient pollinator of *B. campestris* var. toria. Mishra *et al.* (1988) studied the foraging behavior of honeybees on *B. campestris* var. sarson and found that average number of *A. dorsata*, *A. florea*, *A. cerana*, and *A. mellifera* per square meter was 2.5, 9.0, 1.1, and 1.6, respectively. Singh *et al.* (1993) found that *A. cerana* was the most important pollinator rapeseed var. M-27 and constituted 83.14% of the total flower visitors.

While honeybees make up about 90% of the effective insect pollinating fauna (Belozerova, 1960; Mesquida and Renard, 1979), several species of solitary bees (Andrenidae, Halctidae, Megachilidae, and lower Apidae) and bumblebees were also recorded visiting oilseed rape (*B. napus* L. var. *oleifera* Mezter). Winter oilseed rape [*B. napus* L. subsp. *oleifera* (Metzg.), Brassicaceae] is an important crop for the production of oilseed in temperate agricultural regions. Oilseed rape is a predominantly self-pollinated crop with about one-third out crossing (Becker *et al.*, 1992). Pollen transport by insects, wind, or gravity as vectors is necessary (Eisikowitch, 1981; Free, 1993; Westcott and Nelson, 2001; Williams, 1978). Earlier studies have shown that the insect pollination of oilseed rape can lead to higher seed set and yield (Westcott and Nelson, 2001; Williams and Simpkins, 1989). However, these effects are dependent on cultivar, environmental growing conditions, and the compensatory capacity of the crop (Free, 1993; Mesquida *et al.*, 1988; Westcott and

Nelson, 2001; Williams *et al.*, 1987). Further replicated experiments comparing yields with and without bees are still needed (Kevan and Phillips, 2001; Mesquida *et al.*, 1988; Williams *et al.*, 1987). The significant heterosis for seed yield in oilseed rape has created interest in the development of hybrid cultivars (Riaz *et al.*, 2001). Oilseed rape hybrids are based on male sterility mainly using the MSL system (Pinochet and Bertrand, 2000). Several oilseed rape hybrids have been recommended to farmers since 1994 (Pinochet and Bertrand, 2000). Hybrid composites consisting of a male-sterile component and a male-fertile component have been widely used in France and the United Kingdom. However, at several locations in Europe low seed set occurred, presumably due to pollen limitation (Pinochet and Bertrand, 2000). This could have been caused by low temperatures and rain reducing pollen transfer by wind and flight activity of insects. Additionally, the destruction of seminatural habitats results in lower pollinator abundance in intensively managed agricultural landscapes, thereby possibly causing pollinator limitation and reduced seed set (Steffan-Dewenter and Tscharntke, 1999; Steffan-Dewenter *et al.*, 2001, 2002, 2005, 2006). In addition to the honeybee, *A. mellifera* L., which is often advocated as the only practical pollinator, the use of solitary bees and flies has been proposed (Delaplane and Mayer, 2000; Richards, 1993; Torchio, 1990). Yields of MS oilseed rape in isolation cages were increased by honeybee pollination (Mesquida and Renard, 1981) and varied with leaf-cutting bee-stocking rates (Soroka *et al.*, 2001). Pollination of other plant species in closed environments was improved successfully by using the solitary mason bee *Osmia rufa* Losinski and the syrphid fly *E. tenax* (L.) (Jarlan *et al.*, 1997; Schittenhelm *et al.*, 1997). Because most studies compare control versus pollinator treatments, pollinator densities required for optimal seed set are as yet still unknown. Steffan-Dewenter (2003) studied the effects of pollinator using the solitary mason bee *O. rufa* and found that there was significant increase in yield and improvement in quality of the produce. *O. rufa* can be easily established in nesting traps made of wood or common reed (Tscharntke *et al.*, 1998) and has been shown to perform well as a pollinator in closed environments such as glasshouses (Delaplane and Mayer, 2000; Free, 1993; Schittenhelm *et al.*, 1997).

A. Bees Can Discriminate Between Different Lines

Wright *et al.* (2002) found that honeybees like *A. mellifera* use odors to identify and discriminate among flowers during foraging. They found that if a bee had been trained with a variety of canola, it was unable to differentiate the odor of one canola flower from the odor of other canola flowers, but it could differentiate canola from the odor of a snapdragon flower. Bees trained with the odor of snapdragon flowers readily differentiated the odor of one variety of a snapdragon from the odor of other varieties of snapdragons and also canola flowers. Our study suggests that both intensity and odor quality affect the ability of honeybees to differentiate among floral perfumes.

B. NUMBER OF COLONIES REQUIRED FOR POLLINATION

Several investigators have attempted to determine the number of colonies of honeybees required for increased yields and their recommendations vary from place to place and crop, for instance, Hammer (1963, 1966) recommended three colonies per hectare; Radchenko (1964), two; Downey and Bolton (1961), one; White (1970), two; and Vesely (1962) three to four colonies per hectare. Although hoverflies appear to play some role in pollination of rape, we consider the honeybee the more efficient pollinator. White (1970) said that both summer turnip rapes and true rapes depend on bees for maximum production. The data indicate that a heavy bee population on rape would be beneficial, until more concrete data are available, the one to two strong colonies of honeybees per acre would appear to be a logical usage. The ideal pollinator population and proper distribution of colonies for most efficient pollination of rape need to be determined.

C. POLLINATION RECOMMENDATIONS

Honeybees are the most effective agents involved in cross-pollination of rapeseed-mustard crops (Palmer, 1959; Zander, 1952). In modern agriculture, farm mechanization and high-yielding varieties (HYV) are very common. In such a condition, to increase the yield more inputs such as water, fertilizer, and other agrochemicals are in use. Indiscriminate use of pesticides/fungicides often kills a large number of pollinators. In certain cases, a single crop over a vast area is cultivated. This also reduces number of wild honeybee colonies in those areas. Therefore, the importance of bee keeping in the field is being realized as an important input to increase the production of oilseed crops.

1. It is the only pollinating insect, which can be employed in large numbers and distributed at desired places.
2. Honeybees have a flower fidelity to one kind of a plant at a given time than either solitary bees or bumblebees.
3. It works for longer periods than most of the solitary bees.

4. Its body parts are specially modified to pick up large number of pollen grains.
5. It is less affected by adverse weather conditions than most other insect pollinators.

For effective pollination and increased yield of oilseed crops, the efficiency of a bee colony as a pollinator would depend on the following factors:

1. Colony strength

Larger and stronger colonies are five times better than the smaller and weaker ones because the former have a higher percentage of older bees as foragers. Thus, good honey yielding colonies are better pollinators also.

2. Number and time of placement of colonies

This factor depends on density of plant stand, total number of flowers in inflorescence of each plant, duration of flowering, strength of bee colonies, and number of flowers over an area of 1 ha of land. Generally, three to five colonies of *A. mellifera* per hectare of crop in bloom and five strong colonies of *A. cerana* are recommended for sufficient and efficient pollination.

3. Distribution of colonies in the field/orchards

Honeybees primarily visit the nectar flow source which is in the range of 0.3- to 0.5-km radius from the apiary. Beyond 0.5-km range, pollination activity greatly diminishes. For efficient pollination hives should be placed singly than in groups. Bees tend to forage in the area which is closest to their hive, particularly when weather is not favorable.

4. Time and placement of colonies

Generally, colonies are to be introduced when 5–10% of the crop is in bloom. Earlier placements of the bees result in foraging in other weeds and wild plants in the vicinity, but ignore the crop bloom. If bees are moved too late, they can only pollinate late and less vigorous ones.

5. Weather conditions

Failure and success of bee pollination depend on the weather conditions as it affects equally the crop as well as bees. *A. cerana* can forage at a lower temperature than *A. mellifera*. Wind velocity of more than 15 miles/h affects the forage behavior, therefore wind breaks are recommended around the orchard/field. Cool, cloudy weather, and storms greatly reduce bee flights.

D. PROBLEMS ASSOCIATED WITH BEE POLLINATION

However, aphid control becomes a problem during pollination when the crop is in bloom and bees are actively foraging in the field (Abrol, 1997, 1993; Abrol and Andotra, 1998, 2003; Mayer, 1980; Mayer *et al.*, 1984). Insecticides presently registered for use in seed crops are considered highly toxic to bees. Thus, to avoid killing bees, pollination often must be interrupted. The colonies are either temporarily removed from the field or applications must be made at night. In either case, several days can pass before the bees resume normal foraging activity. Repeated applications to control aphid populations may cause a significant disruption in pollination and reduced seed yield and quality.

Flowering Brassicas attract not only a large number of insect pollinators especially the honeybees for nectar and pollen but also other insects which feed on flowers, leaves, fruits, thereby causing serious economic losses (Sihag, 1991). Oilseed crops are attacked by aphids, caterpillars, and bugs during flowering and pod formation stage. This requires the application of insecticides to combat the pest (Sihag, 1985a,b,c, 1986, 1988, 1991, 1995; Sihag *et al.*, 1999a,b) which poses serious problem for the foraging activity of honeybees and developing brood. Evidently, problems of infestation of aphids and other pests and their management during flowering period of rapeseed and mustard crops pose major problem for their utilization as effective pollinators for crop production which also get incidentally killed. Aphid control becomes a major problem during pollination when the crop is in bloom and bees are actively foraging in the field. Insecticides presently registered for use in seed crops are considered highly toxic to bees. Thus, to avoid killing bees, pollination often must be interrupted. The colonies are either temporarily removed from the field or applications must be made at night. In either case, several days can pass before the bees resume normal foraging activity. Repeated applications to control aphid populations may cause a significant disruption in pollination and reduced seed yield and quality.

Pollinator–plant interaction is a very complex phenomenon and is influenced by many overlapping effects. The protection of pollinators, including honeybees, is as essential as the protection of crops from the insect pest damage. The use of pesticides for pest control on the one hand and the role of honeybees (*Apis* spp.) for crop pollination on the other have become essential components of modern agriculture. Unfortunately, these two practices are not always compatible, as honeybees are susceptible to many of commonly used pesticides (Cunningham *et al.*, 2002; Johansen, 1977; Russell *et al.*, 1998), used for the control of insect pests (Poehling, 1989;

Stark *et al.*, 1995). The major constraint confronting pollinator–plant interaction is the indiscriminate and excessive use of pesticides for controlling insect pests (Bisht *et al.*, 1980; Rana and Goyal, 1991; Zhong *et al.*, 2004). The loss of honeybees directly affects beekeeping through loss of honey production and indirectly the crop production due to inadequate pollination. Reduction of population of these beneficial insects due to insecticides, therefore, incurs significant environmental, ecological, and economic costs (Bai and Reddy, 1977; Crane and Walker, 1983; Khan and Dethe, 2004; Pimentel *et al.*, 1980; Prakash and Kumaraswami, 1984).

Honeybees constitute a major group of insect pollinators and their pollinating efficacy is manifested not only through increase in yield but also by the improvement of the crop quality through heterosis breeding (Melnichenko, 1976). The bees earn about rupees 10 million to the national exchequer in terms of honey production and beeswax and it is expected that an additional crop yield worth rupees 90 million could be obtained due to pollination of crops (Sorthia and Chari, 1985). Levin (1984) estimated that the value of crops in the United States that benefited directly from the honeybee pollination approaches $20 billion annually. The annual cost of crop loss due to insecticide poisoning of pollinating honeybees has been estimated at $135 million in the United States (Pimentel, 1980). Notwithstanding with the absence of such estimates, the indispensability for a harmonious compromise between pest management and honeybee pollination of crops in India cannot be gainsaid.

Bee poisoning or killing of bees from pesticides continuous to be a serious problem for beekeepers. Most bee kill occurs when pesticides are applied or allowed to drift on to blooming crops or weeds (Mayer, 2003). Most (99%) bee kills result from bees picking up the pesticides when foraging. The hazards of insecticidal application on flowering crops include direct mortality, fumigative effects, repellent effect, and toxicity of residues present on various floral parts and in nectar to the insect visitor. A highly toxic insecticide generally reduces the field force of a colony within a short period of time. Colonies may be reduced by one-third to half in strength within 24–48 h (Eckert and Tucker, 1954), thus adversely affecting both the production and marketing segments of the honey and beekeeping industry. Generally, fumigative action of insecticides used under field conditions is of much shorter durations than the effect of contact and stomach poison. A prolong repellent effect will deprive the flowers of pollination benefits of insect visits, while a short repellency will deter the insect pollinators from visiting the treated bloom for a brief period and, thereafter, allow them to resume their foraging activity (with minimum residual hazards) without compromising with the yield potential of the crop.

Poisoning of insecticides to honeybees is generally more pronounced because of their long hours of working on the crop flowers for pollen and nectar collection, continuous working nature, and long flights with pollen loads. Bee poisoning is more chronic in areas which lack sufficient wild pollens and nectar plants to sustain the number of colonies required for crop pollination. Most bees are pastured in agricultural areas where they are subjected to killing from pesticides (Mayer and Johansen, 1983). The beekeeper has little or no control of when and what pesticides are applied in the areas of his bee forage (Mayer, 2003).

Further more modern agricultural practices have resulted in the reduction of wild insect pollinators and disturbed insect–flower relationship by way of disappearance of wastelands and uncultivated strips of land, destruction of certain food sources by weed control, and overall changes in the environment. Wild bees are also damaged by pesticides. Poisoning may result from contaminated food as well as from florets, leaves, soil, or other material used by the bees in nesting. The toxicity of a specific insecticide to honeybees and wild bees is not always the same, and even among wild bees some materials are more toxic to one species than to another.

Wild bees are also damaged by pesticides. Poisoning may result from contaminated food as well as from florets, leaves, soil, or other material used by the bees in nesting. The toxicity of a specific insecticide to honey bees and wild bees is not always the same, and even among wild bees some materials are more toxic to one species than to another.

III. PESTICIDE APPLICATION PRACTICES TO REDUCE BEE POISONING

1. Apply pesticides only when needed.
2. Choose the pesticide with the lowest hazard rating for bees, particularly the lowest residual toxic effect, from the list of pesticides available for a particular pest control program.
3. Liquid or granule applications are less hazardous than dusts. Microencapsulated forms of pesticides have a significantly longer residual life than other application forms. The minute capsules can be carried back to the colony in the same manner that pollen is carried, and can kill brood and young adult bees.
4. Ground application is less hazardous than aerial application, particularly when applied in close distances to apiaries. Fine sprays are less toxic than coarse sprays.
5. Where practicable, apply pesticides when bees are not active on the crop. For pesticides considered a low hazard when they have dried, early

morning may be suitable. For pesticides with a residual toxic effect of a few hours, apply in the late afternoon or early evening.

6. The time of day when a pesticide is applied should be chosen to minimize the risk of spray drift occurring either over apiaries or over plants being foraged by bees. Treatments can and should be applied only when bees are not foraging for nectar or pollen. If bee-attractive plants need to be treated while in bloom, they should be treated at night or in the early morning or late evening when the bees are not flying.

7. Where practical, beekeepers should be given prior notice, preferably a minimum of 48 h, of a pesticide application to allow apiaries to be moved from the area.

IV. CONCLUSION

The foregoing discussion reveals that by proper management and supplementing field with honeybee colonies oilseed production can be enormously increased. The honeybee pollination not only increases yield in self-incompatible varieties but even the self-fertile varieties give many times increased yield over those deprived of insect visits. There is also significant increase in quality of the crop and other parameters through the process of heterosis. Honeybee pollination stimulates germination of pollen on stigma of flowers and improves selectivity in fertilization, increases the viability of embryos and seeds, enhances resistance to diseases and other adverse environmental conditions, increases nectar production in the nectaries of plants, results in early and uniform crop set, and increases oil content in oilseed crops. The study suggests that both protective applications of pesticides and use of bees should be integrated in a manner to boost oilseed production and honey production.

REFERENCES

Abrol, D. P. (1993). Insect pollination and crop production in Jammu and Kashmir. *Current Science* **65**(3), 265–269.

Abrol, D. P. (1997). "Bees and Beekeeping in India," p. 450. Kalyani Publishers, Ludhiana.

Abrol, D. P. and Andotra, R. S. (1998). Impact of pesticides on foraging activity of honeybee,*Apis mellifera* L.on *Brassica campestris* L.var.*toria*. *In* "Proceedings of the Fourth International Asian Apicultural Association Conference," p. 123. Kathmandu, Nepal, March 25–27.

Abrol, D. P. and Andotra, R. S. (2003). Relative toxicity of some insecticides to *A. mellifera* L. *Journal of Asia Pacific of Entomology* **6**(2), 235–237.

Abrol, D. P. and Kapil, R. P. (1996). Insect pollinators of some oilseed corps. *Journal of Insect Science* **9**(2), 172–174.

Ahmad, B. (1999). Insect pollinator complex of rapeseed (*Brassica campestris* var. toria) M.Sc. thesis, Assam Agricultural University, Jorhat, Assam 52 p.

Bai, A. R. K. and Reddy, C. C. (1977). Laboratory toxicity of some insecticides to *Apis cerana indica*. *Journal of Apicultural Research* **16**(3), 161–162.

Becker, H. C., Damgaard, C. and Karlsson, B. (1992). Environmental variation for outcrossing rate in rapeseed (*Brassica napus*). *Theoretical and Applied Genetics* **84**, 303–306. Bee World 79 (3), 115–123.

Belozerova, E. L. (1960). Bees increase the seed crop from winter rape. *Pchelovodstvo* **37**, 36–40.

Bhalla, O. P., Vema, A. K. and Dhaliwal, H. S. (1983). Insect visitors mustard bloom, their foraging behaviour under mid hill conditions. *Journal of Entomological Research* **1**(1), 15–17.

Bisht, D. D., Naim, M. and Mehrotra, K. N. (1980). Studies on the role of honeybees in rapeseed production. *In* "Proceedings of Second International Conference of Apiculture in Tropical Climate," pp. 491–496. Indian Agricultural Research Institute, New Delhi.

Chand, H. and Hameed, S. F. (1994a). Population dynamics of honeybees and insect pollinator on Indian mustard, *Brassica juncea* L. *Journal of Entomological Research* **18**(3), 233–239.

Chand, H. and Hameed, S. F. (1994b). Studies on the flowering period, density of Indian mustard and competing flora for honeybees in and around Pusa, Bihar (India). *Journal of Entomological Research* **18**(4), 361–367.

Crane, E. and Walker, P. (1983). "The Impact of Pest Management on Bees and Pollination." p. 107. International Bee Research Association, England.

Cunningham, S. A., Fitzgibben, F. and Heard, T. A. (2002). The future of pollinators for Australian Agriculture. *Australian Journal of Agricultural Research* **53**(8), 893–900.

Delaplane, K. S. and Mayer, D. F. (2000). "Crop Pollination by Bees," CAB International, Wallingford, UK.

Deodikar, G. B. and Suryanaryana, M. C. (1972). Crop yields and bee pollination. *Indian Bee Journal* **34**, 52–63.

Downey, R. K. (1964). Effect of bees on seed yields of rapeseed. *Forage Notes* **10**, 1–5.

Downey, R. K. and Bolton, J. L. (1961). Production of rape in western Canada. Canada Dept. Agr. Res. Br. Pub. **1021**, 19 p.

Downey, R. K., Bolton, J. L., Pawlowski, S. H. and Mcansh, J. (1970). Rapeseed – Canada's "cinderalla," crop. 2nd Edn., Rapeseed Association of Canada Pub 8, 40 p.

Eckert, J. E. and Tucker, K. W. (1954). The toxicity of newer pesticides to honeybees. *Gleanings in Bee Culture* **82**, 529–531, 570.

Eisikowitch, D. (1981). Some aspects of pollination of oil seed rape (*Brassica napus* L.). *Journal of Agricultural Science (Cambridge)* **96**, 321–326.

Free, J. B. (1993). "Insect Pollination of Crops," 2nd Edn., p. 656. Academic press, London.

Free, J. B. and Nuttall, P. M. (1968). The pollination of oilseed rape (*Brassica napus)* and the behaviour of bees on the crop. *Journal of Agricultural Science (Cambridge)* **71**, 91–94.

Free., J. B. and Spencer-Booth, Y. (1963). The pollination of mustard by honeybees. *Journal of Apicultural Research* **2**, 69–70.

Fujita, M. (1939). Influence of honeybees on the fructification of rape. Bull. Imp. Zootech. *Experimental Station Chiba-shi* **34**, 1–6.

Hammer, O. (1952). Rape growing, bees and seed production. *Dansk landbr.* **71**, 67–69.

Hammer, O. (1963). Summer rape as a competitor affecting the pollination of clovers. Dansk Froavl No. 14, p. 7.

Hammer, O. (1966). Some problems of competition between summer rape and clover, in relation to pollination. *In* "2nd International Symposium on Pollination," London, 1964, *Bee World* **47**(Suppl. 1), 99–106.

Hasler, A. and Maurizio, A. (1949). The action of boron on seed-setting and nectar secretion in rape (*Brassica napus* L.). *Phytopathology Zeits.* **15**(2), 193–207.

Hasler, A. and Maurizio, A. (1950). Influence of various nutritive substances on the development of blossom, the secretion of nectar, and the seed-yield, of nectar plants (especially summer rape). *Schweizerische Landwirtschaftliche Monatshefte* **6**, 201–211.

Jarlan, A., De Oliveira, D. and Gingras, J. (1997). Effects of *Eristalis tenax* (Diptera: Syrphidae) pollination on characterisitcs of greenhouse sweet pepper fruits. *Journal of Economic Entomology* **90**, 1650–1654.

Jenkinson, J. G. and Jones, G. D. G. (1953). Observations on the pollination of oil rape (*Brassica napus*) and broccoli (*Brassica oleracea*). *Bee World* **34**, 173–177.

Johansen, C. A. (1977). Pesticides and pollinators. *Annual Review of Entomology* **22**, 172–199.

Kapil, R. P., Grewal, G. S., Kumar, S. and Atwal, A. S. (1969). Insect pollinators of rape seed and mustard. (Abstract) 56th Indian Sci. Cong. Proc., pt. **3**, 509.

Kapil, R. P., Grewal, G. S., Kumar, S. and Atwal, A. S. (1971). Insect pollinators of rapeseed and mustard. *Indian Journal of Entomology* **33**, 61–66.

Kevan, G. and Eisikowitch, D. (1990). The effects of insect pollination on canola (*Brassica napus* L. cv. O.A.C. Triton) seed germination. *Euphytica* **45**, 39–41.

Kevan, P. G. and Phillips, T. P. (2001). The economic impacts of pollinator declines: An approach to assessing the consequences. *Conservation Ecology* **5**, 2.

Khan, B. M. and Chaudhary, M. F. (1991). Comparative assessment of honeybees and other insect visitors with self pollination of sarson in Peshawar. *Pakistan Journal of Forestry* **38**(4), 231–237.

Khan, R. B. and Dethe, M. D. (2004). Median lethal time of new pesticides to foragers of honeybee. *Pestology* **28**(1), 28–29.

Kisselhagen, S. (1977). Seed yield is increased if there are bees on the rape. *Tidsskrift for Biavl* **111**, 66–67.

Koutensky, J. (1958). The results of the pollinating work of bees. *Vcelarstvi* **11**(5), 72–73.

Koutensky, J. (1959). The pollinating effect of the honey bee (*Apis mellifera* L.) On the increase in rape and white mustard yields per hectare. Ceskoslov. Akad. Zemedel. Ved, sborn. *Rostlinna vyroba* **32**(4), 571–582.

Kubisova, S., Nedbalova, V. and Plesnik, R. (1980). The pollinating activity of honeybees on rape. *J. Polnohospodarstvo* **26**, 744–754.

Kumar, J., Rao, K. V. K. and Gupta, J. K. (1994). Pollination efficiency of bees visiting blossom of *Brassica campestris* L. var. toria in mid hills of Himachal Pradesh, India. *Indian Bee Journal* **56**, 202–206.

Langridge, D. F. and Goodman, R. D. (1975). A study on pollination of oilseed rape (*Brassica campestris*). *Australian Journal of Experimental Agriculture and Animal Husbandry* **15**, 285–288.

Langride, D. F. and Goodman, R. D. (1982). Honeybee pollination of oil seed rape, cultivar midas. *Australian Journal of Experimental Agriculture and Animal Husbandry* **22**, 124–126.

Langridge, D. F. and Goodman, R. D. (2003). Honeybee pollination of oilseed rape, cultivar Midas. *Australian Journal of Experimental Agriculture and Animal Husbandry* **22**(115), 124–126.

Latif, A., Qayyum, A. and Abbas, M. (1960). The role of *Apis indica* in the pollination of [oil seeds] "toria" and "sarson" (*Brassica campestris* var. Toria and dichotoma). *Bee World* **41**, 283–286.

Levin, M. D. (1984). Value of honeybee pollination to United States agriculture. *American Bee Journal* **124**, 184–186.

Louveaux, J. and Verge, J. (1952). Researches on the pollination of winter rape. *Apiculteur* **96**(Sect. Sci.), 15–18.

Mahanta, M. and Rahman, A. (1997). Pollination studies of honeybees (*Apis cerana*) on rapeseed, *Brassica campestris* var. *toria*. *Plant Health* **2**, 66–69.

Maisymora, A. K. (1990). Flowering stages of three rapeseed varieties in Siberia. *Crop Science* **31**(1), 921–929.

Manning, W. G. and Boland, W. T. (2000). A preliminary investigation into honey bee (*Apis mellifera*) pollination of canola (*Brassica napus* cv. Karoo) in Western Australia. *Australian Journal of Experimental Agriculture* **40**(3), 439–442.

Masierowska, M. L. (2003). Floral nectaries and nectar production in brown mustard (*Brassica juncea*) and white mustard (*Sinapis alba*) (Brassicaceae). *Plant Systematics and Evolution* **238**(1–4), 97–107.

Mayer, D. F. (1980). Monitoring of honeybee mortality from insecticides. *American Bee Journal* **120**, 352–355.

Mayer, D. F. (2003). Bee kills from pesticides and the expert witness. *American Bee Journal* **143**(7), 563–568.

Mayer, D. F. and Johansen, C. A. (1983). Occurrence of honeybee (Hyonenoptera: Apidae). Poisoning in eastern Washington. *Environmental Entomology* **12**(2), 317–320.

Mayor, D. F., Patten, K. D., Macfarlane, R. P. and Shanks, C. H. (1984). Differences between susceptibility of four pollinators species (Hymenoptera; Apoidea) to filed weathered residue. *Melandria* **50**, 24–27.

McGregor, S. E. (1976). "Insect Pollination of Cultivated Crop Plants." Agricultural Handbook No. 496. Agriculture Research Science, United States Department of Agriculture. p. 411.

Melnichenko, A. N. (1976). Role of insect pollinators in increasing yields of agricultural plants. *In* "Pollination of Entomophilous Agricultural Crops by Bees" (R. B. Kozin, ed.), p. 331. Amerind Publishing Co. Pvt. Ltd., New Delhi.

Mesquida, J. and Revard, M. (1979). Entomolophilous pollination of male sterile strains of winter rape seed (*Brassica napus* L.) Metzger var. *Oleifera* and a preliminary study of alternating devices. *Proceedings of 4th International Symposium on Pollination*, pp. 49–57. Maryland, USA.

Mesquida, J. and Renard, M. (1981). Pollinisation du colza d'hiver mâle-fertile et mâle-stérile (*Brassica napus* L. *var. oleifera* Metzger) par l'abeille domestique (*Apis m. mellifica* L.). Effets sur la phénologie et le rendement., *Apidologie* **12**, 345–362.

Mesquida, J., Renard, M. and Pierre, J. S. (1988). Rapeseed (*Brassica napus* L.) productivity: The effect of honeybees (*Apis mellifera* L.) and different pollination conditions in cage and field tests. *Apidologie* **19**, 51–72.

Meyerhoff, G. (1954). Investigation on the effect of bee visits on rape. *Arch. F. Geflugelzucht und kleintierkunde* **3**(3/4), 259–306.

Meyerhoff, G. (1958). Behavior of bees foraging on rape. *Leipzig. Bienenztng.* **72**(6), 164–165.

Mishra, R. C. and Kaushik, H. D. (1992). Effect of cross-pollination on yield and oil content of *Brassica campestris* L. var. sarson with pollination efficiency of honeybees, *Apis* spp. *Annals of Entomology* **10**(2), 33–37.

Mishra, R. C., Kumar, J. and Gupta, J. K. (1988). The effect of mode of pollination on yield and oil potential of *Brassica campestris* L. var. *sarson* with observation on insect pollinators. *Journal of Apicultural Research* **27**(3), 186–189.

Mohammad, A. (1935). Pollination studies on toria (*Brassica napus* L. var. *dichotoma* Prain) and sarson (*Brassica campestris* L. var. *sarsona* Prain). *Indian Journal of Agricultural Sciences* **5**, 125–154.

Muller, H. (1883). "The Fertilization of Flowers." Macmillan, London.

Nikitina, A. I. (1950). Honeybees raise seed yields of turnips and rutabaga. *Pchelovodstvo* **27**(5), 271–274.

Olsson, G. (1952). Investigations of the degree of cross-pollination in white mustard and rape. *Sverig. Utsadesfpren. Tidskr.* **62**(4), 311–322.

Olsson, G. (1955). Wind pollination of cruciferous oil plants. *Sverig. Utsadesforen. Tidskr.* **65**(6), 418–422.

Olsson, G. (1960). Self-incompatibility and outcrossing in rape and white mustard. *Hereditas* **46**, 241–252.

Palmer, S. (1959). A honey plant par excellence. *Gleanings in Bee Culture* **87**, 460–461.

Panda, P. and Nanda, U. K. (1989). Insect pollination of some oilseed crops in Orissa. *Indian Bee Journal* **51**(3), 97–98.

Pierre, J. (2000). The role of honeybees (*Apis mellifera*) and other insect pollinators in gene flow between oilseed rape (*Brassica napus*) and wild radish (*Raphanus raphanistrum*). *ISHS Acta Horticulturae* 561: VIII International Symposium on Pollination - Pollination: Integrator of Crops and Native Plant Systems.

Pimentel, D. (1980). Environmental and social cost of pesticides: A preliminary assessment. *Oikos* **34**(2), 126–140.

Pinochet, X. and Bertrand, R. (2000). Oilseed rape grain yield productivity increases with hybrid varietal types: A first balance sheet with post registration tests in France and Europe. *OCL-Oleagineux Corps Gras Lipides* **7**, 11–16.

Poehling, H. M. (1989). Selective application strategies for insecticides in agricultural crops. *In* "Pesticides and Non Target Invertebrates," pp. 151–175. Intercept, Wimborne, Dorset, UK Dorset, UK.

Prakash, R. and Kumaraswami, J. (1984). Toxicity of some insecticides to the Indian bee, *Apis cerana* Feb.. *Indian Bee Journal* **46**, 15–17.

Prasad, D., Hameed, S. F., Singh, R., Yazdani, S. S. and Singh, B. (1989a). Effect of bee pollination on the quantity and quality of rai crop (*Brsassica juncea* Coss.). *Indian Bee Journal* **51**(2), 447.

Prasad, D., Hameed, S. F., Singh, R., Yazolam, S. S. and Singh, B. (1989b). Effect of bee pollination on quantity and quality of rai crop (*Brassica juncea* Coss.). *Indian Bee Journal* **51**(2), 45–47.

Pritsch, G. (1965). Increasing the yield of oil plants by using honey bees. *Ved. Prace vyzkam. Ustav. Vcelar csazv* **4**, 157–163.

Radchenko, T. H. (1964). The influence of pollination on the crop and the quality of seed of winter rape. *Bdzhil'nitstvo* **1**, 68–74.

Rahman, K. A. (1940). Insect pollinators of toria (Brassica napus linn., var. Dichotoma prain) and sarson (*B. Campestris* Linn., var. Sarson prain) at lyallpur. *Indian Journal of Agricultural Sciences* **10**(3), 422–447.

Rana, B. S. and Goyal, N. P. (1991). Field toxicity of methyl demeton and dimethoate to the foragers of honeybee (*Apis cerana indica* F.) visitors to *Brassica chinensis* L. *Indian Bee Journal* **53**(1–4), 73–77.

Rao, G. M., Suryanaryana, M. C. and Thakur, C. V. (1980). Bees can boost oil seed production. *Indian Farming* **29**, 25–26.

Von Rhein, W. (1952). Results of training bees by scent during the 1952 rape flow. *Hess. Biene* **88**(8), 192–194; **88**(9), 218–220.

Riaz, A., Li, G., Quresh, Z., Swati, M. S. and Quiros, C. F. (2001). Genetic diversity of oilseed *Brassica napus* inbred lines based on sequence-related amplified polymorphism and its relation to hybrid performance. *Plant Breeding* **120**, 411–415.

Richards, K. W. (1993). Non-Apis bees as crop pollinators. *Revue Suisse de Zoologie* **100**, 807–822.

Russell, D., Meyer, D. and Bukowski, J. (1998). Potential impact of micro encapsulated pesticides on New Jersay Apiaries. *American Bee Journal* **138**(3), 207–210.

Sabbahi, R., De Oliveira, D. and Marceau, J. (2005). Influence of honey bee (Hymenoptera: Apidae) density on the production of canola (*Crucifera:* Brassicacae). *Journal of Economic Entomology* **98**(2), 367–372.

Sampson, D. R. (1957). The genetics of self- and cross-incompatibility in *Brassica oleracea*. *Genetics* **42**, 252–263.

Saran, G. and Prasad, R. (2002). Rapeseed-mustard. *In* "Text Book of Field Crop Production," Rajindera Prasad (ed.) Directorate of Information and Agriculture, New Delhi. pp. 435–461.

Schittenhelm, S., Gladis, T. and Rao, V. R. (1997). Efficiency of various insects in germplasm regeneration of carrot, onion and turnip rape accessions. *Plant Breeding* **116**, 369–375.

Sharma, S. K., Singh, J. R. and Ombir, J. R. (2000). Comparative pollination efficiency of three *Apis* species on *Brassica compestris* var. B.S.H-1. *Crop Research, Hissar Agricultural University* **19**(1), 125–127.

Sihag, R. C. (1985a). Floral biology melittophily and pollination ecology of cultivated cruciferous crops. *In* "Recent Advances in Pollen Spore Research" (T. M. Varghese, ed.), pp. 241–268. Allied Publishers, New Delhi.

Sihag, R. C. (1985b). Floral biology, melittophily and pollination ecology of cultivated umbelliferous crops. *In* "Recent Advances in Pollen Spore Research" (T. M. Varghese, ed.), Vol. 11, pp. 269–277. Allied Publishers, New Delhi.

Sihag, R. C. (1985c). Floral biology, melittophily and pollination ecology of cultivated umbelliferous crops. *In* "Recent Advances in Pollen Spore Research" (T. M. Varghese, ed.), Vol. 11, pp. 278–284. Allied Publishers, New Delhi.

Sihag, R. C. (1986). Insect pollination increases seed production in cruciferous and umbelliferous crops. *Journal of Apicultural Research* **25**(2), 121–126.

Sihag, R. C. (1988). Effect of pesticides and bee pollination on seed yield of some crops in India. *Journal of Apicultural Research* **27**(1), 49–54.

Sihag, R. C. (1991). Ecology of European honeybee (*Apis mellifera* L.) in semi arid and sub-tropical climates. 1. Melliferous flora and over-seasoning of the colonies. *Korean Journal of Apiculture* **5**, 31–43.

Sihag, R. C. (ed.) (1995). "Pollination Biology: Environmental Factors and Pollination," p. 196. Rajendra Scientific Publishers, Hissar, India.

Sihag, R. C., Khatkar, S. and Khatkar, S. (1999a). Foraging pattern of three honeybee species on eight cultivars of oilseed crops. 2. Foraging during the entire blooming period of the crops. *International Journal of Tropical Agriculture* **17**(1–4), 253–261.

Sihag, R. C., Khatkar, S. and Khatkar, S. (1999b). Synchronization of foraging activity of honeybees and flowering/anthesis in oilseeds crops. *Annals of Agriculture and Biological Research* **4**(2), 263–269.

Singh, M. P., Singh, K. I. and Sarangthen, A. K. (1993). Studies on the effect of bee pollination on yield of rapeseed. Proceedings of International Symposium on Pollination in Tropics. Bangalore, pp. 211–212.

Singh, M. P., Singh, K. I. and Devi, C. S. (2000). Role of *Apis cerana himalaya* pollination on yield and quality of rapeseed and sunflower. *In* "Asian Bees and Beekeeping" (M. Matsuka, L. R. Verma, S. Wongsiri, K. K. Shrestha and Uma Pratap, eds.), pp. 186–189. Oxford and IBH Publishing Co., New Delhi.

Soroka, J. J., Goerzen, D. W., Falk, K. C. and Bet, K. E. (2001). Alfalfa leaf cutting bee (Hymenoptera: Megachilidae) pollination of oilseed rape (*Brassica napus* L.) under isolation tents for hybrid seed production. *Canadian Journal of Plant Science* **81**, 199–204.

Sorthia, B. K. and Chari, M. S. (1985). Toxicity of some insecticides to honeybees, *Apis florea* F. and *Apis mellifera* L. *Journal of Entomological Research* **92**(2), 195–197.

Stark, D. J. D., Jepson, P. C. and Mayer, D. F. (1995). Limitation to use of topical toxicity data for production of pesticide side effects in the field. *Journal of Economic Entomology* **88**(5), 1081–1088.

Steffan-Dewenter, I. (2003). Seed set of male-sterile and male-fertile oilseed rape (*Brassica napus*) in relation to pollinator density. *Apidologie* **34**, 227–235.

Steffan-Dewenter, I. and Tscharntke, T. (1999). Effects of habitat isolation on pollinator communities and seed set. *Oecologia* **121**, 432–440.

Steffan-Dewenter, I., Münzenberg, U. and Tscharntke, T. (2001). Pollination, seed set and seed predation on a landscape scale. *Proceedings of the Royal Society of London B* **268**, 1685–1690.

Steffan-Dewenter, I., Munzenberg, U., Burger, C., Thies, C. and Tscharntke, T. (2002). Scale-dependent effects of landscape structure on three pollinator guilds. *Ecology* **83**, 1421–1432.

Steffan-Dewenter, I., Potts, S. G. and Packer, L. (2005). Pollinator diversity and crop pollination services are at risk. *Trends in Ecology and Evolution* **20**, 651–652.

Steffan-Dewenter, I., Klein, A. M., Alfert, T., Gaebele, V. and Tscharntke, T. (2006). Bee diversity and plant–pollinator interactions in fragmented landscapes. *In* Specialization and generalization in plant–pollinator interactions, (N. M. Waser and J. Ollerton, eds.), pp. 387–408. Chicago Press, Chicago, IL.

Torchio, P. F. (1990). Diversification of pollination strategies for US crops. *Environmental Entomology* **16**, 1649–1656.

Tscharntke, T., Gathmann, A. and Steffan-Dewenter, I. (1998). Bioindication using trap nesting bees and wasps and their natural enemies: Community structure and interactions. *Journal of Applied Ecology* **35**, 708–719.

Vesely, V. (1962). The economic effectiveness of bee pollination on winter rape (*Brassica napus* l., var. Oleifera metz.). Min. Zemedel. Lesn. A vodniho hospodar. Ust. Vedtech. Inform. *Zemedel. Ekon.* **8**(9), 659–673.

Westcott, L. and Nelson, D. (2001). Canola pollination: An update. *Bee World* **82**, 115–129.

White, B. (1970). Pollination of commercial rape seed crops. *Australasian Beekeeper* **72**(4), 99–100.

Williams, I. H. (1978). The pollination requirements of swede rape (*Brassica napus* L.) and of turnip rape (*Brassica campestris* L.). *Journal of Agricultural Science (Cambridge)* **91**, 343–348.

Williams, I. H. and Simpkins, J. (1989). Honeybee pollination of the double low oil-seed rape line. *Journal of Applied Biology* **23**, 343–346.

Williams, I. H., Martin, A. P. and White, R. P. (1987). The effect of insect pollination on plant development and seed production in winter oil-seed rape (*Brassica napus* L.). *Journal of Agricultural Science (Cambridge)* **109**, 135–139.

Wright, G. A., Skinner, B. D. and Smith, B. H. (2002). Ability of honeybee, *Apis mellifera*, to detect and discriminate odors of varieties of canola (*Brassica rapa* and *Brassica napus*) and snapdragon flowers (Antirrhinum majus). *Journal of Chemical Ecology* **28**(4), 721–740.

Zander, E. (1952). Rape and bees. *Z. Bienenforsch.* **1**(8), 135–140.

Zhong, H. E., Mark, L., Sheve, P. and Cate, B. (2004). Minimizing the import if the mosquito adulterine noted or honeybees, *Apis mellifera* (Hymenoptera: Apidae): Aerial Ultra volume application using a high-pressure nozzle system. *Journal of Economic Entomology* **97**(1), 1–7.

Genetic Variation and Metabolism of Glucosinolates

NATALIA BELLOSTAS,* ANNE DORTHE SØRENSEN,† JENS CHRISTIAN SØRENSEN* AND HILMER SØRENSEN*

**Department of Natural Sciences, Biochemistry and Natural Product Chemistry, Faculty of Life Sciences, Copenhagen University, Thorvaldsensvej 40, DK-1871, Frederiksberg C, Denmark*
†Department of Human Nutrition, Group for Paediatric and International Nutrition, Faculty of Life Sciences, Copenhagen University, Rolighedsvej 30, DK-1958, Frederiksberg C, Denmark

ABSTRACT

Progress in the production and utilisation of oilseed rape (*Brassica napus* L. and *B. rapa* L.) and their derived products in the past years has resulted in the production of high-yielding double low varieties with low levels of erucic acid and glucosinolates.

Advances in Botanical Research, Vol. 45
Incorporating Advances in Plant Pathology

0065-2296/07 $35.00
DOI: 10.1016/S0065-2296(07)45013-3

In order to meet this objective, extensive knowledge on the occurrence, structure and properties of glucosinolates and myrosinase isoenzymes (thioglucohydrolase; EC 3.2.1.147) co-occurring with glucosinolates is essential. The present chapter focusses on these subjects including heredity, genetic variation and metabolism of glucosinolates.

A number of stages in the glucosinolate biosynthetic pathway are well known, but significant elements still remain to be further explored and elucidated. In recent years, most research has been confined on gene technology using *Arabidopsis thaliana* as a model and hence considering only the glucosinolates present in this particular plant. To date, more than 140 structurally different glucosinolates have been described of which ~30 glucosinolates occur in *Brassica* species. Only relatively few glucosinolates are produced in individual plant genera and/or families. Knowledge on the genetic- and metabolic-based regulation of glucosinolate accumulation in different plants is therefore of great relevance in order to explain the reasons for the chemotaxonomically limited occurrence of structurally different glucosinolates. The metabolic regulation and thus the biological consequences of those changes are however unknown. Similarly, the reasons for the high selectivity of the multi-enzyme complexes towards the precursor amino acids in the biosynthesis of glucosinolates, as well as the specificity of the various enzymes needed for the modification of glucosinolate side chains and the substitution/esterification of the thioglucose residue of the molecules are yet to be clarified. The glucosinolate catabolism, which comprises the transformation of glucosinolates into various bioactive products, is another important area where limited information is available on the oligomeric myrosinase isoenzymes and associated compounds. This is especially relevant regarding the differences in the catalytic properties of the individual isoenzymes as well as the opportunities to modify these properties. New analytical methods allow for the synchronous detection of substrate and products in myrosinase-catalysed reactions, which broadens the perspectives for progress in this field.

I. INTRODUCTION

Glucosinolates in cruciferous oilseed crops have been a hot subject for a long time with research being intensified some decades ago, which resulted in comprehensive plant-breeding programmes directed to the production of double low oilseed crops (less than 0.5% erucic acid in the oil and less than 20 μmol of glucosinolates per gram seed dry matter). The advantages of these double low oilseed rape varieties were the subject of EU supported programmes as presented and discussed in Braunschweig (Bunting, 1981), Copenhagen (Sørensen, 1985) and Gembloux (Wathelet, 1987). One of the main problems in the study of the genetic variation and the metabolism of glucosinolates in cruciferous oilseed crops has been the lack of knowledge on the analytical methods needed for the determination of all of the individual glucosinolates present in the starting material, for example 4-hydroxyindol-3-ylmethylglucosinolate in seeds of oilseed rape (Bjerg *et al.*, 1987a). When this particular glucosinolate is neglected, serious problems with the quality of rapeseed products can be encountered (Jensen *et al.*, 1991) owing to the easy oxidation of the 4-hydroxyindol group (Bjergegaard *et al.*, 2001) and the subsequent formation of dark products, which affect both the colour and other quality parameters of the oil and especially the de-oiled meal. The problems related to this glucosinolate, along with various other reasons, raise concern on the reliability of analytical data based on procedures such as GLC and a number of other simple procedures for total glucosinolate determination, which are based on the measurement of products released by myrosinase-catalysed glucosinolate hydrolysis (Sørensen, 1985).

Since the various effects of glucosinolates and their transformation products depend on their actual structures and concentrations, all of the structurally different glucosinolates need to be taken into account independently. The biological effects of glucosinolate degradation products include adverse effects on internal organs and xenobiotic metabolism as well as detrimental effects on the nutritional value, taste and smell of food and feed containing double low oilseed rape, which otherwise contains high-quality oil/lipids, protein and dietary fibres (Ahlin *et al.*, 1994; Allison *et al.*, 2001; Bell and Shires, 1982; Bille *et al.*, 1983a,b; Bjerg *et al.*, 1987b; Campbell, 1979; Ciska and Pathak, 2004; Cox-Ganser *et al.*, 1994; Danielsen *et al.*, 1987, 1994; Eggum *et al.*, 1985; Hansen *et al.*, 1995, 1997; Henriksen *et al.*, 1987; Jensen *et al.*, 1990, 1991; Martland *et al.*, 1984; Mason and Lucas, 1983; Michaelsen *et al.*, 1994; Morton and Campbell, 1997; Schmid and Schmid, 1992; Schone *et al.*, 1997a,b, 2001, 2002; Spiegel *et al.*, 1993; Taljaard, 1993; Umemura *et al.*, 1977).

During recent decades, focus has increasingly been placed on the health-promoting effects of transformation products of glucosinolates (Bao and Fenwick, 2004; Barillari *et al.*, 2005; Bjergegaard *et al.*, 1994; Bonnesen *et al.*, 1999; Cover *et al.*, 1998; Fahey *et al.*, 1997; Fenwick and Heaney, 1983; Kim *et al.*, 1997; Kuang and Chen, 2004; Loft *et al.*, 1992; Moreno *et al.*, 2006; Rose *et al.*, 2005; Staack *et al.*, 1998; Vang *et al.*, 1995; Verhoeven *et al.*, 1997; Wallig *et al.*, 1998; Wortelboer *et al.*, 1992; Wu *et al.*, 2004; Vang *et al.*, 2001). On the basis of the large amount of data available on both positive and negative effects from glucosinolates and glucosinolate-derived products, it is thus clearly revealed that there is a need for analytical techniques able to separate and quantify the glucosinolates individually. This is now possible by using a combination of methods developed for the determination of intact glucosinolates by HPLC and/or capillary electrophoresis (HPCE-MECC) (Bellostas *et al.*, 2006; Bjerg and Sørensen, 1987a,b; Bjerg *et al.*, 1987a; Bjergegaard *et al.*, 1991, 1995; Feldl *et al.*, 1994; Helboe *et al.*, 1980; Hill *et al.*, 2003; Michaelsen *et al.*, 1992; Rangkadilok *et al.*, 2002; Sørensen, 1990; Sørensen *et al.*, 1999; Tian *et al.*, 2005). Glucosinolate determination based on GLC has several limitations as mentioned above and discussed elsewhere (Olsen and Sørensen, 1981), and this is also the case for the methods based on desulphoglucosinolates (Bjerg and Sørensen, 1987a,b; Sørensen, 1990). However, the methods based on desulphoglucosinolates, which can be either based on HPLC or HPCE-MECC methods (Bjerg and Sørensen, 1987a,b; Bjerg *et al.*, 1987a; Bjergegaard *et al.*, 1995; Sørensen *et al.*, 1999), have a series of advantages regarding the compounds detectable by this technique. For the analytical determination of glucosinolate-derived products, such as volatile isothiocyanates and nitriles, GLC is an efficient analytical method (Bjergegaard *et al.*, 1998; Olsen and Sørensen, 1981). Some more hydrophilic and non-volatile glucosinolate products can be determined with the use of HPLC (Bjergegaard *et al.*, 1998) or HPCE-MECC (Agerbirk *et al.*, 1996; Bellostas *et al.*, 2006; Bjergegaard *et al.*, 1999; Feldl *et al.*, 1994). For oligomeric compounds of low solubility in aqueous systems, for example the compounds produced during the degradation of indol-3-ylmethylglucosinolates, the methods of analysis based on supercritical fluid chromatography (SFC) techniques and enhanced fluid liquid chromatography (EFLC-ELSD) have several advantages (Buskov *et al.*, 2000a,b,c,d,e; Sørensen *et al.*, 1999).

The various effects that glucosinolates and transformation products thereof have on human health and animal feed quality (vide supra) as well as on the environment and as biofumigants (Brown and Morra, 1997; Buskov *et al.*, 2002; Hopkins *et al.*, 1998) have boosted the interest in the analyses, the occurrence, the relation between structure and properties, the genetically defined variation as well as the metabolism of glucosinolates.

II. CHEMICAL STRUCTURE OF GLUCOSINOLATES, NAMES AND PROPOSED NUMBERING SYSTEM FOR KNOWN GLUCOSINOLATES

Glucosinolates are alkyl-*N*-hydroximine sulphate esters with a β-D-thioglucopyranoside group attached to the hydroximine carbon in *Z*-configuration to the sulphate group (Buskov *et al.*, 2000a,b,c,d; Ettlinger and Kjær, 1968; Ettlinger and Lundeen, 1956, 1957; Kjær, 1960; Olsen and Sørensen, 1981; Sørensen, 1990), as shown in Fig. 1.

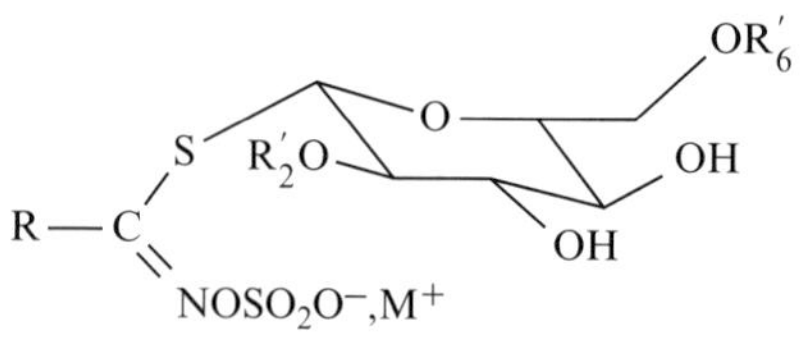

Fig. 1. Structure of glucosinolates. R = side chain with structural resemblance to the parent amino acids; R_2' and R_6' = H or acyl derivatives, especially cinnamoyl derivatives; M^+ = cation.

The structural features common for glucosinolates are the β-D-thioglucopyranoside group attached to C-0 as *cis* or *Z*- in the *N*-hydroximine sulphate ester group and the side chain (R-group) at C-0 (Fig. 1). This gives the total functional group known as the glucosinolate group, with semi-systematic names of individual glucosinolates based on the name for the R- and R′-groups used as a prefix to the word glucosinolate (Kjær, 1960; Olsen and Sørensen, 1981; Sørensen, 1990). The glucosinolate structure was originally confirmed by X-ray investigations on synthesised sinigrin (Ettlinger and Lundeen, 1956, 1957; Marsh and Waser, 1970). This structure is found to be a general feature for all glucosinolates investigated by ^{1}H- and ^{13}C NMR spectroscopy (Buskov *et al.*, 2000a,b; Karcher and El Rassi, 1999; Matthäus and Luftmann, 2000; Olsen and Sørensen, 1981; Prestera *et al.*, 1996; Sørensen, 1990).

Structural differences between individual glucosinolates are due either to substituents on the thioglucoside group (R′ ester groups, Fig. 1) or to structurally different side chains (R-groups). Owing to the strongly acidic sulphate group, glucosinolates occur as salts. This fact, together with the presence of the glucoside group, gives the glucosinolates their hydrophilic properties, and thus polar solvents, such as water, are needed to dissolve these compounds. Ester groups as substituents on the thioglucoside part of glucosinolates were originally described for sinapoyl- and isoferuloyl derivatives (Bjerg and Sørensen, 1987b; Lincheid *et al.*, 1980; Sørensen, 1990). These glucosinolate derivatives are relatively unstable in alkaline solutions, and they are not substrates for the enzymes myrosinases (EC 3.2.1.147) and sulphatases (EC 3.1.6.1). Therefore, such compounds will not be detected when using analytical methods based on desulphoglucosinolates and/or analytical methods based on the determination of products from myrosinase-catalysed hydrolysis of glucosinolates (Bjerg and Sørensen, 1987a,b; EC, 1990; Karcher and El Rassi, 1999; Matthäus and Luftmann, 2000; Prestera *et al.*, 1996; Sørensen, 1990). Analytical methods based on desulphoglucosinolates are most often performed as on-column (anion exchanger) desulphatation, which will also result in the loss of possibilities for the determination of glucosinolates with acidic groups in the side chain (R-group) (Sørensen, 1990). These different types of glucosinolates can, however, be determined when the analytical techniques are based on intact glucosinolates, for example, by HPLC (Bjerg and Sørensen, 1987a,b; Helboe *et al.*, 1980) and especially, capillary electrophoresis (Bjergegaard *et al.*, 1994; Buskov *et al.*, 1997; Karcher and El Rassi, 1999; Michaelsen *et al.*, 1992; Sørensen *et al.*, 1999). When appropriate methods of analysis of glucosinolates are used, it is expected that additional thioglucoside substituted glucosinolates, other than the known sinapoyl- and isoferuloyl derivatives, will be found (Sørensen, 1990).

To facilitate the study of glucosinolates, including their chemotaxonomically defined occurrence, Table I shows structures, names and an applied numbering system. The compounds are arranged according to the expected biosynthetic relation to the amino acid precursors and to side chain modifications (vide infra). The trivial names are those generally accepted (Sørensen, 1990), and for those without trivial names partly self-created names are used, following the accepted system with the last part of the names indicating the plants in which they are found. References to occurrence and the original identification of the glucosinolates are carefully treated elsewhere (Daxenbichler *et al.*, 1991; Ettlinger and Kjær, 1968; Fahey *et al.*, 2001; Halkier, 1999; Kjær, 1960; Rosa *et al.*, 1997; Sørensen, 1990, 2001). On the basis of the counts of glucosinolates presented in these references, more than 140 different compounds are known.

III. GLUCOSINOLATE BIOSYNTHESIS

The general metabolism in all living cells comprises the transformation of carbohydrates in glycolysis/gluconeogenesis and pentose phosphate cycles, which is metabolically linked to fatty acid/lipid metabolism, the TCA cycle and amino acid, protein and nucleic acid metabolism. In addition to these ubiquitous metabolic processes, a great variety of chemotaxonomically defined metabolic processes are the basis for the production and metabolism of allelochemicals. Considering the structures of glucosinolates (Fig. 1; Table I), it is evident that amino acids are the precursors of the side chain structures. However, despite several hundreds of non-protein amino acids are known as plant constituents (Eggum and Sørensen, 1989), only some function as glucosinolates precursors (Table I).

Structural variations in the side chain of glucosinolates (R-group, Fig. 1) are due to the metabolic processes occurring in the glucosinolate biosynthesis from amino acid precursors (Hill *et al.*, 2003, and references cited therein). On the basis of the amino acid precursor (Bjerg *et al.*, 1987c), it is relevant to divide the sequence in the glucosinolate biosynthesis into five main groups corresponding to the amino acid precursor: (1) methionine; (2) tryptophan; (3) phenylalanine/tyrosine; (4) neutral aliphatic amino acids, Ala, Val/Leu, Ile and (5) acidic amino acids, Asp, Glu.

Glucosinolates with Gly, Ser, Thr, Cys and basic amino acids as precursors, for example His, Orn, Lys, Arg, have not yet been described. This may be a result of the specificity of the enzymes transforming amino acids into glucosinolates or it can be due to the analytical techniques frequently used for investigation of glucosinolates, such as in the case of the determination/

TABLE I

Structures and Numbers of Glucosinolates Together with Semi-Systematic Names (Ettlinger and Kjær, 1968; Olsen and Sørensen, 1981; Sørensen, 1990) and Trivial Names

Number	Structure of side chain (or complete glucosinolate)	Semi-systematic name	Trivial name
0.1	$CH_2{=}CH{-}CH_2{-}$	Prop-2-enylglucosinolate	Sinigrin
0.2	$CH_2{=}CH{-}CH_2{-}CH_2{-}$	But-3-enylglucosinolate	Gluconapin
0.3	$CH_2{=}CH{-}CH_2{-}CH_2{-}CH_2{-}$	Pent-4-enylglucosinolate	Glucobrassicanapin
0.4	$CH_2{=}CH{-}CH_2{-}CH_2{-}CH_2{-}CH_2{-}$	Hex-5-enylglucosinolate	Glucowasabiamin
0.5	$CH_2{=}CH{-}CH_2{-}CH_2{-}CH_2{-}CH_2{-}CH_2{-}$	Hept-6-enylglucosinolate	Glucowasajaponicain
1.1	$CH_2{=}CH{-}C(OH)(H){-}CH_2{-}$ (OH above, H below)	(2*R*)-2-Hydroxybut-3-enylglucosinolate	Progoitrin
1.2	$CH_2{=}CH{-}C(H)(OH){-}CH_2{-}$ (H above, OH below)	(2*S*)-2-Hydroxybut-3-enylglucosinolate	Epiprogoitrin
1.3	$CH_2{=}CH{-}CH_2{-}C(OH)(H){-}CH_2{-}$ (OH above, H below)	(2*S*)-2-Hydroxypent-4-enylglucosinolate	Napoleiferin
1.4	$CH_2{=}CH{-}CH_2{-}C(H)(OH){-}CH_2{-}$ (H above, OH below)	(2*R*)-2-Hydroxypent-4-enylglucosinolate	Epinapoleiferin
2.1	$CH_3{-}S{-}CH_2{-}CH_2{-}$	2-Methylthioethylglucosinolate	Elucoamoracia lapathifolin
2.2	$CH_3{-}S{-}CH_2{-}CH_2{-}CH_2{-}$	3-Methylthiopropylglucosinolate	Glucoibervirin
2.3	$CH_3{-}S{-}CH_2{-}CH_2{-}CH_2{-}CH_2{-}$	4-Methylthiobutylglucosinolate	Glucoerucin
2.4	$CH_3{-}S{-}CH_2{-}CH_2{-}CH_2{-}CH_2{-}CH_2{-}$	5-Methylthiopentylglucosinolate	Glucoberteroin

(continues)

TABLE I (*continued*)

Number	Structure of side chain (or complete glucosinolate)	Semi-systematic name	Trivial name
2.5	$CH_3-S-CH_2-CH_2-CH_2-CH_2-CH_2-CH_2-$	6-Methylthiohexylglucosinolate	Glucoarabidopthalianin
2.6	$CH_3-S-CH_2-CH_2-CH_2-CH_2-CH_2-CH_2-CH_2-$	7-Methylthioheptylglucosinolate	Glucoarabishirsutain
2.7	$CH_3-S-CH_2-CH_2-CH_2-CH_2-CH_2-CH_2-CH_2-CH_2-$	8-Methylthiooctylglucosinolate	Glucoarabishirsuin
2.8	$CH_3-S-CH_2-CH_2-CH_2-CH_2-CH_2-CH_2-CH_2-CH_2-CH_2-$	9-Methylthiononylglucosinolate	Glucoarabispurpureain
2.9	$CH_3-S-CH_2-CH_2-CH_2-CH_2-CH_2-CH_2-CH_2-CH_2-CH_2-CH_2-$	10-Methylthiodecylglucosinolate	
2.10	$CH_3-S-CH{=}CH-CH_2-CH_2-$	4-Methylthiobut-3-enylglucosinolate	Glucoraphasatin
2.11	$CH_3-S-CH_2-CH_2-CH(OH)-CH_2-CH_2-$	3-Hydroxy-5-methylthiopentylglucosinolate	
2.12	$CH_3-S-CH_2-CH_2-CH_2-CH(OH)-CH_2-CH_2-$	3-Hydroxy-6-methylthiohexylglucosinolate	
2.13	$CH_3-S-CH_2-CH_2-CH_2-C(=O)-CH_2-CH_2-$	3-Oxo-6-methylthiohexylglucosinolate	
2.14	$CH_3-S-CH_2-CH_2-CH_2-CH_2-C(=O)-CH_2-CH_2-$	3-Oxo-7-methylthioheptylglucosinolate	
2.15	$CH_3-S-CH_2-CH_2-CH_2-CH_2-CH_2-C(=O)-CH_2-CH_2-$	3-Oxo-8-methylthiooctylglucosinolate	
3.1	$CH_3-\ddot{S}(\rightarrow O)-CH_2-CH_2-CH_2-$	(*R*)-3-Methylsulfinylpropylglucosinolate	Glucoiberin
3.2	$CH_3-\ddot{S}(\rightarrow O)-CH_2-CH_2-CH_2-CH_2-$	(*R*)-4-Methylsulfinyl butylglucosinolate	Glucoraphanin
3.3	$CH_3-\ddot{S}(\rightarrow O)-CH_2-CH_2-CH_2-CH_2-CH_2-$	(*R*)-5-Methylsulfinyl pentylglucosinolate	Glucoalyssin

3.4	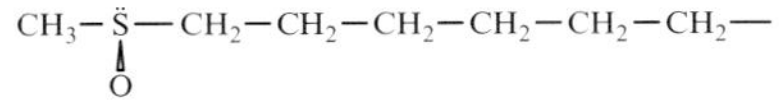	(*R*)-6-Methylsulfinylhexylglucosinolate	Glucohesperalin
3.5	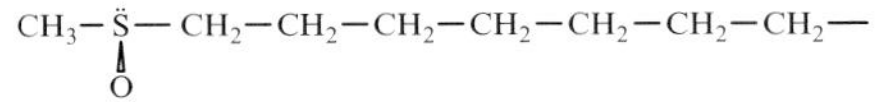	(*R*)-7-Methylsulfinylheptylglucosinolate	Glucoarabidopsithalianin
3.6	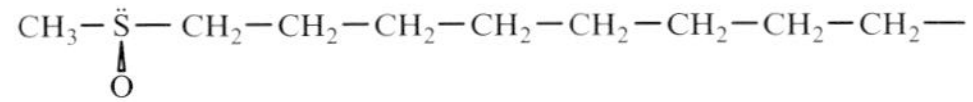	(*R*)-8-Methylsulfinyloctylglucosinolate	Glucohirsutin
3.7	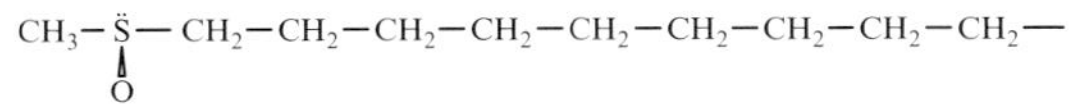	(*R*)-9-Methylsulfinylnonylglucosinolate	Glucoarabin
3.8	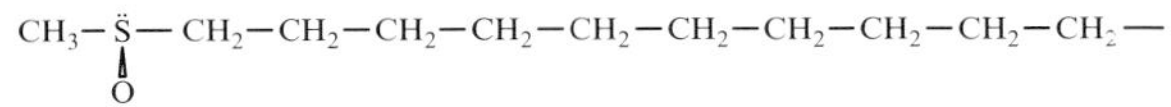	(*R*)-10-Methylsulfinyldecylglucosinolate	Glucocamelein
3.9	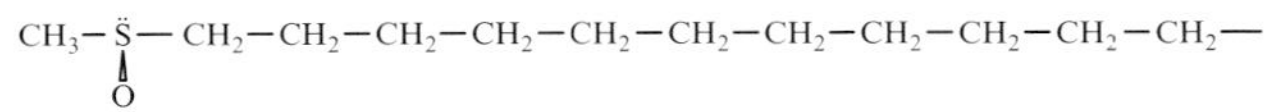	(*R*)-11-Methylsulfinylundecylglucosinolate	Gluconesliapaniculatin
3.10	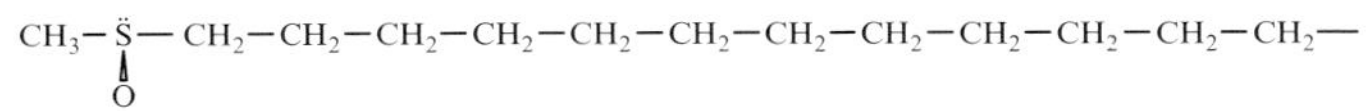	(*R*)-12-Methylsulfinyldodecylglucosinolate	Glucocamelinain
3.11	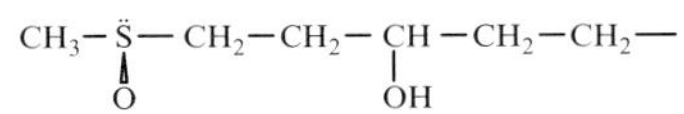	(*R*)-3-Hydroxy-5-methylsulfinyl pentylglucosinolate	Glucoerysimumhieraci folinin
3.12	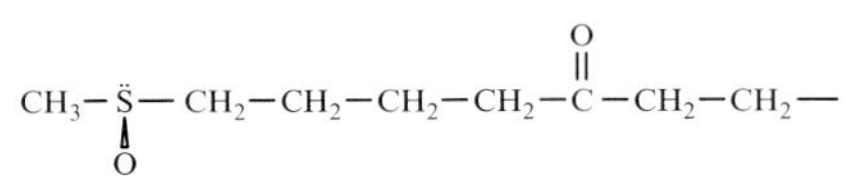	(*R*)-3-Oxo-7-methylsulfinyl heptylglucosinolate	
3.13	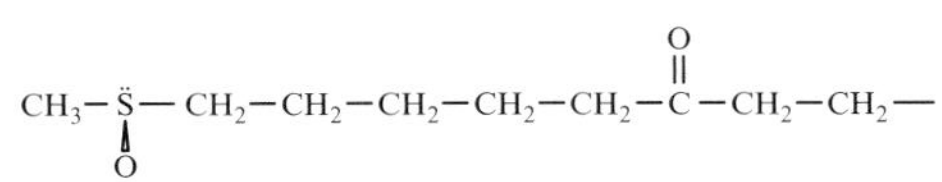	(*R*)-3-Oxo-8-methylsulfinyl octylglucosinolate	Glucoarabishirsutain (skal ændres)
3.14	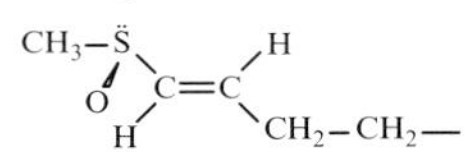	(*R*)-4-Methylsulfinylbut-3-enylglucosinolate	Glucoraphenin

(*continues*)

TABLE I (*continued*)

Number	Structure of side chain (or complete glucosinolate)	Semi-systematic name	Trivial name
3.15	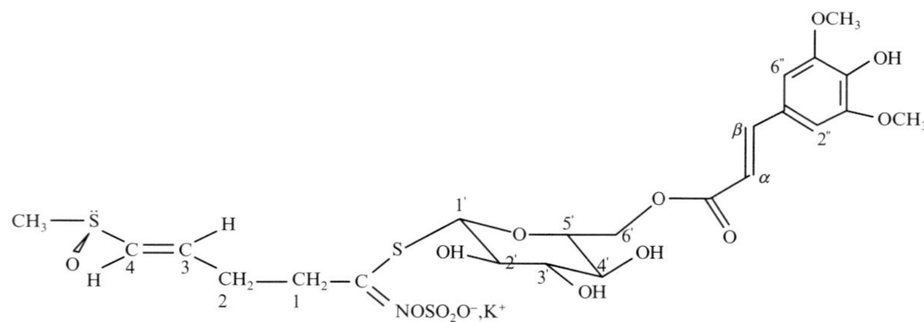	6′-Sinapoyl-(*R*)-4-methylsulfinylbut-3-enylglucosinolate	6′-Sinapoylglucoraphenin
3.16	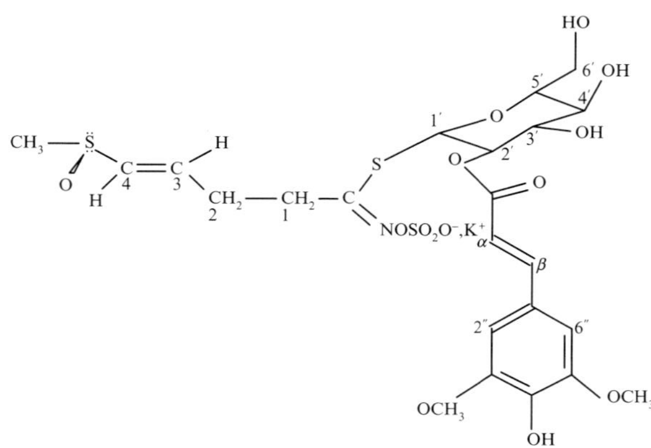	2′-Sinapoyl-(*R*)-4-methylsulfinylbut-3-enylglucosinolate	2′-Sinapoylglucoraphenin

3.17	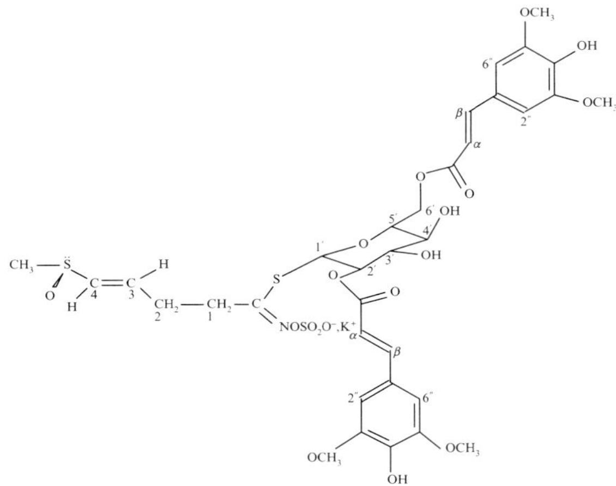	2,6′-Disinapoyl-(*R*)-4-methylsulfinylbut-3-enylglucosinolate	2′,6′-Disinapoylglucoraphenin
4.1	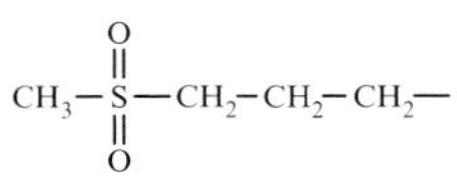	3-Methylsulfonylpropylglucosinolate	Glucocheirolin
4.2	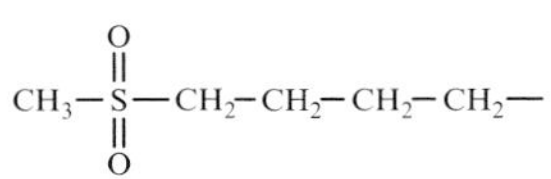	4-Methylsulfonylbutylglucosinolate	Glucoerysolin
4.3	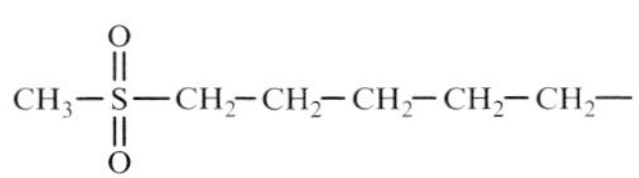	5-Methylsulfonylpentylglucosinolate	Glucoerysihieracifoliumin
4.4	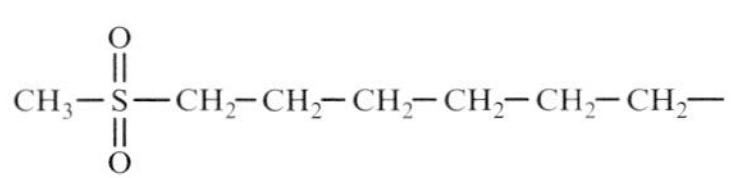	6-Methylsulfonylhexylglucosinolate	

(*continues*)

TABLE I (*continued*)

Number	Structure of side chain (or complete glucosinolate)	Semi-systematic name	Trivial name
4.5	$CH_3-S(=O)_2-CH_2-CH_2-CH_2-CH_2-CH_2-CH_2-CH_2-CH_2-$	7-Methylsulfonylheptylglucosinolate	
4.6	$CH_3-S(=O)_2-CH_2-CH_2-CH_2-CH_2-CH_2-CH_2-CH_2-CH_2-CH_2-$	8-Methylsulfonyloctylglucosinolate	
4.7	$CH_3-S(=O)_2-CH_2-CH_2-CH_2-CH_2-CH_2-CH_2-CH_2-CH_2-CH_2-CH_2-$	9-Methylsulfonylnonylglucosinolate	
4.8	$CH_3-S(=O)_2-CH=CH-CH_2-CH_2-$	4-Methylsulfonylbut-3-enylglucosinolate	
4.9	$CH_3-S(=O)_2-CH_2-CH_2-CH(OH)-CH_2-CH_2-$	5-Methylsulfonyl-3-hydroxypentylglucosinolate	
4.10	$CH_3-S(=O)_2-CH_2-CH_2-CH_2-CH(OH)-CH_2-CH_2-$	6-Methylsulfonyl-3-hydroxyhexylglucosinolate	
5.1	CH_3-	Methylglucosinolate	Glucocapparin
5.2	CH_3-CH_2-	Ethylglucosinolate	Glucoslepidiin
5.3	$CH_3-CH_2-CH_2-$	Propylglucosinolate	
5.4	$CH_3-CH_2-CH_2-CH_2-$	Butylglucosinolate	Glucocapparisflexuosain
5.5	$CH_3-CH_2-CH_2-CH_2-CH_2-$	Pentylglucosinolate	Glucokohlrabiin

5.6	$CH_3-CH_2-CH_2-CH_2-CH_2-CH_2-$	Hexylglucosinolate	Glucoraphasativusain
5.7	$CH_3-CH_2-CH_2-CH_2-CH_2-CH_2-CH_2-$	Heptylglucosinolate	
5.8	$CH_3-CH_2-CH_2-CH=CH-$	Pent-1-enylglucosinolate	
6.1	$HO-CH_2-CH_2-$	2-Hydroxyethylglucosinolate	Glucocapparimasiakain
6.2	$CH_3-CH(OH)-CH_2-$	2-Hydroxypropylglucosinolate	Glucoarabidopsithalin
6.3	$HO-CH_2-CH_2-CH_2-$	3-Hydroxypropylglucosinolate	Glucoerysimumhieracifolium
6.4	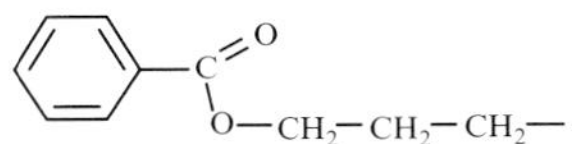	3-Benzoyloxypropylglucosinolate	Glucomalcolmiin
6.5	$CH_3-CH(OH)-CH_2-CH_2-$	3-Hydroxybutylglucosinolate	Glucocappariflexin
6.6	$HO-CH_2-CH_2-CH_2-CH_2-$	4-Hydroxybutylglucosinolate	Glucoarabidopsithalianain
6.7	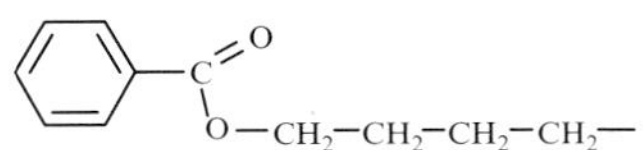	4-Benzoyloxybutylglucosinolate	
6.8	$CH_3-CH_2-CH_2-CH(OH)-CH_2-$	2-Hydroxypentylglucosinolate	Glucoarmoracialapathin
6.9	$HO-CH_2-CH_2-CH_2-CH_2-CH_2-$	5-hydroxypentylglucosinolate	Glucoarabidopsin
6.10	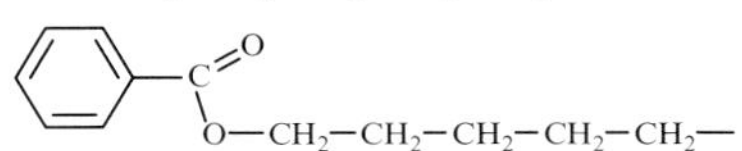	5-Hydroxybenzoyloxy pentylglucosinolate	Glucoarabidopsithain
6.11	$HO-CH_2-CH_2-CH_2-CH_2-CH_2-CH_2-$	6-Hydroxyhexylglucosinolate	
6.12	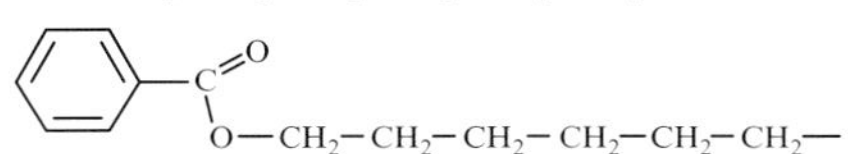	6-Hydroxybenzoyloxy hexylglucosinolate	

(*continues*)

TABLE I (*continued*)

Number	Structure of side chain (or complete glucosinolate)	Semi-systematic name	Trivial name
6.13	$CH_3-C(=O)-CH_2-CH_2-CH_2-$	3-Oxopentylglucosinolate	
6.14	$CH_3-CH_2-CH_2-C(=O)-CH_2-CH_2-CH_2-$	5-Oxoheptylglucosinolate	Glucocapparisangulatain
6.15	$CH_3-CH_2-C(=O)-CH_2-CH_2-CH_2-CH_2-$	6-Oxoheptylglucosinolate	Gluconorcappasalin
6.16	$CH_3-CH_2-CH_2-C(=O)-CH_2-CH_2-CH_2-CH_2-$	5-Oxooctylglucosinolate	Glucocappasalin
6.17	$CH_3-CH_2-CH_2-CH(OH)-CH(OH)-CH(OH)-CH(OH)-CH_2-CH_2-CH_2-$	4,5,6,7-Tetrahydroxydecyl glucosinolate	Glucocapparisgrandisin
7.1	$(CH_3)_2CH-$	Isopropylglucosinolate	Glcuputranjivin
7.2	$(CH_3)_2CH-CH_2-$	Isobutylglucosinolate	Glucoconringianin
7.3	$(CH_3)_2CH-CH_2-CH_2-$	3-Methylbutylglucosinolate	
7.4	$(CH_3)_2CH-CH_2-CH_2-CH_2-$	4-Methylpentylglucosinolate	Glucoraphanusativasin
7.5	$CH_2=C(CH_3)-CH_2-$	2-Methylprop-2-enylglucosinolate	
7.6	$CH_2=C(CH_3)-CH_2-CH_2-$	3-Methylbut-3-enylglucosinolate	Glucocapparilinearisin

7.7	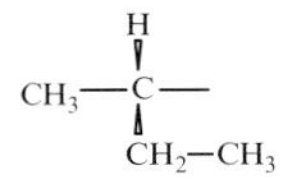	(1*S*)-Methylpropylglucosinolate	Glucocochlearin
7.8	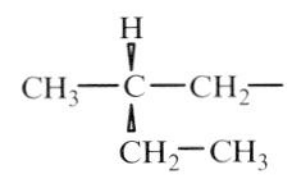	(2*S*)-2-Methylbutylglucosinolate	Glucojiaputin
8.1	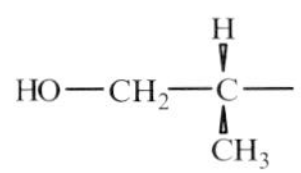	(1*R*)-2-Hydroxy-1-methylethylglucosinolate	Glucosisymbrin
8.2	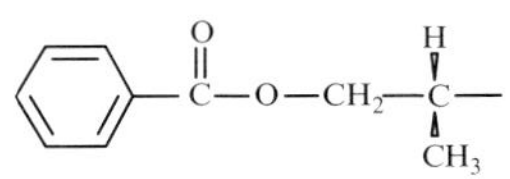	(1*R*)-2-Benzoyloxy-1-methylethylglucosinolate	Glucobenzosisymbrin
8.3	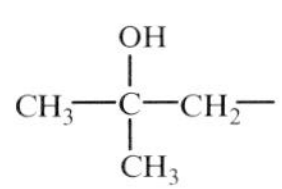	2-Hydroxy-2-methylpropylglucosinolate	Glucoresedalbain
8.4	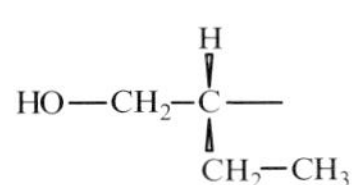	(1*R*)-2-Hydroxy-1-ethylethylglucosinolate	Glucosisaustriacin
8.5	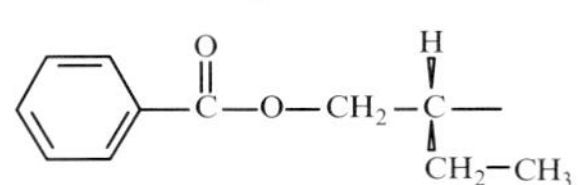	(1*R*)-Benzoyloxy-1-ethylethylglucosinolate	Glucoaustriacuin
8.6	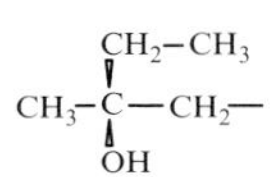	(2*R*)-2-Hydroxy-2-methylbutylglucosinolate	Glucocleomin

(*continues*)

TABLE I (*continued*)

Number	Structure of side chain (or complete glucosinolate)	Semi-systematic name	Trivial name
9.1	$CH_3-O-C(=O)-CH_2-CH_2-CH_2-CH_2-$	4-Methoxy-4-oxobutylglucosinolate	Glucoerypestrin
10.1	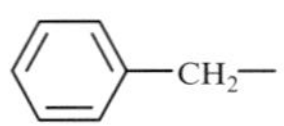	Benzylglucosinolate	Glucotropaeolin
10.2	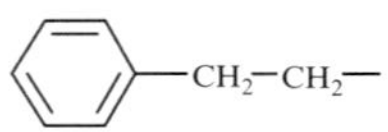	Phenethylglucosinolate	Gluconasturtiin
10.3	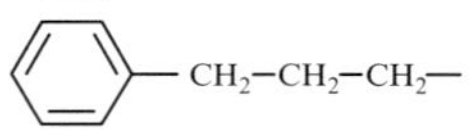	3-Phenylpropylglucosinolate	Glucoarmoracialapicin
10.4	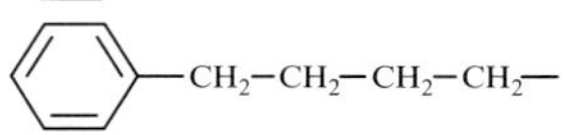	4-Phenylbutylglucosinolate	Glucoarmoracialafolicin
11.1	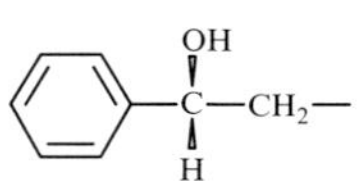	(2*R*)-2-Hydroxy-2-phenylethylglucosinolate	Glucosibarin
11.2	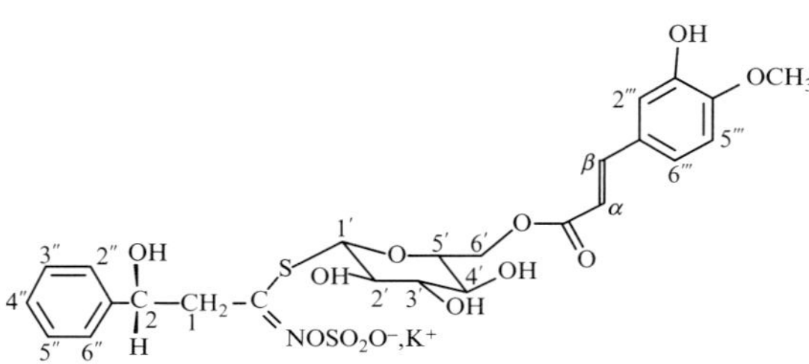	6′-Isoferuloyl-(2*R*)-2-Hydroxy-2-phenylethylglucosinolate	6′-Isoferuloylglucosibarin

11.3	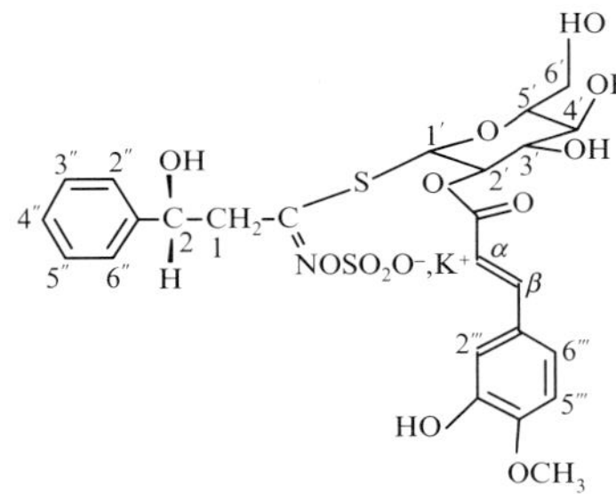	2′-Isoferuloyl-(2*R*)-2-Hydroxy-2-phenylethylglucosinolate	2′-Isoferuloylglucosibarin
11.4	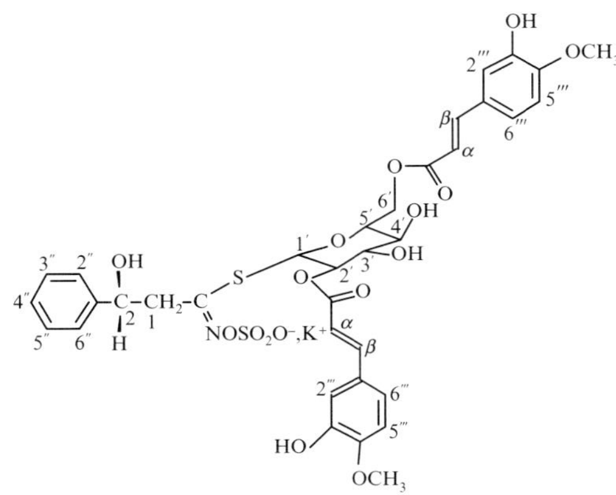	2′,6′-Diisoferuloyl-(2*R*)-2-Hydroxy-2-phenylethylglucosinolate	2′,6′-Isoferuloylglucosibarin

(*continues*)

TABLE I (*continued*)

Number	Structure of side chain (or complete glucosinolate)	Semi-systematic name	Trivial name
11.5	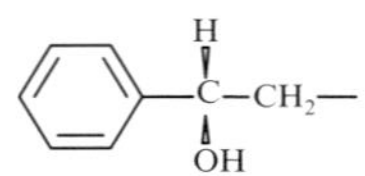	(2*S*)-2-Hydroxy-2-phenylethylglucosinolate	Glucobarbarin
11.6	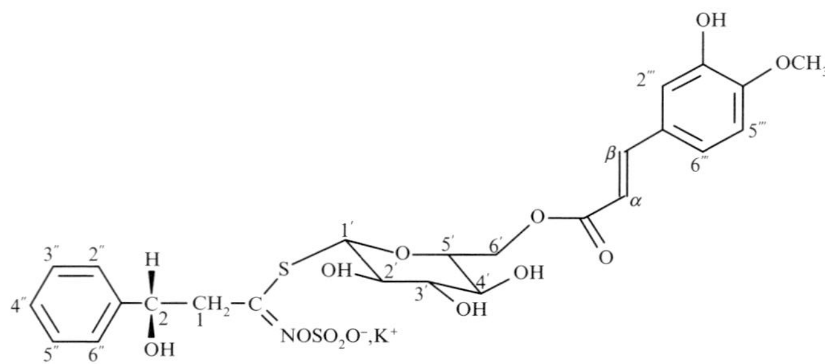	6′-Isoferuloyl-(2*S*)-2-Hydroxy-2-phenylethylglucosinolate	6′-Isoferuloylglucobarbarin
11.7	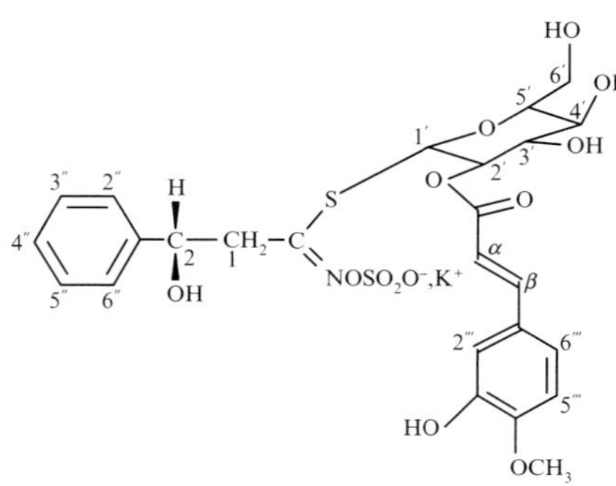	2′-Isoferuloyl-(2*S*)-2-Hydroxy-2-phenylethylglucosinolate	2′-Isoferuloylglucobarbarin

11.8	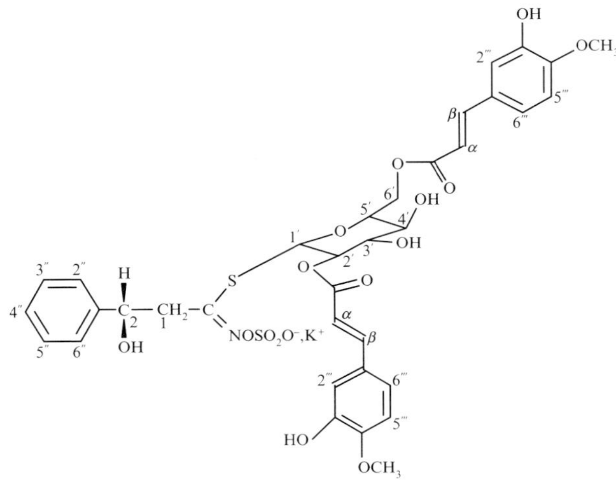	2′,6′-Diisoferuloyl-(2*S*)-2-Hydroxy-2-phenylethylglucosinolate	2′,6′-Diisoferuloylglucobarbarin
11.9	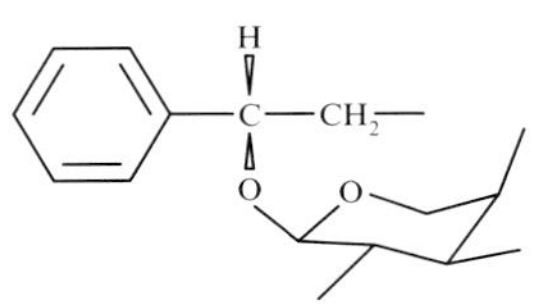	(2*S*)-2-α-L-Arabinopyranosyloxy-2-phenylethylglucosinolate	Glucosesamoidin
11.10	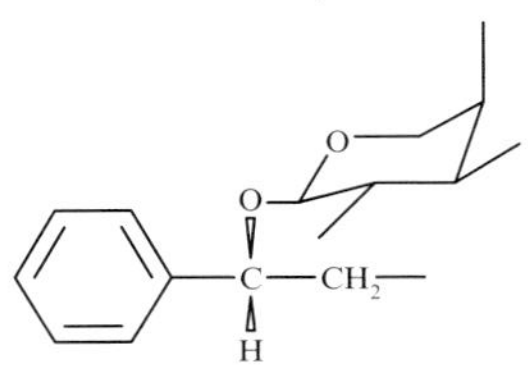	(2*R*)-2-α-L-Arabinopyranosyloxy-2-phenylethylglucosinolate	Glucocanescenein

(*continues*)

TABLE I (*continued*)

Number	Structure of side chain (or complete glucosinolate)	Semi-systematic name	Trivial name
12.1	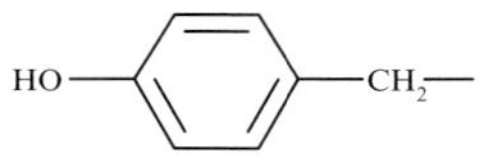	4-Hydroxybenzylglucosinolate	Sinalbin
12.2		4-Methoxybenzylglucosinolate	Glucoaubrietin
12.3	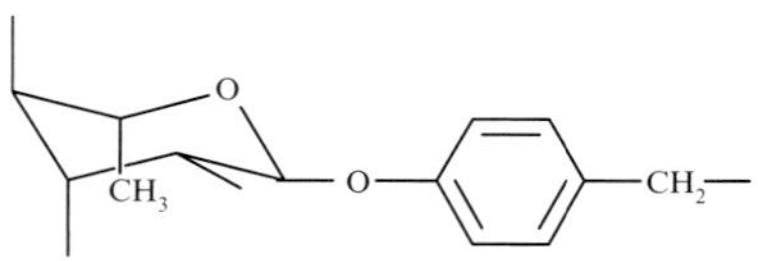	4-α-L-Rhamnopyranolsyloxy benzylglucosinolate	Glucomoringain
12.4	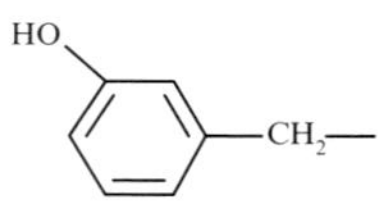	3-Hydroxybenzylglucosinolate	Glucolepigramin
12.5	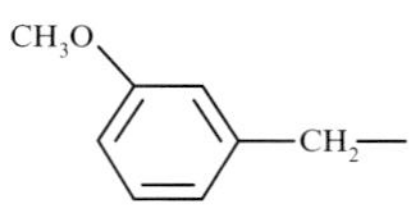	3-Methoxybenzylglucosinolate	Glucolimnanthin
12.6	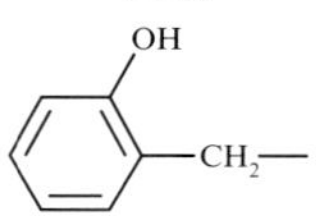	2-Hydroxybenzylglucosinolate	Glucoreluteolain
12.7	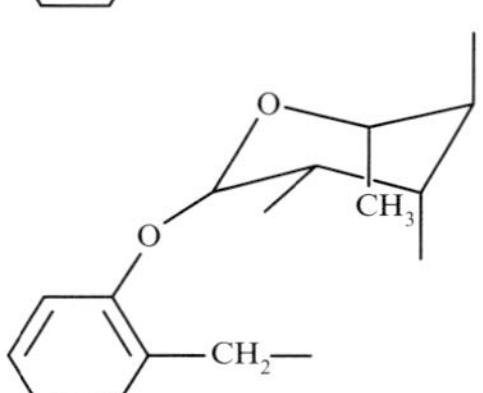	2-Rhamnopyranosyloxybenzyl-glucosinolate	Glucoreodoratain

12.8	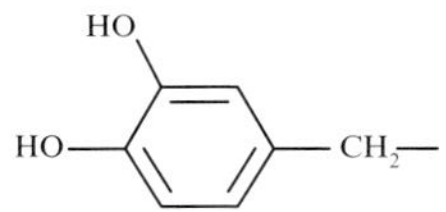	3,4-Dihydroxybenzylglucosinolate	Glucohesmatrolin
12.9	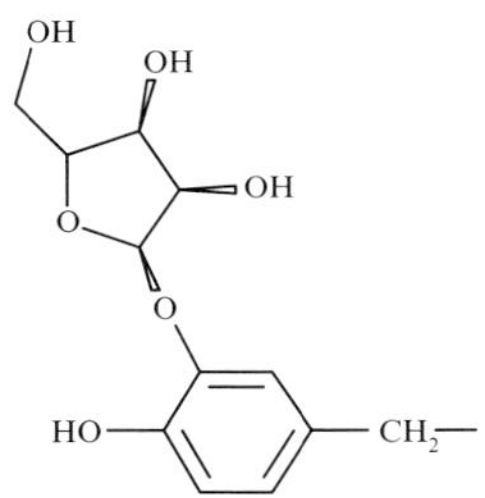	4-Hydroxy-3-apiofuranosyloxybenzylglucosinolate	Glucohesmatrolalin
12.10	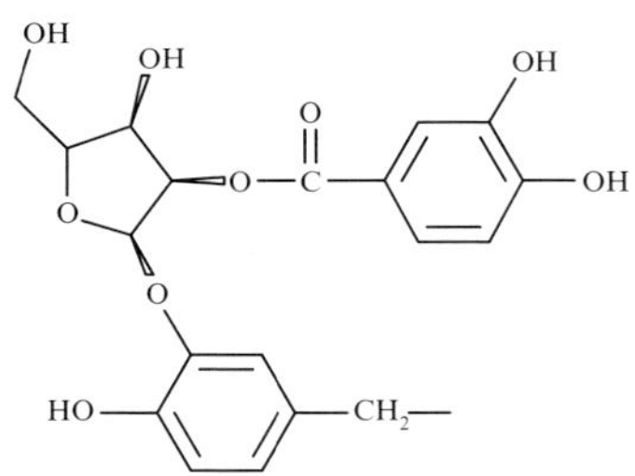	4-Hydroxy-3-[2′-(3,4-dihydroxybenzoyl) apiofuranosyloxy] benzylglucosinolate	
12.11	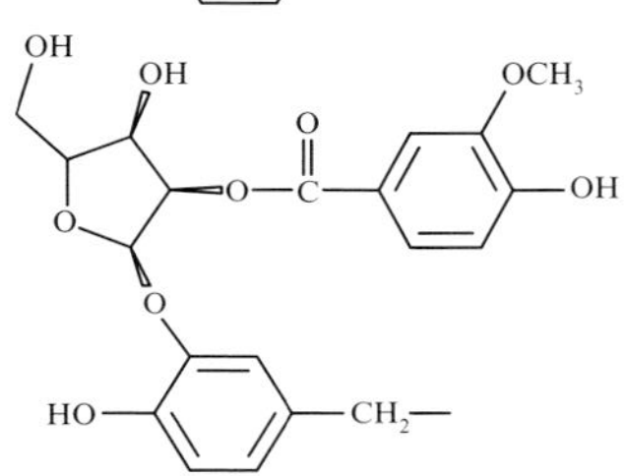	4-Hydroxy-3-[2′-(4-hydroxy-3-methoxybenzoyl)apiofuranosyloxy] benzylglucosinolate	

(*continues*)

TABLE I (*continued*)

Number	Structure of side chain (or complete glucosinolate)	Semi-systematic name	Trivial name
12.12	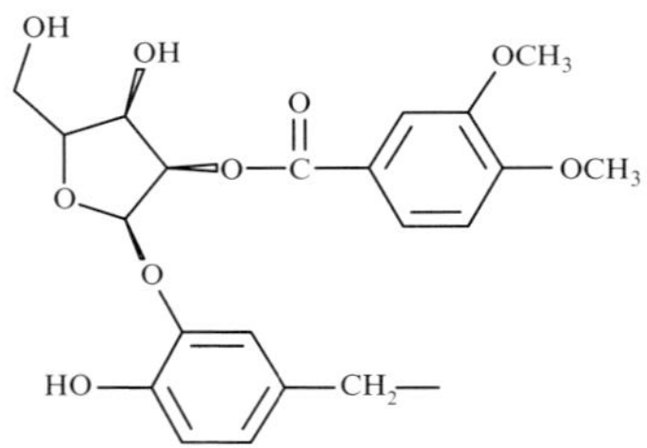	4-Hydroxy-3-[2′-(3,4-dimethoxybenzoyl)apiofuranosyloxy] benzylglucosinolate	
12.13	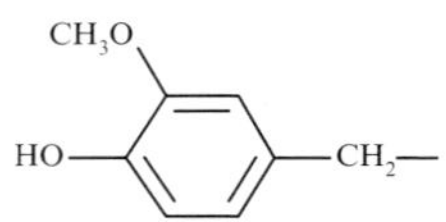	4-Hydroxy-3-methoxybenzylglucosinolate	Glucoelongatin
12.14	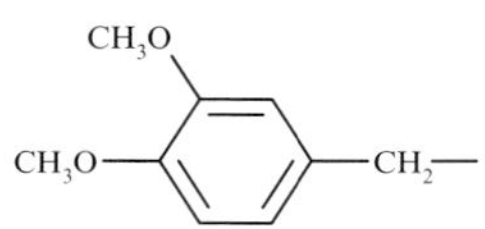	3,4-Dimethoxybenzylglucosinolate	Glucolongifoliain
12.15	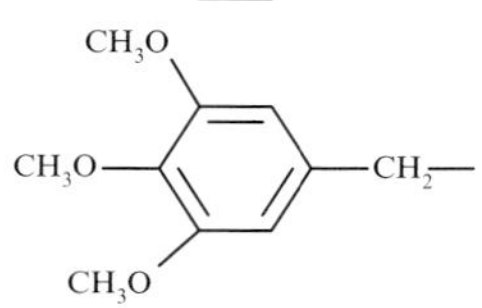	3,4,5-Trimethoxybenzylglucosinolate	Glucolepidiumhyssopifolium
12.16	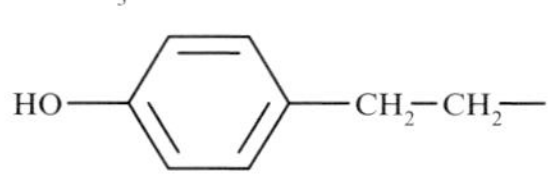	2-(4-Hydroxyphenyl)ethylglucosinolate	
12.17	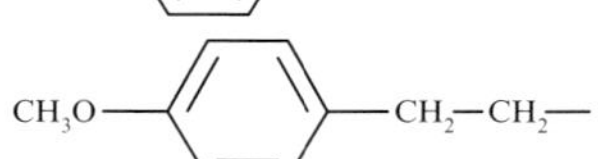	2-(4-Methoxyphenyl)ethylglucosinolate	

12.18	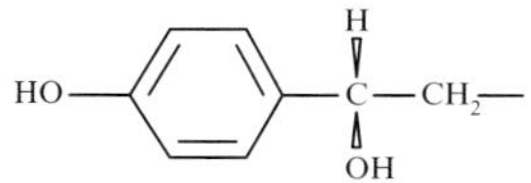	(2*S*)-2-Hydroxy-2-(4-hydroxyphenyl) ethylglucosinolate	Glucoarabihirsuin
12.19	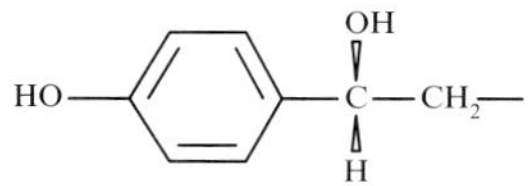	(2*R*)-2-Hydroxy-2-(4-hydroxyphenyl) ethylglucosinolate	Glucoarabihirin
12.20	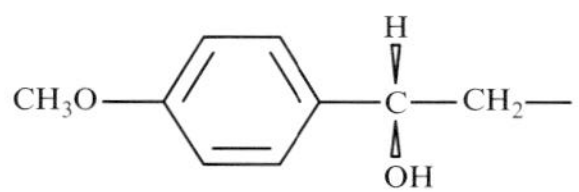	(2*S*)-2-Hydroxy-2-(4-methoxyphenyl) ethylglucosinolate	
12.21	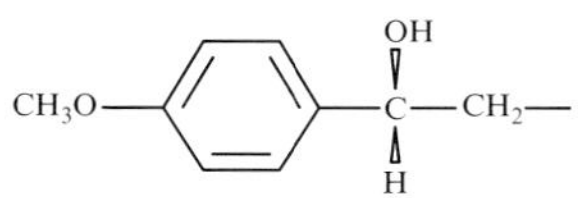	(2*R*)-2-Hydroxy-2-(4-methoxyphenyl) ethylglucosinolate	
13.1	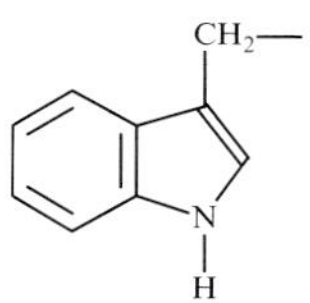	Indol-3-ylmethylglucosinolate	Glucobrassicin
13.2	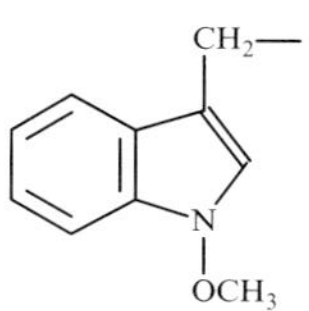	*N*-methoxyindol-3-ylmethylglucosinolate	Neoglucobrassicin
13.3	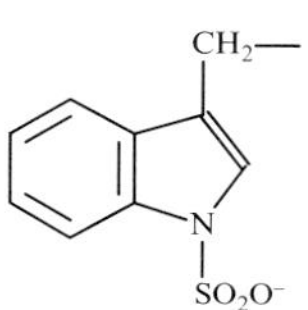	*N*-sulfoindol-3-ylmethylglucosinolate	Sulfoglucobrassicin

(*continues*)

TABLE I (*continued*)

Number	Structure of side chain (or complete glucosinolate)	Semi-systematic name	Trivial name
13.4	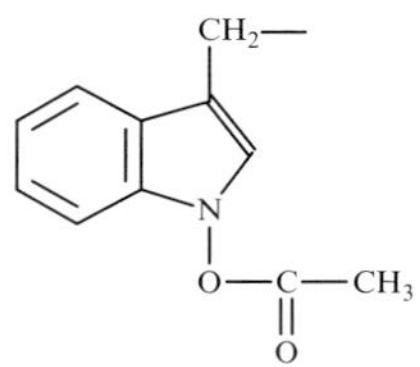	*N*-acetylindol-3-ylmethylglucosinolate	*N*-acetylglucobrassicin
13.5	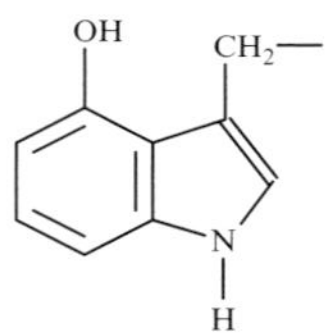	4-Hydroxyindol-3-ylmethylglucosinolate	4-Hydroxyglucobrassicin
13.6	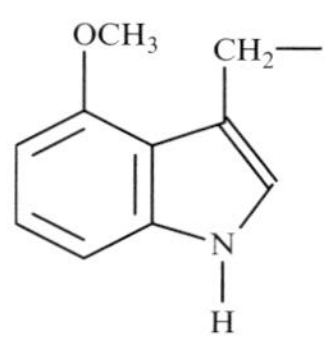	4-Methoxyindol-3-ylmethylglucosinolate	4-Methoxyglucobrassicin
13.7	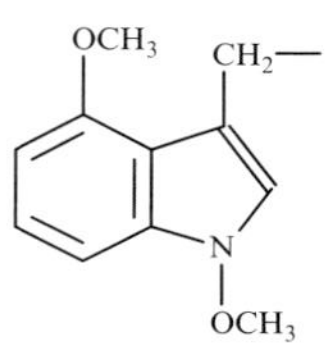	1,4-Dimethoxyindol-3-ylmethylglucosinolate	

identification of glucosinolates with a negative charge in the side chain (Sørensen, 1990). Glucosinolates biosynthetically derived from Met (group 1), such as those accumulated in *Brassica* species as oilseed rape, have all undergone at least one side chain elongation (Hill *et al.*, 2003). No known glucosinolates in group 2 (Trp-derived compounds such as the indol-3-ylmethylglucosinolates) have side chain elongation, whereas glucosinolates in groups 3–5 can have either chain elongation or no chain elongation.

The role of amino acids as precursors of glucosinolates has also been proven by use of *in vivo* ^{14}C, ^{3}H, ^{15}N and ^{35}S tracer experiments (Kutacek *et al.*, 1962; Underhill *et al.*, 1962). The chain elongation steps typical for several glucosinolates (Table I) have been shown to result from the incorporation of acetyl units (Chrisholm and Wetter, 1967), and the use of the ^{14}C acetate technique has also shown that such chain elongations do not occur in biosynthesis of glucosinolate from tryptophan (Bjerg *et al.*, 1987c; Haughn *et al.*, 1991). Tracer experiments with amino acid precursors double labelled with ^{14}C and ^{15}N at carbon C-2 indicate that the chain-elongated amino acids are the glucosinolate precursors (Underhill and Chisholm, 1964). The study of glucosinolate biosynthesis based on radioactivity labelled amino acid precursors was carefully treated in an excellent review by Ettlinger and Kjær (1968). Tracer experiments have also focused on revealing the biosynthetic steps involved in the transformation of amino acids into glucosinolates, where formation of the aldoxim intermediate has been an initial problem (Kindl and Underhill, 1968; Matsuo, 1968; Matsuo and Underhill, 1969; Tapper and Butler, 1967; Underhill, 1967; Underhill *et al.*, 1973). The initial and general steps involved in the glucosinolate biosynthesis including chain elongation of amino acid precursors have been the subject of comprehensive treatments in various review papers (Bennett *et al.*, 1995; Ettlinger and Kjær, 1968; Grubb and Abel, 2006; Halkier and Gershenzon, 2006; Mithen *et al.*, 1995, 2000; Rosa *et al.*, 1997; Rücker and Röbbelen, 1994). The sequence of reactions in glucosinolate biosynthesis seems to occur in multi-enzyme complexes with high specificity for the amino acid precursor since only few of the large number of amino acids in glucosinolate producing plants serve as precursors for glucosinolates (vide supra; Table I). The sequence of reactions is illustrated in Fig. 2 (Bjerg *et al.*, 1987c) where the enzymes and genes responsible for the enzymes are to a great extent known (Halkier and Gershenzon, 2006; Hill *et al.*, 2003). The specificity of the multi-enzyme complexes towards the amino acid precursors and the differences with respect to chain elongation steps can be seen in the chemotaxonomic occurrence of glucosinolates. For example, plants in the Resedaceae family are derived from groups (vide supra) (4) neutral amino acids, (3) phenylalanine/tyrosine and (2) tryptophan, whereas plants in the *Brassica* genera

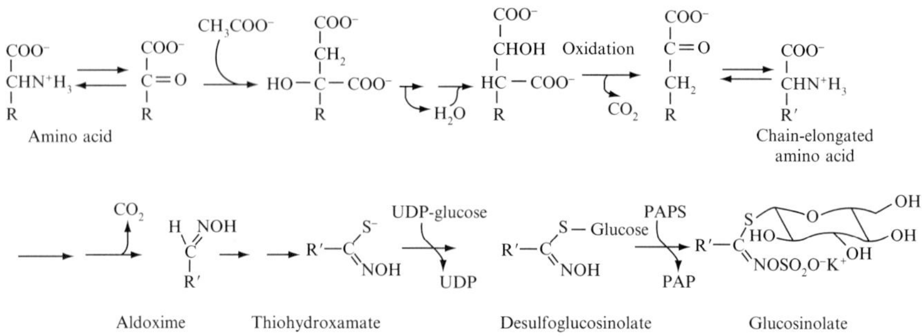

Fig. 2. Illustration of the biosynthetic pathway from amino acid to glucosinolates, with possible chain elongations that occur two times for progoitrin, gluconapin, glucoraphanin, glucoerucin, glucoerysolin, glucoraphenin and glucoraphasatin derived from Met.

accumulate glucosinolates of the groups (1) methionine, (2) tryptophan and (3) phenylalanine. Considering both structural types and relative concentrations of the individual glucosinolates, the determination of the glucosinolate profile in a particular species is a very powerful chemotaxonomic tool.

Following the initial glucosinolate biosynthesis (Fig. 2), various types of additional structural modifications result in the final glucosinolate structure. These modifications are based on specific enzymes, and thereby genes, which results in a large number of genes being involved in glucosinolate biosynthesis (Hill *et al.*, 2003). These final modifications of glucosinolate structures comprise:

Formation of double bonds ± elimination of the methylthio- or methylsulfinyl group in, for example, glucoraphenin
Stereo-specific oxidation of the methylthio-group (Christensen and Kjær, 1963, 1965; Dalgaard *et al.*, 1977; Klyne *et al.*, 1960; Sørensen, 1990)
Stereo-specific hydroxylation to either (*R*) or (*S*) configuration of, for example, C-2 in the R-group (Table I)
Hydroxylation/oxidation of the aromatic side chains (Table I)
Introduction of methoxy-, sulpho- or glycoside groups in the side chains (R-groups)
Acylation of the thioglucopyranoside group or the R-groups (Fig. 1; Table I)

The potentially large number of biosynthetic steps does not occur in all glucosinolate containing plants, but instead it occurs as chemotaxonomically defined processes. These modifications may, for example, involve the introduction of new chiral centres in the glucosinolate structure, which gives a need for the consideration of the stereo-chemical aspects, as a change of chirality can give appreciable changes in physicochemical as well as biochemical–physiological properties of glucosinolates and their hydrolysis products affecting the quality and/or properties of crops/products containing these compounds (Bagger *et al.*, 1999a,b; Bjerg *et al.*, 1989; Palmieri *et al.*, 1998; Sørensen, 1990).

IV. GLUCOSINOLATE STEREO-CHEMISTRY

The stereo-chemistry of glucosinolates is well defined with respect to the thio-β-D-glucopyranoside group and its attachment in *Z* (*cis*) stereo-chemistry in relation to the sulphate ester group at the thiohydroximinsulphate group (Fig. 1). The chiral centres in the side chains originating directly from the amino acid precursor Ile (either with or without chain elongation) are unchanged in all of the glucosinolates investigated (Ettlinger and Kjær, 1968) (Fig. 3).

Methylsulphinyl groups in glucosinolates derived from the precursor Met seem always to have (*R*)-configuration (Fig. 4), which is opposite to the chirality at the methylsulfinyl groups in the majority of other plant-produced compounds derived from Met (Christensen and Kjær, 1963, 1965; Dalgaard *et al.*, 1977; Klyne *et al.*, 1960; Sørensen, 1990).

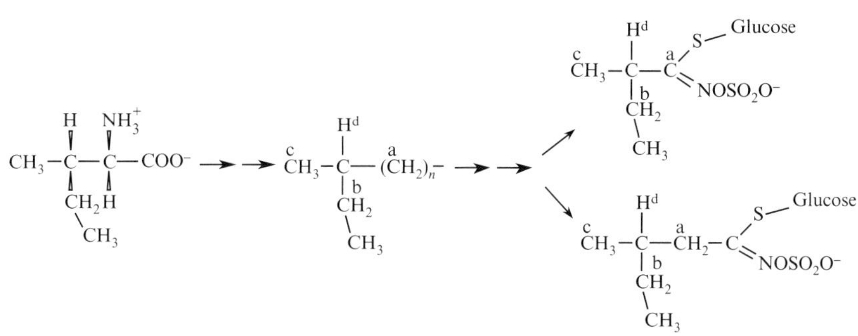

Fig. 3. Isoleucine-derived glucosinolates with (*S*)-configuration at the chiral centre in the side chain.

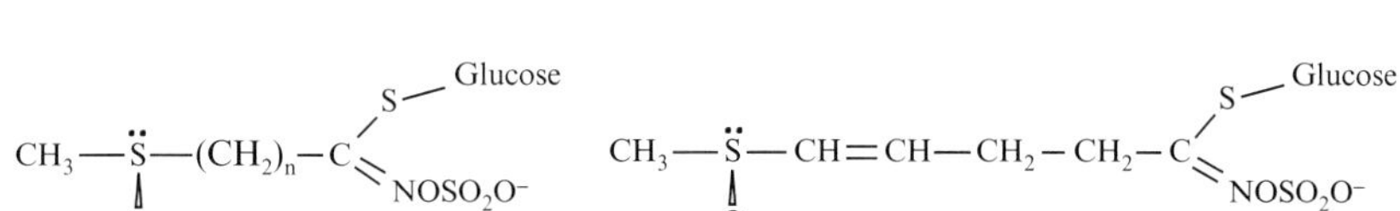

Fig. 4. (*R*)-configuration at the methylsulphinyl group in glucosinolates.

Chiral centres in glucosinolate side chains introduced by oxidation/hydroxylation can both be with (*R*)- and (*S*)-configuration (Fig. 5). These differences in chirality have great influence on the properties of the glucosinolates and especially of the oxazolidine-2-thiones produced in myrosinase-catalysed hydrolysis of 2-hydroxysubstituted glucosinolates (Bjerg *et al.*, 1989; Palmieri *et al.*, 1998).

Introduction of hydroxy groups in glucosinolate side chains most often results in a chiral centre that can be of both (*R*)- and (*S*)-configuration (exceptions are, for example, $(CH_3)_2$-COH-CH_2-Gls). In most cases, when

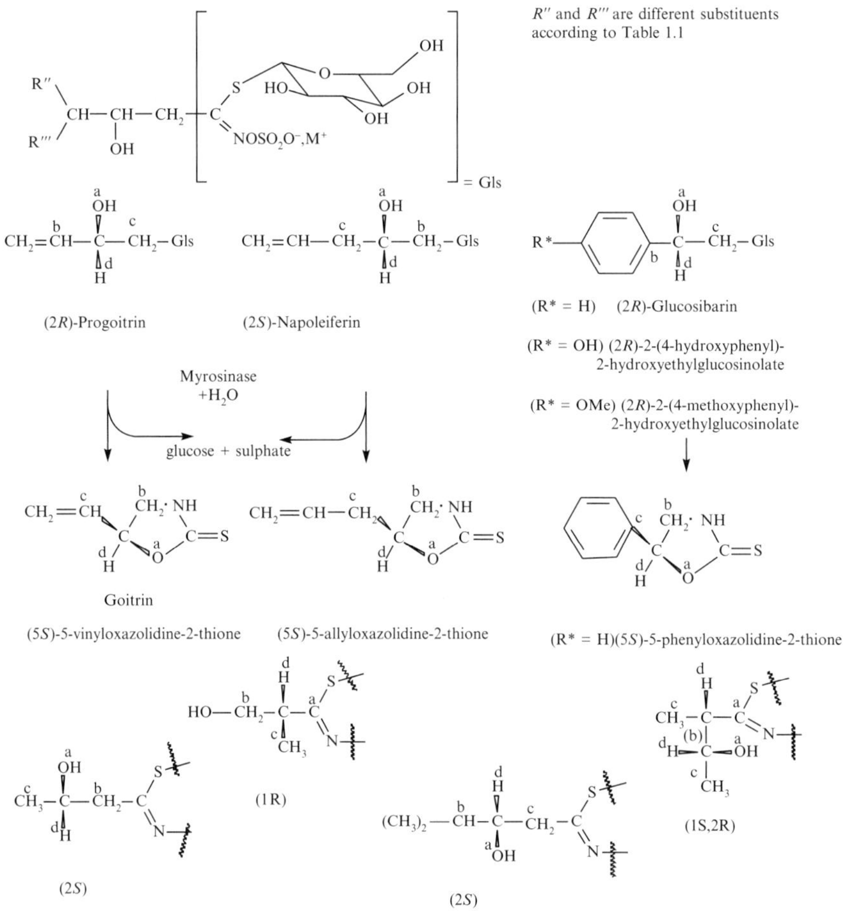

Fig. 5. The chirality introduced by enzyme-catalysed oxidation/hydroxylation of glucosinolates results in both (*R*)- and (*S*)-configuration. The symbol Gls used in the figure represents the general structure for glucosinolates with β-D-thioglucopyranoside group attached to C-0 in the *cis* or *Z*-*N*-hydroximine sulphate ester group.

both (*R*) and (*S*) stereoisomers occur together, one of the diastereoisomeric glucosinolates will most often be far the quantitatively dominating compound. In *Crambe* species, epiprogoitrin is the quantitatively dominating glucosinolate accompanied by minor amounts of progoitrin (Fig. 5). In seeds of oilseed rape (*Brassica napus* L.), progoitrin and napoleiferin (Fig. 5) are the quantitatively dominating 2-hydroxysubstituted glucosinolates, accompanied by minor amounts of the isomeric compounds (2*S*)-epiprogoitrin and (2*R*)-epinapoleiferin, respectively. In the genera *Barbarea*, some species have (2*R*)-glucosibarin co-occurring with minor amounts of the diastereoisomeric (2*S*)-glucobarbarin, whereas for some other *Barbarea* species, the opposite is found. Similar results are found for 2-hydroxy-substituted glucosinolates occurring in *Reseda* species as well as for the diastereoisomers of 2-(4-hydroxyphenyl)-2-hydroxyethylglucosinolate and of 2-(4-methoxyphenyl)-2-hydroxyethylglucosinolate (Fig. 5), which occur in *Arabis* species. There is strong experimental evidence on the specific modification of glucosinolate side chains by oxidation/hydroxylation (vide supra), by formation of (*R*)-configuration in the methylsulphinyl group (Fig. 4) and by oxidation resulting in a carbon–carbon double bond without elimination of the methylthio- or methylsulphinyl group (Sørensen, 1990). These results strongly indicate that all these processes occur as a result of enzyme-catalysed reactions following the general stages of the glucosinolates biosynthesis in multi-enzyme complexes (vide supra).

Interest in glucosinolates is mainly due to the physicochemical-biochemical-biological properties of glucosinolates and especially of their degradation products, properties that often to a great extend depend on the stereo-chemistry of the compounds (Bjerg *et al.*, 1989). As separation of optical antipodes (enantiomers) is generally more difficult than separation of diastereoisomers, proper and reliable glucosinolate identification cannot be based exclusively on determination of degradation products, but needs other determination methods including analysis of intact glucosinolates (Bjergegaard *et al.*, 2000; Buskov *et al.*, 2000a,b,c,d; Olsen and Sørensen, 1981; Sørensen, 1990; Sørensen *et al.*, 1999).

V. OCCURRENCE OF GLUCOSINOLATES

Glucosinolates are ubiquitous in plants of the order Capparales and in some few other plants (Dahlgreen, 1975; Ettlinger and Kjær, 1968; Kjær, 1960; Rodman *et al.*, 1996). A restricted number of other plants have also been mentioned as containing glucosinolates, mostly on the basis of their content of isothiocyanates and nitriles, which are considered to be degradation products of some aliphatic glucosinolates (Bjerg and Sørensen, 1987a; Bjergegaard *et al.*, 1994; Buskov *et al.*, 2000a,b). However, some of these reports indicating glucosinolates as being constituents of *Plantago* species (Plantaginaceae) and of cocoa (*Theobroma cacao* L. Sterculiaceae) have been seriously questioned (Bjerg *et al.*, 1987d; Larsen *et al.*, 1983).

According to (Dahlgreen, 1975; Halkier, 1999), the glucosinolate containing plant order Capparales comprises in the botanical broad sense the following families: Limnanthaceae, Tropaeolaceae, Bretschneideraceae, Salvadoraceae, Tovariaceae, Pentadeplandraceae, Gyrostemonaceae, Bataceae, Moringaceae, Brassicaceae, Capparaceae and Resedaceae. Of these families, especially the Brassicaceae (Cruciferae) has attracted great attention as a large number of cruciferous plants have been developed into important oil- and protein-rich agricultural crops, horticultural crops as well as crops for the production of food rich in oil/lipids, proteins, carbohydrates/dietary fibres and bioactive products (Bagger *et al.*, 1998; Fenwick and Heaney, 1983; Rosa *et al.*, 1997). Production of food and non-food products is mainly based on oil- and protein-rich crops and plant parts, especially seeds, as rapeseed. Cruciferous plants developed for human and animal nutrition belong mainly to the genera *Brassica*, *Sinapis* and *Raphanus*, which for human consumption, include rooty or tuberous vegetables, leafy or flower-head vegetables, and seeds (mustard), flowers (capers) and roots (horseradish) as spices.

VI. SPECIFICITY OF MYROSINASE ISOENZYMES AND INITIAL TRANSFORMATION OF GLUCOSINOLATES

Metabolism and/or transformation of glucosinolates in non-enzymatic and enzyme-catalysed processes result in various bioactive compounds; therefore, information on these processes is of key importance in relation to the control of the quality and safety of glucosinolate-containing foods. An important step in these reactions is the hydrolysis of the thioglucoside bond; reaction① (Fig. 6) catalysed by myrosinase isoenzyme (EC 3.2.1.147).

Myrosinases from different cruciferous plants occur as groups of oligomeric isoenzymes of different molecular weight and composed by different protein subunits, with some of these associated protein being easily released during certain purification procedures (Bellostas *et al.*, 2003; Bjergegaard *et al.*, 2003). Depending on the isoenzymes used, various products are formed in the myrosinase-catalysed transformation of glucosinolates, with the isoenzymes having different specificity towards structurally different glucosinolates (Petersen *et al.*, 2003). Myrosinase isoenzymes have very specific

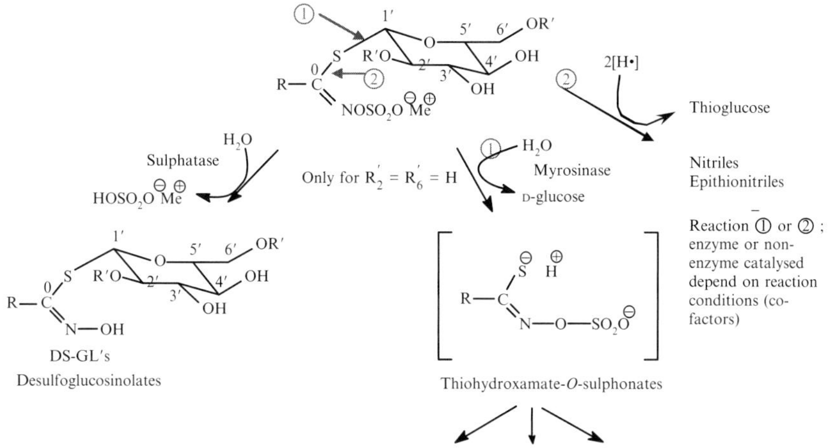

Fig. 6. Glucosinolates with $R_2' = R_6' = H$ are transformed into desulphoglucosinolates by catalysis with sulphatase and into thiohydroxamates by myrosinase-catalysed hydrolysis (reaction ①). Reaction ② resulting in nitriles can occur both as an enzyme catalysed and a non-enzymatic process catalysed by, for example, ferro ion, and in all cases with use of two redox equivalents.

requirements to the glucosinolate substrates with respect both to presence of the sulphate group and to an unsubstituted β-D-thioglucopyranoside group. Hence, myrosinase isoenzymes or thioglucosidases catalyse the hydrolysis of the β-D-thioglucopyranoside bond in glucosinolates where R_2' and R_6' is hydrogen. On hydrolysis glucose is released and the aglucone formed further rearranges into a variety of biologically active products with different structures depending on the parent glucosinolate and the reaction conditions (Bellostas *et al.*, 2006; Bones and Rossiter, 2006; Sørensen, 1990) (Fig. 7).

VII. GLUCOSINOLATE METABOLISM IN UNDAMAGED PLANTS, AUTOLYSIS PROCESSES AND *IN VITRO* SYSTEMS

Glucosinolates and myrosinases are physically separated in undamaged plants (Bones and Rossiter, 2006). Limited information is available concerning glucosinolate metabolism in the undamaged plant tissues, although it is known that only very low concentrations of isothiocyanates occur in undamaged plants (Ettlinger and Kjær, 1968). Among the few examples of glucosinolate metabolism in undamaged plants are the production of amines from glucosinolates and γ-glutamylpeptides of these amines (Eggum and Sørensen, 1989) as well as the active metabolism and selective transport/

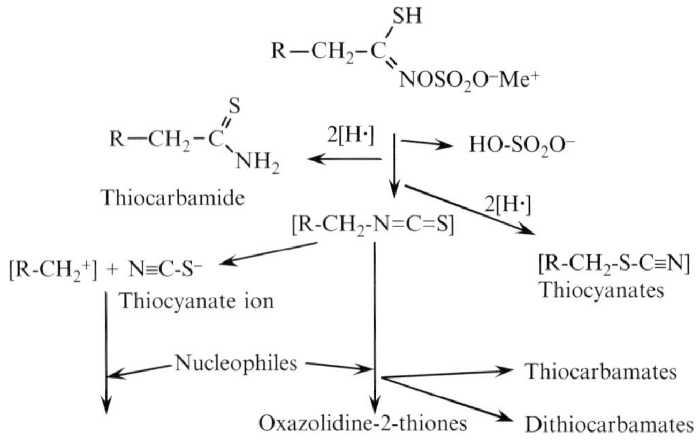

Fig. 7. Transformation of reactive thiohydroxamate-*O*-sulphonates and isothiocyanates into various bioactive products.

localisation of glucosinolates in different plant parts that occur as a function of germination and plant development (Bjerg *et al.*, 1987c). Transformation products of glucosinolates are known to be highly reactive towards nucleophiles as, for example, amines and thiols, which are compounds that occur in all living plants. Hence, autolysis (Olsen and Sørensen, 1981) of glucosinolate-containing plants gives rise to a complex mixture of products (Bjergegaard *et al.*, 1994; Buskov *et al.*, 2000a,b,c,d).

Glucosinolate hydrolysis products have been very broadly studied, initially due to their negative effect in feed for mono-gastric animals (Bille *et al.*, 1983b; Bjerg *et al.*, 1989) and in the last 15 years due to their beneficial effect on human health (Bonnesen *et al.*, 1999; Fahey *et al.*, 1997) and their potential as food and plant protection agents (Brown and Morra, 1997; Rosa *et al.*, 1997). Isothiocyanates have been the most studied glucosinolate products with regard to the biological activity of glucosinolate-containing plants since they are the major initial compounds produced at neutral pH after the Lossen re-arrangement of the aglucone. However, at more acidic pH and in the presence of certain co-factors, the formation of nitriles is favoured (Bones and Rossiter, 2006; Gil and MacLeod, 1980; Hasapis and MacLeod, 1982; Sørensen, 1990). The *in vitro* formation of nitriles is thus influenced by pH (Gil and MacLeod, 1980) and by the presence of certain co-factors such as thiol groups (Tookey and Wolff, 1970; Uda *et al.*, 1986a) and Fe^{2+} (Tookey and Wolff, 1970; Uda *et al.*, 1986b). Non-enzymatic formation of nitriles through the effect of Fe^{2+} has also been reported (Austin and Gent, 1967; Austin *et al.*, 1968a,b; Youngs and Perlin, 1967), and it was suggested to occur through the formation of a glucosinolate-Fe^{2+} complex (Youngs and Perlin, 1967). Although the complex has never been observed, its existence was postulated on the basis of a stable 2:1 complex between methylthiohydroxamic acid and Ni^{2+} (Sato *et al.*, 1966).

The isothiocyanates formed from glucosinolates containing indol-3-ylmethyl groups or ortho- and para-hydroxy/glycoside-/methoxybenzyl *R*-groups release relatively fast the thiocyanate ion and form products corresponding to carbonium ion-derived products (Figs. 8–11). In this case, the solubility is a serious problem, and therefore the application of supercritical fluids is recommendable (Buskov *et al.*, 2000a,b,c).

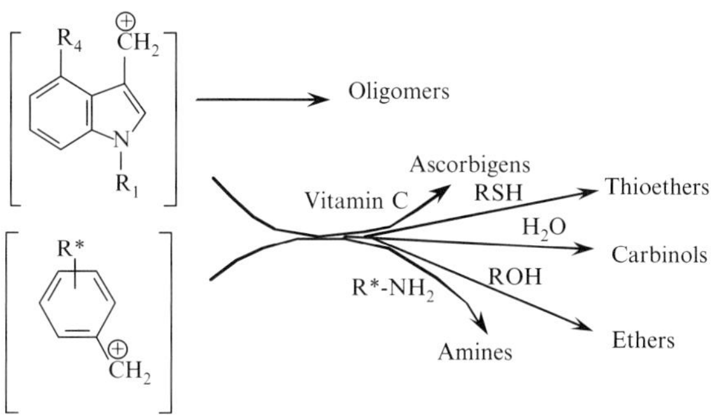

Fig. 8. Stabilised carbonium ions as reactive intermediates in transformation of glucosinolates into more stable products.

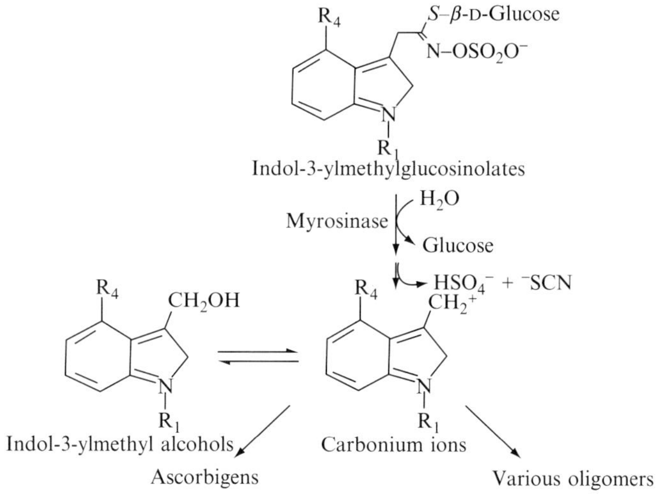

$R_1 = R_4 = H$ (glucobrassicin); $R_1 = H$, $R_4 = OH$ (4-hydroxyglucobrassicin);
$R_1 = H$, $R_4 = OCH_3$ (4-methoxyglucobrassicin); $R_1 = OCH_3$, $R_4 = H$ (neoglucobrassicin)

Fig. 9. Indol-3-ylmethylglycosinolates and transformation products thereof. Adapted from Buskov *et al.* (2000a).

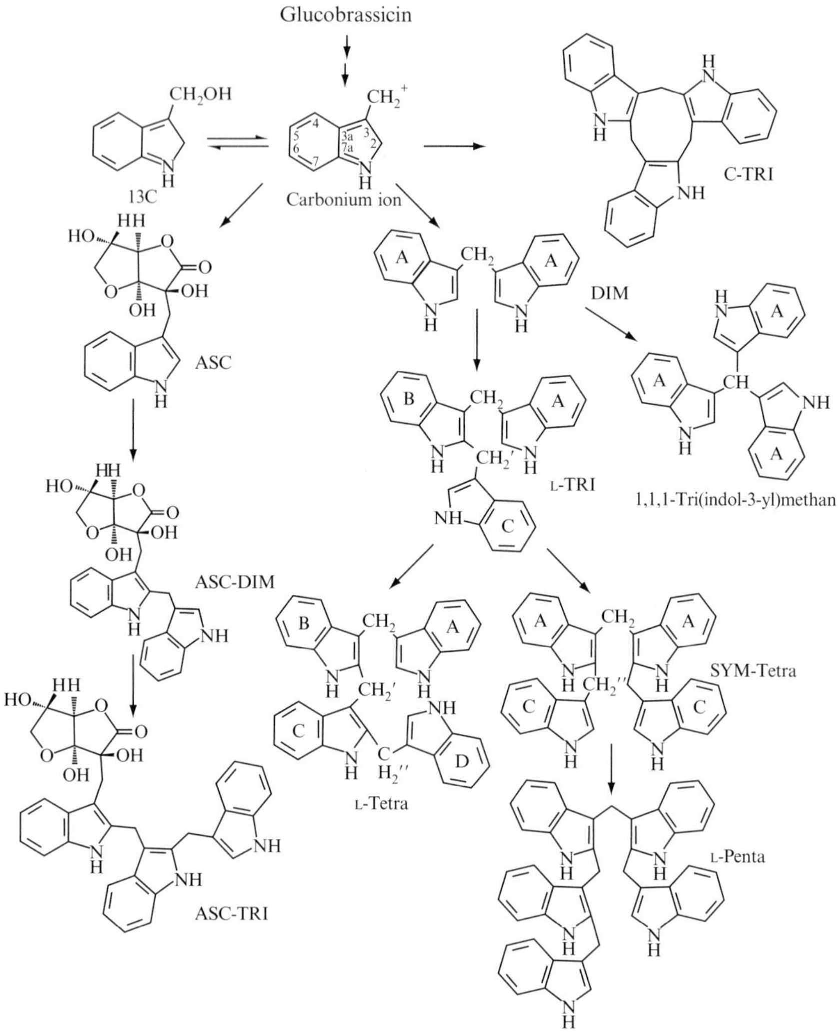

Fig. 10. Oligomers and ascorbigens from glucobrassicin. Adapted from Buskov *et al.* (2000b).

VIII. PERSPECTIVES: CRUCIFEROUS PLANTS (RAPESEED) AS CHEMICAL FACTORIES

The value of cruciferous crops and products obtained from them depend on the amount and type/structure of the glucosinolates present, including their stereo-chemistry. The increasing attention paid to glucosinolates and products thereof in the recent years is due to the rising interest in the production

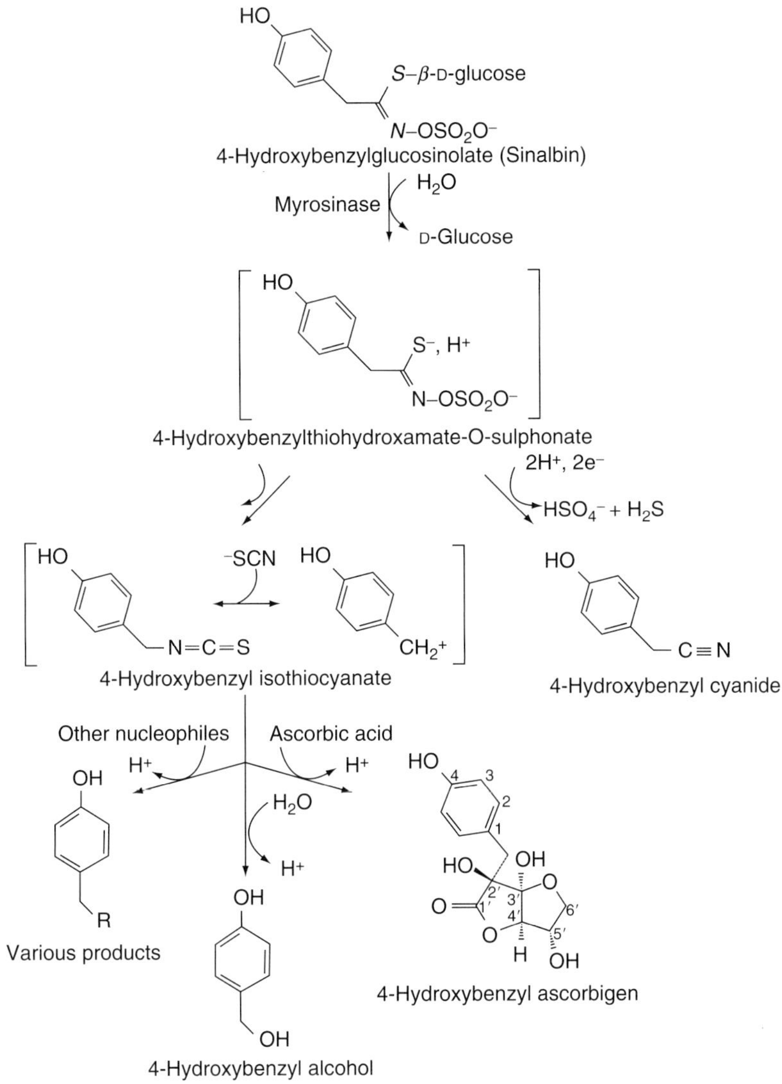

Fig. 11. Structures of sinalbin products formed during myrosinase-catalysed hydrolysis. Adapted from Buskov *et al.* (2000a).

of high-quality (added value) products from oil- and protein-rich cruciferous crops both for food and non-food purposes (Bagger *et al.*, 1998; Bjergegaard and Sørensen, 1996; Bjergegaard *et al.*, 1998; Jensen *et al.*, 1990; Palmieri *et al.*, 1998). Other glucosinolate-containing plants have also been considered for these purposes (Andersen *et al.*, 1999; Bagger *et al.*, 1999a,b). Knowledge on the structure and properties of glucosinolates and products derived from them is thus of outmost importance for the optimal utilisation of glucosinolate-containing plants (Buskov *et al.*, 2000a,b). Similarly, knowledge on glucosinolate heredity and on changes of glucosinolate content and composition and its control by appropriate plant-breeding programmes is also highly relevant (Hill *et al.*, 2003).

The consideration of plants as factories for biologically active phytochemicals as glucosinolates and products thereof is a subject of increasing interest (Sørensen *et al.*, 2001). This is especially the case in many industrialised countries, where focus is placed on healthy and safe bioactives, which can be used as environment friendly biodegradable products (Sørensen, 2001). This comprises both positive and negative effects from the actual components, for example odour and taste, as well as antioxidants, anti-nutrients, anti-cancerogenic and hyper-cholesterolemic compounds. Another area of growing interest is the use of bioactive natural products as feed, food and plant-protection agents, both as biofumigants (Brown and Morra, 1997; Buskov *et al.*, 2002) and in agro-industries (Palmieri *et al.*, 1998).

IX. CONCLUSION

Progress in the production and utilisation of cruciferous oilseed crops including oilseed rape (*B. napus* L. and *B. rapa* L.) and their derived products is closely related to successful plant breeding, which in the past years has resulted in the production of high-yielding double low varieties with low levels of erucic acid and glucosinolates. In order to meet this objective, extensive knowledge on the occurrence, structure and properties of glucosinolates and myrosinase isoenzymes (thioglucohydrolase; EC 3.2.1.147) co-occurring with glucosinolates is essential. The present chapter focusses on these subjects including heredity, genetic variation and metabolism of glucosinolates.

A number of stages in the glucosinolate biosynthetic pathway are well known, but significant elements still remain to be further explored and elucidated. Pioneer work was performed several decades ago, initially based on analytical and tracer experiments. In recent years, most research has been based on gene technology using *A. thaliana* as a model and hence considering only the glucosinolates present in this particular plant. To date, more than 140 structurally different glucosinolates have been described of which $\sim$30 glucosinolates occur in *Brassica* species. Only relatively few glucosinolates are produced in individual plant genera and/or families. Knowledge on the genetic- and metabolic-based regulation of glucosinolate accumulation in different plants is therefore of great relevance in order to explain the reasons for the chemotaxonomically limited occurrence of structurally different glucosinolates. The new technologies used in plant breeding along with gene technology allow modifying the glucosinolate levels and/or including additional or structurally new glucosinolates in cruciferous oilseed crops, which results in changed properties of the crops and their derived products. The metabolic regulation and thus the biological consequences of those changes are however unknown. Similarly, the reasons for the high selectivity of the multi-enzyme complexes towards the precursor amino acids in the biosynthesis of glucosinolates, as well as the specificity of the various enzymes needed for the modification of glucosinolate side chains and the substitution/esterification of the thioglucose residue of the molecules, are yet to be clarified. The glucosinolate catabolism, which comprises the transformation of glucosinolates into various bioactive products, is another important area where limited information is available on, for example, the oligomeric myrosinase isoenzymes and associated compounds. This is especially relevant regarding the differences in the catalytic properties of the individual isoenzymes as well as the opportunities to modify these properties. New analytical methods allow for the synchronous detection of substrate and products in myrosinase-catalysed reactions, which broadens the perspectives for progress in this field.

ACKNOWLEDGMENT

The Commission of the European Union (FP6-Novel Q 015710–2) is gratefully acknowledged for financial support of this work.

REFERENCES

Agerbirk, N., Bjergegaard, C., Olsen, C. E. and Sørensen, H. (1996). Kinetic investigation of the transformations of indol-3-ylcarbinol into oligomeric indolyl compounds based on micellar electrokinetic capillary chromatography. *Journal of Chromatography A* **745**, 239–248.

Ahlin, K. A., Emanuelson, M. and Wiktorson, H. (1994). Rapeseed products from double-low cultivars as feed for dairy cows: Effects of long term feeding on thyroid function, fertility and animal health. *Acta Veterinaria Scandinavia* **35**, 37–53.

Allison, R. D., Moss, A. R. and Blake, J. S. (2001). The effect of feeding rations containing heat treated rapeseed meal, lupins and beans to lactating dairy cows on milk yield and quality. *Proceedings of the British Society of Animal Science* **2001**, 4.

Andersen, A. A. M., Merker, A., Nilsson, P., Sørensen, H. and Åman, P. (1999). Chemical composition of the potential new oilseed crops *Barbarea vulgaris*,

Barbarea verna and *Lepidium campestre*. *Journal of the Science of Food and Agriculture* **79**, 179–186.

Austin, F. L. and Gent, C. A. (1967). (*S*)-3-Hydroxypent-4-enethionamide: A new reation product of a natural thioglucoside. *Chemical Communications* **7**, 1–72.

Austin, F. L., Gent, C. A. and Wolff, I. A. (1968a). Enantiomeric 3-hydroxypent-4-enethionamides from thioglucosides of Crambe and Brassica seeds by action of ferrous salts. *Canadian Journal of Chemistry* **46**, 1507–1512.

Austin, F. L., Gent, C. A. and Wolff, I. A. (1968b). Degradation of natural thioglucosides with ferrous salts. *Journal of Agricultural and Food Chemistry* **16**, 752–755.

Bagger, C. L., Sørensen, J. C., Sørensen, S. and Sørensen, H. (1998). Biorefining of oilseed crops as basis for molecular farming of high quality products. *GCIRC Bulletin No* **15**, 96–99.

Bagger, C. L., Bjergegaard, C., Sørensen, H., Sørensen, J. C. and Sørensen, S. (1999a). High quality oils, proteins and bioactive products for food and non-food purposes produced in pilot plant scale by biorefining of cruciferous crops. *In* "Proceedings of 10th International Rapeseed Congress." Canberra, Australia.

Bagger, C. L., Buskov, S., Hasselstrøm, J. V., Rosa, E., Sørensen, H. and Sørensen, J. C. (1999b). Bioactives from cruciferous crops especially glucosinolate derived products produced in pilot plant scale and used as biocides supplementary to synthetic pesticides. *In* "Proceedings of 10th International Rapeseed Congress." Canberra, Australia.

Bao, Y. and Fenwick, G. R. (2004). Future perspectives in phytochemical and health research. *In* "Phytochemicals in Health and Disease" (Y. Bao and R. G. Fenwick, eds.), pp. 331–342. Marcel Dekker Inc., New York, Basel.

Barillari, J., Canistro, D., Paolini, M., Ferroni, F., Pedulli, G. F., Iori, R. and Valgimigli, L. (2005). Direct antioxidant activity of purified glucoerucin, the dietary secondary metabolite contained in rocket (*Eruca sativa* mill.) seeds and sprouts. *Journal of Agricultural and Food Chemistry* **53**(7), 2475–2482.

Bell, J. M. and Shires, A. (1982). Comparison and digestibility by pigs of hull fractions from rapeseed cultivars with yellow or brown seed coats *Brassica campestris*, *Brassica napus*. *Canadian Journal of Animal Science* **62**, 557–565.

Bellostas, N., Jørgensen, A. L. W., Lundin, N. V. F., Petersen, I. L., Sørensen, H., Sørensen, J. C., Sørensen, R. and Tidmand, K. D. (2003). Comparison of physico-chemical properties of myrosinase isoenzymes occuring in seeds of Brassica species of the U triangle. *In* "Proceedings of 11th International Rapeseed Congress, 6–10 July." Copenhagen, Denmark, ISBN 87-7611-003-6, PP25, 720–723.

Bellostas, N., Sørensen, J. C. and Sørensen, H. (2006). Micellar electrokinetic capillary chromatography – Synchronous monitoring of substrate and products in the myrosinase catalysed hydrolysis of glycosinolates. *Journal of Chromatography A* **1130**, 246–252.

Bennett, R., Ludvig-Muller, J., Kiddle, G., Hilgenberg, W. and Wallsgrove, R. (1995). Developmental regulation of aldoxim formation in seedlings and mature plants of chinese cabbage (*Brassica campestris* ssp. pekinensis) and oilseed rape (*Brassica napus*): Glucosinolate and IAA biosynthetic enzymes. *Planta* **196**, 239–244.

Bille, N., Eggum, B. O., Jacobsen, I., Olsen, O. and Sørensen, H. (1983a). The effects of processing on antinutritional rape constituents and the nutritive value of double low rapeseed meal. *Zeitschrift für Tierphysiologie, Tierernährung und Futtermittelkunde* **49**, 148–163.

Bille, N., Eggum, B. O., Jacobsen, I., Olsen, O. and Sørensen, H. (1983b). Antinutritional and toxic effects in rats of individual glucosinolates (± myrosinases) added to a standard diet. 1. The effects on protein utilization and organ weights. *Zeitschrift für Tierphysiologie, Tierernährung und Futtermittelkunde* **49**, 195–210.

Bjerg, B. and Sørensen, H. (1987a). Isolation of intact glucosinolates by column chromatography and determination of their purity. *In* "World Crops: Production, Utilization, Description. Glucosinolates in Rapeseeds: Analytical Aspects" (J.-P. Wathelet, ed.), Vol. 13, pp. 59–75. Martinus Nijhoff Publishers, Kluwer Academic Publishers Group, Dordrecht/Boston/Lancaster.

Bjerg, B. and Sørensen, H. (1987b). Quantitative analysis of glucosinolates in oilseed rape based on HPLC of desulfoglucosinolates and HPLC of intact glucosinolates. *In* "World Crops: Production, Utilization, Description. Glucosinolates in Rapeseeds: Analytical Aspects" (J.-P. Wathelet, ed.), Vol. 13, pp. 125–150. Martinus Nijhoff Publishers, Kluwer Academic Publishers Group, Dordrecht/Boston/Lancaster.

Bjerg, B., Larsen, L. M. and Sørensen, H. (1987a). Reliability of analytical methods for quantitative determination of individual glucosinolates and total glucosinolate content in double low oilseed rape. *In* "Proceedings of the Seventh International Rapeseed Congress," Vol. 6, pp. 1330–1341. Poznan Poland, IV.

Bjerg, B., Eggum, B. O., Rasmussen, K. W. and Sørensen, H. (1987b). Acceptable concentrations of glucosinolates in double low oilseed rape and possibilities of further quality improvements by processing and plant breeding. *In* "Proceedings of the Seventh International Rapeseed Congress," Vol. 7, pp. 1619–1626. Poznan Poland, VII.

Bjerg, B., Kachlicki, P. W., Larsen, L. M. and Sørensen, H. (1987c). Metabolism of glucosinolates. *In* "Proceedings of the Seventh International Rapeseed Congress," Vol. 2, pp. 496–506. Poznan Poland, 2, I.

Bjerg, B., Fenwick, G. R., Spinks, A. and Sørensen, H. (1987d). Failure to detect glucosinolates in coca. *Phytochemistry* **26**, 567–568.

Bjerg, B., Eggum, B. O., Jacobsen, I., Otte, J. and Sørensen, H. (1989). Antinutritional and toxic effects in rats of individual glucosinolates (+/- myrosinase) added to a standard diet (2). *Journal of Animal Physiology and Animal Nutrition* **61**, 227–244.

Bjergegaard, C. and Sørensen, H. (1996). Antinutritional compounds in feed and food. *In* "Proceedings of Fourth International Feed Production Conference," pp. 105–135. Piacenza Facolta Di Agraria, Italy.

Bjergegaard, C., Michaelsen, S., Møller, P. and Sørensen, H. (1991). High performance capillary electrophoresis: Determination of individual anions, carboxylates, intact- and desulfoglucosinolates. *GCIRC Congress, Saskatoon, Canada* **III**, 822–827.

Bjergegaard, C., Li, P. W., Michaelsen, S., Møller, P., Otte, J. and Sørensen, H. (1994). Glucosinolates and their transformation products - compounds with a broad biological activity. *Bioactive Substances in Food of Plant Origin* **1**, 1–15.

Bjergegaard, C., Michaelsen, S., Møller, P. and Sørensen, H. (1995). Separation of desulphoglucosinolates by micellar electrokinetic capillary chromatography based on a bile salt. *Journal of Chromatography A* **717**, 325–333.

Bjergegaard, C., Jensen, S. K., Quinsac, A. and Sørensen, H. (1998). Analyses of antinutritional factors in rapeseed. Recent advances of research in antinutritional factors in legume seeds and rapeseed. *In* "Proceedings of the Third International Workshop on 'Antinutritional Factors in Legume Seeds and Rapeseed"' (A. J. M. Jansman, G. D. Hill, J. Huisman and A. F. B. van der Poel, eds.), pp. 67–90. Wageningen Pers, Wageningen.

Bjergegaard, C., Møller, P., Sørensen, H., Sørensen, J. C. and Sørensen, S. (1999). Micellar electrokinetic capillary chromatography of thiocarbamoyl derivatives produced in reactions between isothiocyanates and amino acids. *Journal of Chromatography A* **836**, 115–127.

Bjergegaard, C., Buskov, S., Sørensen, H., Sørensen, J. C., Sørensen, M. and Sørensen, S. (2000). Reactions between glucosinolate products and thiol groups in food components Czech. *Journal of Food Science* **18**, 28–30.

Bjergegaard, C., Buskov, S., Ib, K., Sørensen, H., Sørensen, J. C. and Sørensen, S. (2001). Effects of various antioxidants and storage conditions on oxidative degradation of the indole glucosinolate 4-hydroxyglucobrassicin. *In* "Biologically-Active Phytochemicals in Food. Analysis, Metabolism, Bioavailability and Function." The Royal Society of Chemistry, UK. ISBN 0-85404-806-5, 88–90.

Bjergegaard, C., Mortensen, K., Petersen, I. L., Sørensen, H. and Søresnen, J. C. (2003). Isolation and characterisation of myrosinase isoenzymes occurring in *Brassica napus* L. and *Sinapis alba* L. seeds. *In* "Proceedings of 11th International Rapeseed Congress, 6–10 July." Copenhagen, Denmark. ISBN 87-7611-003-6, PP24, 712–715.

Bones, A. M. and Rossiter, J. T. (2006). The enzymic and chemically induced decomposition of glucosinolates. *Phytochemistry* **67**, 1053–1067.

Bonnesen, C., Stephensen, P. U., Andersen, O., Sørensen, H. and Vang, O. (1999). Modulation of cytochrome P-450 and glutathione S-transferase isoform expression *in vivo* by intact and degraded indolyl glucosinolates. *Nutrition and Cancer* **33**(2), 178–187.

Brown, P. D. and Morra, M. J. (1997). Control of soil-borne plant pests using glucosinolate-containing plants. *Advances in Agronomy* **61**, 167–231.

Bunting, E. S. (ed.) (1981). "Production and Utilization of Protein and Oilseed Crops." Martinus Nijhoff Publishers, The Hague/Boston/London. ISBN 90-247-2532-1.

Buskov, S., Mortensen, K., Møller, P., Sørensen, H., Sørensen, J. C. and Sørensen, S. (1997). Comparison of micellar electrokinetic capillary chromatography and high performance liquid chromatography for determination of different hydroxycinnamic acid esters of malic acid occurring in cruciferous seedlings. *In* "Proceedings of Euro Food Chem IX, Authenticity and Adulteration of Food—The Analytical Approach," pp. 632–637. Interlaken, Switzerland.

Buskov, S., Hasselstrøm, J., Olsen, C. E., Sørensen, H., Sørensen, J. C. and Sørensen, S. (2000a). Supercritical fluid chromatography as method of analysis for determination of 4-hydroxybenzylglucosinolate degradation products. *Journal of Biochemical and Biophysical Methods* **43**, 157–174.

Buskov, S., Olsen, C. E., Sørensen, H. and Sørensen, S. (2000b). Supercritical fluid chromatography as basis for identification and quantitative determination of indol-3-ylmethyloligomers and ascorbigens. *Journal of Biochemical and Biophysical Methods* **43**, 175–195.

Buskov, S., Hansen, L. B., Olsen, C. E., Sørensen, J. C., Sørensen, H. and Sørensen, S. (2000c). Determination of ascorbigens in autolysates of various Brassica species using supercritical fluid chromatography. *Journal of Agricultural and Food Chemistry* **48**(7), 2693–2701.

Buskov, S., Sørensen, H. and Sørensen, J. C. (2000d). Oligomeric compounds formed under acidic processing of food containing hydroxybenzyl and indol-3-ylmethylglucosinolates Czech. *Journal of Food Science* **18**, 31–33.

Buskov, S., Hasselstrøm, J., Sørensen, H., Sørensen, J. C. and Sørensen, S. (2000e). Ascorbigens in Brassica and Sinapis based food. Czech. *Journal of Food Science* **18**, 28–30.

Buskov, S., Serra, B., Rosa, E., Sørensen, H. and Sørensen, J. C. (2002). Effects of intact glucosinolates and products produced from glucosinolates in myrosinase-catalysed hydrolysis on the potato cyst nematode (Globodera rostochiensis cv. Wolll). *Journal of Agricultural and Food Chemistry* **50**, 690–695.

Campbell, L. D. (1979). Incidence of liver haemorrhage among white leghorn strains fed on diets containing different types of rapeseed meals. *British Poultry Science* **20**, 239–246.

Chrisholm, M. D. and Wetter, L. R. (1967). The biosynthesis of some isothiocyanates and oxazolidinethiones in rape (*Brassica campestris* L.). *Plant Physiology* **42**, 1726–1730.

Christensen, B. W. and Kjær, A. (1963). The absolute configuration of Cleomin. *Acta Chemica Scandinavica* **17**, 279–280.

Christensen, B. W. and Kjær, A. (1965). The absolute configuration of methionine sulphoxide. *Chemical Communications* **11**, 225–226.

Ciska, E. and Pathak, D. R. (2004). Glucosinolate derivatives in stored fermented cabbage. *Journal of Agricultural and Food Chemistry* **52**, 1938–1943.

Cover, C. M., Hsieh, S. J., Tran, S. H., Hallden, G., Kim, G. S., Bjeldanes, L. F. and Firestone, G. L. (1998). Indole-3-carbinol inhibits the expression of cyclin-dependent kinase-6 and induces a G1 cell cycle arrest of human breast cancer cells independent of estrogen receptor signaling. *Journal of Biological Chemistry* **273**(7), 3838–3847.

Cox-Ganser, J. M., Jung, G. A., Pushkin, R. T. and Reid, R. L. (1994). Evaluation of Brassicas in grazing systems for sheeep: II. Blood composition and nutrient status. *Journal of Animal Science* **72**, 1832–1841.

Dahlgreen, R. (1975). Botanical Notes. **128**, p. 119.

Dalgaard, L., Nawaz, R. and Sørensen, H. (1977). 3-Methylthiopropylamine and (R)-3-methylsulphinylpropylamine in *Iberis amara*. *Phytochemistry* **16**, 931–932.

Danielsen, V., Eggum, B. O., Rasmussen, K. W. and Sørensen, H. (1987). Long-term studies of requirements to the quality of rapeseed meal from double low varieties used in sow diets. *In* "Proceedings of the Seventh International Rapeseed Congress," Vol. 7, pp. 1727–1734. Poznan Poland, V.

Danielsen, V., Eggum, B. O., Jensen, S. K. and Sørensen, H. (1994). Dehulled protein-rich rapeseed meal as protein source for early weaned piglets. *Animal Feed Science and Technology* **46**, 239–250.

Daxenbichler, M. E., Spencer, G. F., Carlson, D. G., Rose, G. B., Brinker, A. M. and Powell, R. G. (1991). Glucosinolate composition of seeds from 297 species of wild plants. *Phytochemistry* **30**, 2623–2638.

EC (1990). Determination of the oilseed glucosinolates content by HPLC. *Official Journal of European Community* L170, 03.07.27-34.

Eggum, B. O. and Sørensen, H. (1989). Chemistry and analysis of amino acids. *In* "Absorption and Utilization of Amino Acids" (M. Friedman, ed.), Vol. III, Chapter 17, pp. 265–295. CRC press, Boca Raton, Florida.

Eggum, B. O., Just, A. and Sørensen, H. (1985). Double low rapeseed meal in diets to growing - finishing pigs. *In* "Advances in the Production and Utilization of

Cruciferous Crops" (H. Sørensen, ed.), Vol. 11, pp. 167–176. Martinus Nijhoff/Dr. W. Junk Publ., Dordrecht/Boston/Lancaster.

Ettlinger, M. and Kjær, A. (1968). Sulfur compounds in plants. *In* "Recent Advances in Phytochemistry" (T. J. Mabry, R. E. Alston and V. C. Runeckles, eds.), Vol. 1, pp. 59–144. North Holland Publishing Co., Amsterdam.

Ettlinger, M. G. and Lundeen, A. J. (1956). The structure of sinigrin and sinalbin; An enzymatic rearrangement. *Journal of the American Chemical Society* **78**, 4172–4173.

Ettlinger, M. G. and Lundeen, A. J. (1957). First synthesis of a mustard oil glycoside: The enzymatic Lossen rearrangement. *Journal of the American Chemical Society* **79**, 1764–1765.

Fahey, J. W., Zhang, Y. S. and Talalay, P. (1997). Broccoli sprouts: An exceptionally rich source of inducers of enzymes that protect against chemical carcinogens. *Proceedings of the National Academy of Sciences of the United States of America* **94**(19), 10367–10372.

Fahey, J. W., Zalcman, A. T. and Talalay, P. (2001). The chemical diversity and distribution of gluocosinolates and isothiocyanates among plants. *Phytochemistry* **56**(1), 5–51.

Feldl, C., Møller, P., Otte, J. and Sørensen, H. (1994). Micellar electrokinetic capillary chromatography for determination of indolyl glucosinolates and tranformation products thereof. *Analytical Biochemistry* **217**, 62–69.

Fenwick, R. G. and Heaney, R. K. (1983). Glucosinolates and their breakdown products in cruciferous crops, foods and feedingstuffs. *Food Chemistry* **11**, 249–271.

Gil, V. and MacLeod, A. J. (1980). The effects of pH on glucosinolate degradation by a thioglucoside glucohydrolase preparation. *Phytochemistry* **19**, 2547–2551.

Grubb, C. D. and Abel, S. (2006). Glucosinolate metabolism and its control. *Trends in Plant Science* **11**(2), 89–100.

Halkier, B. A. (1999). Glucosinolates. *In* "Naturally Occurring Glycosides" (R. Ikem, ed.), pp. 193–223. Wiley, Chichester, UK.

Halkier, B. A. and Gershenzon, J. (2006). Biology and Biochemistry of glucosinolates. *Annual Review of Plant Biology* **57**(1), 303–333.

Hansen, M., Møller, P. and Sørensen, H. (1995). Glucosinolates in broccoli stored under controlled atmosphere. *Journal of American Horticultural Science* **120**(6), 1069–1074.

Hansen, M., Laustsen, A. M., Olsen, C. E., Poll, L. and Sørensen, H. (1997). Chemical and Sensory quality of broccoli (*Brassica oleracea* L. var. italica). *Journal of Food Quality* **20**, 441–459.

Hasapis, X. and MacLeod, A. J. (1982). Effects of metal ions on benzylglucosinolte degradation in *Lepidium sativum* seed autolysates. *Phytochemistry* **21**, 559–563.

Haughn, G. W., Davin, L., Giblin, M. and Underhill, E. W. (1991). Biochemical genetics of plant secondary metabolies in *Arabidopsis thaliana*. The glucosinolates. *Plant Physiology* **97**, 217–226.

Helboe, P., Olsen, O. and Sørensen, H. (1980). Separation of glucosinolates by high-performance liquid chromatography. *Journal of Chromatography* **197**, 199–205.

Henriksen, P., Hillemann, G., Mortensen, K. and Sørensen, H. (1987). Requirements to the quality of oilseed rape which can be used without problems in diets to mink (Mustela vison Screb.). *In* "Proceedings of the Seventh International Rapeseed Congress," Vol. 7, pp. 1817–1824. Poznan Poland, VII.

Hill, J., Lethenborg, P., Li, P. W., Rahman, M. H., Sørensen, H. and Sørensen, J. C. (2003). Inheritance of progoitrin and total aliphalic glucosinolates in oilseed rape (*Brassica napus* L.). *Euphytica* **134**, 179–187.

Hopkins, R. J., Ekbom, B. and Henkow, L. (1998). Glucosinolate content and susceptibility for insect attack of three populations of *sinapis alba*. *Journal of Chemical Ecology* **V24**(7), 1203–1216.

Jensen, S. K., Olsen, H. S. and Sørensen, H. (1990). Aqueous enzymatic processing of rapeseed for production of high quality products. *In* "Rapeseed/Canola: Production, Chemistry, Nutrition and Processing Technology" (F. Shahidi, ed.), Chapter 19, pp. 331–343. Van Nostrand Reinhold Publisher, New york.

Jensen, S. K., Michaelsen, S., Kachlicki, P. and Sørensen, H. (1991). 4-Hydroxyglucobrassicin and degradation products of glucosinolates in relation to unsolved problems with the quality of double low oilseed rape. *GCIRC Congress, Saskatoon, Canada* **V**, 1359–1364.

Karcher, A. and El Rassi, Z. (1999). Capillary electrophoresis of glucosinolates and their degradation products. *Electrophoresis* **20**, 3181–3189.

Kim, D. J., Han, B. S., Ahn, B., Hasegawa, R., Shirai, T., Ito, N. and Tsuda, H. (1997). Enhancement by indole-3-carbinol of liver and thyroid gland neoplastic development in a rat medium-term multiorgan carcinogenesis model. *Carcinogenesis* **18**(2), 377–381.

Kindl, H. and Underhill, E. W. (1968). Biosynthesis of mustard oil glucosides: N-hydroxyphenylalanine, a precursor of glucotropaeolin and substrate fort he enzymatic and nonenzymatic formation of phenylacetaldehyde oxime. *Phytochemistry* **7**, 745–756.

Kjær, A. (1960). Naturally derived isothiocyanates (mustard oils) and their parent glucosides. *Fortschritte der Chemie organischer Naturstoffe* **18**, 122–176.

Klyne, W., Day, J. and Kjær, A. (1960). isoThiocyanates XXXVII. Rotatory dispersion studies of optically active sulphoxides derivable from isothiocyanates of natural origin. *Acta Chemica Scandinavica* **14**, 215–216.

Kuang, Y. F. and Chen, Y. H. (2004). Induction of apoptosis in a non-small cell human lung cancer cell line by isothiocyanates is associated with P53 and P21. *Food and Chemical Toxicology* **42**(10), 1711–1718.

Kutacek, M., Prochazka, A. and Veres, K. (1962). Biogenesis of glucobrassicin, the in vitro precursor of ascorbigen. *Nature* **194**, 393–394. PL.

Larsen, L. M., Olsen, O. and Sørensen, H. (1983). Failure to detect glucosinolates in *Plantago* species. *Phytochemistry* **2**, 2314–2315.

Lincheid, M., Wendisch, D. and Strack, D. (1980). The structures of sinapic acid esters and their metabolism in cotelydons of *Raphanus sativus*. *Zeitschrift für Naturforschung* **35C**, 907–914.

Loft, S., Otte, J., Poulsen, H. E. and Sørensen, H. (1992). Influence of intact and myrosinase-treated indolyl glucosinolates on the metabolism *in vivo* of metronidazole and antipyrine in the rat. *Food and Chemical Toxicology* **30**(11), 927–935.

Marsh, R. E. and Waser, J. (1970). Refinement of the crystal structure of sinigrin. *Acta Crystallographica Section B* **20**, 1030–1037.

Martland, M. F., Butler, E. J. and Fenwick, G. R. (1984). Rapeseed induced liver haemorrhage reticulolysis and biochemical changes in laying hens: The effects of feeding high and low glucosinolate meals. *Research in Veterinary Science* **63**, 298–309.

Mason, R. W. and Lucas, P. (1983). Acute poisoning in cattle after eating old non-viable seed of chou moellier (*Brassica oleracea* convar. *Acephala*). *Australian Veterinary Journal* **60**, 272–273.

Matsuo, M. (1968). Biosynthesis of sinigrin. VII. Incorporation of 4-methylthiobutyraldoxime-1–14C, 15N into sinigrin. *Tetrahedron Letters* **38**, 4101–4104.

Matsuo, M. and Underhill, E. W. (1969). A UDP-glucose: Thiohydroximate glucosyltransferase from Tropaeolum majus L. *Biochemical and biophysical research communications* **36**, 18–23.

Matthäus, B. and Luftmann, H. (2000). Glucosinolates in members of the family Brassicaceae: Separation and identification by LC/ESI-MS-MS. *Journal of Agricultural and Food Chemistry* **48**, 2234–2239.

Michaelsen, S., Møller, P. and Sørensen, H. (1992). Factors influencing the separation and quantitation of intact glucosinolates and desulfoglucosinolates by micellar electrokinetic capillary chromatography. *Journal of Chromatography* **608**, 363–374.

Michaelsen, S., Otte, J., Simonsen, L.-O. and Sørensen, H. (1994). Absorption and degradation of individual intact glucosinolates in the digestive tract of rodents. *Acta Agriculturae Scandinivica Section A* **44**, 25–37.

Mithen, R., Clarke, J., Lister, C. and Dean, C. (1995). Genetics of aliphatic glucosinolates. III. Side chain structure of aliphatic glucosinolates in Arabidopsis thalisma. *Heredity* **74**, 210–215.

Mithen, R. F., Dekker, M., Verkerk, R., Rabot, S. and Johnson, I. T. (2000). The nutritional significance, biosynthesis and bioavailability of glucosinolates in human food. *Journal of the Science of Food and Agriculture* **80**, 967–984.

Moreno, D. A., Carvajal, M., Lopez-Berenguer, C. and Garcia-Viguera, C. (2006). Chemical and biological characterisation of nutraceutical compounds of broccoli. *Journal of Pharmaceutical and Biomedical Analysis* **41**(5), 1508–1522.

Morton, J. M. and Campbell, P. H. (1997). Disease signs reported in south-eastern Australian dairy cattle while grazing Brassica species. *Australian Veterinary Journal* **75**, 109–113.

Olsen, O. and Sørensen, H. (1981). Recent advances in the analysis of glucosinolates. *Journal of American Oil and Chemical Society* **58**, 857–865.

Palmieri, S., Rollin, P., Sørensen, H. and Sørensen, S. (1998). Enzyme technology for a potential glucosinolate utilization in Agro-Industry. *Agro Food Industry High Tech* **9**(1), 24–27.

Petersen, I. L., Sørensen, H., Sørensen, J. C. and Sørensen, S. (2003). Kinetic parameters of myrosinase isoenzymes isolated from Brassica napus L. and Sinapis alba L. seeds. *In* "Proceedings of 11th International Rapeseed Congress, 6–10 July." Copenhagen, Denmark, ISBN 87-7611-003-6, PP24, 716–719.

Prestera, T., Fahey, J. W., Holtzclaw, W. D., Abeygunawardana, C., Kachinski, J. L. and Talalay, P. (1996). Comprehensive chromatographic and spectroscopic methods for the separation and identification of intact glucosinolates. *Analytical Biochemistry* **239**, 168–179.

Rangkadilok, N., Nicolas, M. E., Bennett, R. N., Premier, R. R., Eagling, D. R. and Taylor, P. W. J. (2002). Determination of sinigrin and glucoraphanin in Brassica species using a simple extraction method combined with ion-pair HPLC analysis. *Scientia Horticulturae* **96**(1–4), 27–41.

Rodman, J. E., Kard, K. G., Price, R. A. and Sytsma, K. J. (1996). Molecules, morphology and Dahlgreens expanded order Capparales. *Systematic Botany* **21**, 289–307.

Rosa, E. A. S., Heaney, R. K., Fenwick, G. R. and Portaz, C. A. M. (1997). Glucosinolates in crop plants. *Horticultural Reviews* **19**, 99–215.

Rose, P., Armstrong, J. S., Chua, Y. L., Ong, C. N. and Whiteman, M. (2005). [beta]-Phenylethyl isothiocyanate mediated apoptosis; contribution of Bax and the mitochondrial death pathway. *The International Journal of Biochemistry and Cell Biology* **37**(1), 100–119.

Rücker, B. and Röbbelen, G. (1994). Inheritence of total and individual glucosinolate contents in seeds of winter oilseed rape (*Brassica napus* L.). *Plant Breeding* **113**, 206–216.

Sato, T., Nagata, K., Shiro, M. and Koyama, H. (1966). Structure of the methylthiohydroxamic acid complex of nickel. *Chemical Communications* **192**.

Schmid, A. and Schmid, H. (1992). Rapeseed poisoning of wild herbivores. *Tierarztliche Praxis* **20**, 321–325.

Schone, F., Leiterer, M., Jahreis, G. and Rudolph, B. (1997a). Effect of rapeseed feedstuff with different glucosinolate conaten and iodine administration on gestating and lactating sow. *Zentralblatt für Veterinärmedizin Reihe A* **44**, 325–339.

Schone, F., Rudolph, B., Kirchheim, U. and Knapp, G. (1997b). Counteracting the negative effects of rapeseed and rapeseed press cake in pig diets. *British Journal of Nutrition* **78**, 947–962.

Schone, F., Leiterer, M., Hartung, H., Jahreis, G. and Tischendorf, F. (2001). Rapeseed glucosinolates and iodine in sows affect the milk iodinie concentration and the iodine status of piglets. *British Journal of Nutrition* **85**, 659–670.

Schone, F., Tischendorf, F., Kirchheim, U., Reichardt, W. and Bargholz, J. (2002). Effects of high fat rapeseed press cake on growth, carcass, meat quality and body fat composition of leaner and fatter pig croβ breeds. *Animal Science* **74**, 285–297.

Sørensen, H. (ed.) (1985). "Advances in the Production and Utilization of Cruciferous Crops." Mortinus Nijhoff Publishers, Dordrecht/Boston/Lancaster. ISBN 90-247-3196-8.

Sørensen, H. (1990). Glucosinolates: Structure-properties-function. *In* "Rapeseed/Canola: Production, Chemistry, Nutrition and Processing Technology," Chapter 9, pp. 149–172. Van Nostrand Reinhold Publisher, New York.

Sørensen, H., Sørensen, S., Bjergegaard, C. and Michaelsen, S. (1999). "Chromatography and Capillary Electrophoresis in Food Analysis," p. 470. Royal Society of Chemistry, UK. ISBN 85404-561-9.

Sørensen, H., Sørensen, J. C. and Sørensen, S. (2001). Phytochemicals in food: The plant as chemical factories. *In* "Biologically-Active Phytochemicals in Food. Analysis, Metabolism, Bioavailability and Function," pp. 3–12. The Royal Society of Chemistry, UK. ISBN 0-85404-806-5.

Sørensen, J. C. (2001). "Biorefined Oilseed Products." Ph.D. Thesis, Chemistry Department, The Royal Veterinary and Agricultural University, Frederiksberg, Denmark.

Spiegel, C., Bestetti, G. E., Rossi, G. L. and Blum, J. W. (1993). Normal circulating triiodothyronine concentration are maintained despite severe hypothyroidism in growing pigs fed rapeseed presscake meal. *Journal of Nutrition* **123**, 1554–1561.

Staack, R., Kingston, S., Wallig, M. A. and Jeffery, E. H. (1998). A Comparison of the individual and collective effects of four glucosinolate breakdown products from brussels sprouts on induction of detoxification enzymes. *Toxicology and Applied Pharmacology* **149**(1), 17–23.

Taljaard, T. L. (1993). Cabbage poisoining in ruminants. *Journal of the South African Veterinary Association* **64**, 96–100.

Tapper, B. A. and Butler, G. W. (1967). Conversion of oximes to mustard oil glucosides (Glucosinolates). *Archives of Biochemistry and Biophysics* **20**, 719–721.

Tian, Q., Roselot, R. A. and Schwartz, S. J. (2005). Quantitative determination of intact glucosinolates in broccoli, broccoli sprouts, Brussels sprouts, and cauliflower by high-performance liquid chromatography-electrospray ionization-tandem mass spectrometry. *Analytical Biochemistry* **343**, 93–99.

Tookey, H. L. and Wolff, I. A. (1970). Effect of organic reducing agents and ferrous ion on thioglucosidase activity of *Crambe abyssinica* seed. *Canadian Journal of Biochemistry* **48**, 1024–1028.

Uda, Y., Kurata, T. and Arakawa, T. (1986a). Effects of pH and ferrous ion on the degradation of glucosinolates by myrosinase. *Agricultural and Biological Chemistry* **50**, 2735–2740.

Uda, Y., Kurata, T. and Arakawa, T. (1986b). Effects of thiol compounds on the formation of nitriles from glucosinolates. *Agricultural and Biological Chemistry* **50**, 2741–2746.

Umemura, T., Yamashiro, S., Bhatnagar, M. K., Moody, D. L. and Slinger, S. J. (1977). Liver fibrosis of the turkey on rapeseed products. *Research in Veterinary Science* **23**, 139–145.

Underhill, E. W. (1967). Biosynthesis of mustard oild blucosides: Conversion of phenylacetaldehyde oxime and 3-phenylpropionaldehyde oxime to glucotropaeolin and gluconasturtiin. *European Journal of Biochemistry* **2**, 61–63.

Underhill, E. W. and Chisholm, M. D. (1964). Biocynthesis of mustar oil glucosides. III. Formation of glucotropaeolin form L-phenylalanin-C^{14}-N^{15}. *Biochemical and Biophysical Research Communications* **14**, 425–430.

Underhill, E. W., Chisholm, M. D. and Wetter, L. R. (1962). Biosynthesis of mustard oil glucosides. I. Administration of C14-labelled compounds to horseradish, nasturtium and watercress. *Canadian Journal of Biochemistry and Physiology* **40**, 1505–1514.

Underhill, E. W., Wetter, L. R. and Chisholm, M. D. (1973). Biosynthesis of glucosinolates. *Biochemical Society Symposium* **38**, 303–326.

Vang, O., Rasmussen, B. F., Sørensen, H., Clausen, J. and Andersen, O. (1995). Effects of dietary broccoli on antioxidant enzymes. *Clinical Chemistry* **41**, 1910–1911.

Vang, O., Frandsen, H., Hansen, K. T., Sørensen, J. N., Sørensen, H. and Andersen, O. (2001). Biochemical effects of dietary intakes of different broccoli samples. I. Different modulation of cytochrome P-450 activities in rat liver, kidney and colon. *Metabolism. Clinical and Experimental* **50**(10), 1123–1129.

Verhoeven, D. T. H., Verhagen, H., Goldbohm, R. A., van den Brandt, P. A. and van Poppel, G. A. (1997). Rreview of mechanisms underlying anticarcinogenicity by brassica vegetables. *Chemico-Biological Interactions* **103**, 79–129.

Wallig, M. A., Kingston, S., Staack, R. and Jeffery, E. H. (1998). Induction of rat pancreatic glutathioneS-transferase and quinone reductase activities by a mixture of glucosinolate breakdown derivatives found in brussels sprouts. *Food and Chemical Toxicology* **36**(5), 365–373.

Wathelet, J. P. (ed.) (1987). "Glucosinolates in Rapeseed: Analytical Aspects." Mortinus Nijhoff Publishers, Dordrecht/Boston/Lancaster. ISBN 90-247-3525-4.

Wortelboer, H. M., Van der Linden, E. C. M., De Kruif, C. A., Noordhoek, J., Blaauboer, B. J., Van Bladeren, P. J. and Falke, H. E. (1992). Effects of indole-3-carbinol on biotransformation enzymes in the rat: *In vivo* changes in liver and small intestinal mucosa in comparison with primary hepatocyte cultures. *Food and Chemical Toxicology* **30**(7), 589–599.

Wu, L., Noyan Ashraf, M. H., Facci, M., Wang, R., Paterson, P. G., Ferrie, A. and Juurlink, B. H. J. (2004). Dietary approach to attenuate oxidative stress, hypertension, and inflammation in the cardiovascular system. *Proceedings of the National Academy of Sciences of the United States of America* **101**(18), 7094–7099.

Youngs, C. G. and Perlin, A. S. (1967). Fe (II)-catalyzed decomposition of sinigrin and related thioglycosides. *Canadian Journal of Chemistry* **45**, 1801–1804.

Mutagenesis: Generation and Evaluation of Induced Mutations

SANJAY J. JAMBHULKAR

Nuclear Agriculture and Biotechnology Division, Bhabha Atomic Research Centre, Mumbai, India

ABSTRACT

Mutation breeding is one of the approaches to enhance the spectrum of beneficial alleles of agronomic and economic significance in crop improvement program. It has been demonstrated in oleiferous *Brassica* species too. Physical and chemical mutagens were used to isolate mutations for reduced plant height and yield-contributing characters. Wide range of variability for flower morphology has been isolated. Yellow seed coat mutants and their use in crossbreeding program have resulted in the development of high oil content lines. Ethyl methane sulfonate (EMS) has played an important

Advances in Botanical Research, Vol. 45
Incorporating Advances in Plant Pathology

0065-2296/07 $35.00
DOI: 10.1016/S0065-2296(07)45014-5

role to tailor fatty acid composition of rapeseed-mustard. Mutations for reduced erucic acid and increased oleic acid have been achieved using EMS. Use of mutations for modified fatty acids in crossbreeding has improved oil quality and seed yield potential. The haploid system was found to be beneficial in *Brassica napus* for the induction of mutation for herbicide resistance, disease resistance, and oil and meal quality. Mutation breeding has been successfully employed to develop high-yielding varieties. So far 31 high-yielding varieties have been developed and released for cultivation all over the world. Compared to important cereal and oilseed crops, limited work on mutation breeding has been undertaken in rapeseed-mustard. Therefore, efforts need to be strengthened to enhance the wide range of variability for agronomical, morphological, biochemical, and biotic and abiotic stress-tolerant characters and to use them in breeding program to break the yield barriers in *Brassica* crop species. Analysis of mutant at molecular level would reinforce the understanding on mechanism of mutations in the era of genomics.

I. INTRODUCTION

Brassica species are characterized by extensive morphological diversity and the ability to adapt to a wide range of habitats and growing environments. Environmental fluctuations, sensitiveness of crop plants to various biological and nonbiological stress factors, and increasing health consciousness have imposed more responsibility on crop improvement program. Crop improvement is a key route to ensuring continued benefits arises from food and plant products. Wide range of genetic, biochemical, and metabolic variation needs to be generated to utilize beneficial allele for effective crop improvement. Mutation breeding is one of the approaches to enhance the spectrum of variability for characters of agronomic and economic significance, which is a prerequisite in crop improvement program. It has been successfully employed to enhance the production and productivity of crop plants (Bhatia *et al.*, 1999; Chopra, 2005; Maluszynski *et al.*, 1995). Mutagenesis has also been successfully employed in genetic improvement of qualitative and quantitative traits in oleiferous *Brassica* species (Bhatia *et al.*, 1999; Robbelen, 1990). Overview on the modifications in morphological, biochemical, and yield attributes through mutagenesis, their direct and indirect use to develop high yielding varieties, and molecular mechanism of selected mutations have been mentioned in this chapter.

A. MORPHOLOGICAL CHARACTERS

Morphological characters play an important role as phenotypic markers to identify and maintain the purity of genotype as well as in linkage mapping. *Brassica* germplasm consists of large spectrum of variation for morphological characters. However, desirable changes in a specific character without affecting the rest of the genome could be induced and utilized in basic studies and for the development of high-yielding varieties.

B. DWARFS

Use of dwarfing genes was a major factor in the success of green revolution (Khush, 2001). Dwarfing genes could also be useful in *Brassica* species to increase seed yield by reducing lodging and increasing harvest index. Dwarf mutants compared to their parents have been isolated (Chauhan and Kumar, 1986; Das and Rahman, 1988; Javed *et al.*, 2003; Rai and Singh, 1993; Shah *et al.*, 1999; Tyagi *et al.*, 1983; Zanewich *et al.*, 1991) in rapeseed-mustard using physical and chemical mutagens, whereas Devlin *et al.* (1992) isolated elongated internode (*ein*) mutation in *Brassica rapa*.

C. FLOWER MUTATIONS

Rapeseed-mustard in general possesses yellow-colored flowers. Rai and Jacob (1956) using X-rays isolated white flower mutation. A novel yellow–white flower color mutation in the male-sterile progenies derived from commercial *B. napus* hybrid CO22 was isolated by Yu *et al.* (2004). The exploitation of male sterility for the production of hybrids is one of the approaches to break the yield barrier in crop plants. Male-sterile mutants in *B. juncea* have been isolated using chemical mutagens like ethyl methane sulfonate (EMS), ENU, Ebr (Bhat *et al.*, 2001), and γ-rays and EMS (Chauhan and Singh, 1998). Transient male sterility can also be induced using nontraditional mutagenic agent called chemical-hybridizing agent (CHA). Various types of CHAs have been used as gametocide in rapeseed-mustard (Banga and Labana, 1983; Guan, 1995; Guan and Stringam, 1998; Liu *et al.*, 1999; Singh and Chauhan, 2001, 2004; Yu *et al.*, 2006; Zhang *et al.*, 1999).

Apetalous *Brassica* plant showed higher photosynthetic efficiency, less evaporation, higher root activity, and probable avoidance for diseases, and therefore development of apetalous *Brassica* plants could be beneficial. Apetalous trait is a kind of floral abnormality that was first reported in India. Mutants for apetalous flower were discovered in *B. rapa* (Buzza, 1983; Cours and Williams, 1977; Singh, 1961a,b) and *B. carinata* (Rana, 1985).

D. SILIQUA CHARACTERS

Siliqua of rapeseed-mustard stands $\sim$50° angle to the raceme branch. Reduction of angle would cut down the vigorous growth of aphids. In *B. juncea*, X-ray-induced appressed pod mutants were isolated from var. Rai5 (Rai, 1958) and RL 9 (Nayar and George, 1969). Lavania (1979) obtained four-siliqua mutant from T-25 using EMS. Bhat *et al.* (2001) treated the seeds of var. Pusa Jaikisan with various chemical mutagens like EMS, ENU, Ebr, and isolated tri- and tetralocular siliqua and nonshattering mutations.

Singh and Sareen (2004) reported bunching and appressed pod mutants in combination treatment of γ-rays and EMS to var. RH 30.

Among other morphological characters, improved root morphology may help to develop drought tolerant genotypes. Basak and Prasad (2004) isolated mutants for high root biomass on treatment of "Rose Red" variety of turnip (*B. rapa* L.) by γ-rays and EMS.

E. YELLOW SEED COAT

In general, seeds of *Brassica* species are brown/black in color. Yellow-seeded rapeseed-mustard are more desirable than brown/black seed because it has thinner seed coat, higher oil content, high protein, and lower fiber content than brown-seeded varieties within same genetic background (Shirzadegan and Robbelen, 1985; Stringam *et al.*, 1974; Woods, 1980; Xiao, 1982). It has improved nutritive value of the meal after oil extraction (Simbaya *et al.*, 1995; Slominski *et al.*, 1999). Yellow-seeded genotypes are available in *B. rapa*, *B. juncea*, and *B. carinata*. No natural or induced mutations have been reported in *B. napus*, and therefore interspecific hybridization was extensively used to develop yellow seed coat genotypes (Rahman, 2001).

Till late 1960s, all *B. juncea* genotypes available in the germplasm collection had brown or black seed coat. Induced mutations to isolate yellow seed coat were initiated at BARC, Mumbai, India. Using ^{35}S radioisotope, two yellow seed coat mutants (*YSM1* and *YSM2)* were isolated from blackish brown seed var. Rai5 but could not be published. New yellow-seeded mutant was isolated from the same var. Rai5 using ^{32}P radioisotope (Nayar, 1968) and named as Trombay Mustard 1 (TM1). Using this mutant in crossbreeding program, improved high-yielding genotypes were developed (Nayar, 1976, 1979). The yellow seed coat mutants and their derivatives were extensively used in crossbreeding program throughout the India and large number of high-yielding bold-seeded genotypes has been developed (Abraham and Bhatia, 1986). In recent years, we have also isolated yellow seed coat mutant from most popular var. "Varuna" using γ-rays (data unpublished). These mutants are being used in crossbreeding to develop high-yielding varieties (Jambhulkar *et al.*, 2005).

F. INSECT PEST

Resistance to insect pests is one of the major constraints in realizing the yield potential of rapeseed-mustard. Among various insect pests, aphid is important. Systematic breeding approaches are required (Singh and Rana, 2002). Mutation breeding stands better chances because limited or no variability in germplasm has been observed. However, information on this aspect is scanty. Waxy leaves were found to be susceptible to aphids. Using γ-rays, ethylene imine (EI) and hydrazine, a nonwaxy mutant was isolated in var. Pusa Sweti and PTW Globe of *B. rapa*, which had shown resistance to aphids (Srinivasachar and Malik, 1972).

II. YIELD-CONTRIBUTING CHARACTERS

Various yield components directly or indirectly contribute to seed and oil yields. Induced mutations have been successfully employed to isolate mutants for desirable economic characters such as plant height, number of siliquae per plant, number of seeds per siliqua, seed weight, seed yield, and oil content (Chauhan and Kumar, 1986; Javed *et al.*, 2003; Mahla *et al.*, 1990, 1991; Rehman, 1996; Rehman *et al.*, 1987; Robbelen, 1990; Shah *et al.*, 1990, 1998, 1999). The most important yield-contributing characters in rapeseed-mustard are numbers of primary and secondary branches, siliquae per plant, siliquae on main fruiting axis, number of seeds per siliqua, and seed weight. Mutants with more branches and siliquae per plant have been reported (Chauhan and Kumar, 1986; Javed *et al.*, 2003; Naz and Islam, 1979; Shah *et al.*, 1990) in rapeseed-mustard.

Reduction in time to flowering could well be at the expense of yield potential. However, mutation for early flowering in the agronomically superior lines could be useful in the various cropping patterns. Early flowering mutants were obtained in *B. napus* (Shah *et al.*, 1999; Thurling and Depittayanan, 1992) and *B. juncea* (Nayar and George, 1969).

III. OIL QUALITY AND FATTY ACIDS

Edible oil is an important component of the human diet and provides energy. In vegetable oils, the major part of the fatty acids is represented by oleic acid (C18:1), linoleic acid (C18:2), and α-linolenic acid (C18:3) along with other fatty acids like palmitic and stearic acids. Traditional rapeseed-mustard oil contains high proportion ($\sim$50%) of erucic acid (C22:1), and thus it is an exception to other vegetable oils. Mutation breeding has been successful in tailoring oil crops for desirable fatty acid composition (Robbelen, 1990) because oil crop plants tolerate wide range of fluctuation in fatty acid composition without loosing viability and single mutation can result into desirable oil composition. Liho the first zero erucic acid mutant was reported by Stefansson *et al.* (1961) in *B. napus*, which opened the era of mutant-assisted quality improvement in oilseed crops. Downey (1964) suggested that erucic acid-free oil should also occur in *B. campestris* as it is one of the parents in *B. napus* and reported zero erucic acid natural mutant/variability in *B. campestris*. Two zero erucic acid natural mutants/variability were found in Chinese accession of *B. juncea* (Kirk and Oram, 1981) and termed them as *zem1* and *zem2*. Olsson (1984) also reported natural mutation/variability for low erucic acid (LEA) in *B. juncea*. High level of erucic acid also has industrial applications; however, mutation for high erucic acid has not yet reported.

Linolenic acid is easily oxidized and oil cannot be stored for long period. However, no variety or species of Cruciferae was found free from linolenic acid. In view of this, Robbelen and Rakow (1970) in Germany started mutation breeding program and isolated reduced linolenic acid mutant in *B. napus* (Rakow, 1973; Robbelen and Nitsch, 1975). High oleic acid in oil is considered as nutritionally desirable for human health. Auld *et al.* (1992) and Rucker and Robbelen (1997) isolated high oleic acid mutant in *B. napus*. This initial success of mutant isolation has laid foundation stone for improvement of oil quality in *Brassica* crops. Mutation for various fatty acids has been listed in Table I.

IV. *IN VITRO* CULTURE AND MUTAGENESIS

The development of efficient *in vitro* culture method has facilitated the use of mutation technique for crop improvement (Maluszynski *et al.*, 1995). The haploid system can be successfully used for induction of genetic variation through mutagenesis. It has advantages like immediate fixation of mutated genotypes in the first generation, *in vitro* selection, and increased selection efficiency. Mutagenic treatment can be given to spikes, buds, anthers, microspores, and haploid calli; embryos or protoplasts. The dose of mutagen needs to be adjusted with sensitivity of material used. Such studies for chemical mutagen, like EMS (Barro *et al.*, 2001; Beversdorf and Kott, 1987), ENU (Swanson *et al.*, 1988, 1989), NaN_3 (Polsoni *et al.*, 1988), MNH (MNU) (Jedrzejaszek *et al.*, 1997), and physical mutagen, like γ-rays (Beversdorf and Kott, 1987; McDonald *et al.*, 1991; Swanson *et al.*, 1988), X-rays (McDonald *et al.*, 1991), and ultraviolet (UV) rays (Jedrzejaszek *et al.*, 1997; McDonald *et al.*, 1991), were undertaken in various *Brassica* species to isolate mutation. It was opined that microspore mutagenesis has been most successful in rapeseed to isolate desirable stable mutants.

Combining mutagenic treatment with *in vitro* selection at haploid level can improve recovery of desired mutants. Direct application of herbicide as

TABLE I

Mutations for Various Fatty Acids in Brassica *Species*

Fatty acids	*Brassica* species	Parent with fatty acid (%)	Mutagen	Fatty acid in mutant (%)	References
Palmitic acid	*B. napus*	Wotan (4.7)	EMS	9.0	Rucker and Robbelen, 1997
Oleic acid	*B. napus*	Cascade (64.8)	EMS	80.2	Auld *et al.*, 1992
		Wotan (60.3)	EMS	71.0–80.3	Rucker and Robbelen, 1997
	B. carinata	C-101 (9.4)	EMS	23.3 and 15.3	Velasco *et al.*, 1997
Linoleic acid	*B. napus*	Oro (21.5)	EMS	22.40–37.92	Robbelen and Nitsch, 1975
	B. rapa	R-500 (11.9)	EMS	2.1	Auld *et al.*, 1992
	B. carinata	C-101 (18.3)	EMS	8.3 and 9.1	Velasco *et al.*, 1997
Linolenic acid	*B. napus*	Cascade (17.6)	EMS	3–5	Auld *et al.*, 1992
		Wotan (9.9)	EMS	3.1 and 6.2	Rucker and Robbelen, 1997
		Oro	EMS	3.24–8.42	Robbelen and Nitsch, 1975
		M57 (5.6)	EMS	3.2	Rakow, 1973; Robbelen and Nitsch, 1975
	B. rapa	R-500 (8.6)	EMS	3.0	Robbelen and Nitsch, 1975
	B. juncea	Zem1 (15.0)	EMS/γ-rays	9.0	Auld *et al.*, 1992
	B. carinata	C-101 (12.9)	EMS	6.3, 6.6, and 9.1	Velasco *et al.*, 1997
Erucic acid	*B. carinata*	C-101 (44.0)	EMS	5–10	Velasco *et al.*, 1995
	B. carinata	N2-6230		54.9	Velasco *et al.*, 1998
	B. napus		Natural mutant/variability	0	Stefansson *et al.*, 1961
	B. campestris		Natural mutant/variability	0	Downey, 1964
	B. juncea		Natural mutant/variability	0	Kirk and Oram, 1981

Sources: Bhatia *et al.* (1999) and Robbelen (1990).

the selection agent to the culture medium enforces immediate expression of tolerance in the tissue. Selection for herbicide resistance is likely to be the simplest test for this system since a single point mutation at any one of the number of genes could effectively interfere with the uptake, assimilation, or translocation of the herbicide, resulting in a resistant/tolerant plant. *In vitro* mutagenesis and selection have resulted in the development of mutants for herbicide resistance (Ahmad *et al.*, 1991; Beversdorf and Kott, 1987; Kott, 1995, 1998; McDonald *et al.*, 1991; Palmer *et al.*, 1996; Polsoni *et al.*, 1988; Swanson *et al.*, 1988, 1989), disease resistance (Ahmad *et al.*, 1991; Bansal *et al.*, 1998; Liu *et al.*, 2005; McDonald and Ingram, 1986; Newsholme *et al.*, 1989; Sacriston, 1982), and long pod and short plant (Shi *et al.*, 1995) in *B. napus*.

Kott *et al.* (1996) and Kott (1998) have demonstrated that microspore mutagenesis can be successfully used to improve oil and meal quality of rapeseed. Double haploid mutations for increased level of oleic acid and reduction of linolenic acid have been found by Auld *et al.* (1992) and Wong and Swanson (1991). Similarly, Turner and Facciotti (1990) and Huang (1992) isolated mutants for high oleic acid and decreased level of saturated fatty acids from mutagenized microspore culture. Barro *et al.* (2001) reported high and low levels of erucic acid microspore-derived mutations. Beaith *et al.* (2005) treated microspore with UV light isolated mutants for reduced palmitic and stearic acids content in *B. napus*.

V. MOLECULAR BASIS OF MUTATIONS

Phenotypic expression is governed by biochemical processes that are controlled by genes. Mutations like deletion, duplication, and substitutions of base pairs in deoxyribo nucleic acid (DNA) impose alteration in the biochemical products that result into the mutant phenotype. Characterization of mutations at protein and DNA levels strengthens the basic understanding of mutations.

A. FATTY ACIDS

LEA is characterized by a near absence of very long-chain fatty acids (VLCFAs) in the seed oil, which has been correlated with lack of acyl-CoA elongation activity (Roscoe *et al.*, 2001). To validate the speculation that presence of serine at position 282 in all functional proteins of HEA is replaced by phenylalanine in LEA *B. napus* under the control of *FAE1* enzyme, Katavic *et al.* (2002) used site directed mutagenesis. The phenylalanine 282 residue was substituted with a serine residue in the *FAE1* polypeptide from

LEA *B. napus* cv. Westar. This mutated gene was expressed in yeast and GC analysis revealed the presence of very long-chain monounsaturated fatty acids (VLCMFAs), indicating that the elongase activity was restored in the LEA *FAE1* enzyme by the single amino acid serine. Thus, LEA trait in LEA *B. napus* can be attributed to phenylalanine, which prevents the biosynthesis of eicosenoic and erucic acids.

Synthesis of oleic acid is under the influence of *fad2* gene that codes an enzyme, endoplasmic delta-12 oleate desaturase, responsible for desaturation of oleic acid (C18:1) into linoleic acid (C18:2). Tanhuanpaa *et al.* (1998) reported that single nucleotide mutation in the *fad2* gene of *B. rapa* caused an increase in the C18:1 content. This mutation substituted leucine with proline. Hu *et al.* (2006) also identified single nucleotide mutation from C to T in the gene *fad2* of *B. napus*. This mutation generated stop codon (TAG) leading to premature termination of the peptide chain during translation. As a result, only 185 amino acids were incorporated into the polypeptide instead of all 384 amino acids representing the full-length polypeptide. Thus, truncated polypeptide may not function as an active desaturase for the desaturation of C18:1 to C18:2 and, therefore will result in the accumulation of C18:1 in the seeds of the mutant line.

B. DWARFS

Mutations in the biosynthetic pathways of gibberellin (GA) can cause dwarfing phenotype in the plants (Hedden, 2003; Peng *et al.*, 1999; Sun and Gubler, 2004). GA promotes stem growth by causing degradation of DELLA protein in the ubiquitine proteosome pathway. The most widely utilized semidwarf wheat cultivars have *Rht 131b* or *Rht D1b* allele that encodes a mutant form of a DELLA protein that is GA-signaling repressor (Hedden, 2003; Peng *et al.*, 1999). Five DELLA protein genes (*GA1*, *RGA*, *RGL1*, *RGL2*, and *RGL3*) are responsible for GA-regulated plant growth and development in *Arabidopsis*. Muangprom and Osborn (2004) found that *dwf2* mutation in *B. rapa* was insensitive to exogenous GA_3 for both plant height and flowering time and inferred that it is not a mutation in the GA biosynthetic pathway. Muangprom *et al.* (2005) reported that mutation of *B. rapa* (Brrga1-d) is caused by substitution of conserved amino acid in the C-terminal domain of a DELLA protein. Brrga1-d retains its repressor function in the presence of GA and does not interact with a protein component required for degradation, suggesting that the mutated amino acid causes dwarfism by preventing an interaction needed for degradation. The elongated internode (*ein*) mutation of *B. rapa* leads to a deficiency in immunochemically detectable phytochrome B (Devlin *et al.*, 1997). Molecular analysis of

TABLE II

Released Varieties in Rapeseed-Mustard Using Mutation Breeding

Variety	Country and year of release	Mutagen/mutant in cross	Improved characters
B. juncea (Indian mustard)			
Shambal (BAU-/248)	Bangladesh, 1984	EMS 0.64%	Early, seed yield
Agrani	Bangladesh, 1991	γ-Rays 700 Gy	Early, blight resistance
Safal	Bangladesh, 1991	γRays 700 Gy	Seed yield, oil content, blight resistance
RLM 198	India, 1975	γ-Rays	Seed yield, oil content, moderately tolerant to aphid and leaf minors
RLM 514	India, 1980	γ-Rays	Early, bold seed, seed yield, oil content, moderately resistant to aphid and shattering
RLM 619	India, 1983	γ-Rays	Seed yield, oil content, bold seed, moderately tolerant to aphid, downey mildew and white rust
RLM 185	India, 1983	γ-Rays	Seed yield, oil content, tolerant to drought
RL 1359	India, 1987	γ-Rays-induced mutant RLM 514 used in cross (RLM 514 $\times$ Varuna)	Seed yield, bold seed, comparatively dwarf, tolerant to aphids
TM 2	India, 1978	X-rays 750 Gy	Seed yield, appressed pod
TM 4	India, 1987	^{32}P-induced yellow seed coat mutantTM1 used in cross (Varuna $\times$ TM1)	Seed yield, oil content, yellow seed, less susceptible to powdery mildew
NIFA-Mustard canola	Pakistan, 2003	γ-Rays, 14 Gy	Seed yield
TPM1	India, 2006	^{32}P	Seed yield, early, oil content, tolerant to powdery mildew, reduced erucic acid

Brassica sp. (white mustard)			
Svalof's Primex	Sweden, 1950	X-rays 350 Gy	Seed yield, oil content
Seco	Sweden, 1961	X-ray-induced mutant Primex used in cross (Primex × Rumanian white mustard)	Seed yield, early, stalk stiffness, resistant to shattering
Trico	Sweden, 1967	Selection from Primex	Seed yield, oil content
B. napus			
Binasharisha-3	Bangladesh, 1997	γ-Rays 800 Gy	Seed and oil yields, early
Binasharisha-4	Bangladesh, 1997	γ-Rays 700 Gy	Seed and oil yield, early
Stellar	Canada, 1987	EMS-induced mutant Mil used in cross (Mil × Regent) × Regent	Seed yield, low linolenic and erucic acids, low glucosinolates
Apollo	Canada, 1993	–	Seed yield, low linolenic acid
Ganyou No.5	China, 1984	γ-Rays 1400 Gy	Seed yield, short plant height, early, long siliqua, resistant to white rust and *Scerotinia*
Huyou No.4	China, 1970	γ-Rays, 600 Gy	Seed yield, cold tolerant
Xinyou No.1	China, 1979	γ-Rays 700 Gy	Seed yield, tolerant to cold, drought, saline and alkaline soils
Xiuyou No.1	China, 1979	γ-Rays, 80Gy, F_1 seeds	Seed yield, early, cold resistant
Hua-Yellow No.1	China	γ-Rays	Seed yield, seed viability
Abasin 95	Pakistan, 1996	γ-Rays 1400 Gy	Seed yield, early
Regina varraps elite A	Sweden, 1953	X-rays, 350 Gy	Seed yield, oil content
Regina varraps elite F	Sweden, 1962	X-rays, 450 Gy	Seed yield, oil content
Tismenitskii	USSR, 1989	DMS 0.006%	Seed yield, oil content
Ivana	USSR, 1990	MNH 0.0025%	Seed yield, oil content, insect resistance
B. campestris			
Haya-natane	Japan, 1961	Colchicine	Seed yield
Hanakkori	Japan, 1999	Colchicine	Seed yield, plant architecture

Source: http://www-mvd.iaea.org/MVD; Kumar *et al.* (2000).

the PHYB gene from *ein* indicates deletion in the flanking DNA 5′ of the ATG start codon, which could interfere either with PHYB transcription or processing of the PHYB transcript. Restriction fragment length polymorphism and inverse PCR fragments generated from the PHYB gene of wild type and *ein* seedlings demonstrated the deletion to be 500 bp in length.

VI. CONCLUSION

Utilization of induced mutations in crop improvement program has resulted in the development of more than 2541 high-yielding varieties in different crop plants. Among them, 163 varieties belong to oilseed crops. The maximum number of varieties has been released in soybean (58), followed by groundnut (44), sesame (16), linseed (15), castor (4), and sunflower (1). Mutation breeding in rapeseed-mustard has also resulted in the development of total 31 high-yielding varieties comprising 12 in *B. juncea*, 14 in *B. napus*, 2 in *B. rapa*, and 3 in white mustard. Fifteen varieties have been developed using γ-rays and four by X-rays. Remaining varieties were developed using chemical mutagen and mutants used in hybridization. In *B. juncea*, Labana (1976) obtained a γ-ray-induced mutant RLM 198 with 25% higher seed yield, moderately resistant to aphids and leaf miners, higher oil content, and 5–6 days earlier maturity. Labana (1981) evolved another mutant RLM 214 with 22% more seed yield with high oil content and shattering resistant from parent RL 18 using X-rays. Mutations for increased seed yield resulting into the development of high-yielding varieties have also been reported by Rehman (1996) and Shah *et al.* (1999). Backcrossing of M-11, a low linolenic acid mutant with "Regent," resulted into the development of high-yielding, low linolenic acid var. "Stellar" (Scrath *et al.*, 1988) and Apollow (Scrath *et al.*, 1995). The details of the released varieties are presented in Table II.

REFERENCES

Abraham, V. and Bhatia, C. R. (1986). Development of strains with yellow seed coat in Indian mustard (*Brassica juncea* Czern. & Coss.). *Plant Breeding* **97**, 86–88.

Ahmad, I., Day, J. P., McDonald, M. V. and Ingram, D. S. (1991). Haploid culture and UV mutagenesis in rapid-cycling *Brassica napus* for the generation of resistance to chlorsulfuron and *Alternaria brassicicola*. *Annals of Botany* **67**, 521–525.

Auld, D. L., Heikkinen, M. K., Erickson, D. A., Sernyk, J. L. and Romero, J. E. (1992). Rapeseed mutants with reduced level of polyunsaturated fatty acids and increased levels of oleic acid. *Crop Science* **32**, 657–662.

Banga, S. S. and Labana, K. S. (1983). Production of F_1 hybrid using etherel induced male sterility in indian mustard (*B. Juncea*). *Journal of Agricultural Science Cambridge* **101**, 453–455.

Bansal, V. K., Thiagarajah, M. R., Stringam, G. R. and Hardin, R. T. (1998). Haploid plantlet screening in the development of blackleg resistant DH lines of *Brassica napus*. *Plant Breeding* **117**, 103–106.

Barro, F., Fernandez-Escobar, J., De La Vega, M. and Martin, A. (2001). Double haploid lines of *Brassica carinata* with modified erucic acid content through mutagenesis by EMS treatment of isolated microspores. *Plant Breeding* **120**, 262–264.

Basak, S. and Prasad, C. (2004). Screening for root mutant in turnip (*Brassica rapa* L.). *Cruciferae Newsletter* **25**, 9–10.

Beaith, M. E., Fletcher, R. S. and Kott, L. S. (2005). Reduction of saturated fats by mutagenesis and heat selection in *Brassica napus* L. *Euphytica* **144**, 1–9.

Beversdorf, W. D. and Kott, L. S. (1987). An *in vitro* mutagenesis/selection system for *Brassica napus*. *Iowa state Journal of Research* **64**, 435–443.

Bhat, S. R., Haque, A. and Chopra, V. L. (2001). Induction of mutations for cytoplasmic male sterility and some rare phenotypes in Indian mustard (*Brassica juncea* L.). *Indian Journal of Genetics* **61**, 335–340.

Bhatia, C. R., Nichterlein, K. and Maluszynski, M. (1999). Oilseed cultivars development from induced mutations and mutation altering fatty acid composition. *Mutation Breeding Review* **11**, 1–36.

Buzza, G. C. (1983). The inheritance of an apetalous character in canola (*B. napus* L.). *Cruciferae Newsletter* **8**, 11–12.

Chauhan, S. V. S. and Singh, N. K. (1998). Gamma ray and EMS induced male sterility mutants in *Brassica juncea*. *Journal of Cytology and Genetics* **33**, 29–34.

Chauhan, Y. S. and Kumar, K. (1986). Gamma ray induced chocolate B seeded mutant in *Brassica campestris Current Science* **55**, 410.

Chopra, V. L. (2005). Mutagenesis: Investigating the process and processing the outcome for crop improvement. *Current Science* **89**, 353–359.

Cours, B. J. and Williams, P. H. (1977). Genetic studies in *Brassica campestris* L. *Plant Breeding Abstract* **51**, 1533.

Das, M. L. and Rahman, A. (1988). Induced mutagenesis for the development of high yielding varieties in mustard. *Journal of Nuclear Agriculture and Biology* **17**, 1–4.

Devlin, P. F., Rood, S. B., Somers, D. E., Quail, P. H. and Whitelam, G. C. (1992). Photophysiology of the elongated internode (*ein*) mutant of *Brassica rapa: Ein* mutant lacks a detectable phytochrome B-like polypeptide. *Plant Physiology* **100**, 1442–1447.

Devlin, P. F., Somers, D. E., Quail, P. H. and Whitelam, G. C. (1997). The *Brassica rapa* elongated internode (*EIN*) gene encodes phytochrome B. *Plant Molecular Biology* **34**, 537–547.

Downey, R. K. (1964). Selection of *Brassica campestris* L. containing no erucic acid in its seed oil. *Canadian Journal of Plant Science* **44**, 295.

Guan, C. Y. (1995). Chemical induced male sterility and other ways of heterosis breeding. *In* "Breeding and Utilization of Hybrid Rapeseed" (T. D. Fu, ed.), pp. 124–135. Hubei Science and Technology Press, Wuhan, China.

Guan, C. Y. and Stringam, G. R. (1998). The effect of ZMA on inducing male sterility on spring canola. *Cruciferae Newsletter* **20**, 55–56.

Hedden, P. (2003). The genes of the green revolution. *Trends Genetics* **19**, 5–9.

Hu, X., Sullivan-Gilbert, M., Gupta, M. and Thompson, S. A. (2006). Mapping of loci controlling oleic and linolenic acid contents and development of *fad2* and *fad3* allele-specific marker in canola. *Theoretical and Applied Genetics* **113**, 497–501.

Huang, B. (1992). Genetic manipulation of microspores and microspore derived embryos. *In Vitro Cellular and Developmental Biology* **28**, 53–58.

Jambhulkar, S. J., Shitre, A. S. and Pathak, B. (2005). Heterosis in Indian mustard using induced mutations. *In* "Conference on Resource Development and Marketing Issues in Rapeseed-Mustard, March 28–29," pp. 27–28. Jaipur, India.

Javed, M. A., Siddiqui, M. A., Khan, M. K. R., Khatri, A., Khan, I. A., Dahar, N. A., Khanzada, M. H. and Khan, R. (2003). Development of high yielding mutants of *Brassica campestris* L. cv. Toria selection through gamma rays irradiation. *Asian Journal of Plant Sciences* **2**, 192–195.

Jedrzejaszek, K., Kruczkowska, H., Pawlowska, H. and Skucinska, B. (1997). Stimulating effect of mutagens on *in vitro* plant regeneration. *Mutation Breeding Newsletter* **43**, 10–11.

Katavic, V., Mietkiewska, E., Borton, D. L., Giblin, E. M., Reed, D. W. and Taylor, D. C. (2002). Restoring enzyme activity in nonfunctional low erucic acid *brassica napus* fatty acid elongase 1 by single amino acid substitution. *European Journal of Biochemistry* **269**, 1925–1931.

Kirk, J. T. O. and Oram, R. N. (1981). Isolation of erucic acid free lines of *Brassica juncea*: Indian mustard now a potential oilseed crop in Australia. *Journal of the Australian Institute of Agricultural Science* **47**, 51–52.

Kott, L. (1995). Production of mutants using the rapeseed double haploid system. *In* "Induced Mutations and Molecular Techniques for Crop Improvement," pp. 505–515. IAEA, Vienna.

Kott, L. (1998). Application of double haploid technology in breeding of oilseed *Brassica napus*. *AgBiotech News and Information* **10**, 69N–74N.

Kott, L., Wong, R., Swanson, E. and Chen, J. (1996). Mutation and selection for improved oil and meal quality in *Brassica napus* utilizing microspore culture. *In* "*In-Vitro* Haploid Production in Higher Plants" (S. M. Jain, S. K. Sopory and R. E. Veilleux, eds.), Vol. 2, pp. 151–167. Springer, Berlin Heidelberg, New York.

Kumar, P. R., Chauhan, J. S., Singh, A. K. and Yada, S. K. (2000). "Rapeseed-Mustard Varieties of India." pp. 1–112. NRCRM, Sewar, Bharatpur, India.

Khush, G. S. (2001). Green revolution: The way forward. *Nature Reviews Genetics* **2**, 815–822.

Labana, K. S. (1976). Release of mutant variety of raya (*Brassica juncea*). *Mutation Breeding Newsletter* **7**, 11.

Labana, K. S. (1981). *Mutation Breeding Newsletter* **17**, 13.

Lavania, U. C. (1979). Interphase sensitivity and specificity of siliqua mutations induced by EMS in Indian mustard. *Current Science* **48**, 101.

Liu, S., Wang, H., Zhang, J., Fitt, B. D. L., Xu, Z., Evans, N., Liu, Y., Yang, W. and Guo, X. (2005). *In vitro* mutation and selection of double haploid *Brassica napus* lines with improved resistance to sclerotinia aclerotiorum. *Plant Cell Reports* **24**, 133–144.

Liu, X. X., Dong, Z. S., Liu, C. S., Dong, J. G. and Qiao, Y. Q. (1999). A preliminary study on the effect and its application of a new chemical inducing male sterility agent EN to *B. napus* L. *Acta Agric. Boreal Occident. Sinica.* **8**, 60–62.

McDonald, M. V., Ahmad, I., Menten, J. O. M. and Ingram, D. S. (1991). Haploid culture and *in-vitro* mutagenesis (UV light, X-rays and gamma rays) of rapid cycling *Brassica napus* for improved resistance to diseases. *In* "Plant Mutations Breeding for Crop Improvement," Vol. 2, pp. 129–138. IAEA, Vienna.

McDonald, M. V. and Ingram, D. S. (1986). Towards the selection *in vitro* for resistance to *Alternaria brassicicola* (Schw) Welts., in *Brassica napus* ssp. Oleifera (Metzg.) Sinsk., winter oilseed rape. *New Phytologist* **104**, 621–629.

Mahla, S. V. S., Mor, B. R. and Yadava, J. S. (1990). Effect of mutagen on yield and its componenet characters in mustard. *Haryana Agriculture University Journal Research* **20**, 259–264.

Mahla, S. V. S., Mor, B. R. and Yadava, J. S. (1991). Mutagen induced polygenic variability in some mustard (*Brassica juncea* L.) varieties and their hybrids. *Journal of Oilseed Research* **8**, 173–177.

Maluszynski, M., Ahloowalia, B. S. and Sigurbjornsson, B. (1995). Application of *in vivo* and *in vitro* mutation techniques for crop improvement. *Euphytica* **85**, 303–315.

Muangprom, A. and Osborn, T. C. (2004). Characterisation of dwarf gene in *Brassica rapa*, including the identification of a candidate gene. *Theoretical and Applied Genetics* **108**, 1378–1384.

Muangprom, A., Thomas, S. G., Sun, T. P. and Osborn, T. C. (2005). A novel dwarfing mutation in a green revolution gene from *Brassica rapa*. *Plant Physiology* **137**, 931–938.

Nayar, G. G. (1968). Seed colour mutation in *Brassica juncea* Hook. F. and Thomas induced by radioactive phosphorus 32P. *Science and Culture* **34**, 421–422.

Nayar, G. G. (1976). Studies on radiation induced mutants and their recombinant types in mustard. *Oilseed Journal* January–June, 46–47.

Nayar, G. G. (1979). Breeding strategy for the improvement of mustard (*Brassica juncea*) coss. *In* "Proc. Symp. The Role of Induced Mutation in Crop Improvement," Sept. 10–13, pp. 258–270. Osmania University, Hyderabad, India.

Nayar, G. G. and George, K. P. (1969). X-ray induced early flowering, appressed pod mutant in *Brassica juncea* Coss. *In* "Proc. Symp. Radiation and Radiomimetic Substances in Mutation Breeding," Sept. 26–29, pp. 409–413. Bombay, India.

Naz, R. M. A. and Islam, R. Z. (1979). Effect of irradiation of *Brassica trilocularis* (yellow sarson). *Pakistan Journal of Agricultural Research* **17**, 87–93.

Newsholme, D. M., MacDonald, M. V. and Ingram, D. S. (1989). Studies of selection *in-vitro* for novel resistance to phytotoxic products of *Leptospheria maculans* (Desm.) Ces. & De Not in secondary embryogenic lines of *Brassica napus* ssp. Oleifera (Metzg.) Sinsk., winter oilseed rape. *New Phytologist* **113**, 117–126.

Olsson, G. (1984). Selection for low erucic acid in *Brassica juncea*. *Sveriges Utsadesf. Tidskr.* **94**, 187.

Palmer, C. E., Keller, W. A. and Arnison, P. G. (1996). Utilization of *Brassica* haploids. *In* "*In Vitro* Haploid Production in Higher Plants" (S. M. Jain, S. K. Sopory and R. E. Veilleux, eds.), pp. 173–192. Kluwer Academic Publisher, Dordrecht.

Peng, J., Richards, D. E., Hartley, N. M., Murphy, G. P., Devos, K. M., Flintham, J., Beales, J., Fish, J., Worland, A. J. and Pelica, F. (1999). 'Green revolution' genes encode mutant gibberellin response modulators. *Nature* **400**, 256–261.

Polsoni, L., Kott, L. S. and Beversdorf, W. D. (1988). Large scale microspore culture techniques for mutation/selection studies in *Brassica napus*. *Canadian Journal of Botany* **66**, 1681–1685.

Rahman, M. H. (2001). Production of yellow-seeded *Brassica napus* through interspecific crosses. *Plant Breeding* **120**, 463–472.

Rai, B. and Singh, D. (1993). A note on potential sources of dwarfing genes in Indian rapeseed (*Brassica campestris* L. Prain). *Indian Journal of Genetics* **53**, 153–156.

Rai, U. K. (1958). X-ray induced appressed pod mutant in *Brassica juncea*. *Science and Culture* **24**, 46–47.

Rai, U. K. and Jacob, K. T. (1956). Induced mutation studies in sesamum and mustard. *Science and Culture* **22**, 344–346.

Rakow, G. (1973). Selektion auf Linol- und Linolensauregehalt in rapssamen nach mutagener behandlung. *Z Pflanzenzuchtg* **69**, 205–209.

Rana, M. A. (1985). Developmental morphology of flower and inheritance of an apetalous mutant in *Brassica carinata*. Dissertation, University of California Davis. *Abstract International B (Science and Engineering)* **45**, 3425B.

Rehman, A. (1996). New mutant cultivar. *Mutation Breeding Newsletter* **42**, 27.

Rehman, A., Das, M. L., Hawlidar, M. A. R. and Mansure, M. A. (1987). Promising mutants in *Brassica campestris*. *MBNL* **29**, 14–15.

Robbelen, G. (1990). Mutation breeding for quality improvement a case study for oilseed crops. *Mutation Breeding Review* **6**, 1–43.

Robbelen, G. and Nitsch, A. (1975). Genetical and physiological investigations on mutants for polyeonic fatty acid in rapeseed, *Brassica napus* L. I. Selection and description of new mutants. *Z Pflanzenzuchtg* **75**, 93–105.

Robbelen, G. and Rakow, G. (1970). Selection for fatty acids in rapeseed. *In* "Proceedings of International Conference Rapeseed. Set. Adele." pp. 476–490.

Roscoe, T. J., Lessire, R., Puyaubert, J., Renard, M. and Delseny, M. (2001). Mutation in the fatty acid elongation 1 gene are associated with a loss of beta-ketoacyl-CoA synthase activity in low erucic acid rapeseed. *FEBS Letters* **492**, 107–111.

Rucker, B. and Robbelen, G. (1997). Mutants of *Brassica napus* with altered seed lipid fatty acid composition. *In* "Proceedings of 12th International Symposium Plant Lipids," pp. 316–318. Kluwer Academic Publisher, Dordrecht.

Sacriston, M. D. (1982). Resistance response to *Phoma lingam* of plants regenerated from selected cell and embryogenic cultures of haploid *Brassica napus*. *Theoretical and Applied Genetics* **61**, 193–200.

Scrath, R., McVetty, R. B. E., Rimmer, S. R. and Stefansson, B. R. (1988). Stellar low linolenic-high linoleic acid summer rape. *Canadian Journal of Plant Science* **68**, 509–511.

Scrath, R., Rimmer, S. R. and McVetty, P. B. E. (1995). Apollo low linolenic summer rape. *Canadian Journal of Plant Science* **75**, 203–204.

Shah, S. A., Ali, I. and Rahman, K. (1990). Induction and selection of superior genetic variables of oilseed rape, *Brassica napus* L. *The Nucleus* **7**, 37–40.

Shah, S. A., Ali, I. and Rahman, K. (1998). Use of induced genetic variability for the improvement of oilseed rape (*Brassica napus* L.). *In* "Proceedings of 2nd International Symposium New Genetical Approaches to Crop improvement-II Nuclear Institute of Agriculture," pp. 229–237. Tandojam, Pakistan.

Shah, S. A., Ali, I., Iqbal, M. M., Khattak, S. U. and Rahman, K. (1999). Evolution of high yielding and early flowering variety of rapeseed (*Brassica napus* L.) through *in-vivo* mutagenesis. *In* "Proceedings of 3rd International Symposium New Genetical Approaches to Crop Improvement-III Nuclear Institute of Agriculture," pp. 47–53. Tandojam, Pakistan.

Shi, S., Wu, J. and Liu, H. (1995). *In-vitro* mutation for long pod and short plant. *Nucleic Agriculture* **9**, 252–253.

Shirzadegan, M. and Robbelen, G. (1985). Influence of seed colour and hull proportions of seeds in *Brassica napus* L. *Fette Seifen Anstrichmittel* **87**, 235–237.

Simbaya, J., Slominski, B. A., Rakow, G., Campbell, L. D., Downey, R. K. and Bell, J. M. (1995). Quality characteristics of yellow seeded *Brassica* seed meals: Protein, carbohydrates and dietary fibre components. *Journal of Agricultural and Food Chemistry* **43**, 2062–2066.

Singh, D. (1961a). Heredity changes in number of petals in brown sarson. *Indian Oilseed Journal* **5**, 190–193.

Singh, D. (1961b). An apetalous mutant in toria brown sarson (*B. campestris* var. brown sarson) and its inheritance. *Current Science* **30**, 62–63.

Singh, D. and Rana, J. S. (2002). Insect-pest resistance breeding in oilseeds *Brassica*. *Brassica* **4**, 52–57.

Singh, K. P. and Sareen, P. K. (2004). Induced podding mutants of Indian mustard (*Brassica juncea* L. Czern & Coss.). *Cruciferae Newsletter* **25**, 17.

Singh, V. and Chauhan, S. V. S. (2001). Benzotriazole- A new chemical hybridizing agent for *Brassica juncea* L. *Journal of Cytology and Genetics* **2**, 81–83.

Singh, V. and Chauhan, S. V. S. (2004). Bud pollination and hybrid seed production in detergent-induced male sterile plants of *Brassica juncea*. *Plant Breeding* **122**, 421–425.

Slominski, B. A., Simbaya, J., Campbell, L. D., Rakow, G. and Guenter, W. (1999). Nutritive value for broilers of meals derived from newly developed varieties of yellow- seeded canola. *Animal Feed Science and Technology* **78**, 249–262.

Srinivasachar, D. and Malik, R. S. (1972). An induced aphid reistant non-waxy mutant in turnip, *Brassica rapa*. *Current Science* **41**, 820–821.

Stefansson, B. R., Hougen, F. W. and Downey, R. K. (1961). Note on the isolation of rape plants with seed oil free from erucic acid. *Canadian Journal of Plant Science* **41**, 218–219.

Stringam, G. R., McGregor, D. I. and Pawlowski, S. H. (1974). Chemical and morphological characteristics associated with seed coat colour in rapeseed. *In* "Proceedings of 4th International Rapeseed Congress," pp. 99–108. Giessen, West Germany.

Sun, T. P. and Gubler, F. (2004). Molecular mechanism of gibberellin signalling in plants. *Annual Review of Plant Biology* **55**, 197–223.

Swanson, E. S., Coumans, M. P., Brown, G. L., Patel, J. D. and Beversdorf, W. D. (1988). The characterization of herbicide tolerant plants in *Brassica napus* L. after *in-vitro* selection of microspores and protoplasts. *Plant Cell Reports* **7**, 83–87.

Swanson, E. S., Herrgesell, M. J., Arnoldo, M., Sippell, D. and Wong, R. S. C. (1989). Microspore mutagenesis and selection: Canola plants with field tolerance to the imidazolinones. *Theoretical and Applied Genetics* **78**, 525–530.

Tanhuanpaa, P., Vikki, J. and Vihinen, M. (1998). Mapping and cloning of *FAD2* gene to develop allele-specific PCR for oleic acid in spring turnip rape (*Brassica rapa* ssp.oleifera). *Molecular Breeding* **4**, 543–550.

Thurling, N. and Depittayanan, V. (1992). EMS induction of early flowering mutants in spring rape (*Brassica napus*). *Plant Breeding* **108**, 177–184.

Turner, J. and Facciotti, D. (1990). High oleic acid *Brassica napus* from mutagenized microspores. *In* "Proceedings of the 6th Crucifer Genetics Workshop" (J. R. McFerson, S. Kresovich and S. G. Dwyer, eds.), pp. 24. Geneva, New York.

Tyagi, D. V. S., Rai, B. and Verma, R. B. (1983). A note on the bunching dwarf mutant in toria. *Indian Journal of Genetics* **43**, 374–377.

Velasco, L., Fernandez-Martinez, J. M. and De Haro, A. (1995). Isolation of induced mutants in ehtiopean mustard (*Brassica carinata* Braun). *Plant Breeding* **116**, 396–397.

Velasco, L., Fernandez-Martinez, J. M. and De Haro, A. (1997). Selection for reduced linolenic acid content in ethiopian mustard (*Brassica carinata* Braun) with low levels of erucic acid. *Plant Breeding* **114**, 454–456.

Velasco, L., Fernandez-Martinez, J. M. and De Haro, A. (1998). Increasing erucic acid content in ethiopian mustard through mutation breeding. *Plant Breeding* **117**, 85–87.

Wong, R. S. C. and Swanson, E. (1991). Genetic modification of canola oil: High oleic acid anola. *In* "Fat and Chlesterol Reduced Food" (C. Haberstroh and C. E. Morris, eds.), pp. 154–164. Gulf, Houston, Texas.

Woods, D. L. (1980). The association of yellow seed coat with other characters in mustard *Brassica juncea*. *Cruciferae Newsletter* **5**, 23–24.

Xiao, D. (1982). Analysis of the correlation between seed coat colour and oil contents of *Brassica napus* L. *Acta Agronomica Sinica* **8**, 24–27.

Yu, C., Hu, S., He, P., Sun, G., Zhang, C. and Yu, Y. (2006). Inducing male sterility in *Brassica napus* L. by a sulphonylurea herbicide, tribenuron-methyl. *Plant Breeding* **125**, 61–64.

Yu, C. Y., Hu, S. W., Zhang, C. H. and Yu, Y. J. (2004). The discovery of novel flower color mutation in male sterile rapeseed (*Brassica napus* L.). *Yi Chuan* **26**, 330–332.

Zanewich, K. P., Rood, S. B., Southworth, C. E. and Williams, P. H. (1991). Dwarf mutant of *Brassica*: Responses to applied gibberellins and gibberellin content. *Plant Growth Regulation* **10**, 121–127.

Zhang, X. K., Li, J. N. and Tang, Z. L. (1999). Effect of the chemical hybridization agent on the fertility of *Polima* male sterile lines of rape (*Brassica napus* L). *Journal of Southwest Agricultural University* **21**, 140–143.

Rapeseed Biotechnology

VINITHA CARDOZA AND C. NEAL STEWART, JR.

Department of Plant Sciences, University of Tennessee, Knoxville, Tennessee 37996

ABSTRACT

Biotechnology has been used for rapid improvement of germplasm for a number of traits in rapeseed and has been crucial to gain a better understanding of its biology. Much has been accomplished in the cellular and molecular biology of canola in the past few years. Plant regeneration has been increasingly optimized via organogenesis and somatic embryogenesis from various explants. The improvement of tissue culture techniques has been exceptionally important. The production of haploids and doubled haploids using microspores has accelerated the production of homozygous lines in canola with uses in breeding and genetic mapping. Crop improvement using somaclonal variation and mutational breeding has also been achieved. The use of

Advances in Botanical Research, Vol. 45
Incorporating Advances in Plant Pathology

0065-2296/07 $35.00
DOI: 10.1016/S0065-2296(07)45015-7

molecular markers in marker-assisted selection and breeding, transformation technology for the introduction of desirable traits, and a comparative analysis of these as well as their future prospects are important parts of current research that are reviewed. Since canola is one of the few crops with wild relatives that it can hybridize and partner with gene introgression, special considerations must be made in mitigating and monitoring transgene flow from crop to related weeds.

I. INTRODUCTION

Rapeseed is a member of the family Brassicaceae (syn. Cruciferae) and is amenable to growth in cooler climates. Once considered a specialty crop for Canada, it is now a global oilseed crop. Many other countries including the United States, Australia, and those in Europe also grow canola, but Canada and the United States are the largest producers.

As with most crops, rapeseed issue culture, transformation, and molecular genetics have made great strides in the past 20 years. Compared with many other dicots, canola is relatively amenable for tissue culture and transformation. Since it is in the same family with the ultimate model plant, *Arabidopsis thaliana*, there is considerable genomics information that is transferable between the two species. Here we will review the state of canola biotechnology.

II. ORGANOGENESIS

Organogenesis is the formation of organs, either shoots or roots from undifferentiated tissue. Organogenesis *in vitro* depends on the amount and balance of phytohormones in media, especially auxins and cytokinins, and the ability of the tissue to respond to the phytohormone species during culture. Organogenesis is an indispensable tool for plant regeneration using tissue culture techniques and for plant transformation. Because of the everincreasing consumer demand for canola oil, and its relative ease of genetic manipulation, the species has been widely exploited for tissue culture purposes. Regeneration protocols have been developed for various canola cultivars. Organogenesis has been the most widely used means for regeneration in canola. Regeneration of plants via organogenesis has been accomplished from various explants such as cotyledons (Moloney *et al.*, 1989; Ono *et al.*, 1994), hypocotyl segments (Cardoza and Stewart, 2003; De Block *et al.*, 1989), peduncle segments (Stringham, 1977), leaves (Radke *et al.*, 1988), flower stalks (Boulter *et al.*, 1990), thin cell layers of epidermal and subepidermal cells (Klimaszewska and Keller, 1985; Shu and Loh, 1991), roots (Sharma and Thorpe, 1989; Xu *et al.*, 1982), and protoplasts (Hu *et al.*, 1999; Spangenberg *et al.*, 1986) and leaves (Kennedy *et al.*, 2005). However, hypocotyl segments remain the most desirable explant for tissue culture in canola because of relatively high efficiency for regeneration.

There are several genetic, developmental, and environmental factors in organogenesis. The first is age of the explant. Young explants respond better and some genotypes respond better than the others. There was a huge variation ranging from 0 to 91% in the 100 cultivars tested (Ono *et al.*, 1994). The second, as in other species, is genotype. We do not know the exact number of unresponsive genotypes as the data are often not published. However, one standard cultivar for tissue culture and transformation is "Westar," a spring-type canola. However, *Brassica napus* cv. GSL-1 had better regeneration efficiency than Westar (Phogat *et al.*, 2000). Third, organogenesis from leaf explants in canola was enhanced by the addition of silver nitrate (Kennedy *et al.*, 2005), a very commonly used ethylene inhibitor in other *Brassica* species. Perhaps capacity for organogenesis in canola could be enhanced by the use of other ethylene inhibitors such as aminovinylglycine (AVG) or silver thiosulphate.

III. SOMATIC EMBRYOGENESIS

Somatic embryogenesis, which has been the subject of increasing research in the genus, has become one of the most desired pathways in the regeneration of plants via tissue culture for several reasons. First, it bypasses the necessity of timely and costly manipulations of individual explants to obtain organogenesis. Second, it does not require the time-consuming subculture steps to increase clonal stock. Third, somatic embryogenesis may also overcome difficulties with micropropagation procedures in species that are difficult to root. Fourth, it can accelerate the introduction of improved clones into commercial production as somatic embryos can be encapsulated and handled as seeds. Finally, it may provide a regeneration system amenable to gene transfer techniques. Though somatic embryogenesis has been used in transformation and regeneration systems in many plant species, there are only a few reports in canola probably because of the tractability and advanced state of organogenesis techniques in the species. Somatic embryos have been obtained from explants from hypocotyl tissues (Kohlenbach *et al.*, 1982; Moghaieb *et al.*, 2006), protoplast-derived colonies (Kranz, 1988), and immature cotyledons (Turgut *et al.*, 1998).

A. MICROSPORE CULTURE AND DOUBLED HAPLOIDS

One of the most useful developments in biotechnology has been the production of haploid and doubled haploid plants for advanced breeding. Haploids and doubled haploids have been produced in canola using anther culture or isolated microspores, thus providing a tool for the rapid production of homozygous lines for the production of hybrid seeds. This technology serves as a valuable tool in breeding self-incompatible, outcrossing lines.

Microspores have been used for the production of plantlets via the somatic embryogenesis pathway in *B. napus* as early as 1977 (Keller and Armstrong, 1977) and ever since it has been exploited for various purposes. Microspores have been used for an array of studies. Micropore-derived embryos of *B. napus* have been used to investigate biochemical pathways and screen end products (Wiberg *et al.*, 1991). The amount of glucosinolate has been studied in microspore-derived embryos (Iqbal *et al.*, 1995), leading to the selection of desirable *B. napus* lines. Microspore-derived somatic embryos have also been used for the study of the production of very long-chain monounsaturated fatty acids (VLCMFAs) (Qi *et al.*, 1998). It has also been possible to transfer genes into microspores and obtain transgenic plants in *B. napus* (Fukuoka *et al.*, 1998; Nehlin *et al.*, 2000).

B. FACTORS AFFECTING MICROSPORE CULTURE

Various factors affect the regeneration efficiency of microspores in culture. A few important factors have been described. The developmental stage of microspores is crucial where late uninucleate to early binucleate stage has been found to be ideal in stimulating embryogenesis in canola (Hansen and Svinnset, 1993; Pechan and Keller, 1988). To improve embryogenesis, there is a need of fresh culture medium in *B. napus*. Discharged toxic substances in the first day of culture and addition of fresh media at different ratios can help remove harmful substances that may hinder embryo development (Kott *et al.*, 1988). Addition of polyethylene glycol (PEG) stimulated microspore embryogenesis (Ilic-Grubor *et al.*, 1998). Embryogenesis decreased with the age of the inflorescence (Chuong *et al.*, 1988). Microspores isolated from older *B. napus* plants had a higher embryogenic frequency than very young plants (Takahata *et al.*, 1991).

However, as in organogenesis, microspore culture in *B. napus* is genotype dependent. Hence, there is still a need to extend this technology to nonresponsive cultivars before it becomes an efficient tool in plant breeding.

C. PRODUCTION OF DOUBLED HAPLOIDS

After the production of haploid plants using microspores, it is possible to obtain doubled haploids either by using colchicine on haploid plants or by the process of spontaneous doubling of chromosomes. In Brassicas, spontaneous doubling depends on the genotype used, the microspore stage, and the culture conditions, especially subjecting to low temperature (Charne *et al.*, 1988; Chen and Beversdorf, 1992; Keller and Armstrong, 1978). Spontaneous double haploids have been produced in *B. napus* cv. Topas (XuHan *et al.*, 1999), and doubled haploids using colchicine have been reported by Murphy and Scarth (1998). In *B. napus* (Zhao *et al.*, 1996), colchicine simultaneously induced microspores embryogenesis and doubled the ploidy level. Microspore-derived doubled haploids prove to be promising lines that could be used by breeders. In the recent years, breeders have been using doubled haploids since they are able to produce homozygous lines in the *B. napus* species, in a relatively short time and without the use of laborious conventional breeding methods.

IV. SOMATIC HYBRIDIZATION

Somatic hybridization is the fusion of two or more somatic cells to form hybrids or cybrids. Using protoplast fusion, it has been possible to create hybrid and cybrid combinations of species that are sexually incompatible. This method is of importance for transfer of valuable traits, which are present in a related, but sexually incompatible, species, where the genes have not been identified, isolated, and cloned. Development in somatic cell fusion has led to the production of hybrids and cybrids not only within the *Brassica* genus but also to the production of intergeneric hybrids. Various desirable traits from the parents have been transferred to the hybrids and cybrids.

One success of protoplast fusion has been the production of hybrids that are disease resistant. Somatic hybrids that were resistant to *Leptosphaeria maculans* were produced by fusing the protoplasts of *B. napus* and *Sinapis arvensis*, which is a wild relative of *B. napus* (Hu *et al.*, 2002), in addition these hybrids also showed a high fertility rate. Novel rapid cycling *B. napus* lines have been produced by protoplast fusion between rapid cycling *B. oleracea* and *B. rapa*. The seeds of the somatic hybrid *B. napus* possessed novel fatty acid compositions (Hansen and Earle, 1994). Somatic hybridization between *B. napus* and *Thlaspi caerulescens*, a heavy metal hyperaccumulator, produced hybrids that tolerated high levels of zinc (Brewer *et al.*, 1999). Somatic hybrids between *B. napus* and *Lesquerella fendleri* were produced by fusing mesophyll protoplasts (Skarzhinskaya *et al.*, 1996).

Another important development in protoplast fusion is the production of male-sterile lines. Male-sterile, cold-tolerant *B. napus* somatic hybrids were produced by crossing an Ogura male-sterile cold-sensitive cauliflower inbred (*B. oleracea* var. *botrytis* inbred NY7642A) and a cold-tolerant, fertile canola-type *B. rapa* (cv. Candle) (Heath and Earle, 1996). Male-sterile cybrids have been produced by the fusion of protoplasts of *B. napus* and *B. tournefortii* (Liu Clarke *et al.*, 1999).

Protoplast fusion between *A. thaliana* and *B. napus* has resulted in the production of asymmetric hybrids of which included three male-sterile hybrids (Yamagishi *et al.*, 2002). The male-sterile plants would be excellent candidates for the study of genes involved in CMS. With the sequencing of the *A. thaliana* genome, the production of somatic hybrids and cybrids between *Arabidopsis* and canola may help unravel the functions of many more genes.

V. INDUCED MUTATIONS

In breeding programs, induced mutations have extensively been used to select desired traits. In *Brassica*, mutation studies have been carried out using both chemical and physical mutagens. By using ethyl methane sulfonate (EMS) in *B. napus*, polyunsaturated linolenic acid was considerably reduced and oleic acid was increased (Auld *et al.*, 1992). This trait was found to be heritable. The M_3 and M_4 generations of the mutant lines also had reduced polyunsaturated linolenic acid and increased oleic acid demonstrating that mutagenized traits were stably inherited. As in other crops, work has been accomplished to produce herbicide-tolerance in canola: sulfonylurea resistance was developed using chemical mutagenesis (Tonnemaker *et al.*, 1992). In addition, early flowering mutations were induced using EMS in *B. napus* (Thurling and Depittayanan, 1992).

A popular target for mutagenesis has been the microspore. Microspores that have been exposed to γ-irradiation and the chemical mutagens sodium azide and EMS were used to develop a system for mutation selection using microspore-derived embryos (Beversdorf and Kott, 1987). Mutagenesis of microspores has led to the development of *B. napus* lines with thinner seed coats, higher oil and protein contents, and low fiber content (Henderson and Pauls, 1992). Canola plants with field tolerance to imidazolinones were developed using a microspore mutagenesis system (Swanson *et al.*, 1989). These mutants were tolerant to the imidazolinone herbicides Scepter, Pursuit, and Assert. Two of these semidominant mutants that contained two unlinked mutated genes were combined to produce an F_1 hybrid, which was superior in imidazolinone tolerance to either of the heterozygous mutants alone. Glyphosate-resistant canola was developed by an *in vitro* mutagenesis system using microspores (Kott, 1996). Rapid cycling *B. napus* resistant to chlorosulfuron has been developed using ultraviolet rays as the mutagen (Ahmad *et al.*, 1991).

VI. SOMACLONAL VARIATION

Genetic variation is very important in crop improvement and forms the basis of development of new varieties. Somaclonal variation is a valuable tool in plant breeding wherein variation in plants regenerated from somatic cells grown using tissue culture can be used in the development of crops with novel traits. It has been used by plant breeders to develop new varieties of crops. By applying selection pressure during tissue culture, it is also possible to develop somaclones resistant to biotic and abiotic stresses (Jain, 2001). Somaclonal variation has been associated with changes in chromosome number and structure, point mutations, and DNA methylation (Brown, 1991; Brown *et al.*, 1993). Somaclonal variants have been developed using anther culture-derived double haploids of *B. napus* (Hoffmann, 1978; Hoffmann *et al.*, 1982; Wenzel *et al.*, 1977).

VII. MOLECULAR MARKERS AND THEIR USE

Molecular markers have been used for various purposes in canola breeding. It has been used for mapping of agronomically important genes, to assess genetic variability, identify genotypes, perform phylogenetic analysis, and establish the true-to-type nature of clonally propagated or micropropagated plants. Molecular markers have mostly been used for genetic mapping using various DNA marker systems in single segregating populations developed for specific investigations of particular traits of interest (Snowdon and Friedt, 2004). With the sequencing of the *A. thaliana* genome complete, it has been possible to conduct comparative studies of *Brassica* genomes with its close relative. Molecular markers linked to various agronomically important traits have been reported and many are integrated into oilseed-breeding programs (Snowdon and Friedt, 2004).

The development of markers and identification of QTL controlling various traits such as oil content (Badani *et al.*, 2002), disease resistance (Manzanares-Dauleux *et al.*, 2000; Pilet *et al.*, 1998; Zhao and Meng, 2003), flowering time (Ferreira *et al.*, 1995), and fertility restoration (Hansen *et al.*, 1997; Jean *et al.*, 1997) are positive steps toward marker-assisted breeding of *B. napus*. Molecular markers have helped in understanding that the spring and winter rapeseed genotypes constitute two genetically different groups (Becker *et al.*, 1995; Diers *et al.*, 1996).

VIII. TRANSFORMATION

Although canola transformation technologies are well developed, efforts are still being continued to increase transformation efficiency and develop novel transformation systems. *Agrobacterium tumefaciens*-mediated transformation is most widely used in canola as it is generally quite efficient and practical for most species in the genus. However, there is still a need for developing efficient transformation methods to overcome genotype dependency. There have been reports on increasing transformation efficiencies in canola (Cardoza and Stewart, 2003, 2004).

Transformation has been used in improving many traits, but the most prominent trait has been for herbicide resistance (HR), and HR canola is the fourth most planted transgenic crop in the world. In Canada, in the year 2003, 3.15 million ha of transgenic canola were planted, about two-third of the total canola cultivation in that country. Canola varieties that are tolerant to herbicides, such as imidazoline, glufosinate, and glyphosate, are now available commercially in the United States and Canada. Other examples of HR include sulfonylurea and bromoxynil resistance in *B. napus* (Blackshaw *et al.*, 1994; Zhong *et al.*, 1997).

Oil-quality improvement has been another target for canola transformation. Canola oil is in great global demand and technology is available to custom-tune fatty acid profiles in seeds. Some of the very interesting features are explained below. Silencing the endogenous oleate desaturase increased oleic acid levels in (Stoutjesdijk *et al.*, 2000). Canola with high γ-linolenic acid was produced by transformation of δ12-desaturase genes from the fungus *Mortierella alpina* (Liu *et al.*, 2001). The overexpression of the Garm FatA1, an acyl–acyl carrier protein (ACP) thioesterase, isolated from *Garcinia mangostana* in canola (Hawkins and Kridl, 1998) have been able to produce canola with elevated levels of stearate. By site-directed mutagenesis, Facciotti *et al.* (1999) have engineered the Garm FatA1 gene for increased activity and have been able to generate seven mutants that conferred a 13-fold increase in enzyme activity toward ACP *in vitro*. The seed-specific expression of the mutant gene resulted in transgenic plants that accumulated 55–58% more stearate than plants expressing the wild-type gene. By crossing transgenic bay thioesterase (BTE) canola with transgenic canola expressing a coconut lyphosphatidic acid acyltransferase (CLP) cDNA, plants containing both transgenes (BTE and CLP plants) were produced. Seed oil from these lines showed an increase in laurate at sn-2 position (Knutzon *et al.*, 1999).

In addition to oil improvements, transformation can convert a crop to biochemical factories for the production of pharmacological and industrial products such as the biodegradable polymer poly(B-hydroxybutyrate) (PHB). Canola is an ideal candidate for the commercial production of PHB since acetyl-CoA, the substrate required for the first step of PHB biosynthesis, is prevalent during oilseed fatty acid biosynthesis. Three genes (*phbA* or *bktB, phbB*, and *phbC*) that are required for the production of the biopolymer were engineered in *B. napus* and the polymer was produced in the leucoplasts (Houmiel *et al.*, 1999). *B. napus* has also been used in the production of carotenoids (Shewmaker *et al.*, 1999), which act as antioxidants in the human body.

Insect resistance is a target trait as there are increasing pest problems in canola cultivation. Since canola is very susceptible to the diamondback moth, a good approach is to control it to overproduce a *Bacillus thuringiensis* endotoxin crystal protein such as Bt *cry*1A(c). This gene has been introduced in *B. napus* (Halfhill *et al.*, 2001; Stewart *et al.*, 1996).

IX. CONCLUSION

Because of the desirability of its oil, ease of culture, and ease of genetic transformation, transgenic canola that contains new traits such as disease resistance, custom oils, improved nutritional qualities, and improved stress tolerance should continue to be adopted by growers. There is one potential hurdle in the adoption of novel traits in canola—biosafety considerations. Canola is the only current major transgenic row crop with sexually compatible wild relatives. Among the numerous wild relatives, the one that can outcross at highest frequency is wild turnip (*B. rapa*) (Stewart *et al.*, 2003). Wild turnip × transgenic, herbicide-resistant canola hybrids have been found in commercial production in Canada (Warwick *et al.*, 2003), and backcrossed hybrids have been discovered as well (Warwick *et al.*, personal communication). The major route of transgene escape will likely be through canola pollen. Thus, adoption of transgenic canola could be slowed depending on the severity of risk of the consequences of gene flow endowed by the trait itself and changes in regulatory environments. Even though transgenic backcrossed hybrids with weedy *B. rapa* were found to be less fit than their nontransgenic counterparts (Halfhill *et al.*, 2005), there remains serious concerns about gene flow in canola (Stewart, 2004). Thus, a biocontainment system might eventually be required for the release of certain traits in transgenic canola. One such biocontainment system that has been recently described, "GM-gene deletor," which excises transgenes out of pollen using a derived site-specific recombination system could be adopted for canola biocontainment (Luo *et al.*, submitted for publication). In any case, it remains to be seen if transgenic plants with any biosafety failsafe systems can become widely adopted in Europe, where the regulatory systems have been mired in politics. More important than Europe, however, is India for *Brassica* cultivation. Transgenic traits can help Indian farmers provide more and better food for the people of this rapidly developing country. For the technology to fully realize the potential illustrated in this chapter, biosafety and biocontainment must be addressed in sustainable fashion.

REFERENCES

Ahmad, I., Day, J. P., MacDonald, M. V. and Ingram, D. S. (1991). Haploid culture and UV mutagenesis in rapid-cycling *Brassica napus* for the generation of resistance to chlorosulfuron and *Alternaria brassicicola*. *Annals of Botany* **67**, 521–525.

Auld, D. L., Heikkinen, M. K., Erickson, D. A., Sernyk, J. L. and Romero, J. E. (1992). Rapeseed mutants with reduced levels of polyunsaturated fatty acids and increased levels of oleic acid. *Crop Science* **32**, 657–662.

Badani, A. G., Koch, M., Baetzel, R., Luehs, W., Friedt, W. and Renate, H. (2002). Molecular mapping of loci controlling yellow seed colour in oilseed rape (*Brassica napus*) (Jan 12–16). *In* "Plant Animal and Microbe Genomes X Conference," Abstract, pp. 468. San Diego, CA.

Becker, H. C., Engqvist, G. M. and Karlsson, B. (1995). Comparison of rapeseed cultivars and resynthesized lines based on allozyme and RFLP markers. *Theoretical and Applied Genetics* **91**, 62–67.

Beversdorf, W. D. and Kott, L. S. (1987). An *in vitro* mutagenesis/selection system for *Brassica napus*. *Iowa State Journal of Research* **61**, 435–443.

Blackshaw, R. E., Kanashiro, D., Moloney, M. M. and Crosby, W. L. (1994). Growth, yield and quality of canola expressing resistance to acetolactate synthase inhibiting herbicides. *Canadian Journal of Plant Science* **74**, 745–751.

Boulter, M. E., Croy, E., Simpson, P., Shields, R., Croy, P. R. D. and Shirsat, A. H. (1990). Transformation of *Brassica napus* L. (oilseed rape) using *Agrobacterium tumefaciens* and *Agrobacterium rhizogenes* a comparison. *Plant Science* **70**, 91–99.

Brewer, E. P., Saunders, J. A., Angle, J. S., Chaney, R. L. and McIntosh, M. S. (1999). Somatic hybridization between the zinc accumulator *Thlaspi caerulescens* and *Brassica napus*. *Theoretical and Applied Genetics* **99**, 761–771.

Brown, P. T. H. (1991). The spectrum of molecular changes associated with somaclonal variation. *Newsletter IAPTC* **66**, 14–25.

Brown, P. T. H., Lange, F. D., Kranz, E. and Lorz, H. (1993). Analysis of single protoplasts and regenerated plants by PCR and RAPD technology. *Molecular and General Genetics* **237**, 311–317.

Cardoza, V. and Stewart, C. N., Jr. (2003). Increased *Agrobacterium* mediated transformation and rooting efficiencies in canola (*Brassica napus* L.) from hypocotyl explants. *Plant Cell Reports* **21**, 599–604.

Cardoza, V. and Stewart, C. N., Jr. (2004). *Agrobacterium*-mediated transformation of canola. *In* "Transgenic Crops of the World–Essential Protocols" (I. S. Curtis, ed.), pp. 379–387. Kluwer Academic Publishers, The Netherlands.

Charne, D. G., Pukacki, P., Kott, L. S. and Beversdorf, W. D. (1988). Embryogenesis following cryopreservation in isolated microspores of rapeseed (*Brassica napus* L.). *Plant Cell Reports* **7**, 407–409.

Chen, J. L. and Beversdorf, W. D. (1992). Production of spontaneous diploid lines from isolated microspores following cryopreservation in spring rapeseed (*Brassica napus* L.). *Plant Breeding* **108**, 324–327.

Chuong, P. V., Deslauriers, C., Kott, L. S. and Beversdorf, W. D. (1988). Effects of donor genotype and bud sampling on microspore culture of *Brassica napus*. *Canadian Journal of Botany* **66**, 1653–1657.

De Block, M., De Brower, D. and Tenning, P. (1989). Transformation of *Brassica napus* and *Brassica oleracea* using *Agrobacterium tumefaciens* and the expression of the *bar* and *neo* genes in the transgenic plants. *Plant Physiology* **91**, 694–701.

Diers, B. W., McVetty, P. B. E. and Osborn, T. C. (1996). Relationship between heterosis and genetic distance based on restriction fragment length polymorphism markers in oilseed rape (*Brassica napus* L.). *Crop Science* **36**, 79–83.

Facciotti, M. T., Bertain, P. B. and Yuan, L. (1999). Improved stearate phenotype in transgenic canola expressing a modified acyl-acyl carrier protein thioesterase. *Nature Biotechnology* **17**, 593–597.

Ferreira, M. E., Satagopan, J., Yandell, B. S., Williams, P. H. and Osborn, T. C. (1995). Mapping loci controlling vernalization requirement and flowering time in *Brassica napus*. *Theoretical and Applied Genetics* **90**, 727–732.

Fukuoka, H., Ogawa, T., Matsuoka, M., Ohkawa, Y. and Yano, H. (1998). Direct gene delivery into isolated microspores of rapeseed (*Brassica napus* L.) and the production of fertile transgenic plants. *Plant Cell Reports* **17**, 323–328.

Halfhill, M. D., Richards, H. A., Mabon, S. A. and Stewart, C. N., Jr. (2001). Expression of GFP and Bt transgenes in *Brassica napus* and hybridization with *Brassica rapa*. *Theoretical and Applied Genetics* **103**, 659–667.

Halfhill, M. D., Moon, H. S., Sutherland, J. P., Poppy, G. M., Warwick, S. I., Rufty, T. W., Weissinger, A. K., Raymer, P. L. and Stewart, C. N., Jr. (2005). Growth, productivity, and competitiveness of introgressed weedy *Brassica rapa* hybrids selected for the presence of Bt *cry1Ac* and *gfp* Transgenes. *Molecular Ecology* **14**, 3177–3189.

Hansen, L. N. and Earle, E. D. (1994). Novel flowering and fatty acid characters in rapid cycling *Brassica napus* L. synthesized by protoplast fusion. *Plant Cell Reports* **14**, 151–156.

Hansen, M. and Svinnset, K. (1993). Microspore culture of Swede (*Brassica napus* ssp. *rapifera*) and the effects of fresh and conditioned media. *Plant Cell Reports* **12**, 496–500.

Hansen, M., Hallden, C., Nisson, N. O. and Sall, T. (1997). Marker-assisted selection of restored male-fertile *Brassica napus* plants using a set of dominant RAPD markers. *Molecular Breeding* **6**, 449–456.

Hawkins, D. and Kridl, L. (1998). Characerization of acyl-ACP thioesterase of mangosteen (*Garcinia mangosteena*) seed and high levels of state production in transgenic canola. *Plant Journal* **13**, 743–752.

Heath, D. W. and Earle, E. D. (1996). Synthesis of Ogura male sterile rapeseed (*Brassica napus* L.) with cold tolerance by protoplast fusion and effects of atrazine resistance on seed yield. *Plant Cell Reports* **15**, 939–944.

Henderson, C. A. P. and Pauls, K. P. (1992). The use of haploidy to develop plants that express several recessive traits using light-seeded canola (*Brassica napus*) as an example. *Theoretical and Applied Genetics* **83**, 476–479.

Hoffmann, F. (1978). Mutation and selection of haploid cell culture systems of rape and rye. *In* "Production of Natural Compounds by Cell Culture Methods" (A. W. Alfermann and Reinhard, eds.), pp. 319–329. Gesellschaft für Strahlen-and Umweltforschung, München.

Hoffmann, F., Thomas, E. and Wenzel, G. (1982). Anther culture as a breeding tool in rape. II. Progeny analysis of androgenetic lines and induced mutants from haploid cultures. *Theoretical and Applied Genetics* **61**, 225–232.

Houmiel, K. L., Slater, S., Broyles, D., Casagrande, L., Colburn, S., Gonzalez, K., Mitsky, T. A., Reiser, S. E., Shah, D. and Taylor, N. B. (1999). Poly (beta-hydroxybutyrate) production in oilseed leukoplasts of *Brassica napus*. *Planta* **209**, 547–550.

Hu, Q., Anderson, S. B. and Hanson, L. N. (1999). Plant regeneration capacity of mesophyll protoplasts from *Brassica napus* and related species. *Plant Cell Tissue and Organ Culture* **59**, 189–196.

Hu, Q., Anderson, S. B., Dixelius, C. and Hansen, L. N. (2002). Production of fertile intergeneric somatic hybrids between *Brassica napus* and *Sinapis arvensis* for the enrichment of the rapeseed gene pool. *Plant Cell Reports* **21**, 147–152.

Ilic-Grubor, K., Attree, S. M. and Fowke, L. C. (1998). Induction of microspore-derived embryos of *Brassica napus* L. with polyethylene glycol (PEG) as osmoticum in a low sucrose medium. *Plant Cell Reports* **17**, 329–333.

Iqbal, M. C. M., Röbbelen, G. and Möllers, C. (1995). Biosynthesis of glucosinolates by microspore derived embryoids and plantlets *in vitro* of *Brassica napus* L. *Plant Science* **112**, 107–115.

Jain, S. M. (2001). Tissue cultured-derived variation in crop improvement. *Euphytica* **118**, 153–166.

Jean, M., Brown, G. G. and Landry, B. S. (1997). Genetic mapping of nuclear fertility restorer genes for the 'polima' cytoplasmic male sterility in canola (*Brassica napus* L.) using DNA markers. *Theoretical and Applied Genetics* **95**, 321–328.

Keller, W. A. and Armstrong, K. C. (1977). Embryogenesis and plant regeneration in *Brassica napus* anther cultures. *Canadian Journal of Botany* **55**, 1383–1388.

Keller, W. A. and Armstrong, K. C. (1978). High frequency production of microspore-derived plants from *Brassica napus* anther cultures. *Z Pflanzenzüecht* **80**, 100–108.

Kennedy, Y. A., Yoshida, H. and Takahata, Y. (2005). Efficient plant regeneration from leaves of rapeseed (*Brassica napus* L): The influence of $AgNO_3$ and genotype. *Plant Cell Reports* **24**, 649–654.

Klimaszewska, K. and Keller, K. (1985). High frequency plant regeneration from thin cell layer explants of *Brassica napus*. *Plant Cell Tissue and Organ Culture* **4**, 183–197.

Knutzon, D. S., Hayes, T. R., Wyrick, A., Xiong, H., Davies, H. M. and Voelker, T. A. (1999). Lysophosphatidic acid acyltransferase from coconut endosperm mediates the insertion of laurate at the sn-2 position of triacylglycerols in lauric rapeseed oil and can increase total laurate levels. *Plant Physiology* **120**, 739–746.

Kohlenbach, H. W., Wenzel, G. and Hoffmann, F. (1982). Regeneration of *Brassica napus* plantlets in cultures from isolated protoplasts of haploid stem embryos as compared with leaf protoplasts. *Z Pflanzenphysiol* **105**, 131–142.

Kott, L. (1996). Doubled haploid technology accelerates canola breeding. *Agri-Food Research in Ontario* **19**, 16–18.

Kott, L. S., Polsoni, L., Ellis, B. and Beversdorf, W. D. (1988). Autotoxicity in isolated microspore cultures of *Brassica napus*. *Canadian Journal of Botany* **66**, 1665–1670.

Kranz, E. (1988). Somatic embryogenesis in stationary phase suspension cultures derived from hypocotyl protoplasts of *Brassica napus* L. *Plant Cell Tissue and Organ Culture* **12**, 141–146.

Liu, J. W., DeMichele, S., Bergana, M., Bobik, E., Hastilow, C., Chuang, L. T., Mukherji, P. and Huang, Y. S. (2001). Characterization of oil exhibiting high gamma-linolenic acid from a genetically transformed canola strain. *Journal of the American Oil Chemists Society* **78**, 489–493.

Liu Clarke, J. H., Chevre, A. M., Landgren, M. and Glimelius, K. (1999). Characterization of sexual progenies of male-sterile somatic cybrids between *Brassica napus* and *Brassica tournefortii*. *Theoretical and Applied Genetics* **99**, 605–610.

Luo, K., Duan, H., Zhao, D., Zheng, X., Deng, W., Chen, Y., Stewart, C. N., Jr., Wu, Y., Jiang, X., He, A., McAvoy, R., Pei, Y., *et al.* (2007). "GM-gene-deletor": Fused loxP-FRT recognition sequences dramatically improves efficiency of FLP or Cre recombinase on transgene excision from pollen and seed of tobacco plants. *Plant Biotechnology Journal.* **5,** 263–274.

Manzanares-Dauleux, M. J., Delourme, R., Baron, F. and Thomas, G. (2000). Mapping of one major gene and of QTLs involved in resistance to clubroot in *Brassica napus*. *Theoretical and Applied Genetics* **101**, 885–891.

Moghaieb, R. E. A., El Awady, M. A., El Mergawy, R. G., Youssef, S. S. and El Sharkawy, A. M. (2006). A reproducible protocol for regeneration and transformation in canola (*Brassica napus* L.). *African Journal of Biotechnology* **5**, 143–148.

Moloney, M. M., Walker, J. M. and Sharma, K. K. (1989). High efficiency transformation of *Brassica napus* using *Agrobacterium* vectors. *Plant Cell Reports* **8**, 238–242.

Murphy, L. A. and Scarth, R. (1998). Inheritance of the vernalization response determined by doubled haploids in spring oilseed rape (*Brassica napus* L.). *Crop Science* **38**, 1463–1467.

Nehlin, L., Möllers, C., Bergman, P. and Glimelius, K. (2000). Transient *β-gus* and *gfp* expression and viability analysis of microprojectile bombarded microspores of *Brassica napus* L. *Journal of Plant Physiology* **156**, 175–183.

Ono, Y., Takahata, Y. and Kaizuma, N. (1994). Effect of genotype on shoot regeneration from cotyledonary explants of rapeseed (*Brassica napus* L.). *Plant Cell Reports* **14**, 13–17.

Pechan, P. M. and Keller, W. A. (1988). Identification of potentially embryogenic microspores in *Brassica napus*. *Physiologia Plantarum* **74**, 377–384.

Phogat, S. K., Burma, P. K. and Pental, D. (2000). High frequency regeneration of *Brassica napus* varieties and genetic transformation of stocks containing fertility restorer genes of two cytoplasmic male sterility systems. *Journal of Plant Biochemistry and Biotechnology* **9**, 73–79.

Pilet, M. L., Delourme, R., Foisset, N. and Renard, M. (1998). Identification of loci contributing to quantitative field resistance to blackleg disease, casual agent *Leptosphaeria maculans* (Desm.) Ces. et de Not., in winter rapeseed (*Brassica napus* L.). *Theoretical and Applied Genetics* **96**, 23–30.

Qi, Q., Rose, P. A., Abrams, G. D., Taylor, D. C., Abrams, S. R. and Cutler, A. J. (1998). (+)-Abscisic acid metabolism, 3-ketoacyl-coenzyme A synthase gene expression, and very-long-chain monounsaturated fatty acid biosynthesis in *Brassica napus* embryos. *Plant Physiology* **117**, 979–987.

Radke, S. E., Andrews, B. M., Moloney, M. M., Crouch, M. L., Krid, J. C. and Knauf, V. C. (1988). Transformation of *Brassica napus* L. using *Agrobacterium tumefaciens*: Developmentally regulated expression of a reintroduced napin gene. *Theoretical and Applied Genetics* **75**, 685–694.

Sharma, K. K. and Thorpe, T. A. (1989). *In vitro* regeneration of shoot buds and plantlets from seedling root segments of *Brassica napus* L. *Plant Cell Tissue and Organ Culture* **18**, 129–141.

Shewmaker, C. K., Sheehy, J. A., Daley, M., Colburn, S. and Ke, D. Y. (1999). Seed-specific overexpression of phytoene synthase: Increase in carotenoids and other metabolic effects. *Plant Journal* **20**, 401–412.

Shu, W. and Loh, C. S. (1991). Secondary embryogenesis from thin cell layers of *Brassica napus ssp. oleifera*. *New Phytologist* **119**, 427–432.

Skarzhinskaya, M., Landgren, M. and Glimelius, K. (1996). Production of intertribal hybrids between *Brassica napus* L. and *Lesquerella fendleri* (Gray) Wats. *Theoretical and Applied Genetics* **93**, 1242–1250.

Snowdon, R. J. and Friedt, W. (2004). Molecular markers in Brassica oilseed breeding: Current status and future possibilities. *Plant Breeding* **123**, 1–8.

Spangenberg, G., Koop, H. U., Lichter, R. and Schweiger, H. G. (1986). Microculture of single protoplasts of *Brassica napus*. *Physiologia Plantarum* **66**, 1–8.

Stewart, C. N., Jr. (2004). "Genetically Modified Planet: Environmental Impacts of Genetically Engineered Plants," p. 240. Oxford University Press, New York.

Stewart, C. N., Adang, M. J., All, J. N., Raymer, P. L., Ramachandran, S. and Parrott, W. A. (1996). Insect control and dosage effects in transgenic canola containing a synthetic *Bacillus thuringiensis cry1Ac* gene. *Plant Physiology* **112**, 115–120.

Stewart, C. N., Jr., Halfhill, M. D. and Warwick, S. I. (2003). Transgene introgression from genetically modified crops to their wild relatives. *Nature Reviews Genetics* **4**, 806–817.

Stoutjesdijk, P. A., Hurlestone, C., Singh, S. P. and Green, A. G. (2000). High-oleic acid Australian *Brassica napus* and *B. juncea* varieties produced by co-suppression of endogenous delta 12-desaturases. *Biochemical Society Transaction* **28**, 938–940.

Stringham, G. R. (1977). Regeneration in stem explants of haploid rapeseed (*Brassica napus* L.). *Plant Science Letters* **9**, 115–119.

Swanson, E. B., Hergesell, M. J., Arnoldo, M., Sippell, D. W. and Wong, R. S. C. (1989). Microspore mutagenesis and selection: Canola plants with field tolerance to the imidazolinones. *Theoretical and Applied Genetics* **78**, 525–530.

Takahata, Y., Brown, D. C. W. and Keller, W. A. (1991). Effect of donor plant age and inflorescence age on microspore culture of *Brassica napus* L. *Euphytica* **58**, 51–55.

Thurling, N. and Depittayanan, V. (1992). EMS induction of early flowering mutants in spring rape (*Brassica napus*). *Plant Breeding* **108**, 177–184.

Tonnemaker, K. A., Auld, D. L., Thill, D. C., Mallory-Smith, C. A. and Erickson, D. (1992). Development of sulfonylurea-resistant rapeseed using chemical mutagenesis. *Crop Science* **32**, 1387–1391.

Turgut, K., Barghchi, M. and Scott, R. (1998). Efficient shoot regeneration and somatic embryogenesis from immature cotyledons of *Brassica napus* L. *Plant Breeding* **117**, 503–504.

Warwick, S. I., Simard, M. J., Légère, A., Braun, L., Beckie, H. J., Mason, P., Zhu, B. and Stewart, C. N., Jr. (2003). Hybridization between *Brassica napus* L. and its wild relatives: *B. rapa* L., *Raphanus raphanistrum* L., *Sinapis arvensis* L., and *Erucastrum gallicum* (Willd.) O. E. Schulz. *Theoretical and Applied Genetics* **107**, 528–539.

Wenzel, G., Hoffman, F. and Thomas, E. (1977). Anther culture as a breeding tool in rape. 1. Ploidy level and phenotype of androgenetic plants. *Z Pflanzenzücht* **78**, 149–155.

Wiberg, E., Rahlen, L., Hellman, M., Tillberg, E., Glimelius, K. and Stymne, S. (1991). The microspore derived embryo of *Brassica napus* L. as a tool for studying embryo-specific lipid biogenesis and regulation of oil quality. *Theoretical and Applied Genetics* **82**, 515–520.

Xu, Z. H., Davey, M. R. and Cocking, E. C. (1982). Plant regeneration from root protoplasts of *Brassica*. *Plant Science Letters* **24**, 117–121.

XuHan, X., Jing, H. C., Cheng, X. F., Iwanowska, A., Kieft, H., Bergervoet, J. H. W., Groot, S. P. C., Bino, R. J. and van Lammeren, A. A. M. (1999). Polyploidization in embryogenic microspore cultures of *Brassica napus* L. cv. Topas enables the regeneration of double haploid clones by somatic embryogenesis. *Protoplasma* **208**, 240–247.

Yamagishi, H., Landgren, M., Forsberg, J. and Glimelius, K. (2002). Production of asymmetric hybrids between *Arabidopsis thaliana* and *Brassica napus* utilizing an efficient protoplast culture system. *Theoretical and Applied Genetics* **104**, 959–964.

Zhao, J. and Meng, J. (2003). Genetic analysis of loci associated with partial resistance to *Sclerotinia sclerotiorum* in rapeseed (*Brassica napus* L.). *Theoretical and Applied Genetics* **106**, 759–764.

Zhao, J., Simmonds, D. H. and Newcomb, W. (1996). High frequency production of doubled haploid plants of *Brassica napus* cv. Topas derived from colchicine-induced microspore embryogenesis without heat shock. *Plant Cell Reports* **15**, 668–671.

Zhong, R., Zhu, F., Liu, Y. L., Li, S. G., Kang, L. Y. and Luo, P. (1997). Oilseed rape transformation and the establishment of a bromoxynil-resistant transgenic oilseed rape. *Acta Botanica Sinica* **39**, 22–27.

Oilseed Rape: Co-existence and Gene Flow from Wild Species

RIKKE BAGGER JØRGENSEN

Risø National Laboratory, Biosystems Department, Denmark

ABSTRACT

Oilseed rape (*Brassica napus*) is partly outcrossing. This gene flow between oilseed rape and the environment may have consequences such as effects on purity and quality of the oilseed rape harvest and abundance of wild relatives. The rape pollen is transferred by insects and wind, and seeds are spilled before and after harvest. Hybridization between different oilseed rape fields and between fields and volunteers will take place. Especially in relation to dispersal of transgenes from GM oilseed rape, this intraspecific gene flow is unwanted. Impurities from unintended mixture of seeds of different varieties during harvest or transportation is another source of adventitious presence. To limit dispersal of genes through pollen and seed, a number of measures could be taken. Among the most effective are physical separation of fields, effective control of volunteers, testing the purity of the certified seed, and cleaning of agricultural machinery. Hybridization between oilseed rape and wild or cultivated relatives can also occur. This interspecific gene transfer is most frequent between

Advances in Botanical Research, Vol. 45
Incorporating Advances in Plant Pathology

0065-2296/07 $35.00
DOI: 10.1016/S0065-2296(07)45016-9

oilseed rape and the species *B. rapa*. After the formation of hybrids between the crop and the relative, survival of the hybrids in the ecosystem will depend on the hybrid-fitness. Wild species that receive transgenes from GM oilseed rape may experience decrease as well as increase in their abundance depending on the selection for the transgene.

I. INTRODUCTION

The transfer of genes into oilseed rape may be deliberate or spontaneous. The deliberate gene flow is a consequence of controlled crossings between oilseed rape and donor plants. This type of hybridization is most often part of breeding activities or a product of developing specific plant material for research purposes. A comprehensive overview of the wild gene resources that can be used in oilseed rape breeding is given at www.brassica.info and Scheffler and Dale (1994). Chèvre *et al.* (2004) reported the possibilities of gene transfer from controlled crosses. As several chapters in this volume deal with breeding of oilseed rape including deliberate introgression of genes, this chapter is devoted to the spontaneous transfer of genes to oilseed rape. The spontaneous transfer results from natural hybridization between plants.

The spontaneous transfer of genes can take place between oilseed rape plants (intraspecific transfer) and may result in unwanted admixture of genes from other varieties. The strategies to avoid this unwanted spontaneous transfer of genes between oilseed rape lines are named coexistence measures. Spontaneous transfer also takes place between oilseed rape and related species within the *Brassica* genus (interspecific transfer) or between oilseed rape and species from other genera (intergeneric transfer). Interspecific and intergeneric transfer is expected to be relatively rare.

A number of different issues must be addressed when assessing the likelihood of spontaneous gene flow between a crop and its related varieties or wild relatives: (1) the existence of close contact between recipient and donor plants resulting in overlapped flowering in time and space, (2) the production of fertile F_1 hybrids and their survival, (3) the transmission of genes through successive backcross generations or selfings, (4) stable gene introgression through recombination between the genomes of recipient and donor, and (5) maintenance of the introgressed genes in the recipient. Generally for spontaneous crosses between different oilseed rape types, items two to five do not present hindrances to the transfer. However, for the interspecific and intergeneric transfer, all items can represent bottlenecks in the gene transfer process.

II. INTRASPECIFIC TRANSFER OF GENES: FACTORS AFFECTING COEXISTENCE

As GM-types of oilseed rape have long been on the market and are increasing their global distribution, the spontaneous transfer of transgenes from GM oilseed rape varieties to non-GM oilseed rape types has become a focus point. Many consumers especially in Europe are requesting GM-free oilseed rape products; also some organic farmers' economy is dependent on the production of GM-free oilseed rape, and it can become crucial if specific GM-traits from varieties bred for different usage are intermixed. Therefore, coexistence measures are now in practice in many countries. Figure 1 gives the possible routes of unwanted admixture of transgenes in the oilseed rape harvest. The admixture takes place through both pollen transfer of transgenes and seed transfer. Problems with coexistence first and foremost relate to GM crops, but could as well be seen as the general problem of separating genes (transgenes as well as genes from conventional breeding) where mixing would be a problem.

A. POLLEN DISPERSAL

Oilseed rape is both self- and cross-pollinated. The frequency of outcrossing is dependent on the environment and the genotype of the variety. Self-pollination is assumed to be most prevalent, for example, between 53% and 88% for the var. Topas under Scandinavian conditions (Becker *et al.*, 1992). Cross-pollination with foreign pollen mainly occurs between an oilseed rape

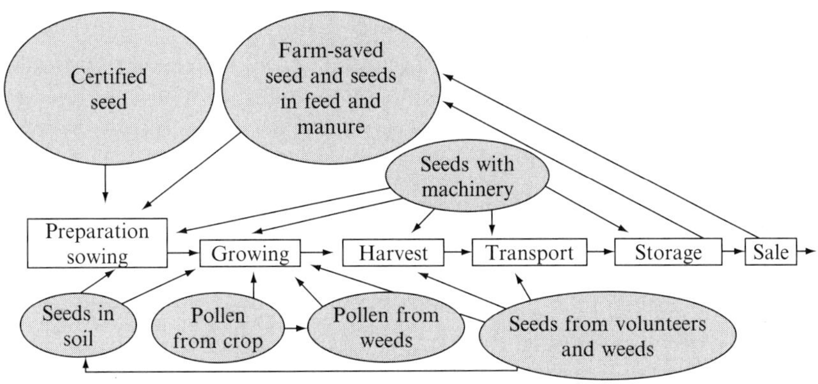

Fig. 1. Dispersal routes for possible adventitious presence of transgenes in the oilseed rape harvest. Modified after Tolstrup *et al.* (2003).

crop and a pollen donor field in the neighborhood, although long-distance dispersal also occurs (Devaux *et al.*, 2005; Rieger *et al.*, 2002). Quite high frequencies have been reported. Devaux *et al.* (2005) found that 13% of the seeds on male sterile bait plants in a French oilseed rape production area were fertilized by pollen from fields more than 1000 m away. Tolstrup *et al.* (2003) summarized a number of different studies where the frequency of gene flow from pollen varied from 0 to 21%. In an attempt to predict the pollen dispersal, several models have been developed (Cresswell, 2005; Damgaard and Kjellsson, 2005; Klein *et al.*, 2006).

Pollen transfer can also take place from oilseed rape volunteers or ferals found at the edge of fields, along roads, at ruderal places, and so on. Factors especially important for the amount of pollen dispersal are the size and form of the oilseed rape fields, volunteer and feral populations, the distance to neighboring fields and ferals, the landscape topology, and frequency of pollinators (Tolstrup *et al.*, 2003). The mitigating measures necessary to reduce pollen flow will of course depend on the degree of dispersal tolerated, and the frequency of donor and recipient in the region (Tolstrup *et al.*, 2003).

B. Seed Dispersal

Oilseed rape has a high frequency of seed shattering before harvest. The amount of seed spilled to the environment is normally between 5% and 10% of the total seed set, but occasionally, the spillage can be larger (Price *et al.*, 1996). If these seeds are incorporated in the soil seed bank, they may survive for several years (Jørgensen *et al.*, 2007; Lutman *et al.*, 2005; Pessel *et al.*, 2001). The resulting volunteers contaminate the harvest of subsequent crops through both seed and pollen. In Danish settings, the seed soil bank of oilseed rape was reported to be minimum 50–100 seeds/m^2 (Jørgensen *et al.*, 2007), and a substantial amount of volunteers of other varieties were observed in oilseed rape fields (Devaux *et al.*, 2005; Jørgensen *et al.*, 2007). The long persistence of seeds in the soil seed bank can locally contribute to hybrids between oilseed rape and related species. These hybrid populations may act as gene reservoirs for oilseed rape genes (Hansen *et al.*, 2001, 2003). Other sources of adventitious presence of transgenes are the certified seed. If impurities are found in the certified seed, they will add to the adventitious admixture of other varieties in the resulting harvest (Devaux *et al.*, 2005; Jørgensen *et al.*, 2007). This of course also applies to impurities in farm-saved seeds. Other possible sources of seed dispersal and harvest impurities are insufficiently cleaned machinery and seed loss during storage. As the seeds of *Brassica napus* are very small and round, they have a potential for dispersal during transportation. Likewise, kilos of oilseed rapeseeds (1 kilo $\sim$ 200,000 seeds) can be spread, if combines are not cleaned between the harvests of different oilseed rape fields. The rotation practice and the soil management are factors especially important for the amount of control of volunteers (Gruber *et al.*, 2005; Lutman *et al.*, 2005; Tolstrup *et al.*, 2003).

Many reports have been published on adventitious mixture of transgenes in the oilseed rape harvest; however, the extent to which this adventitious presence of transgenes will take place is still being evaluated. To estimate the gene flow from pollen, seeds, and volunteers, a landscape model, GENESYS, has been developed (Colbach *et al.*, 2001a,b) that can account for gene flow from the different admixture sources to a number of fields in a region. Damgaard *et al.* (2006) also predicted the combined effects of different GM contamination sources to the seed production of oilseed rape. They found that volunteers and contamination of the certified seed were the sources contributing the most to adventitious presence.

III. INTERSPECIFIC AND INTERGENERIC GENE TRANSFER

Oilseed rape (*B. napus* L., AACC, $2n = 38$) is a natural hybrid between the diploid species, *B. rapa* L. (AA, $2n = 20$) and *B. oleracea* L. (CC, $2n = 18$). In spite of the presence of wild populations of the progenitors, wild oilseed rape populations have not been reported, but feral populations of *B. napus* occur frequently in anthropogenically affected environments. In oilseed rape fields, genes may be transferred spontaneously between *B. napus* and co-occurring and cross-compatible wild or weedy relatives, as there is normally a considerable overlap in flowering time.

The first task in evaluating the extent and consequences of the gene transfer is to establish whether hybrids are formed under natural conditions, in what frequencies, and if they are fertile. This undertaking is relatively simple in cases where hybrids are rather numerous but becomes increasingly laborious as hybrids become scarcer (Jørgensen and Wilkinson, 2005). The next task is to analyze the fitness of these hybrids, for example to what extent they will establish, compete, and produce offspring. The final task is the consequence evaluation of this gene flow.

Many of the wild relatives of oilseed rape are abundant in cultivated fields, and therefore potential donors to oilseed rape (Chèvre *et al.*, 2004; Scheffler and Dale, 1994). However, among 240 species belonging to the *Brassiceae* tribe (Warwick *et al.*, 2000), only 4 wild species are reported to hybridize spontaneously with *B. napus*. They are *B. rapa* (synonym *B. campestris*), *Raphanus raphanistrum*, *Sinapis arvensis*, and *Hirschfeldia incana* (synonym

B. adpressa). Table I summarizes the potential of spontaneous and controlled hybridization between some of the wild relatives and oilseed rape. Besides, *B. napus* can hybridize spontaneously with the cultivated species brown mustard, *B. juncea* (Frello *et al.*, 1995) and cultivated radish, *R. sativus* (Ammitzbøll and Jørgensen, 2006).

A. *B. NAPUS–B. RAPA*

1. Frequency of F_1 hybrids

B. rapa (synonym *B. campestris*, $2n = 20$, genomes AA) is one of the parental species of oilseed rape. *B. rapa* is found as an annual weedy ecotype, which is a common weed in agricultural fields worldwide. There are opportunities for hybridization with the weedy *B. rapa* especially in oilseed rape fields, where weed control is difficult (Jørgensen *et al.*, 2004). There is another ecotype of *B. rapa* found along streams in United Kingdom (Davenport *et al.*, 2000).

Harberd (1975) reported the spontaneous occurrence of the *B. napus* × *B. rapa* hybrid (*B.* × *harmsiana*) in oilseed rape fields. Different frequencies of F_1 hybrids between oilseed rape and the weedy *B. rapa* have also been reported from fields, field experiments, and natural mixed populations of the species (Johannessen *et al.*, 2006a,b; Jørgensen and Andersen, 1994; Jørgensen *et al.*, 1998; Landbo *et al.*, 1996; Pertl *et al.*, 2002; Scott and Wilkinson, 1999; Warwick *et al.*, 2003). Heenan and Dawson (2005) found hybrid frequencies between 0 and 69% of the seeds. The frequency of hybrids is, as expected, highly dependent on environmental and genetic factors, such as density and proportions of plants of the parental species, agricultural practices, and parental genotypes (Jørgensen and Andersen, 1994; Johannessen *et al.*, 2006a,b; Pertl *et al.*, 2002). With oilseed rape as the mother, less interspecific hybrids are produced compared to when *B. rapa* functions as maternal parent (Hauser *et al.*, 1997; Johannessen *et al.*, 2006a; Jørgensen and Andersen, 1994; Jørgensen *et al.*, 1998). This has been shown both in spontaneous crosses in the field and in the mixed pollinations, where the style received pollen from both *B. napus* and *B. rapa*. The hybrid seeds have no dormancy (Landbo and Jørgensen, 1997). This makes hybrids quite vulnerable to weed control but still adult hybrids can be found in agricultural settings, especially in organic fields where herbicides are not used (Hansen *et al.*, 2001).

2. Frequency of backcrossing

The F_1 hybrids have reduced pollen fertility (Jørgensen and Andersen, 1994; Pertl *et al.*, 2002) and often but not always have poor seed production (Hauser *et al.*, 2003; Johannessen *et al.*, 2006a,b). Hybrid plants may have

TABLE I

Current Base Line Available Concerning the Different Combinations

Female–Male	Interspecific hybrid production		Production of subsequent generations	Introgression into the genome of the wild relative	Spontaneous introgression in wild populations
	Hand pollination	Field conditions			
B. napus–B. rapa	Yes	Yes	Yes	Yes	Yes
B. rapa–B. napus	Yes	Yes	Yes	Yes	Yes
B. napus–B. oleracea	Yes	?	?	?	?
B. oleracea–B. napus	Yes	?	?	?	?
B. napus–B. nigra	Yes	?	Yes	?	?
B. nigra–B. napus	Yes	?	?	?	?
B. napus–R. raphanistrum	Yes	Yes	Yes	No	?
R. raphanistrum–B. napus	Yes	Yes	No	No	No
B. napus–S. arvensis	Yes	Yes¤	?	?	?
S. arvensis–B. napus	Yes	No	No	No	No
B. napus–H. incana	Yes	Yes¤	Yes	No	?
H. incana–B. napus	Yes	Yes	Yes	No	?
B. napus–E. gallicum	Yes	No	Yes	No	?
E. gallicum–B. napus	No	No	No	No	No

Yes: data provided; No: data provided but negative results; ¤: hybrid obtained only on male sterile oilseed rape; ?: no experimental data.
Modified from Chèvre *et al.* (2004).

more flowers and pods compared to the parental species, and thereby overcompensate for the reduced number of seeds/pod. Even though fertility is often reduced, spontaneous backcrossing to *B. napus* and the wild relative does take place. *B. rapa* and interspecific hybrids (*B. napus* ♀ × *B. rapa* ♂) with a transgene providing herbicide tolerance were sown together in field experiments to assess the extent of backcrossing to the wild parent (Mikkelsen *et al.*, 1996). Seed set per pod on interspecific hybrids was low (~2.5%) compared to seed set on the parental species (16–23%). Sixty-seven percent of the offspring were herbicide resistant. Among the offspring, 0.5% were almost identical to *B. rapa* having a chromosome number $2n = 20$, high pollen fertility, a normal number of seeds in crosses with genuine *B. rapa* (Mikkelsen *et al.*, 1996), but they retained the herbicide tolerance. The reciprocal cross, *B. rapa* ♀ × hybrid ♂, was not observed among more than 2000 offsprings from 30 *B. rapa* plants. The backcross with *B. napus* as recurrent parent (F_1 hybrid × *B. napus*) was found to be very common (80–94% of the offspring) in a 1:1:1 proportion of *B. napus*, *B. rapa*, and their interspecific hybrid (Johannessen *et al.*, 2006b).

Several studies have reported on spontaneous introgression between oilseed rape and *B. rapa* in natural or naturalized populations (Hansen *et al.*, 2001, 2003; Heenan and Dawson, 2005; Norris *et al.*, 2004). For example, in a large population of *B. napus* and *B. rapa* occurring as weeds, morphologically deviating *Brassica* plants were observed and analyzed together with weedy oilseed rape and *B. rapa*. AFLP analysis revealed that 44% were introgressed having both oilseed rape and *B. rapa* specific markers 7% had only *B. napus* markers and 49% had only *B. rapa* markers (Hansen *et al.*, 2001). The analysis also revealed that the majority of introgressed plants were introgressed beyond the BC_1 generation and that introgression brought about incorporation of *B. napus* C genome DNA into the *B. rapa* genome and exchange of chloroplast DNA producing *B. rapa*-like plants with oilseed rape chloroplasts (Hansen *et al.*, 2003).

3. *Fitness of hybrids and backcross plants*

The fitness of different generations of introgressed plants (F_1, F_2, BC_1, BC_2, BC_3, and so on) in natural populations, fields, and growth chamber experiments have been measured (Allainguillaume *et al.*, 2006; Hauser *et al.*, 1998, 2003; Johannessen *et al.*, 2006a,b; Pertl *et al.*, 2002; Snow *et al.*, 1999). The overall conclusion from the fitness data is that in some environments and for some of the introgressed plants, their fitness will be comparable to their wild parent and occasionally to the oilseed rape crop. A transgene transferred from oilseed rape may of course increase or decrease the fitness of the receiving hybrid/backcross plants. Ammitzbøll *et al.* (2005) found no differences in gene expression in hybrids compared to their transgenic oilseed rape parent.

B. *B. NAPUS–R. RAPHANISTRUM*

R. raphanistrum is one of the most widely distributed weedy relatives of oilseed rape. Field trials performed under agronomic conditions were conducted in France (Chèvre *et al.*, 2000), Canada (Warwick *et al.*, 2003), and Australia (Rieger *et al.*, 2002) with a herbicide resistant oilseed rape variety and various densities of *R. raphanistrum.* In all cases, the frequency of interspecific F_1 hybrids on either parent was very low, ranging from 10^{-5} to 10^{-7}. Gueritaine *et al.* (2003) and Ammitzbøll and Jørgensen (2006) showed that the frequency of hybrids depends on the genotype of the parents. Fertility of hybrids is greatly reduced (Ammitzbøll and Jørgensen, 2006; Chèvre *et al.*, 1997; Warwick *et al.*, 2003). For example, the pollen fertility was never observed to be higher than 15%. Five successive generations were derived in field plots where an equal number of herbicide resistant hybrids and *R. raphanistrum* were planted (Chèvre *et al.*, 1997). The chromosome number of the hybrids decreased during the generations. The percentage of herbicide resistant plants decreased, while male and female fertility increased. Some herbicide resistant plants had a female fertility equivalent to that of *R. raphanistrum*, but none had 18 chromosomes as in *R. raphanistrum*, suggesting that the transgene had not recombined into the *R. raphanistrum* genome. Herbicide resistant hybrids with the *R. raphanistrum* cytoplasm always carried at least one additional oilseed rape chromosome and had a growth pattern similar to that of *R. raphanistrum*, but with male and female fitness values several times lower than that of *R. raphanistrum* (Gueritaine *et al.*, 2002).

C. *B. NAPUS–S. ARVENSIS*

F_1 hybrids have also been produced using male sterile oilseed rape with *S. arvensis* as the pollinator. Here 0.18 hybrid seeds were produced per 100 flowers (Chèvre *et al.*, 1996). Few seeds were obtained on these hybrids after backcrossing to *S. arvensis* in open pollinations (Chèvre *et al.*, 1996).

In contrast, no hybrids have been detected under field conditions (Lefol *et al.*, 1996a; Moyes *et al.*, 2002; Warwick *et al.*, 2003). A 2-year field survey to detect hybridization between *S. arvensis* and glyphosate-resistant *B. napus* crops was conducted in Saskatchewan. A total of 79 populations of *S. arvensis* were analyzed but no interspecific hybrids were detected

(Warwick *et al.*, 2003). Corresponding results were found from survey of *S. arvensis* populations in United Kingdom (Moyes *et al.*, 2002).

D. *B. NAPUS–H. INCANA*

When *H. incana* (synonym *B. adpressa*) was used as the female parent, Lefol *et al.* (1996b) showed that interspecific hybridization occurred spontaneously in field experiments. On average over 3 years, 0.6 hybrids were produced per isolated plant of *H. incana* and represented 0.4% of the seed. No gene introgression from oilseed rape into *H. incana* was detected after five generations of backcrossing to *H. incana* (Darmency and Fleury, 2000).

IV. CONCLUSION

No doubt that the intraspecific gene flow between different plants of oilseed rape will be much more abundant than the interspecific and intergeneric gene flow between species or genera. However, the consequences of the two types of gene flow are not likely to be different. Possible scenarios of the gene transfer are (1) the crop or wild relative may increase its persistence and invasiveness due to the new genes received, with the effect that agricultural practice has to be changed or that the species composition and balance of natural ecosystems are changed, (2) the products of the transferred genes may affect the environment, for example through toxic effects, which may affect the biodiversity. Besides, the intraspecific transgene flow may have economic consequences, if the harvest cannot be sold as GM-free (~GM contents below a given threshold). To what extent consequences of gene transfer will occur of course depends on the traits transferred, the recipient, and the environment. That gene transfer between species has already had consequences like development of new weeds or endangering rare species has already been documented for many crop plants (Ellstrand *et al.*, 1999).

REFERENCES

Allainguillaume, J., Alexander, M., Bullock, J. M., Saunders, M., Allender, C. J., King, G., Ford, C. S. and Wilkinson, M. J. (2006). Fitness of hybrids between rapeseed (*Brassica napus*) and wild *Brassica rapa* in natural habitats. *Molecular Ecology* **15**, 1175–1184.

Ammitzbøll, H. A. and Jørgensen, R. B. (2006). Hybridization between oilseed rape (*Brassica napus*) and different populations and species of *Raphanus*. *Environmental Biosafety Research* **5**, 3–13.

Ammitzbøll, H. A., Mikkelsen, T. and Jørgensen, R. B. (2005). Environmental effects of transgene expression on hybrid fitness—a case study on oilseed rape. *Environmental Biosafety Research* **4**, 3–12.

Becker, H. C., Damgaard, C. and Karlsson, B. (1992). Environmental variation for outcrossing in rapeseed (*Brassica napus*). *Theoretical and Applied Genetics* **84**, 303–306.

Chèvre, A. M., Eber, F., Baranger, A., Kerlan, M. C., Barret, P., Vallee, P. and Renard, M. (1996). Interspecific gene flow as a component of risk assessment for transgenic *Brassicas*. *Acta Horticulturae* **407**, 169–179.

Chèvre, A. M., Eber, F., Baranger, A. and Renard, M. (1997). Gene flow from transgenic crops. *Nature* **389**, 924.

Chèvre, A. M., Eber, F., Darmency, H., Fleury, A., Picault, H., Letanneur, J. C. and Renard, M. (2000). Assessment of interspecific hybridization between transgeneic oilseed rape and wild radish under normal agronomic conditions. *Theoretical and Applied Genetics* **100**, 1233–1239.

Chèvre, A. M., Ammitzbøll, H., Breckling, B., Dietz-Pfeilstetter, A., Frédérique Eber, F., Fargue, A., Gomez-Campo, C., Jenczewsk, E., Jørgensen, R. B., Lavigne, C. Meier, M. S., *et al.* (2004). A review on interspecific gene flow from oilseed rape to wild relatives. *In* "Introgression from Genetically Modified Plants into Wild Relatives and Its Consequences" (H. Nijs, D. Bartsch and J. Sweet, eds.), pp. 235–251. CABI Publishing, UK.

Colbach, N., Clermont-Dauphin, C. and Meynard, J. M. (2001a). GENESYS A model of the influence of cropping systems on gene escape from herbicide tolerant rapeseed crops to rape volunteers. I. Temporal evolution of a population of rapeseed volunteers in a field. *Agriculture, Ecosystems and Environment* **83**, 235–253.

Colbach, N., Clermont-Dauphin, C. and Meynard, J. M. (2001b). GENESYS A model of the influence of cropping systems on gene escape from herbicide tolerant rapeseed crops to rape volunteers. II. Genetic exchanges among volunteer and cropped populations in a small region. *Agriculture, Ecosystems and Environment* **83**, 255–270.

Cresswell, J. E. (2005). Accurate theoretical prediction of pollinator-mediated gene dispersal. *Ecology* **86**, 574–578.

Damgaard, C. and Kjellsson, G. (2005). Gene flow of oilseed rape (*Brassica napus*) according to isolation distance and buffer zone. *Agriculture, Ecosystems and Environment* **108**, 291–301.

Damgaard, C., Kjellsson, G. and Haldrup, C. (2006). Prediction of the combined effect of various GM contamination sources of seed: A case study of oilseed rape under Danish conditions. *Acta Agriculturae Scandinavica Section B – Soil and Plant Science* DOI: 10.1080/09064710600914801.

Darmency, H. and Fleury, A. (2000). Mating system in *Hirschfeldia incana* and hybridization to oilseed rape. *Weed Research* **40**, 231–238.

Davenport, I. J., Wilkinson, M. J., Mason, D. C., Charters, Y. M., Jones, A. E., Allainguillaume, J., Butler, H. T. and Raybould, A. F. (2000). Quantifying gene movement from oilseed rape to its wild relatives using remote sensing. *International Journal of Remote Sensing* **21**, 3567–3573.

Devaux, C., Lavigne, C., Falentin-Guyomarc'h, H., Vautrin, S., Lecomte, J. and Klein, E. K. (2005). High diversity of oilseed rape pollen clouds over an agro-ecosystem indicates long-distance dispersal. *Molecular Ecology* **14**, 2269–2280.

Ellstrand, N. C., Prentice, H. C. and Hancock, J. F. (1999). Gene flow and introgression from domesticated plants into their wild relatives. *Annual Review of Ecology and Systematics* **30**, 539–563.

Frello, S., Hansen, K. R., Jensen, J. and Jørgensen, R. B. (1995). Inheritance of rapeseed (*Brassica napus*) specific RAPD markers and a transgene in the cross *B. juncea* × (*B. juncea* × *B. napus*). *Theoretical and Applied Genetics* **91**, 236–241.

Gruber, S., Pekrun, C. and Claupein, W. (2005). Life cycle and potential gene flow of volunteer oilseed rape in different tillage systems. *Weed Research* **45**, 83–93.

Gueritaine, G., Bazot, S. and Darmency, H. (2002). Emergence and growth of hybrids between *Brassica napus* and *Raphanus raphanistrum*. *New Phytologist* **158**, 561–567.

Gueritaine, G., Bonavent, J. F. and Darmency, H. (2003). Variation of prezygotic barriers in the interspecific hybridization between oilseed rape and wild radish. *Euphytica* **130**, 349–353.

Hansen, L. B., Siegismund, H. R. and Jørgensen, R. B. (2001). Introgression between oilseed rape (*Brassica napus* L.) and its weedy relative *B. rapa* L. in a natural population. *Genetic Resources and Crop Evolution* **48**(6), 621–627.

Hansen, L. B., Siegismund, H. R. and Jørgensen, R. B. (2003). Progressive introgression between *Brassica napus* (oilseed rape) and *B. rapa*. *Heredity* **91**, 276–283.

Harberd, D. J. (1975). *Brassica* L. *In* "Hybridization and the Flora of the British Isles" (C. A. Stace, ed.), pp. 137–139. Academic Press, London.

Hauser, T. P., Jørgensen, R. B. and Østergård, H. (1997). Preferential exclusion of hybrids in mixed pollinations between oilseed rape (*Brassica napus*) and weedy *B. campestris* (*Brassicaceae*). *American Journal of Botany* **84**, 756–762.

Hauser, T. P., Jørgensen, R. B. and Østergård, H. (1998). Hybridization between weedy populations of *Brassica campestris* and varieties of oilseed rape (*B. napus*): II. Fitness of backcross and F_2 progeny. *Heredity* **81**, 436–443.

Hauser, T. P., Damgaard, C. and Jørgensen, R. B. (2003). Frequency dependent fitness of hybrids between oilseed rape (*Brassica napus*) and weedy *B. rapa* (*Brassicaceae*). *American Journal of Botany* **90**, 571–578.

Heenan, P. B. and Dawson, M. I. (2005). Spontaneous hybrids between naturalised populations of pak choi (*Brassica rapa* var. *chinensis*) and wild turnip (*B-rapa* var. *oleifera*) from near Ashburton, Canterbury, New Zealand. *New Zealand Journal of Botany* **43**, 817–824.

Johannessen, M. M., Andersen, B. A. and Jørgensen, R. B. (2006a). Competition affects gene flow from oilseed rape to *Brassica rapa*. *Heredity* **96**, 360–367.

Johannessen, M. M., Andersen, B. A. and Jørgensen, R. B. (2006b). Competition affects the production of first backcross offspring on F1 hybrids *Brassica napus* × *B. rapa*. *Euphytica*; Euphytica DOI: 10.1007/s10681–006–9083-x.

Jørgensen, R. B. and Andersen, B. (1994). Spontaneous hybridization between oilseed rape (*Brassica napus*) and weedy *B. campestris* (*Brassicaceae*). *American Journal of Botany* **81**, 1620–1626.

Jørgensen, R. B. and Wilkinson, M. J. (2005). Rare hybrids and methods for their detection. *In* "Gene Flow from GM Plants" (M. Guy, G. M. Poppy and J. M. Wilkinson, eds.), pp. 113–142. Blackwell Publishing.

Jørgensen, R. B., Andersen, B., Hauser, T. P., Landbo, L., Mikkelsen, T. and Østergård, H. (1998). Introgression of crop genes from oilseed rape (*Brassica napus*) to related wild species—an avenue for the escape of engineered genes. *Acta Horticulturae* **459**, 211–217.

Jørgensen, R. B., Ammitzbøll, H., Hansen, L. B., Johannesen, M., Andersen, B. and Hauser, T. P. (2004). Gene introgression and consequences in *Brassica*. *In* "Introgression from Genetically Modified Plants into Wild Relatives"

(H. C. M. Nijs, D. Bartsch and J. Sweet, eds.), pp. 253–262. CABI publishing, UK.

Jørgensen, T., Hauser, T. P. and Jørgensen, R. B. (2007). Adventitious presence of other varieties in oilseed rape (*Brassica napus*) from seed banks and certified seed. *Seed Science Research* (in press).

Klein, E. K., Lavigne, C., Picault, H., Michel, R. and Gouyon, P. H. (2006). Pollen dispersal of oilseed rape: Estimation of the dispersal function and effects of field dimension. *Journal of Applied Ecology* **43**, 141–151.

Landbo, L. and Jørgensen, R. B. (1997). Seed germinationin in weedy *Brassica campestris* and its hybrids with *B. napus*; implications for risk assessment of oilseed rape. *Euphytica* **97**, 209–216.

Landbo, L., Andersen, B. and Jørgensen, R. B. (1996). Natural hybridisation between oilseed rape and a wild relative: Hybrids among seeds from weedy *B. campestris*. *Hereditas* **125**, 89–91.

Lefol, E., Danielou, V. and Darmency, H. (1996a). Predicting hybridization between transgenic oilseed rape and wild mustard. *Field Crops Research* **45**, 153–161.

Lefol, E., Fleury, A. and Darmency, H. (1996b). Gene dispersal from transgenic crops II. Hybridization between oilseed rape and the wild hoary mustard. *Sexual Plant Reproduction* **9**, 189–196.

Lutman, P. J. W., Berry, K., Payne, R. W., Simpson, E., Sweet, J. B., Champion, G. T., May, M. J., Wightman, P., Walker, K. and Lainsbury, M. (2005). Persistence of seeds from crops of conventional and herbicide tolerant oilseed rape (*Brassica napus*). *Proceeding of the Royal Society B – Biological Sciences* **272**, 1909–1915.

Mikkelsen, T. R., Andersen, B. and Jørgensen, R. B. (1996). Spread of transgenes. *Nature* **380**, 31.

Moyes, C. L., Lilley, J. M., Casais, C. A., Cole, S. G., Haeger, P. D. and Dale, P. J. (2002). Barriers to gene flow from oilseed rape (*Brassica napus*) into populations of *Sinapis arvensis*. *Molecular Ecology* **11**, 103–112.

Norris, C., Sweet, J., Parker, J. and Law, J. (2004). Implications for hybridization and introgression between oilseed rape (*Brassica napus*) and wild turnip from an agricultural perspective. *In* "Introgression from Genetically Modified Plants into Wild Relatives and its Consequences" (H. Nijs, D. Bartsch and J. Sweet, eds.), pp. 107–123. CABI publishing, UK.

Pertl, M., Hauser, T. P., Damgaard, C. and Jørgensen, R. B. (2002). Male fitness of oilseed rape *Brassica napus*, weedy *B. rapa* and their F1 hybrids in mixed populations. *Heredity* **89**, 212–218.

Pessel, F. D., Lecomte, J., Emeriau, V., Krouti, M., Messean, A. and Gouyon, P. H. (2001). Persistence of oilseed rape (*Brassica napus* L.) outside of cultivated fields. *Theoretical and Applied Genetics* **102**, 841–846.

Price, J. S., Neale, M. A., Hobson, R. N. and Bruce, D. M. (1996). Seed losses in commercial harvesting of oilseed rape. *Journal of Agricultural Engineering Research* **80**, 343–350.

Rieger, M. A., Lamond, M., Preston, C., Powles, S. B. and Roush, R. T. (2002). Pollen-mediated movement of herbicide resistance between commercial canola fields. *Science* **296**, 2386–2388.

Scheffler, J. A. and Dale, P. J. (1994). Opportunities for gene transfer from transgeneic oilseed rape (*Brassica napus*) to related species. *Transgeneic Research* **3**, 263–278.

Scott, S. E. and Wilkinson, M. J. (1999). Low probability of chloroplast movement from oilseed rape (*Brassica napus*) into wild *Brassica rapa*. *Nature Biotechnology* **17**, 390–392.

Snow, A. A., Andersen, B. and Jørgensen, R. B. (1999). Costs of transgenic herbicide resistance introgressed from *Brassica napus* into weedy *Brassica rapa*. *Molecular Ecology* **8**, 605–615.

Tolstrup, K., Andersen, S. B., Boelt, B., Buss, M., Gylling, M., Holm, P. B., Kjellsson, G., Pedersen, S., Østergård, H. and Mikkelsen, S. A. (2003). Report from the Danish working group on the co-existence of genetically modified crops with conventional and organic crops. DIAS Report no. 94, Frederiksberg Bogtryk.

Warwick, S. I., Francis, A. and La Fleche, J. (2000). Guide to wild germplasm of *Brassica* and allied crops (Tribe *Brassiceae*, *Bressicaceae*), 2nd Edn. Agriculture and Agri-food Research Branch Publication, ECORC Ottawa, Canada. Contribution no. 991475. Available at http://www.brassica.info.

Warwick, S. I., Simard, M. J., Legere, A., Beckie, H. J., Braun, L., Zhu, B., Mason, P., Seguin-Swartz, G. and Stewart, C. N. (2003). Hybridization between transgenic *Brassica napus* L. and its wild relatives: *Brassica rapa L., Raphanus raphanistrum L., Sinapis arvensis L.*, and *Erucastrum gallicum* (Willd.) OE Schulz. *Theoretical and Applied Genetics* **107**, 528–539.

Evaluation, Maintenance, and Conservation of Germplasm

RANBIR SINGH AND S. K. SHARMA

National Bureau of Plant Genetic Resources, New Delhi, India

ABSTRACT

Germplasm resources constitute the building blocks of crop improvement program. These provide basic raw material and reservoir of genes for breeding high-yielding and disease-resistant cultivars/varieties/hybrids. It is believed that the genus *Brassica* has Mediterranean origin from where it spread to African and Asian countries. Globally, there are more than 74,000 accessions of oilseeds and vegetable *Brassica* germplasm available, with five countries namely China, India, United Kingdom, United States, and Germany holding more than 60% of total world's rapeseed-mustard germplasm. Indian subcontinent is a rich center of diversity for *Brassica juncea* and different morphotypes of *B. rapa*. Efforts have been made to strengthen genetic resources of oilseed Brassicas through conducting *Brassica* explorations and collections within the country as well as through introductions of trait-specific germplasm from rapeseed-mustard growing countries which have been maintained at different research centers in India. Besides cultivated species, a large collection of wild relatives belonging to 15 genera and 46 species have been maintained. During last two decades, over 9500 accessions have been evaluated and characterized for important morphoagronomic traits at various centers under different agroclimatic

Advances in Botanical Research, Vol. 45
Incorporating Advances in Plant Pathology

0065-2296/07 $35.00
DOI: 10.1016/S0065-2296(07)45017-0

conditions. On the basis of evaluation studies, promising donor genotypes have been identified for important agronomic traits and utilized in development of rapeseed-mustard varieties. To meet the challenges ahead, priority areas for future research on rapeseed have been mentioned.

I. INTRODUCTION

To support the national crop improvement programs, a broad genetic base with germplasm of diverse genetic origin is very necessary, essentially from the centers of crop diversity. Thus, to strengthen genetic resources of rapeseed-mustard, germplasm is being continuously augmented through plant explorations and collections as well as through plant introduction activities (Table I).

The Indian subcontinent is an important source of diversity for Brassicas. The northeastern and northwestern regions hold promising genetic variability in oleiferous Brassicae. Early and dwarf types of *Brassica rapa*, particularly cultivars of toria and yellow sarson, are distributed in plains of Uttar Pradesh, Bihar, and West Bengal. Local types of brown sarson have been collected from the hills of Jammu and Kashmir and Himachal Pradesh. Indian mustard from northeastern parts of India, particularly Arunachal Pradesh, Mizoram, Tripura, and Sikkim, are mainly leafy types with long vegetative phase, small seed, and late maturity. Local types of black mustard (*B. nigra*) are grown in small pockets as condiments all over the *Brassica*-growing areas in India. Wild forms of *B. tournefortii* are even now available in drier tracts of Rajasthan, Punjab, and Haryana.

To collect indigenous variability of oilseed Brassicas, systematic efforts were made by erstwhile Plant Introduction Division of the Indian Agriculture Research Institute (now National Bureau of Plant Genetic Resources), New Delhi, under PL-480 scheme on "Collection, Evaluation and Maintenance of *Brassica* Germplasm" during 1960s. Over 2000 accessions of rapeseed-mustard and allied genera were collected from northeastern plains and hills, and northwestern plains and central plateau (Rana and Singh, 1992). Sporadic efforts were made during 1970s to collect *Brassica* germplasm under multicrop explorations. With the establishment of the Germplasm Management Unit of the All India Coordinated Research Project on Rapeseed-Mustard under the aegis of Indian Council of Agricultural Research (ICAR) at the CCS Haryana Agricultural University, Hissar in 1981 and its subsequent shifting to the National Research Centre on Rapeseed-Mustard (NRCRM) at Bharatpur in 1993, efforts were made to strengthen germplasm through indigenous collections as well as exotic introduction of *Brassica* germplasm and its wild allies. These collections have been

TABLE I
Promising Germplasm in Brassica Introduced Through NBPGR

Species	Traits	Cultivar/accession	Source
B. juncea	High oil content	EC-264486	Canada
		EC-303460-463	United Kingdom
		Donskaja	Russia
	Low erucic acid	ZEM 1 (EC-223759)	Australia
		ZEM 2 (EC-223760)	Australia
		EC-322090-093	China
		EC-367880-885	Canada
	Low glucosinolate	EC-346016	Canada
		EC-287711	Sweden
		BJ 1058	Canada
	Alternaria tolerant	EC-206712	France
	White rust resistant line	EC-264487	Canada
	Frost tolerant	EC-333583	Sweden
	Drought tolerant	EC-333584-585	Sweden
B. rapa	High-yielding lines	EC-312379-386	Canada
		EC-333596-97	Sweden
	Low erucic acid	EC-226807	Sweden
		EC-226808	Sweden
	Double zero lines	EC-302478	Sweden
		EC-242690–691	Sweden
	High oil content	EC-232318	Canada
	Multidisease resistance	EC-201539-544	United States
		EC-362816-830	United Kingdom
B. napus	Double zero lines	EC-271577-581	United Kingdom
		EC-200831-832	Canada
		EC-242685	Sweden
		EC-279846	United Kingdom
		EC-362816	United States
	Drought tolerant	EC-333586-587	Sweden
B. carinata	Improved cultivars	PGR 13221	Canada
		PGR 13222	Canada
		EC-223405	Pakistan

Adopted from Singh (1996).

maintained and studied in detail (Kumar *et al.*, 2000). Systematic, planned, and coordinated efforts have resulted in augmentation of over 17,439 accessions of cultivated, wild, and allied species, which are being maintained at different All India Coordinated Research Project on Oilseeds (AICORPO) centers in the country. Several rapeseed-mustard-specific explorations were undertaken in the drier parts of Gujarat, Rajasthan, Bundelkhand region of Uttar Pradesh, parts of Bihar, West Bengal, Orissa, hilly areas of Jammu and Kashmir, Himachal Pradesh, and the northeastern Himalayas under joint programs of National Bureau of Plant Genetic Resources (NBPGR), New Delhi and AICORPO.

As a result of germplasm explorations, local landraces of *B. juncea*, such as *jata rai*, *desi rai*, and *maghi rai*, were collected from farmer's fields in the areas bordering Bangladesh. In yellow sarson, dwarf and early types with drooping siliqua were collected from Indo–Bangladesh border whereas tall, robust, multilocular types were mainly collected from eastern Uttar Pradesh. Diversity of *B. tournefortii*, *B. nigra*, and *Eruca sativa* was collected from drier parts of Haryana, Punjab, and Rajasthan. Explorations for wild crucifers in Pauri Garwal hills of Uttranchal added 22 accessions of *Capsella*, *Crambe*, *Lepidium*, and *Sisymbrium* spp. (Singh, 1996). The collection of germplasm of rapeseed-mustard gained further momentum after the inception of World Bank funded National Agricultural Technological Project (NATP) on "Sustainable Management of Plant Biodiversity" in October 1999. Under this project, several exploration trips were conducted and over 2987 accessions (Table II) were collected from different parts of country, including Arunachal Pradesh, Assam, Bihar, Chattisgarh, Haryana, Madhya Pradesh, Rajasthan, Uttar Pradesh, and West Bengal. Some of the unique collections were made, which include yellow-seeded toria, dwarf and very early toria, long siliqua, bold-seeded and dwarf mustard, white flowered yellow sarson, and so on.

During the last three decades, NBPGR has introduced over 3950 accessions of rapeseed-mustard from more than 25 countries, under strict phytosanitary conditions, with specific traits like high yield, high oil content, resistance to biotic and abiotic stresses, as well as desirable quality traits. Most of the *Brassica* germplasm lines introduced from Europe and Canada are of *B. napus*. These lines are very tall, late in maturity, and poor yielding under Indian conditions. However, some of the lines are having superior oil quality with low erucic acid and low glucosinolate, and have been introgressed into Indian cultivars.

Besides cultivated forms, germplasm of crop plants includes primary and secondary gene pools of wild species, which constitute source for resistance to biotic and biotic stresses (Harlan, 1984). Although wild germplasm usually contains many undesirable traits and are difficult to cross, but through conventional as well as biotechnological methods, a few improved cultivars and breeding lines with desirable traits have been developed. So far, only a limited use of wild species have been made for improvement of cultivated species, but with advancement in biotechnology and molecular techniques, there will be greater use of wild species. A large collection of wild species germplasm belonging to 15 genera and 46 species are maintained at National Research Centre on Plant Biotechnology (NRCPB), New Delhi.

TABLE II

Rapeseed-Mustard Germplasm Collected Under NATP During 1999–2005

Year	Crops	Accessions collected	Areas/state surveyed
1999–2000	Rapeseed-mustard	540	Himachal Pradesh (Bilaspur, Sirmour), Rajasthan, Uttar Pradesh, Andhra Pradesh, Assam, Arunachal Pradesh, and Meghalaya
2000–2001	Rapeseed-mustard Taramira	677 + 49	Rajasthan (Udaipur, Jodhpur, Sirohi, Sikar, Alwar, Bharatpur, Jaipur, Tonk, Ajmer), Assam, Arunachal, Gujarat, Himachal Pradesh, Uttranchal (Champawat, US Nagar)
2001–2002	Rapeseed-mustard Taramira	479 + 34	Uttar Pradesh (Baghpat), Bihar (Vaishali, Patna, Chapra, Mazzafarpur, Siwan), Haryana (Gurgaon, Rewari, Jhajjar), Rajasthan (Pali, Udaipur, Chittorgarh), Jharkhand (East and West Singhbhum, Gumla, Ranchi), West Bengal, Himachal Pradesh (Kangra, Mandi, Kullu, Sirmour), Meghalaya, Arunachal (East–west Siang, Dibang valley, Tawang, and west Kamang)
2002–2003	Rapeseed-mustard Taramira	538 + 10	Himachal Pradesh (Mandi, Hamirpur, Kangra, Una, Kullu, Palampur, Chamba, Sirmour), Haryana (Sirsa, Fatehabad, Bhawani, Mahendergarh), Assam (Sonitpur, Cachar, Hailakandi), Uttranchal (Pithoragarh, Bagheshwar), Jharkhand (Girid, Ranchi, Hazaribagh, East and West Singhbhum), Bihar (Rohtas, Bhojpur), Meghalaya (west Garo hills), Rajasthan (Navgaon, Hanumangarh, Sri Ganganagar)
2003–2004	Rapeseed-mustard	543	Rajasthan (Ajmer, Jalore, Sirohi, Pali, Udaipur), Arunachal (Dibang valley), Assam (Lakhimpur, Golaghat), Uttranchal (Nainital, Champawat, Bagheshwar, Tehri, Uttarkashi, Pithoragarh), Uttar Pradesh (Basti, Deoria, Gorakhpur), Jammu and Kashmir (Kargill, Baramula), Bihar (Gopalganj, Saran, Siwan), Jharkhand (Jamtara, Dhanbad)
2004–2005	Rapeseed-mustard	117	Uttaranchal (Pauri, Rudarprayag, Rishikesh, Garwal), Uttar Pradesh (Bijnor, Sultanpur), Jharkhand (Chamaparan, Hazaribagh), Assam (Sonitpur, Darang, west Kemang, Tawang), Assam (Golaghat, Ananglong, Nagaon)

These include: *Brassica* (10), *Coincya* (2), *Camelina* (1), *Diplotaxis* (11), *Erucastrum* (5), *Enarthrocarpus* (1), *Erucaria* (3), *Eruca* (1), *Hutera* (1), *Hirschfeldia* (1), *Moricandia* (1), *Raphanus* (1), *Rhynchosinapis* (2), *Sinapis* (5), and *Trachystoma* (1).

II. GERMPLASM CHARACTERIZATION AND EVALUATION

To identify important genes for economic traits, germplasm needs to be evaluated for important morphoagronomic descriptors under different agroclimatic conditions and preferably evaluated for disease and pest reactions at the hot spot areas. For characterization of rapeseed-mustard, germplasm is grown in paired rows and representative plants are bagged prior to flower initiation. On the basis of reproductive behavior, the self-pollinated (self-compatible) species are maintained by selfing, whereas cross-pollinated (self-incompatible) species are maintained by sib-mating. The reproductive behavior in genus *Brassica* varies from total cross-pollinated (resulting from sporophytic type of self-incompatibility) to predominantly self-pollinated type. The diploid species, viz., *B. oleracea*, *B. nigra*, and *B. rapa* (var. toria and *lotni* type in brown sarson) are self-incompatible, whereas amphidiploid species, viz., *B. carinata*, *B. juncea*, and *B. napus* as well as *B. rapa* (var. yellow sarson and *tora* type in brown sarson) are generally self-pollinated species. Data is recorded for a set of qualitative and quantitative traits in 5–10 randomly selected plants in self-pollinated species and preferably 15–20 plants in cross-pollinated species. On the basis of evaluation data, crop catalogues are compiled for utilization of desirable genotypes by the crop breeders.

During last two decades, over 9550 accessions have been evaluated and characterized at NBPGR (New Delhi), NRCRM (Bharatpur), and at different AICORPO centers for various agromorphological traits and biotic stresses. A range of values for various agromorphological traits in different *Brassica* spp. have been given in Table III. Information collected on germplasm of rapeseed-mustard demonstrates the availability of valuable genetic reservoir, which could be exploited for improving the existing cultivars. Genetic stocks were subsequently classified into different maturity groups, viz., early, medium, and late types. Under AICORPO, the sources for resistance to pests, diseases, salinity, frost, and drought were identified and utilized in resistance-breeding programs. Improved or local cultivars, which had promising traits, were further utilized for crop improvement programs. Crosses developed through interspecific as well as intraspecific hybridization

TABLE III
Range of Values of Agromorphological Traits of Elite Rapeseed-Mustard Accessions

Characters	Crops				
	Indian mustard	Yellow sarson	Brown sarson	Toria	Gobhi sarson
Days to first flowering	26–87	41–78	38–69	23–65	28–113
Days to 50% flowering	38–96	48–96	46–82	32–73	43–118
Days to maturity	108–190	105–175	110–160	84–110	126–162
Plant height (cm)	70–285	53–209	57–192	36–155	80–220
Primary branches	2.6–14	2.4–16.8	2.4–18.2	2.0–16.0	2.4–12.4
Main shoot length (cm)	18.8–86.4	24.0–68.2	25.2–64.6	16.0–45.4	34.4–96.8
Siliquae on main shoot	10–76	10–78	15–83	13–77	15–107
Siliqua length (cm)	2.2–5.4	2.6–5.6	2.4–5.2	2.4–5.2	2.6–7.2
Seeds per siliqua	6–23	8–44	8–26	8–24	5–30
1000 seed weight (g)	1.1–6.8	1.4–6.0	1.2–6.2	1.0–4.8	2.0–5.8
Oil content (%)	16.1–45.5	30.1–48.0	31.4–47.5	33.5–46.9	33.2–46.2

Adopted from Kumar *et al.* (2004) and Singh (1996).

were used for the transfer of economic characters. The systematic evaluation for various morphoagronomic, quality traits, biotic, and abiotic stresses, resulted in the identification of donor genotypes for use in the varietal development programs (Table IV). Among the exotic introductions, "Midas" and "Tower" of *B. napus*, "Candle," "Span," "Torch," and "Tobin" varieties of *B. rapa* have been found to be promising. The variety "Span" possessing low erucic acid has also been found to be tolerant to frost injury. The sources for frost tolerance, viz. RW 175, RH 781 in *B. juncea*, and DBS 1 in *B. rapa*, are being used in hybridization program. Likewise, the sources for resistance/tolerance to *Alternaria* blight (PAB 9511 and PAB 9534), white rust diseases (Jawahar Mustard 1, JMMWR 93–37, and WR 9541) and aphid tolerance (T 6342) and cream white flower and glossy stem identified from the genetic stock of *B. juncea* are being extensively used in hybridization program and several varieties/strains were developed. *B. carinata*, by and large, has been found to be free from white rust and downy mildew infection, and thus can be used to transfer disease resistance in *B. juncea*.

III. GERMPLASM MAINTENANCE AND CONSERVATION

Increasing population pressure along with adoption of modern agricultural practices has rendered the major crops genetically vulnerable to biotic and abiotic stresses, resulting in narrowing of genetic base. Large-scale cultivation

TABLE IV
Promising Donors Genotypes Identified for Different Agromorphological and Quality Traits

Traits	Species	Genotype
Resistance to biotic stress		
Alternaria blight	*B. juncea*	PBR2012, PAB 9534, PAB2002, KRV tall, SR142-2, PHR-1, Jatai sarson, 38–39, KYM95–5
	B. rapa var. yellow sarson	PYS 16
	B. rapa var. toria	PT 77, Bhawani
	B. napus	PBN 2001, PBN 2002
White rust	*B. juncea*	BIO YSR, BIO 467-95, JYM 10, PAB 2002, JGM 9011, TM 20, DLM 81, PBR 9, RN 505, RN 510 NDR 9902, RK 9902
	B. rapa var. yellow sarson	YSB9, RAURDYS 9702. YS7, YS8, YS9, YS11, YS12
	B. rapa var. brown sarson	Tobin, KBS 3
Downy mildew	*B. juncea*	PYSR5, PYSR7, PYSR10, PBR 91
	B. napus	PBN 2001, PBN 2002
	B. carinata	PPCS1, PBC9427
Powdery mildew	*B. juncea*	TM18, RN505
Sclerotinia wilt	*B. juncea*	RN 505, RN 510, PCR 10, PWR 9541, PW 4810
	B. rapa var. yellow sarson	YS 841
	B. napus	WBBN1, WBBN2
Club root	*B. juncea*	EC-126743, EC-126745, EC-126746, ZEM-1, ZEM-2, Domo, Lethbridge-22, Cutlass
Alternaria blight, white rust, and downy mildew	*B. rapa*	Span
	B. napus	Tower
Orabanche	*B. juncea*	Durgamani
Mustard aphid	*B. juncea*	T-6342, B-85 (glossy), PR-52, B-85 (white flower)
	B. rapa	IB-787, Chamba-1, Chamba-2
	B. napus	GSL-1, Karab, Guliver, PI-171538
Tolerance to abiotic stress		
Drought	*B. juncea*	RH 781, RH 785, RH 819, RH 7363, RH 7513, Vardan, Durgamani, Pusa Barani
Frost	*B. juncea*	RH 781, RW 32–2, PR-52, PR-2030, RH 851, DIRA-343, NDR-8604
	B. rapa	Candle, Span, FR-80, KBS 3
Salt	*B. juncea*	CS 52
Quality traits		
High oil	*B. juncea*	EC-126743-1, EC-126743, Donskaja, NDYR-8, NDYR-10, PR-36
	B. rapa	KS-101, TK-8503, PT-5, PT-25, PT-48

(*continues*)

TABLE IV (*continued*)

Traits	Species	Genotype
Low erucic acid	*B. juncea*	ZEM1, ZEM2, EC-287711, TERI(OE) M21, TERI(OE)M08, PBCM1150-6
Low glucosinolate	*B. juncea*	BJ1058, RC247, RC277, RS79, RS81, EC-287711
	B. rapa	Tobin, Parkland, Candle
Yellow seed coat	*B. juncea*	TM 2, TM 4, TM 17, TM 21, YRT 1, RIK1–1, Domo, Basanti
	B. rapa	Tobin, Parkland, Bele, Torch
Economic agronomic traits		
Earliness	*B. juncea*	IB-1755, IB-1886, Seeta, BRYO, BR-46
Nonshattering	*B. juncea*	RH30, RIK 8, P26–31, PR-10, Pusa Bold, Rajat
Bold seed	*B. juncea*	RH 30, Pusa Bold, R-74-3, Pusa 1002, NDR 8501

of advanced varieties and hybrids is causing erosion of traditional landraces, weedy, and wild relatives of crop plants. Their replacement with advanced varieties has been completed in countries like Korea, Japan, North and South Americas, and several European countries. Significant losses have occurred during last three decades in China, one of the important centers of genetic diversity for oriental Brassicas. Similarly, the wild species *B. tournefortii* ($2n = 20$), a close relative of *B. rapa*, was very common weed in northern India particularly in Punjab, Haryana, Delhi, and Rajasthan during twentieth century (Rana and Singh, 1992), but has become almost extinct. Therefore, there is an urgent need for *in situ* and *ex situ* conservation of wild, weedy, and landraces of *Brassica* species.

In general, seeds of oilseed crops have short viability in comparison to cereals and pulses. Thus, maintenance of rapeseed-mustard requires frequent regeneration, which is very expensive and laborious particularly for allogamous species. It is very important to maintain genetic integrity of germplasm, which depends on the mode of reproduction. Maintenance is not difficult in self-pollinated crops but cross-pollinated crops require special attention. In self-pollinated Brassicas, good healthy seeds can be obtained by enclosing inflorescence in bags (Downey and Robbelen, 1989). In *B. juncea* and *B. napus*, although both are mainly self-pollinated species, some self-incompatible lines have been reported from breeding population. Even under open field conditions, out-crossing by insects and wind pollination has been estimated to range from 5% to 15% in *B. napus* and *B. juncea*.

Maintenance of genetic integrity of germplasm can be easily carried out on buds in *Brassica* species through controlled pollination. For maintaining

TABLE V
Brassica Germplasm Registered During 1997–2006

Species	Germplasm (registration No.)	Traits
B. juncea	CMS trachy (INGR-97006)	CMS line
	CMS trachy (INGR-97007)	CMS line
	Swarna (INGR-98001)	Zero erucic acid, yellow seeded, early (117 DAS)
	NDYR-8 (INGR-98006)	Yellow seeded, high oil content (45.69%)
	NDYR-10 (INGR-98007)	Yellow seeded, high oil content (45.7%)
	CMS oxy (INGR-98020)	CMS line
	CMS mori (INGR-98021)	CMS line
	CMS mori (INGR-98022)	CMS line
	CMS ogu (INGR-98024)	CMS line
	CMS mori (INGR-98025)	CMS line
	Raya Bawal-1 (INGR-99019)	Tetralocular siliquae
	CMS diplotaxis (INGR-99024)	CMS line
	CMS siifolia (INGR-99027)	CMS line
	TM 1 (YSM 2)	Yellow seed
	Heera	Double zero
	NUDH-YJ-5	Double zero
B. napus	Phaguni (INGR-98002)	Zero erucic acid and early (136 DAS)
	Shyamali (INGR-98005)	Zero erucic acid, high oleic acid (70.1%) and oil (50.2%)
	CMS oxy (INGR-98023)	CMS Line
	Gaurav (INGR-99007)	Zero erucic, low glucosinolate, early (125 DAS)
	Garima (INGR-99008)	Zero erucic, low glucosinolate, high oleic (57%)
	CMS erucastrum (INGR-99025)	CMS, Early flowering (31 DAS)
	NUDB 38	Double zero
	NUDB 42	Double zero
B. carinata	*B. carinata* variant (INGR-99026)	Branching like *B. juncea*
B. rapa var. sarson	IC 296707 (INGR-98026)	CMS line carrying cytoplasm of *B. oxyrrhina*

genotypes of individual plant, bud pollination is the best way. However, for maintaining pure populations, insect-proof cages are necessary. Pollination is achieved by keeping honeybee colonies and central rows are harvested for seed purposes. Varieties of rapeseed-mustard can be maintained by growing at appropriate isolation distance, which varies from species to species.

Therefore, to avoid frequent regeneration and to check genetic drift, national facilities have been developed for long-term storage of seed in gene-bank at NBPGR, New Delhi. For long-term storage of germplasm, IBPGR (1981) has recommended storage of rapeseed-mustard at −20 °C. The seed sample size should have minimum of 12,000 seeds per accession, which corresponds to about 50 g of seed for *Brassica* species. So far, 7995 accessions of oilseed Brassicas have been conserved in National Genebank. Beside these, 247 seed samples and 49 pollen samples of rapeseed-mustard and wild relatives have been conserved at −196 °C under liquid nitrogen in cryobanks of NBPGR, New Delhi.

IV. GERMPLASM REGISTRATION

With a view to give due recognition to the plant breeders associated with development/identification of novel genotypes/germplasm/variety, the ICAR has initiated registration of elite germplasm and varieties, and this responsibility was assigned to NBPGR, New Delhi in 1997. In this endeavor, so far 33 germplasm lines have been registered for different traits. A summary of rapeseed-mustard germplasm lines registered during 1997–2005 has been given in Table V (Singh, 2005).

V. GERMPLASM UTILIZATION

For any crop improvement program, a broad range of genetic resources is a very necessary. In the past, a wide spectrum of genetic stocks of Brassicas have been assembled, which include landraces as well as different varieties and related species from indigenous as well as exotic sources. On the basis of preliminary evaluation and characterization data, promising donor genotypes have been identified. Under All India Coordinated Research Project on Rapeseed-Mustard and NRCRM (ICAR), a varietal development program is in progress. A large number of promising varieties have been released (Table VI) for cultivation for different agroclimatic regions and some are in the pipeline for release in the coming future. In rapeseed-mustard, breeders have been working with limited germplasm. This has been one of the major bottlenecks in *Brassica* improvement. The sizeable collection of germplasm available in the country does not represent the entire variability available in the indigenous material.

VI. CONCLUSION

To meet the domestic requirements, the country will require about 24 million ton of rapeseed-mustard oil by 2020 AD. Keeping in view this target and the changing scenario under World Trade Organization (WTO), there is an urgent need to intensify genetic resources programs of oleiferous Brassicae in the country. The following priority areas of research have been identified:

TABLE VI
Released Varieties of Rapeseed-Mustard in India

S. No	Variety/hybrid	Year of release	Areas/states of adaptation
Toria (*B. rapa* L. var. toria)			
1	Agrani (B 54)	1980	West Bengal
2	Anuradha	2002	Orissa
3	Bhawani (TK 8401)	1985	Uttar Pradesh
4	BR 23	Prior to 1951	Bihar
5	DK 1	1970	Himachal Pradesh
6	ITSA	1961	Haryana and Punjab
7	Jawahar Toria (JMT)	1996	Madhya Pradesh
8	M 2	Prior to 1951	Assam
9	M 3	Prior to 1951	Assam
10	M 18	Prior to 1952	Assam
11	M 27	1956	Assam, Orissa, and all NEH States
12	Panchali (TWC 3)	1986	West Bengal
13	PBT 37	1993	Punjab
14	PT 30	1985	Uttar Pradesh
15	PT 303	1985	Assam, Haryana, Punjab, Orissa, Uttar Pradesh, and West Bengal
16	PT 507	1990	Eastern states of the country
17	RAUTS 17	1986	Bihar
18	Sangam	1974	Haryana
19	Suman	1990	Haryana
20	TH 68	1985	Haryana
21	T 9	1961	Arunachal Pradesh, Madhya Pradesh, Manipur, Meghalaya, Rajasthan, Tripura, and Uttar Pradesh
22	T 36	1963	Uttar Pradesh
23	TL 15	1978	Himachal Pradesh and Punjab
24	TLC 1	1983	Punjab
25	TS 29	1980	Assam
26	TS 36	2000	Assam
27	TS 38	2000	Assam
28	TS 46	2000	Assam
29	TWC 2	1986	Eastern India
Yellow Sarson (*B. rapa* L. var. yellow sarson)			
1	66-197-3	1980	Bihar, Central, and Western India
2	Benoy (B 9)	1980	West Bengal
3	Gujarat Sarson 1	1992	Gujarat
4	Indira (PYS 6)	1984	Eastern India
5	Jhumka (NC 1)	1992	West Bengal
6	K 88	1973	Central and Western Uttar Pradesh

(*continues*)

TABLE VI (*continued*)

S. No	Variety/hybrid	Year of release	Areas/states of adaptation
7	Narendra Sarson 2 (NDYS 2)	1996	Madhya Pradesh and Uttar Pradesh
8	Patan Sarson 2 (PS 66)	1966	Northern Gujarat
9	Pusa Gold	1997	Delhi
10	Ragini (MYSL 203)	2000	Central Uttar Pradesh, Bihar, and West Bengal
11	Rajendra Sarson 1 (RAUDYS 89-115)	1998	Bihar
12	Subinoy (YSB 19-7-C)	1982	West Bengal
13	Swarna (RAUDYS 89–07)	1998	Bihar
14	T 10	1959	Uttar Pradesh
15	T 42	1960	Uttar Pradesh
16	T 151	1957	Uttar Pradesh
17	YS Pb 24	1966	Haryana, Punjab, Rajasthan, and Himachal Pradesh
18	YST 151	1957	Uttar Pradesh
Brown Sarson (*B. rapa* L. var. brown sarson)			
1	BS 2	1966	Southwestern Uttar Pradesh
2	BS 70	1966	Eastern Uttar Pradesh
3	BSH 1	1966	Himachal Pradesh, Haryana, Punjab, Rajasthan, and Uttar Pradesh
4	KS 101	1997	Kashmir Valley
5	KBS 3	1996	Himachal Pradesh
6	KOS 1 (Gulachin)	1978	Kashmir Valley
7	Pusa Kalyani (PN 15–5/19)	1973	Assam, Bihar, Haryana, Punjab, Rajasthan, Uttar Pradesh, and West Bengal
8	Suphala (DS 17 M)	1971	Bihar, Haryana, Punjab, Rajasthan, Uttar Pradesh, and West Bengal
Taramira (*Eruca sativa* Mill)			
1	ITSA	1961	Haryana, Punjab, Rajasthan, and parts of Uttar Pradesh
2	Karan Tara (RTM 314)	1995	Haryana, Punjab, Rajasthan, and Uttar Pradesh
3	T 27	1974	Haryana
4	TMC 1	1980	Punjab
5	TMLC 2	1990	Punjab
Gobhi Sarson (*B. napus* L.)			
1	GSL 1	1985	Punjab
2	GSL 2	1994	Punjab
3	Hyola 401	1996	Haryana, Himachal Pradesh, and Punjab
4	Neelam (HPN 3)	1998	Himachal Pradesh
5	PGSH 51	1994	Punjab

(*continues*)

TABLE VI (*continued*)

S. No	Variety/hybrid	Year of release	Areas/states of adaptation
6	Sheetal 1 (HPN 1)	1994	Low and mid-hills of Himachal Pradesh
7	TERI Unaat (TERI 0E-R03)	2001	Madhya Pradesh, Rajasthan, and Uttar Pradesh
Karan Rai (*B. carinata* A Braun)			
1	JTC-1	2001	Haryana, Punjab, Himachal Pradesh
2	Kiran (PBC 9221)	1997	Uttar Pradesh
3	PC 5	1994	Punjab
4	PC 5–17	2001	Rainfed
5	Pusa Gaurav (DLSC 1)	1995	Delhi
6	Pusa Swarnim (IGC 01)	2002	Delhi, Haryana, Himachal Pradesh, and Uttar Pradesh
Indian mustard (*B. juncea* L. Czern. and Coss.)			
1	Aravali (RN 393)	1998	Haryana, Punjab, and Rajasthan
2	Basanti	2000	Uttar Pradesh
3	Bhagirathi (RW 351)	1984	West Bengal
4	BR 13	1965	Bihar
5	BR 40	1960	Gangetic alluvial soils of south Bihar
6	CS 52	1997	Sodic and saline soils of Haryana, Rajasthan, and Uttar Pradesh
7	Durgamani	1968	Rajasthan
8	Geeta (RB 9901)	2002	Delhi, Haryana, Punjab, and Rajasthan
9	GM 1	1989	Dry sandy to medium black soil of Gujarat
10	GM 2	1996	Gujarat
11	Jagannath (VSL 5)	1998	Madhya Pradesh, Uttar Pradesh, and parts of Rajasthan (Bharatpur, Kota)
12	Jawahar Mustard 1 (JMMWR 93-39)	1999	Madhya Pradesh
13	Kanti (RK 9807)	2002	Uttar Pradesh
14	Karishma (PR 8988)	1995	Uttar Pradesh
14	Kranti (PR 45)	1982	Bihar, Delhi, Haryana, Madhya Pradesh, Orissa, Punjab, Rajasthan, Uttar Pradesh, and West Bengal
15	Krishna (PR 18)	1983	Bihar, Delhi, Haryana, Madhya Pradesh, Orissa, Punjab, Rajasthan, and Uttar Pradesh
16	KRV 74	1984	Madhya Pradesh and Uttar Pradesh
17	Laha 101	1947	Uttar Pradesh
18	Laxmi	1996	Haryana
19	Maya	2002	Madhya Pradesh, Uttar Pradesh, and Rajasthan
20	Narendra Ageti Rai 4	1999	Uttar Pradesh
21	Narendra Rai (NDR 8501)	1990	Madhya Pradesh and Uttar Pradesh

(*continues*)

TABLE VI (*continued*)

S. No	Variety/hybrid	Year of release	Areas/states of adaptation
22	Patan Mustard 67 (PM 67)	1967	Mehsana and Banaskantha districts of Gujarat
23	PBR 91	1994	Punjab
24	PBR 97	1996	Haryana, Himachal Pradesh, Jammu, Punjab, and Rajasthan
25	Parkash	1974	Haryana
26	Pusa Agrani (SEJ 2)	1997	Delhi, Haryana, Punjab, and Rajasthan
27	Pusa Bahar (DIRA 247)	1989	Assam, Bihar, Orissa, and West Bengal
28	Pusa Basant (DIRA 367)	1989	Assam, Bihar, Orissa, and West Bengal
29	Pusa Bold (PR 45)	1984	Uttar Pradesh, Rajasthan, Madhya Pradesh, and Delhi
30	Pusa Jaikisan (BIO 902)	1993	Gujarat, Maharashtra, and Western Rajasthan
31	Pusa Kisan (T 6342)	1972	Uttar Pradesh
32	Rajat (PCR 7)	1993	Gujarat, Maharashtra, and Western Rajasthan
33	RCC 4	1996	Himachal Pradesh
34	RGN 13	2002	Rajasthan
35	RGN 48	2004	Rajasthan
36	RH 30	1983	Haryana, Jammu, Punjab, Northern Rajasthan, and Western Uttar Pradesh
37	RH 781	1990	Haryana
38	RH 785	1983	Orissa and West Bengal
39	RH 819	1990	Haryana
40	RH 7361	1984	Haryana
41	RK 01–03	2004	Uttar Pradesh
42	RK 9	1985	Madhya Pradesh and Uttar Pradesh
43	RK 14	1984	Madhya Pradesh and Uttar Pradesh
44	RL 18	1937	Haryana, Punjab, and Rajasthan
45	RL 99-27	2004	Punjab
46	RL 1359	1987	Punjab
47	RLM 185	1983	Madhya Pradesh and Uttar Pradesh
48	RLM 198	1981	Punjab
49	RLM 514	1980	Punjab
50	RLM 619	1985	Punjab
51	Rohini	1995	Madhya Pradesh and Uttar Pradesh
52	Sanjucta Asech (4C-6-3/II)	1988	West Bengal
53	Sarama	1984	West Bengal
54	Saurabh (RH 8113)	1985	Haryana
55	Seeta (B 85)	1980	West Bengal
56	Shekhar (KR 5610)	1978	Uttar Pradesh
57	Shivalik	2002	Delhi, Haryana, Punjab, and Rajasthan
58	Swaran Jyoti (RH 9501)	2002	Haryana, MP, Rajasthan, and Uttar Pradesh

(*continues*)

TABLE VI (*continued*)

S. No	Variety/hybrid	Year of release	Areas/states of adaptation
59	T 11	1936	Uttar Pradesh
60	T 16	1956	Uttar Pradesh
61	TM 2	1987	Assam
62	TM 4	1987	Assam
63	Urvashi (RK 9501)	1999	Uttar Pradesh
64	Vaibhav (RK 1418)	1985	Madhya Pradesh and Uttar Pradesh
65	Vardan (RK 1467)	1985	Madhya Pradesh and Uttar Pradesh
66	Varuna (T 59)	1975	Entire mustard-growing areas of the country
67	Vasundhra	2002	Haryana, Madhya Pradesh, Rajasthan, and Uttar Pradesh
68	YRN 6	2005	Rajasthan
Black mustard (*B. nigra* L. Koch.)			
1	Surya	2002	Andhra Pradesh

Adopted from Kumar *et al.* (2004).

To broaden the genetic base, there is an urgent need for extensive exploration and collection from regions of high genetic diversity/unexplored areas/hot spots.

Presently available resources of wild and weedy relatives of rapeseed-mustard are very limited. There is an urgent need for collection and introduction of germplasm of wild/weedy relatives as well as cultivated species of rapeseed-mustard from the centers/areas of rich genetic diversity.

For selection of widely adaptable material and donor genotypes, multilocation evaluation and characterization of germplasm and subsequently their proper documentation is required to be strengthened.

Taxonomic status of *Brassica* species is still very confusing. By utilization of modern analytical/biochemical tools, germplasm should be properly classified and characterized.

With increasing number of germplasm holdings, there is an urgent need to develop a core set of the germplasm for different traits for efficient handling and utilization of germplasm.

Yellow-seeded varieties of rapeseed-mustard are in great demand. Efforts should be made for developing yellow-seeded varieties since they have high oil and protein content as well as less fibre in meal than the brownseeded varieties.

Developing double or triple "zero" lines (low erucic acid in oil, low glucosinolates, and low fibre content in seed meal) should be breeding objective. There is an increasing demand for double-low varieties in view of tremendous export potential of seed meal.

Exotic germplasm introduced from Canada, Europe, and Australia with double-low traits has not performed well under Indian conditions because of their late maturity. Efforts should be made to develop locally adapted double-low cultivars by using exotic germplasm.

High oleic acid lines of oilseeds are known to be thermostable and thus have better shelf life and keeping quality. Therefore, efforts should be made to select lines with high oleic acid.

Oilseed Brassicas being rich source of erucic acid have great potential for use as industrial raw material for plastic manufacturing. Thus, besides selecting rapeseed-mustard with low erucic acid for edible purposes, germplasm with high erucic acid (>55.0%) should also be selected for industrial uses and biofuels.

REFERENCES

Downey, R. K. and Robbelen, G. (1989). Brassica species. *In* "Oil Crops of the World" (G. Robbelen, R. K. Downey and A. Ashri, eds.), pp. 339–362. McGraw-Hill, New York.

Harlan, J. R. (1984). Evaluation of wild relatives of Crop plants. *In* "Crop Genetic Resources-Conservation and Evaluation" (J. H. W. Holden and S. J. T. Williams, eds.), pp. 212–222. George Allen and Unwin, London.

International Board for Plant Genetic Resources (1981). "Genetic Resources of Cruciferous Crops," pp. 1–48. AGP:IBPGR/BO/100. IBPGR, Rome.

Kumar, P. R., Chauhan, J. S. and Singh, A. K. (2000). Rapeseed mustard genetic resources: Status and priorities. *Indian Journal of Plant Genetic Resources* **13**, 207–218.

Kumar, P. R., Singh, R. and Mishra, A. K. (2004). Rapeseed–mustard. *In* "Plant Genetic Resources-Oilseeds and Cash Crops" (B. S. Dhillon, R. K. Tyagi, S. Saxena and A. Agrawal, eds.), pp. 20–44. Narosa Publishing House, New Delhi.

Rana, R. S. and Singh, R. (1992). Present status of rapeseed and mustard germplasm in India. *In* "Advances in Oilseeds Research, Vol. I" (D. Kumar and M. Rai, eds.), pp. 189–200. Scientific Publishers, Jodhpur, Rajasthan.

Singh, R. (1996). Germplasm Resources. *In* "Oilseed and Vegetable Brassica: Indian Perspective" (V. L. Chopra and S. Prakash, eds.), pp. 279–291. Oxford & IBH Publishing Company, New Delhi.

Singh, R. (2005). Management of rapeseed mustard genetic resources at NBPGR during 2004–05. Paper presented in Annual Rapeseed Workers' Group Meeting on Rapeseed mustard, held at G.B. Pant University of Agriculture & Technology Pantnagar (Uttranchal) from 11–13 August 2005.

Oil Technology

BERTRAND MATTHÄUS

Federal Research Center for Nutrition and Food, Institute for Lipid Research, Piusallee 68/76, 48147 Münster, Germany

Advances in Botanical Research, Vol. 45
Incorporating Advances in Plant Pathology

0065-2296/07 $35.00
DOI: 10.1016/S0065-2296(07)45018-2

ABSTRACT

The production of oil is the main driving force for the cultivation of rapeseed. For this, different methods are in use to separate the oil from the oil-bearing material. By far the major amount of rapeseed oil is produced in huge oil mills, which extract the oilseeds, by pressing and extraction by solvent followed by refining steps. However, a small amount of rapeseed oil is produced in small- and medium-sized facilities, which only use a screw press for the extraction and sedimentation or filtration for purification of the oil from turbid matter. The chapter reviews the different methods of oil processing and the different requirements necessary for the production of high-quality rapeseed oil. Difficulties arising from both methods are also discussed. For rapeseed oil produced in huge oil mills, the purification of the crude oil is an important process because otherwise the use of oil for different applications is not possible. The different steps of the refining process, degumming, alkali or physical refining, bleaching, and deodorization are also described besides the effect of the refining process on the oil. The purification of crude rapeseed oil from small- and medium-sized facilities is much simpler, but the particularities are described. In this context, the difficulties of a water steam treatment of native rapeseed oil are also discussed. Stress is also laid on the description of the modification of rapeseed oil, which is often necessary to match the demands for the preparation of appropriate products or to achieve a desired product quality.

I. INTRODUCTION

The main reason for the cultivation and harvest of rapeseed is the production of oil, which is used as food or in several technical applications, like fuel, lubricants, or raw material for the production of chemical products. For this, it is necessary to separate the oil from the oil-bearing material. The aim of the extraction method is to optimize the oil yield with simultaneous maintenance of the oil quality, whereas the method used for this aim depends on the application for which the oil is provided.

One of the oldest methods for the oil production is the mechanical extraction of the seeds by means of a screw press. Nowadays, this method is used in several small- and medium-sized facilities for the production of cold pressed, native rapeseed oil with a following cleaning step by sedimentation or centrifugation. More efficient is the extensive process in large-scale facilities using mechanical and chemical means which result in raw oil that has to run through different steps of a refining process to make the oil usable. A third method for the production of oil from rapeseed is the extraction by high pressure carbon dioxide. This method is relatively new, but not very common for the production of rapeseed oil, since it is expensive and only suitable for the production of valuable oils or constituents.

In addition to oil, the meal or press cake is an important product used as source of protein in animal feeding. Normally, meal is the by-product of the oil processing, but depending on the price for oil and meal, this relation can change. This chapter describes the different methods of the oil processing and explains the differences between the oil qualities obtained by the different methods.

II. PRETREATMENT

Oil processing not only starts with the oil extraction from the seeds, but for the aim to get high-quality oil, the production has to start with the harvest of high-quality seeds. Only then, it is possible to produce high-quality oil. This is important for large-scale facilities, although they have the possibility to improve oil quality by the refining process, but it is essential for small- and medium-sized plants, which use screw presses for the oil processing and sedimentation or filtration for oil cleaning only (Fig. 1). The producer of

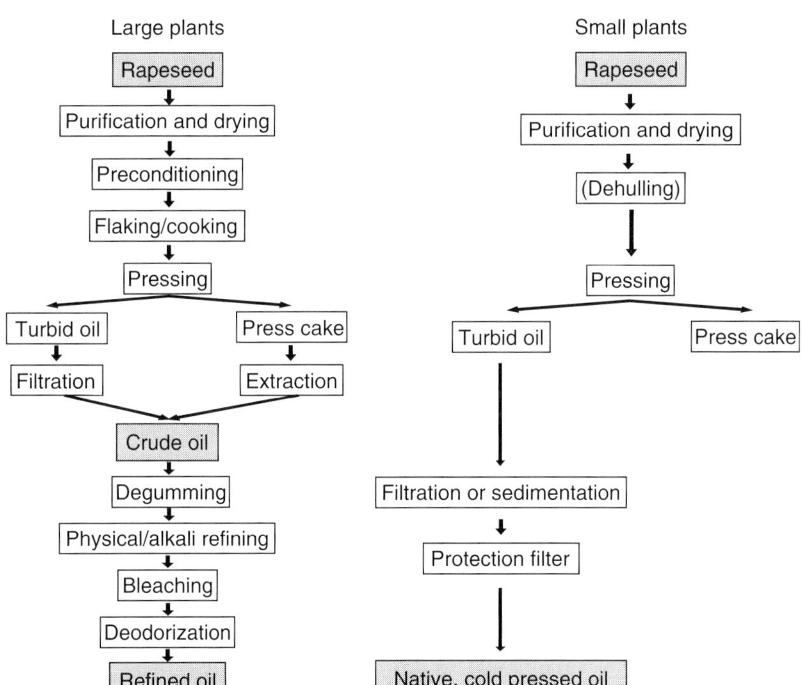

Fig. 1. Oil processing in large and small plants.

this type of oil has no chance to improve the oil quality after the pressing process. Purification by sedimentation or filtration only removes turbid matter, but no components that impair the quality of oil such as free fatty acids or products of oxidative degradation. Therefore, small- and medium-sized plants depend for better or worse on the quality of the raw material, while large-scale facilities with an extensive oil extraction can remove free fatty acids or chlorophyll with some losses from oil. To obtain a high-quality product with a constant quality, the quality of small- and medium-sized plants depend strongly on the quality of the seeds. The quality the producer fills into his screw press influences the quality of oil he gets.

Looking the market of native, cold pressed rapeseed oil, these difficulties become blatantly obvious. The oils show large variations in the quality not only for edible oils but also for oil used as fuel. On one side, it is possible to produce high-quality native rapeseed oils by the simple extraction process, which correspond to the demands of the market, and on the other side, a relatively large number of oils show clear defects. For native edible oils, these defects are primarily detectable in the sensory evaluation of the oil, which is the most important criterion for this type of oil. But large differences can also be found for other parameters, such as oxidative stability, peroxide value, content of free fatty acids, or total amount of impurities, also important for the use of rapeseed oil as fuel.

The composition of rapeseed points out why is it so difficult to handle oilseeds in general and rapeseed especially without loss of quality. Rapeseed contains 45% fats in comparison to cereals that have only 2% fat. On one hand, fat is very susceptible against oxidation and the degradation products of the oxidative process are already noticeable in very low concentrations. On the other hand, fat is an excellent carrier for aroma components, which retains and concentrates volatile compounds. This implicates that the high oil content of rapeseed acts as the sensory memory of the seed. Everything what results in an impairment of the sensory evaluation during pretreatment and storage of the seeds up to the processing is memorized by the oil. In native cold pressed oils, these failures are noticeable directly by sensory evaluation, but also producers of refined rapeseed oils have difficulties with low-quality raw material.

Basis for harvesting high-quality seeds is the optimum date of harvest. Only in the last few days of maturation, the oil content of rapeseed increases. The content of chlorophyll and free fatty acids decreases and the moisture content of the seeds reaches an optimum between 7% and 8%. The oil content increases by about 6% from 35% to 41% oil within the last 14 days before the optimum date of harvest (Fig. 2). The changes in the content of free fatty acids and chlorophyll are still more dramatic in the last few days before

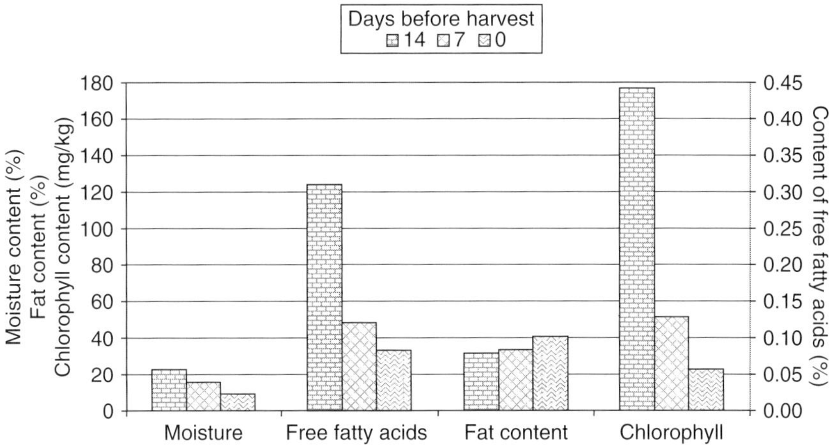

Fig. 2. Influence of the date of harvest on moisture and fat content of the seeds and content of free fatty acids and chlorophyll of the resulting oil.

harvest. These constituents are important for the oxidative stability of native rapeseed oil and also responsible for losses during the refining process. Thus, the content of these compounds should be as low as possible. The moisture content of the seeds decreases from more than 20% to an optimum of 8% within the same period of time. Also important for the production of native cold pressed rapeseed oil is the development of the aroma components within the last 14 days before harvest. While negative sensory attributes like *musty* and *fusty* disappear during this time, the typical positive attributes for native rapeseed oil, *seedlike* and *nutty*, are being developed within the last period of maturation (Fig. 3).

Since rapeseed is not processed directly after harvest, but the processing takes place continuously over at least 1 year, it is necessary to store the seeds after harvest under appropriate conditions. The aim of an optimal storage of rapeseed is to maintain the good quality of the seeds until further processing without losses in quality during storage. But the problem is that the seeds are living organisms. On one side, they have the natural desire to develop new plants, which results in degradation of storage compounds like lipids, carbohydrates, or proteins, and on the other side, only the viable and germinable seeds are optimally storable. Additionally, the ingredients of the seeds are livelihood for microorganisms or insects, which also leads to a degradation of storage compounds.

After harvest, cleaning, and if necessary, drying of seeds is one of the most important steps. Both, the amount of foreign matter and the moisture content are basis for the settlement of the oil mills with the farmers when

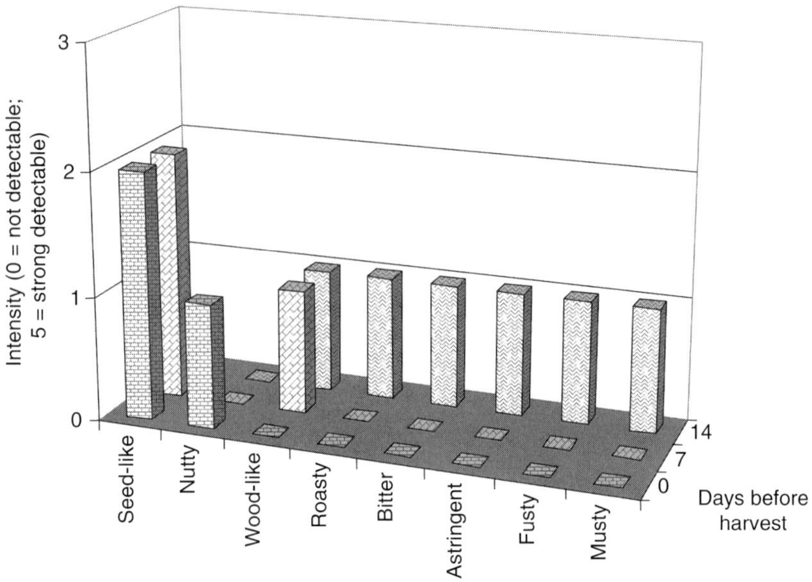

Fig. 3. Influence of the date of harvest on the sensory evaluation of native, cold pressed rapeseed oil.

seeds are delivered. In Canada, the canola standard established by the Canadian Grain Commission defines the maximum amount of foreign matter in the seeds as 2% with moisture content of 6–9%. In Germany, the oil mills have similar conditions except low moisture content (below 9%) otherwise the farmer has to accept deductions. These limits ensure a high-quality oil and meal because both are negatively correlated and hence influence the quality of the seeds during storage.

The moisture content of 9% is sufficient for large plants that have the possibility to improve oil quality by refining, but for producer of native, cold pressed rapeseed oil, this value is too high. The respiration activity of the seeds increases drastically at moisture content higher than 8%, which results in a fast degradation of storage compounds and appropriate reduction of weight during storage. Storage of rapeseed at 9% moisture or higher results in a continual increase in free fatty acids during a storage period of 9 months, whereas no change in the content of free fatty acids is detectable at moisture of 7% (Attenberger *et al.*, 2005).

If rapeseed is stored under wet conditions, then products of degradation glucosinolates, nitriles, and isothiocyanates can be found in the oil, indicating an improper storage of the raw material before processing. The appearance of the above products in the oil with a fusty and musty taste and aroma of the oil further make it impossible to use in human nutrition.

Besides foreign matter, the pods influence the oil yield since it presents noticeable lower oil contents. The objective of seed cleaning is the removal of foreign matter from seed material. This can be done directly before processing, but as described above it is much better to remove weed seeds, pods, dust, broken and sprouted seeds directly after harvest and before storage to keep a high-seed quality for a longer time. For the separation of foreign material from rapeseed, an aspirator, grader, and sieves (Figs. 4 and 5) for removal of over and undersized particles are in use. Often also a combination

Fig. 4. Grader for the separation of foreign matter from rapeseed.

Fig. 5. Sieve for the separation of foreign matter from rapeseed.

Fig. 6. Two types of seed cleaner with two and four trays, respectively (by courtesy of MUEZ-HEST Process Technologies Pvt. Ltd., Mumbai, India).

of these methods is utilized in one unit (Fig. 6). The separation of metal fragments by magnets is recommended to spare the screw press since these fragments can damage the press, if they pass through the newel.

The goal of seed purification by sieves, grader, and aspirator is mainly the removal of foreign matter from rapeseed, but other effects of this procedure are significant drying of the seed surface and decrease of the seed temperature, which result in worse conditions for fungus spores and a delay of metabolic processes.

The result of the cleaning procedure should be to remove as much foreign matter as possible with justifiable efforts. Thus, rapeseed should be cleaned to less than 1% foreign material to maintain the high quality of the seeds during storage.

III. DEHULLING

Rapeseed consists of about 13–19% of hulls (Matthäus, 1998), which contain major parts of fiber and a group of pigments. Contrary to sunflower seeds, dehulling of rapeseed is not a common used process, since it is expensive, due to the small size of the seeds. For the production of refined oils, this type of pretreatment is not economical, although the resulting meal could have some advantages with regard to the use in animal feeding. Phenolic compounds and also fiber are reduced in the resulting meal, which is favorable for the animals from a nutritional point of view. The use of dehulled seeds also has some advantages concerning the reduction of processing costs because the energy effort during further processing is lower. But dehulling has the disadvantages that the amount of antinutritive compounds, like sinapine, glucosinolates, and inositol phosphates, is enriched in the resulting dehulled material (Matthäus, 1998) and percolation of solvent is impaired, since hulls increase the porosity of the material being extracted.

Dehulling is only interesting for some small- and medium-sized plants that produce cold pressed rapeseed oil by screw pressing with an added value. An advantage of dehulling before screw pressing in this type of plants is the possibility to keep the press temperature below 40 °C, which limits the activity of seed-own enzymes and reduces the transfer from undesired components of the seeds into the oil (Piva *et al.*, 1985; Schneider, 1979a,b; Schneider and Khoo, 1986). As a result of dehulling, the taste and smell of this type of oil should be milder than that of conventional native, cold pressed rapeseed oils with a less intensity of the typical sensory attribute *seedlike*.

Different techniques are conceivable for the process of dehulling: pneumatic impact on a rebound plate, splitting in a gap between rollers, and a designed deformation of the seed material. As a result of the designed deformation, the hulls flake from the meal and the meal consisting of germ as well as internal and external part of cotyledons fall to pieces. It is important that the seed material is not completely pulverized and a certain amount of hulls is necessary to enable a successful pressing process. The separation of the hulls from the dehulled material takes place by air floating and sieving (Fig. 7).

To maintain the oil quality from dehulled rapeseeds, it is necessary to work continuously without any delay because dehulled material is remarkably more susceptible to settlement by microorganisms and degradation by enzymes such as myrosinase or lipases. The enzymes being separated in compartments prior dehulling come together with the substrates, forming isothiocyanates and nitriles from glucosinolates and free fatty acids from triglycerides. This results in a fast degradation of the dehulled seed material, and the formation of degradation products strongly influences the oil quality.

Fig. 7. Dehulled seeds in comparison to crushed whole seeds and hulls. (See color Plate 22.)

The following steps, preconditioning, flaking, and cooking, are not necessary and not allowed for the production of native, cold pressed oils. For the production of this type of oil, the purified rapeseed flows directly from the store into the screw press without further treatment.

IV. PRECONDITIONING

Before the extraction process starts, seeds are often heated to about 30–40°C to prevent shattering during the flaking process. Another important point is that this preconditioning helps to improve screw pressing capacity, cake formation, extractability, and hexane recovery from the extracted rapeseed flakes (Unger, 1990). This step is also important for plants in colder climates. Preheating of the whole seeds is performed either by direct contact of the seed material with hot air or indirect heating.

V. FLAKING

The next step is the flaking operation. The aim of this step is to rupture as many cell walls as possible to improve the release of oil from the seeds by destruction of the seed coat and the oil cells. In these cells, lipid particles are surrounded by cell membranes or cell walls, which avoid delivery of oil from the intact seeds. After the flaking process, about 80% of these membranes or walls is destroyed resulting in the unimpeded drain of oil. Another important point is that the surface of rapeseed is drastically enlarged by flaking, which results in an easier contact of the lipid particles among each other by which they easily migrate from the seed material. Additionally, solvent can easily dissolve the lipid fraction from the seeds, which leads to a more comprehensive extraction of the material. The extraction of the seeds needs less thermal energy input during cooking and less mechanical energy during extraction by screw presses.

In general, flaking is carried out by using two cast-iron rolls with a smooth surface, which are set at an appropriate gap to control the thickness of the flakes. During running, one roll revolves at 2–5% higher revolutions per minute with the result that the roll surfaces wipe each other. Distribution of seeds on the whole width of the rolls is achieved by using vibrating feeders, which results in an even spreading. While preheated seeds pass through the rolls, the seed coat and cell walls are destroyed as a result of the pressure occurring during this way. Often flaking is carried out in a two-step process in order to ensure a high grade of flaked seeds. In the first step, the seeds are cracked to about 0.4- to 0.7-mm thickness. The result of this step is that hull or seed coat of the seeds is broken. In the second step, the thickness of the flakes is reduced to about 0.2–0.4 mm, whereas the optimum is between 0.30 and 0.38 mm. Flakes thinner than 0.2 mm are very fragile, while flakes thicker than 0.4 result in lower oil yield (Carr, 1995). The problem of thin flakes is that they contain a large part of small particles that can get into the oil resulting in problems during filtration of the oil.

After flaking, the oil processing has to be carried out without any delay because comparable to dehulled seeds, the protective seed coat is destroyed resulting in the beginning of the work of the appropriate enzymes and attended microorganisms. The result is drastic impairment of the flakes and subsequently of the oil. Therefore, storage of the flakes before further processing is not advisable.

VI. COOKING

Cooking or conditioning of the flaked seeds has different advantages not only for the following process but also for the product quality itself. First of all, intact cells that overcame the flaking process are disrupted during the thermal treatment with temperatures between 75 and 85°C. This results in an easier extractability of oil from the solid residue.

Thus, as a result of higher temperature during cooking, viscosity of the oil droplets is reduced facilitating the mergence of small lipid particles coming from the oil-bearing cells to form larger oil droplets. This relieves the drainage of the oil from the solid during the extraction. A successful cooking process can be seen by the oily appearance of the flakes as a result of the larger oil droplets, whereas the appearance of uncooked flakes is dry resulting in no oil.

Another important point is that the moisture content of the material can be adjusted by the cooking process before starting the extraction. Most screw presses need moisture contents between 3% and 6% for an optimum job because within this range the seeds have the correct elasticity necessary for the use of screw presses.

Most enzymes have a temperature optimum between 50 and 60°C. Within this range, the conversion rates of the enzymes with its substrates are highest. Therefore, it is important to pass through this range as fast as possible. At higher temperature, the process of cooking results in deactivation of the enzymes. This deactivation is important since breakdown products of glucosinolates, such as isothiocyanates or nitriles, partially remain in the solid residue after the extraction process. They are suspected to be harmful to animal when using the residue as fodder. Another point is that glucosinolates consist of sulfur, which can be found in the oil as a result of degradation. Sulfur poisons catalysts and therefore influences their activity during a possible hydrogenation process.

The inactivation of lipases by heat treatment is important since these enzymes lead to the degradation of triglycerides and phospholipids, resulting in higher amounts of free fatty acids responsible for a reduction of the shelf life of the oil.

In general, conditioning of flakes takes place by using drum- or stack-type cookers. The stack-type cooker is a vertical vessel consisting of several steam-heated floors. The flaked material is given on the first floor on the top of the cooker and a stirrer mixes the material for the desired time on the floor to ensure that a uniform uptake of temperature takes place by the flakes. The temperature run of the plate is adjusted so that the desired temperature is achieved as quickly as possible to avoid enzymatic degradation reactions in the material. The resting time of the flakes on each plate depends on the properties of the flakes before cooking, but in most cases, a time between 20 and 40 min is used. After this time, the material falls down on the next floor, and so on. By this procedure, the flakes run progressively through each plate of the cooker. As a result of temperature and time control as well as mixing of the material on each plate, a uniform heating is possible. During the process, all enzymes are deactivated and by the use of hot steam or hot air it is possible to adjust the final water content of the seeds to optimal conditions for the pressing process.

Drum cookers consist of a number of individual horizontal tubes that can be heated. Each tube is equipped with a stirrer to enable the equal motion of the seeds in the cooker. The material is filled into the first tube by screw conveyers, and depending on the properties of the seeds, a certain number of tubes are used for cooking.

The temperature used for cooking should be higher than 80°C to deactivate enzymes, but at temperatures higher than 100°C changes in aroma and taste of the resulting oil arise. Additionally higher cooking temperatures result in an increase in the amount of phosphorus and free fatty acids in the oil. For example, rise in temperature from 80 to 100°C results in an increase in phosphorus from 125 mg/kg to more than 300 mg/kg, and at the same time, the amount of free fatty acids increases from 0.5% to about 0.6% (Unger, 1990). A further aspect of the use of high temperatures during cooking over a longer period of time is the damage of protein components, which can impair the protein quality of the residue.

VII. PRESSING

The aim of the pressing process is to separate the oil from the solid phase of the seed material. In large-scale facilities, two different pressing techniques are used for the separation of oil from the seed. The first one is prepressing of the flaked seeds, which leaves about 15–20% oil in the press cake. Today, this is the commonly used technique, which is followed by the extraction of the press cake by solvent. The advantage of prepressing is that a press cake is formed from the small flakes that allow a good solvent contact and percolation in the extractor. The result is an optimized removal of oil from the seeds (Unger, 1990). Considering both methods, extraction by screw press and solvent extraction combines the economical advantages of both methods, low costs using a screw press and higher oil yield using solvent extraction.

The other technique is a more extensive pressing of the oilseed without further extraction by solvent. This straight pressing is interesting for oilseeds with high oil content, where it is no problem to get over the release of some percent of oil in the residue. Nowadays, this method is used only barely in large-scale facilities since it is not possible to reduce the oil content to less than 5% oil in the press cake by this technique. From an economical point of view, this is not sufficient for this type of oil mill. Nevertheless, this technique has the advantages of lower investment and maintenance cost of achieving higher oil quality (Niewiadomski, 1990).

Different manufacturers for oil presses in the market and their design functionality vary from one manufacturer to another, but in general the principle is nearly the same for all presses. Most of them consist of a rotating screw shaft (Figs. 8 and 9) within a horizontal barrel. This screw shaft transports the

Fig. 8. Screw press without barrel cage to show worm sections.

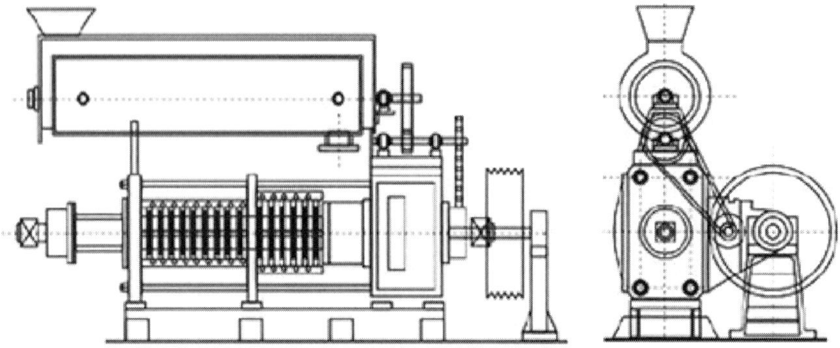

Fig. 9. Sectional drawing of a screw press (by courtesy of MUEZ-HEST Process Technologies Pvt. Ltd., Mumbai, India).

Fig. 10. Barrel of a screw press surrounded by flat bars during pressing of rapeseed.

seeds through the screw flight and in order to increase the pressure onto the material during the passage, the volume of the screw increases or the diameter of the screw flight decreases in direction of the seed flow resulting in less space for the material and increasing pressure onto the material. In dependence on the needs for the seed material, it is possible to vary the slope of the screw which results in a variation of the pressure increase within the screw press. The barrel is surrounded by flat bars made of hardened steel, which form a cage around the screw (Fig. 10). There is some space between the bars, just as much that the oil can flow out of the barrel, while most of the solid material leaves the screw press at the end. It is possible to change and adjust the pressure within the barrel by changing the space between the bars. Another possibility to influence the pressure within the barrel is the size of the press cake exit at the end of the barrel and the slope of the screw. As a result of high pressures, up to 1600 bar in screw presses used in large plants, temperature up to 170°C can occur since kinetic energy of the rotating screw shaft is changed into heat. The result of the pressing process is the reduction of the oil content from 42% in rapeseed to about 16% in the press cake that is used for further solvent extraction.

VIII. OIL EXTRACTION IN SMALL- AND MEDIUM-SIZED FACILITIES

A number of small- and medium-sized facilities produce native, cold pressed rapeseed oil by using only screw pressing and purification of the oil by filtration or sedimentation. This procedure is not new, and the extraction of oilseeds by small presses, going back several hundred years, is rediscovered by the farmers as one possibility to produce products with an added value instead of delivery of the seeds after harvest to the large oil mills. The extraction of rapeseed is remarkably less efficient, but such facilities work and economize in regional material cycles which lead to an increased acceptance of the products by the consumers. So, it is possible to get an added value for the product, which is necessary from an economical point of view. In contrast to large plants, small- and medium-sized facilities have short routes of transport and therefore small costs for transportation. Thus, these plants are able to supply a manageable and quantitatively limited market with high-quality cold pressed rapeseed oils. These oils are mainly used as fuel for agricultural machines and cars, but a small percentage is supplied as native, edible rapeseed oil comparable to native olive oil.

In general, the extraction method by screw presses is comparable to prepressing in large facilities (Fig. 11), but it is necessary to carry out the process gently because these plants have no chance to improve the quality of the oil by a refining process after the extraction. Nevertheless, the oil yield after pressing in small- and medium-sized facilities is higher than during prepressing in large plants. Depending on the type of press and the settings, about

Fig. 11. Screw press of a small plant for the production of native, cold-pressed rapeseed oil.

10–15% oil remains in the press cake, whereas after prepressing the press cake contains 15–20% oil.

A gentle pressing process is recommended because the content of chlorophyll and phosphorus as well as the amount of free fatty acids are strongly influenced by the pressing conditions. Especially, an increase in the temperature within the compressed material leads to an increase in these components in the oil. The amount of free fatty acids, chlorophyll, and phosphorous is of special interest because they negatively influence the stability of the oil and additionally phosphorous is an important parameter if the oil is used as fuel. In that case, the amount should be below 12 mg/kg (Anonymous, 2006). Therefore, it is necessary to avoid the formation of these components during production to improve the quality as well as the shelf life of the oil. But if high-quality seed material is used for the pressing process, the settings of the screw press can be varied in a wide range without running the risk of changing the quality of the resulting oil.

IX. EXTRACTION BY SOLVENT

Extraction by solvent is recommended if it is necessary to reduce the oil content in the raw material lower than 2%. That means the aim of extraction by solvent is to remove as much oil as possible from the oilseed. From an economical point of view, preextraction by screw presses in advisable because equipment and maintenance for solvent extraction is very costly. In most cases, hexane is used as solvent because it is cheap, has good oil solubility at relatively low temperature, has an appropriate boiling temperature, is non-corrosive to metal, does not react chemically with the oil, is stable under the process conditions, and is not mixable with water, which eases the separation of water from the seeds. It is easy to remove from the residue with low-energy input and without impairment of the material. Nevertheless, this solvent also has some disadvantages with regard to its potential danger. Hexane is highly flammable, and mixtures of air and hexane are explosive, which entail the risk of fire and explosion if the plant does not consider appropriate precautionary measures. From this, it is recommended to use explosion-proof electrical equipments to avoid the possibility of any accident. Despite such measures, every year some solvent-extraction plants explode somewhere in the world. Often plants purge all vessels with an inert gas to reduce the risk of fire or explosion.

During solvent extraction of the rapeseed oil, an intensive contact between solvent and press cake is necessary to achieve an exhaustive removal of oil. The solvent is heated to 50–60 °C, but it is important to avoid boiling when hot press cake comes into contact with the solvent. Since press cake becomes hot from the pressing process, further temperature supply is only necessary when the extractor starts working. Temperature is important for the extraction since viscosity of the solvent reduces and the solubility of the extract increases with higher temperatures. Rapeseed oil mainly consists of triglycerides, but also minor components such as phospholipids, chlorophyll, free fatty acids, color pigments, and degradation products of oxidative reactions are coextracted by the solvent. Increase in the temperature from 40 to 58 °C raises the content of phospholipids from 0.2 to 0.8% (Kock, 1981). This should be taken into consideration because removal of minor components during refining is costly.

Another important factor influencing the result of solvent extraction is moisture, which can come from the surface of the press cake or from poor water/hexane separation after distillation. This moisture can avoid an optimal penetration of the press cake by the solvent, resulting in low extraction rates and high residual solvent in the meal. In some cases, "blinding off" of the perforated cell floors occurs resulting in the necessity of a complete shutdown of the plant and cleaning of the floors.

The extraction rate does not only depend on temperature and moisture but also on adjustment of equilibrium between diffusion of solvent into cake particles and the diffusion of oil from within the particles out into the solvent. This time is needed for an extensive extraction of the press cake. Balance between time for the extraction and the oil yield is necessary for each plant.

The highest amount of oil is extracted within the first few minutes of the process (75% in the first 25 min), whereas the total oil yield depends on the further number of extraction steps and the total extraction time. The extraction of the remaining 25% oil from the press cake takes disproportionately more time.

The structure and the behavior of the cake particles are significant for the result of the extraction process. Fine press cake resulting from excessive handling of the cake before processing or poor settings of the screw press during pressing negatively influences the percolation rate of the solvent through the cake. Soybean flakes are stable enough to survive the mechanical and hydraulic strain during the extraction process, whereas rapeseed flakes are more fragile and susceptible. The small particles with a worse surface: volume ratio lead to a remarkably worse percolation of the solvent through the cake bed. This again results in poor extraction efficiency, higher residual oil content in the press cake (Unger, 1990), and also high levels of solvent in the meal. Thus, more extraction steps and a higher solvent requirement are necessary to achieve a sufficient oil yield.

In most cases, solvent extraction of the press cake is carried out by using a continuous countercurrent extraction process moving the press cake and the solvent in opposite directions (Fig. 12). This method is most efficient in

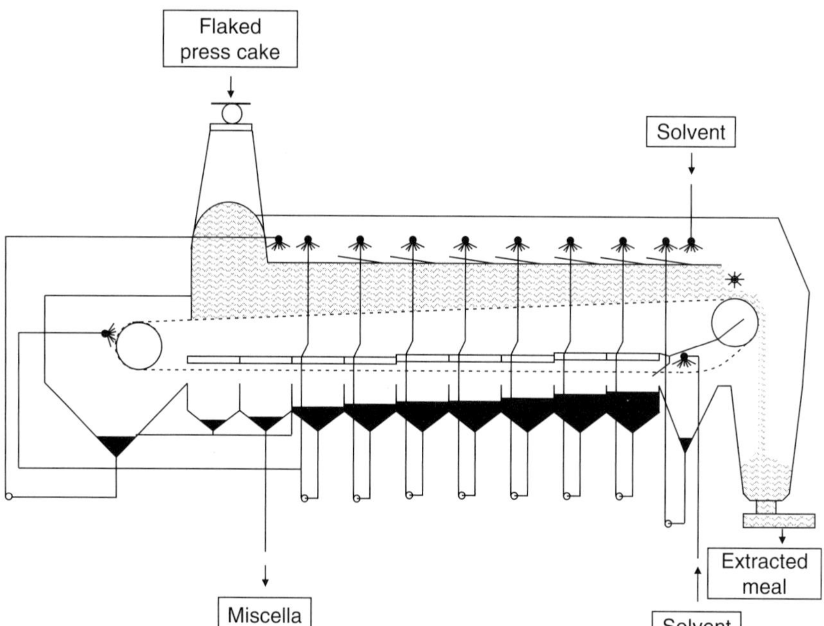

Fig. 12. Extractor.

reducing the oil content of the press cake to a minimum, and the need of solvent remains comparably low. During the process, the cake is brought in contact with solvent several times, and after a certain time of adjustment of equilibrium, the solvent–oil mixture called as miscella is allowed to drain away from the cake. With each step, the amount of oil in the cake is reduced, but it is not possible to remove all of the solvent from the cake. Thus, a certain amount remains together with oil in the cake. Therefore, theoretically the extraction process is an infinite process and the processor has to make a balance between economical aspects and an extraction as extensive as possible (Fig. 13).

The number of extraction steps depends on the performance of the extractor. Often 7–10 successive wash stages have been used. However, an extensive extraction with less steps is possible if the cake is saturated with solvent allowing it to soak over a sufficient time followed by flushing the enriched miscella from the surface of the particles and replacing it with a leaner miscella. In that case, only two or three extraction steps are necessary with a reduced number of pumps, reducing costs of energy, capital, and maintenance (Beach, 1983).

For the extraction of the press cake, a mixture of n-hexane and oil is used and the concentration of oil in hexane decreases with each extraction step. In the last step, before the meal leaves the extractor, it should be washed with pure solvent.

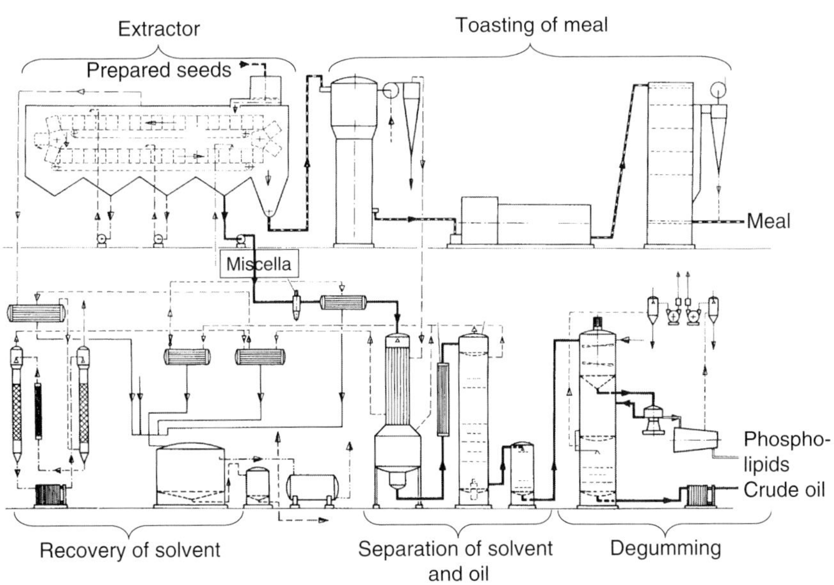

Fig. 13. Extraction plant (Fa. Lurgi).

Several different types of extractors are in use, including rotary, shallow-bed or deep-bed, horizontal belt, and continuous loop extractors. The principle of all these extractors is comparable, but there are some differences in the technical equipment and performance. Some extractors immerse the solids in the solvent, comparable to a Soxhlet extractor in laboratory use, some percolate the solvent through the solids, and some use a combination of both immersion and percolation. Disadvantages of percolation extractors are higher requirement of solvent and problems during transport of material. Therefore, today percolation extractors are the method of choice in industry.

Another important difference is the depth of the cake bed. It is about 20–50 cm in the case of shallow-bed extractors, which allow a fast percolation by the solvent. Using deep-bed extractors, the depth of the cake bed is much higher and the percolation takes much more time, but the extractor is more compact and provides much more capacity per unit of building volume (Beach, 1983). Good extractors offer the possibility to modify the height of the cake bed in order to consider different percolation parameters of different materials.

In most cases, the percolation extractors consist of compartments or cells in which solids are washed with successively less concentrated miscella and finally with fresh solvent. The final miscella contains 20–30% oil. The flakes, given a final wash with raw solvent, have had about 95% of their oil extracted. Another type of extractor is the horizontal belt extractor that conveys solids through a series of solvent sprays. After percolating through the bed, the miscella is collected in a hopper below the conveyor. Continuous loop extractors are shallow-bed extractors that carry the solid material through an enclosed vertical loop. The miscella, low in oil and high in n-hexane at the beginning, is circulated countercurrent to the flow of the cake until the cake leaves the extractor. In a last step, pure solvent extracts the cake just before it leaves the extractor. The solids go through both percolation and immersion and are completely turned over, allowing the solvent to contact flakes from both sides.

X. ENZYME PRETREATMENT

An interesting method to enhance the oil recovery from the seeds is their treatment by enzymes. The aim of this procedure is to modify the seed coat and cell walls to increase the oil yield; as a result, the disruption of cell walls by mechanical treatment during flaking is reduced. The extraction time decreases, which increases the efficiency of the pressing process up to 92%. This procedure demands a lower energy requirement and results in less losses of hexane during solvent extraction. Normally rapeseed meal tends to retain twice the level of residual hexane found in soybean meal. This hexane has to be removed from the meal by more heat resulting in an increase in the overall energy requirements of the process. During the removal of hexane, the meal attains dark color.

Different enzymes of the carbohydrase type are in use such as, β-glucanase, pectinase, hemicellulase, cellulose, or mixed enzyme preparations. The enzymes degraded the cell walls of the seeds which mainly consists of carbohydrates. In this process, the different enzymes show the following relative order of efficiency in enhancement of oil extraction: mixed enzymes > β-glucanase > pectinase > hemicellulase > cellulose (Sosulski *et al.*, 1988). Investigations showed that treatment with enzymes improved throughput of the screw press and increased oil flow rate and oil recovery.

XI. SEPARATION OF SOLVENT AND OIL

After the solvent extraction process, the resulting miscella contains 20–30% oil in hexane. For an economical procedure, the recovery and reutilization of the solvent is necessary. Since recovery of solvent needs energy and equipment, the aim must be to obtain as much oil as possible from the miscella with an economically justifiable need of energy.

After leaving the solvent extractor, finest particles of the seed material resulting from the treatment during solvent extraction, but especially from the treatment during screw pressing, have to be separated from the miscella because they can influence the heat transfer during distillation. The part of particles depends on the equipment used for the extraction. While direct solvent extraction without prepressing results in comparable higher amounts of particles, the amount is much lower if prepressing is used before solvent extraction.

The separation of solvent and oil is achieved by conventional distilling methods by combining a series of sills, stripping columns, and condensers. In general, a two- or three-stage process is carried out. In the first step, hexane and steam vapors from the desolventizer-toaster used for the removal of hexane from the meal and at this step the mixture attains temperatures of about 50°C. Afterward, the resulting oil-enriched liquid is treated in a steam-heated exchanger, where the concentration of hexane decreases to about 5% of the miscella mass at a temperature of about 80°C.

Finally, the concentration of hexane in oil is reduced to less than 1000 ppm by passing the miscella through an oil stripper tower under pressure of 13.3 kPa and temperature of about 100°C. Temperature higher than 105°C adversely affects the quality of the crude oil. The oil is heated at 90–100°C in a preheater and steam is injected from the bottom to increase the effectiveness of the process. For a well-operated extraction plant, it is possible to maintain hexane losses during the process lower than 3 liter/MT of rapeseed processed (Unger, 1990).

XII. TOASTING OF THE MEAL

For the further commercialization of rapeseed meal, it is necessary to remove all hexane from the solvent extracted cake in order to observe the limits of the feeding stuff regulations and to avoid the danger of explosions. After the extraction process, about 25–35% solvent remains in the cake, which has to be removed by evaporation in a desolventizer-toaster. Another effect of this procedure is drying and crisping of the meal and the recovery of the solvent for further use.

Older equipments consisted of several superimposed horizontal tubes with a jacket. The meal is filled into the first tube, and a screw conveyer transports the product to the end of the tube where it falls into the next tube. Each tube is heated by steam to ensure a temperature of the meal of about 100°C.

Today most of the desolventizer-toasters consist of an enclosed vertical vessel equipped with several steam-heated horizontal floors. The solvent extracted cake enters the first floor by a screw conveyer, and a rotating sweep arm ensures a uniform heat transfer by mixing the cake on the floor. The rotating sweep arm transports the cake down onto the next floor. The cake enters the apparatus at about 60°C and heated during the process at 105°C. To improve the recovery of solvent, the addition of direct steam is used which ensures the recovery of hexane absorbed by the protein and "trapped" within the cell walls. Steam also acts as carrier to transport the solvent through the bed of extracted cake. Furthermore, steam effects the degradation of antinutritive compounds, such as glucosinolates, and the inactivation of enzymes. During the toasting process in large oil mills, about 1/3 to 2/3 of the glucosinolates in rapeseed meal are removed depending on the setting of the process (Schumann, 2005). To avoid the degradation of amino acids resulting in lowering the feeding value of the meal, a strict control of temperature, moisture, and retention time in the desolventizer-toaster is important.

XIII. OTHER EXTRACTION METHODS

One alternative to common methods used is the extraction of rapeseed by carbon dioxide under supercritical conditions. Under these conditions (temperature >31.1°C and pressure >73 atm) carbon dioxide adopts properties midway between a gas and a liquid, with comparable dissolving power as hexane. Carbon dioxide has the advantage of not being toxic, easy to remove from the oil, and being nonflammable. Thus, the method is very gentle to minimize deteriorations of the product during extraction, and the amount of minor compounds coextracted with the oil is reduced since carbon dioxide is highly selective for triacylglycerols. The disadvantage of carbon dioxide is that the application is very expensive, which makes the use only interesting for speciality oils produced in low amounts and sold at a high price. Another drawback is the difficulty to use supercritical carbon dioxide in a continuous extraction procedure, which would be preferred for large-scale processing (Booth, 2004).

XIV. PURIFICATION OF THE CRUDE OIL

Additionally to ninety-eight percent triacylglycerols, crude oil contains phospholipids, free fatty acids, colored pigments, sterols, waxes, oxidation products, moisture, aroma components as well as plant parts, and dirt in varying amounts depending on the conditions of the extraction process. The extensive the extraction conditions, the higher the amount of minor compounds, which strongly impair the quality and the shelf life of the oil. Refining involves different steps because otherwise the oil is not worth consumable. Native, cold pressed rapeseed oils are only freed by filtration, sedimentation, or centrifugation of particles from the disrupted seed material, which come during screw pressing into the oil. A separation of undesired minor compounds does not take place. Some cold pressed oils are also washed by hot water steam with the aim to improve the taste and the smell of the oil.

XV. COLD PRESSED OILS

Purification of native, cold pressed rapeseed oil means separation of solid impurities, mainly particles of the seeds. This is an important step to ensure the quality of the oil for a long storage. Disrupted particles of the seeds contain enzymes and also adherently microorganisms that can metabolize the oil to form degradation and metabolism products impairing especially the sensory quality of the oil. Depending on the settings of the screw press, speed of the rotating screw shaft, size of the press cake exit, temperature during pressing, and moisture of the seed, the amount of solid components in the oil ranges between 1% and 13%. The main aim of this type of purification is the separation of the two-phase system consisting of oil (liquid phase) and seed particles (solid phase). In general, two different methods, sedimentation and filtration, are in use, depending on the performance of the plant. A third method, centrifugation of the crude oil, is not widespread but seems to be very gentle.

Sedimentation uses the different density of the liquid and solid phase which leads to a slow settling of the solid particles as a result of gravity. There are two possibilities to use this method. The first one can be used as a batch system, at which a container is filled with crude oil and seed particles are allowed to sediment within 15–30 days. The method is only suitable for small plants with a low production of oil up to 50 kg/h. A drawback of this type of purification is the long contact of oil with the disrupted seed material. Investigations have shown that sedimentation in the batch system over a period of 21 days impair the oil quality significantly in comparison to other types of purification (Attenberger *et al.*, 2005). From this point of view, a second method seems to be more practical. This method is working continuously using a multilevel sedimentation tank, at which addition of crude oil, removal of purified oil, and elimination of enriched solid particles take place simultaneously (Fig. 14).

Filtration is the mechanical separation of solid and liquid phase, which allows removing insoluble solids or suspended material from a liquid by passing it through a porous medium only permeable for the liquid phase. In small- and medium-sized plants, this is mostly achieved by means of the formation of a filter cake from the seed particles between porous filter materials within a pressure gradient. The filter cake improves the result of the filtration. Sometimes filter aids can be used to increase the effect of the filter materials. These are inert cellulose materials that improve the formation of a filter cake for the filtration process. Filtration in small- and medium-sized plants is carried out by use of chamber filter presses or vertical pressure plate filters (Fig. 15), which enable a much higher output in comparison to sedimentation.

The last step of the purification process in small- and medium-sized plants is passing the oil through a fine pored protection filter to ensure a defined clarity

Fig. 14. Multilevel sedimentation system for the purification of rapeseed oil (Technologie und Förderzentrum, Straubing, Germany).

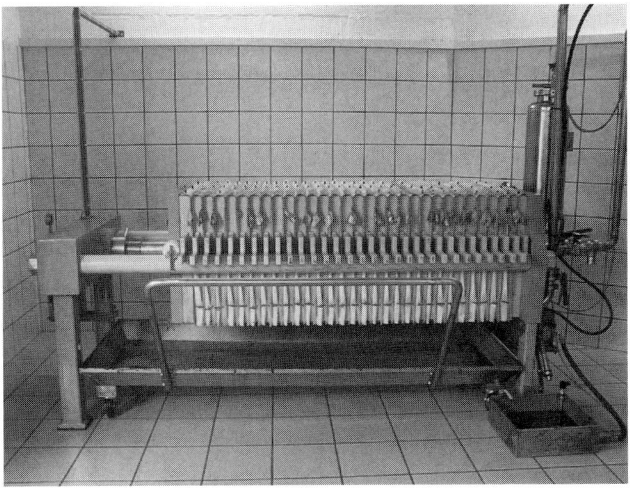

Fig. 15. Chamber filter press for the purification of native, cold pressed rapeseed oil in small- and medium-sized plants.

of the oil. This is necessary since according to DIN V 51605 (Anonymous, 2006), the amount of total impurities should be below 25 mg/kg for the use of pure rapeseed oil as fuel in engines.

Native, cold pressed rapeseed oil as edible oil is also in the market as natural turbid oil, which is not or only insufficient purified. The intensity of the sensory attribute *seedlike*, typical for native rapeseed oil, is stronger than in purified native oil. However, this type of oil has some problems with the formation of free fatty acids as a result of the formation of metabolism and degradation products from enzymes and microorganisms located on the seed particles (Attenberger *et al.*, 2005). Therefore, a longer storage of these oils is not advisable.

Some producers of cold pressed edible rapeseed oil use the treatment of oil with hot water steam after removal of seed particles. The aim of this treatment is to remove undesired volatile compounds, which negatively influence the sensory quality of the oil. In that case, purified oil is treated with hot water steam at about 100–120°C over a period of 1–4 h in a batch system. Under these conditions, no changes in the fatty acid composition or in the composition and the amount of tocopherols are noticeable. Nevertheless, the appearance of oligomer triglycerides or stigmasta-3,5-diene, as degradation product of β-sitosterol, a phytosterol, is a good indicator for a hot water steam treatment.

The refining process takes place in refinery, often directly connected to the oil mill, but sometimes also located at separate places. In general, refining is a multistage process that is carried out either as physical or chemical

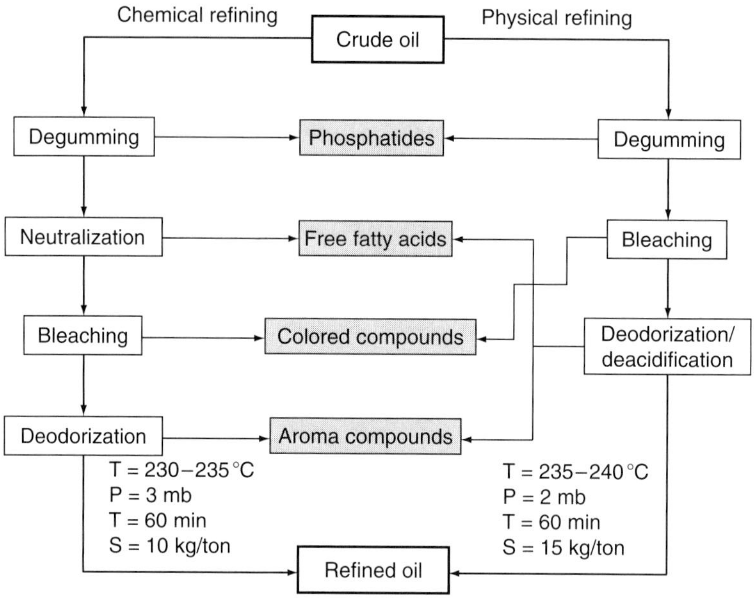

Fig. 16. Different steps of the refining process.

refining (Fig. 16). Using physical refining, free fatty acids are removed by hot water steam, while chemical refining uses alkaline solutions to neutralize free fatty acids. Oil refining involves the steps degumming, neutralization, bleaching, and deodorization in different variations and performances, whereas in most cases the oil mills carry out degumming.

XVI. DEGUMMING

Oils contain high amounts of phospholipids, which should be eliminated because they are jointly responsible for deterioration of the quality during storage. Autoxidation reactions during storage and processing lead to the formation of dark-colored substances from phospholipids and these substances are called melanophosphatides.

The content of phospholipids varies from 1 to 1.5% in the crude oil, corresponding to 400 and 600 mg/kg calculated as phosphorous. Two forms of phospholipids exist, hydratable and nonhydratable phospholipids, which often found together with different cations like magnesium, calcium, or iron. The ratio of both forms depends on the processing conditions and nonhydratable phospholipids usually amount between 20% and 45% of the total phospholipids in rapeseed oil. The removal of hydratable phospholipids is relatively easy by using water to form hydrated gums, which can be further removed by centrifugation. These hydrated gums obtained from crude rapeseed oil, mainly consist of lecithin (phosphaditylcholine) and cephalin (phosphatidylethanolamine).

Much more difficult is the elimination of nonhydratable phospholipids, which requires the use of mild organic acids such as citric and malic acid or phosphoric acid. These acids act as chelating agents to the nonhydratable phospholipids resulting in oil-insoluble metals, salts, and hydratable phospholipids, which can be easily removed.

Water degumming is carried out by using water at about 80°C in an amount of about 2% and intensive mixing with the oil. Hydratable phospholipids become insoluble within 5–45 min in the oil and form a precipitate that is kept in suspension by gentle agitation of the reaction vessel. The mixture consisting of oil and hydrated phospholipids is separated by using a centrifuge (Fig. 17). This results in the reduction of the amount of phosphorous to

Fig. 17. Centrifuge for the separation of oil/water mixtures (by courtesy of MUEZ-HEST Process Technologies Pvt. Ltd., Mumbai, India).

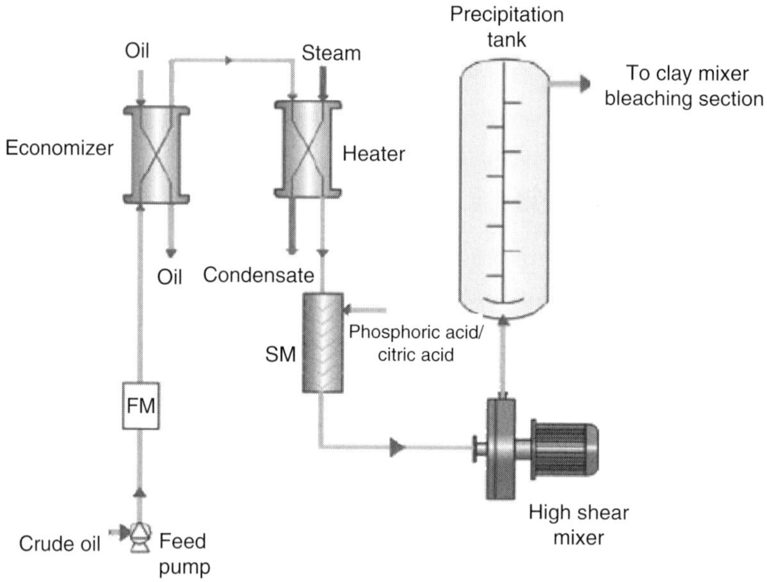

Fig. 18. Degumming (by courtesy of MUEZ-HEST Process Technologies Pvt. Ltd., Mumbai, India).

100 and 200 mg/kg. It is also possible to use hot steam instead of hot water, which improves the contact between oil and water resulting in a better separation of the phospholipids.

During water degumming, it is not possible to remove nonhydratable phospholipids. For this, the oil is mixed with small amounts of acid, for example 0.1–0.5% of a 50% solution of citric or malic acid or 0.1% of 75% phosphoric acid at temperatures between 60 and 70°C (Fig. 18). An intensive mixing of the solution is necessary to ensure an appropriate contact between oil and acidic solution, which influences the reaction time from some seconds to 20 min. During this time, nonhydratable phospholipids are changed to hydratable ones that can be removed from the oil by further addition of about 2% of water. After mixing the mixture for 30 min to 3 h for the formation of large enough agglomerates, the oil is separated from the phospholipids by centrifugation. The resulting oil contains less than 25 mg/kg phospholipids, measured as phosphorous (Mag, 2001), in the range between 5 and 25 mg/kg.

XVII. ALKALI AND PHYSICAL REFINING

Before further processes like bleaching, hydrogenation, or deodorization can be carried out efficiently free fatty acids and phospholipids have to be removed to a very low level. This is achieved either as alkali or physical

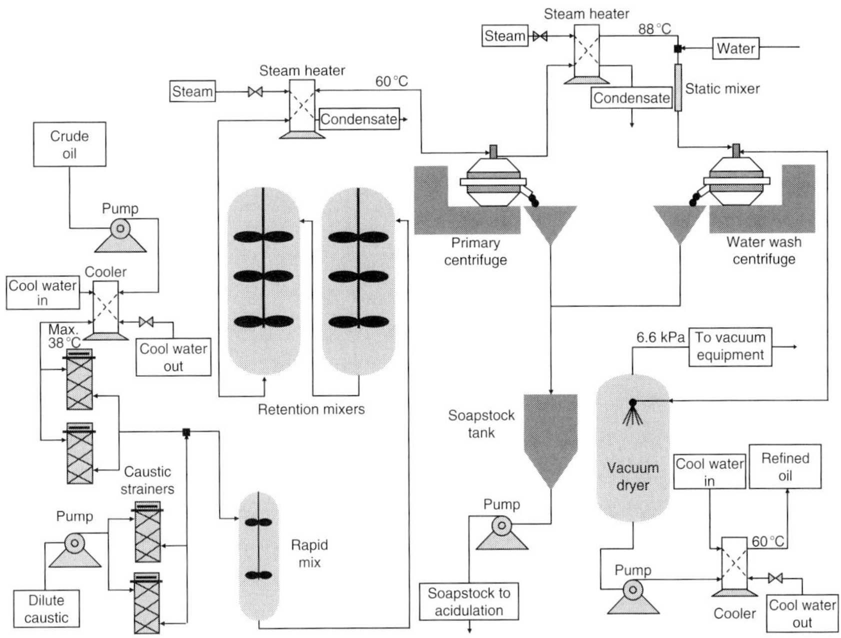

Fig. 19. Refining.

refining, depending on the realized degumming process. If the oil is water degummed or the quality of seeds processed in the oil mill is variable, alkali refining is the appreciable procedure since heat-sensitive compounds in the oil strongly influence the result of physical refining. On the other hand, if the oil is well degummed, physical refining has some advantages from an economical point of view: losses of neutral oil are reduced, less number of processing steps are necessary, no problems with the use of environmentally harmful alkaline solution, and lower capital and operation costs. Especially environmental problems arise from huge amounts of acid water resulting from the neutralization process. This water is highly acidic (pH 2–3) and contains fatty matters from the oil and other organic and inorganic materials that are considered as harmful to environment. Thus, before the water can be disposed to municipal sewers, pH-value has to be neutralized to pH 6–9 and the main part of the oil phase must be separated (Fig. 19).

XVIII. ALKALI REFINING

The first step of alkali refining is adding about 0.05–0.1% of concentrated phosphoric acid to the oil in a high-intensity mixer to precipitate phospholipids and especially chlorophyll and related compounds. Traces of prooxidant metals, if any, are removed. This results into fewer losses of neutral oil, less need of bleaching earth during the further process and better flavor and flavor stability of the oil after deodorization. Other organic acids like citric acid are not found to be as effective as phosphoric acid (Ohlsen and Svensson, 1976).

The free fatty acids are neutralized with a 2- to 3-N sodium hydroxide solution under continuously mixing to ensure contact between oil and alkaline solution, and heating. The amount of sodium hydroxide solution depends on the content of free fatty acids and should be in excess to neutralize phosphoric acid too. The need of sodium hydroxide solution can be determined by using a specific gravity hydrometer calibrated in degrees Baumé (Be).

The soapstock as a result of this procedure is removed together with precipitate phospholipids and other compounds either by continuously centrifuging or in smaller plants by gravity settling in a tank. Small amounts of soaps are removed by washing the oil with water, sometimes hot, which results in soap concentrations less than 50 mg/kg. In the last step, water is removed by drying the oil in a vacuum dryer. In large plants, with high throughput this process is carried out continuously, which has high capital costs because of the use of centrifuges. Small plants also can use batch alkali refining, which is much cheaper, because the separation of the soaps is achieved by gravity only. But the method has disadvantages with regard to the oil yield because of higher losses during the process. By alkali refining, the amounts of free fatty acids are reduced to <0.05%, phosphorous <5 mg/kg, chlorophyll <30 mg/kg, and sulfur <10 mg/kg.

XIX. PHYSICAL REFINING

Using physical refining, free fatty acids are removed from the oil by steam distillation. Therefore, this step in combination with the removal of volatile compounds during deodorization is to obtain bland and odorless oil. Another effect is the degradation of color compounds like carotenoids. The requirement for a successful physical refining is a low phosphorous concentration of the oil (<5 mg/kg) after degumming and bleaching of the oil in order to remove compounds susceptible to heat treatment. During the process, the oil is treated with 0.05–0.1% of 85% phosphoric acid and then further with 1.0% acid-activated bleaching clay at 95–105°C. After about 15 min, the oil is separated from the clay and other impurities, like precipitated phospholipids, chlorophyll, and carotenoids, absorbed onto the clay and transferred into the deodorizer. Depending on the performance, the deodorizer is used at temperatures between 220 and 260°C under 1- to 3-mbar vacuum for 1–3 h.

Another important factor, affecting the result of physical refining, is the amount of stripping steam used for the process. For physical refining, at least 15-kg steam/ton oil is recommended, while the steam consumption during chemical refining is about 10 kg/ton. For a successful process, the interplay between temperature, pressure, and amount of stripping steam is important for the result. The lower pressure and higher temperature result into higher volume of steam resulting from 1 kg of water since the amount of volatile compounds removed from the oil during the process depends on the volume of stripping steam produced.

During this process, most of free fatty acids and volatile compounds responsible for the odor and taste of the oil are removed which may further result into almost tasteless and odorless oil used universally in kitchen.

XX. BLEACHING

Bleaching of the oil by use of adsorbents is necessary, not only as preparation for hydrogenation or physical refining but also for the processing of bottled edible rapeseed oil. The objective of this step is to remove not only colored compounds like chlorophylloids and carotenoids but also soaps, oxidation products, and other minor compounds that retain on the absorptive material. Bottled edible rapeseed oil often is presented in clear glass bottles in the stores, which give no protection against light. Thus, a low concentration of chlorophyll is important to enlarge the shelf life of the oil. More effective is hydrogenation or deodorization whereby the compounds are destroyed by heat treatment and the oils become brighter and less reddish-yellow colored.

For a successful and economically useful bleaching process, it is necessary that the oils fulfill some requirements regarding the content of free fatty acids, soaps, phosphorous, and water. In the case of alkali refining the following limits should be kept: free fatty acids $<0.05\%$, soaps <50 mg/kg, phosphorous <5mg/kg, and water $<0.1\%$, while the soaps and phosphorous should be below 50 and 20 mg/kg, respectively, for oil provided for physical refining (Zschau, 1999). Notably, the amount of soaps and phosphorous is critical since these compounds inactivate the active centers of the clays, which results in worse quality of oil. A higher content of water influences the speed of bleaching concerning drying and adsorption. An excessive amount of water supports hydrolysis of oil during processing (Bockisch, 1993).

The process of bleaching is based on the interaction between the absorptive material (clay) and the solute, which has to be removed from the oil by formation of van der Waals bindings. In the case of soaps and phosphatides, the removal takes place by adsorption of the compounds in pores of the clays.

Usually bleaching is carried out under vacuum to reduce oxidative reactions as a result of high temperature and large surface of the bleaching material. The most important factors in bleaching are the temperature, the period of contact between oil and clay, the amount of clay, and the degree of drying of the oil–clay slurry (Fig. 20).

The oil is heated in the bleacher (Fig. 21) to about 70–80°C before the acid-activated clay is added directly or as a slurry (premix of oil and clay). Then heating continues to about 95–105°C, while the oil–clay slurry is continuously dried to about 0.1% moisture. The amount of clay depends on the oil quality (e.g., chlorophyll content) and the product specification expected as a result of the process. Usually the amount of clay is between 0.5% and 2.5% in the oil. One important point is the contact time between oil and clay, which should be between 5 and 30 min, depending on the quality of oil and the amount of clay used for the processing. Within this time, an equilibrium between coloring compounds adsorbed by the clay and coloring compounds remaining in the oil should be reached, whereas the ratio should be moved in the direction of coloring compounds adsorbed by the clay. By intensive mixing of the slurry, which improves the contact between oil and clay, this equilibrium can be reached within a relatively short time. After reaching this point, no more coloring compounds are absorbed from the oil by the clay. Therefore, a multistep countercurrent process seems to be more effective with regard to clay consumption. In two or more consecutive steps, clay is added to the oil which improves the result of bleaching with a lower consumption of clay.

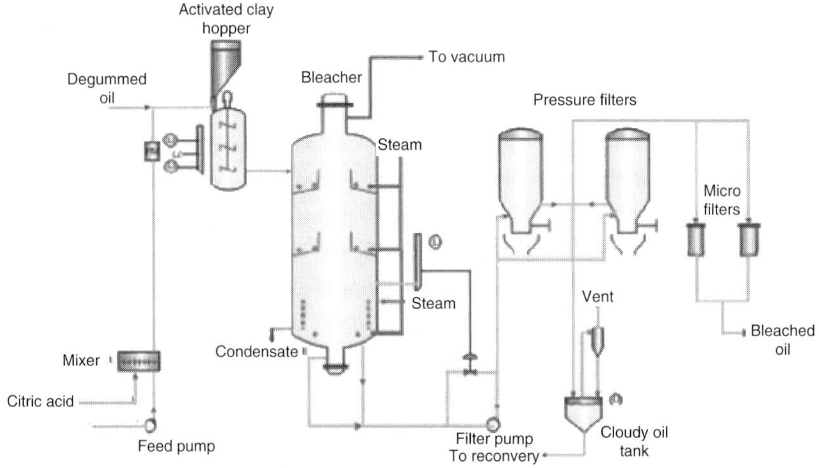

Fig. 20. Bleaching (by courtesy of MUEZ-HEST Process Technologies Pvt. Ltd., Mumbai, India).

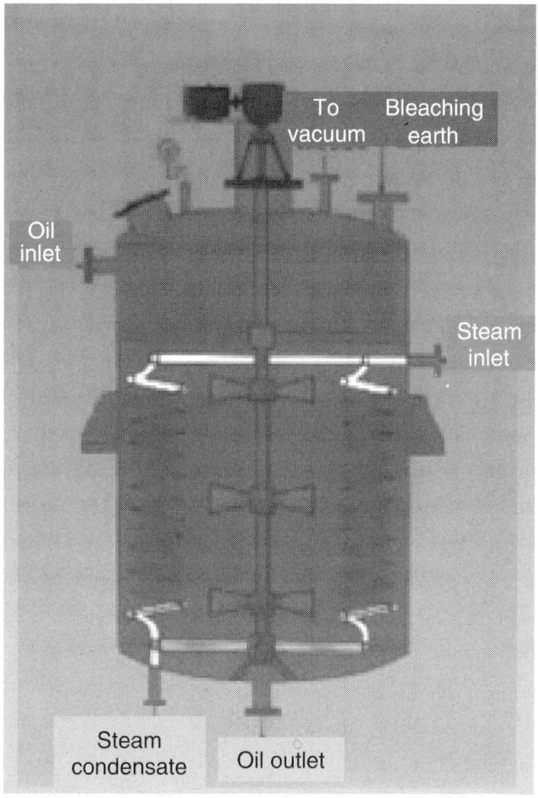

Fig. 21. Bleacher (by courtesy of MUEZ-HEST Process Technologies Pvt. Ltd., Mumbai, India).

The separation of oil and clay takes place by filtration. In most cases, horizontal or vertical leaf filters are in use and the mixture pumped through the filter returns to the bleaching vessel in order to build up a filter cake. At the end of filtration, the cake is dried by steam blowing or nitrogen to lower the amount of remaining oil in the bleaching earth. Nevertheless 20–40% of the cake consists of oil.

XXI. DEODORIZATION

The last step in purification of crude oil in large-scale plants is deodorization. The main objective of this step is to remove all compounds which negatively influence the taste and the odor of the oil. Besides, products formed during the processing, for example bleaching and hydrogenation, are also removed. The resulting oil is neutral in taste and smell, only light yellow colored, and low in free fatty acids with a good stability. Therefore, deodorization often is carried out after hydrogenation if the fatty acid composition of the oil should be changed. Another effect of deodorization is heat bleaching of colored carotenoid compounds at higher temperatures as mentioned above. Last but not least, deodorization is an important step in physical refining to remove free fatty acids.

The principle of deodorization is steam distillation at high temperature (220–250 and 270 °C for physical refining) and low pressure (0.3–0.6 kPa). By the use of low pressure, the volatility of compounds to be removed increases and the compounds are kept away from the oil. Under these conditions, the volatility of compounds to be removed is high enough to leave the deodorizing vessel with steam as carrier except triglycerides. The most important point during the process is that sufficient contact between oil and steam has to be achieved to ensure the removal of all undesired compounds.

Normally double bonds in a fatty acid molecule from crude vegetable oil are arranged in a *cis*-configuration, which is energetically more inconvenient than the *trans*-configuration. The hydrogen atoms are located on the same side of the double bond, which results in a bending of the molecule. The more double bonds in the molecule, the more the molecule is bended. In *trans*-fatty acids, the hydrogen atoms are on opposite sides of the double bond resulting in a more convenient state without bending of the molecule. Thus, the molecule is straight, comparable to saturated fatty acids like stearic acid. Similar to saturated fatty acids, *trans*-fatty acids are found to be responsible for adverse effects on serum lipoproteins and apolipoproteins, which may cause cardiovascular diseases (Precht and Molkentin, 1995; Stender *et al.*, 1995). Since the energy barrier to form *trans*-fatty acids from *cis*-ones is low (125 kJ/mol), conditions occurring during deodorization are sufficient to increase the amount of *trans*-fatty acid in refined oil up to 2% depending on the conditions during processing (Fig. 22). In comparison, native, cold pressed rapeseed oils do not contain more than 0.2% *trans*-fatty acids, otherwise a higher amount is a strong indication of a heat treatment of the oil during production. In gentle steamed oils, notable higher amounts of *trans*-fatty acids can be found as a result of a treatment with hot steam.

Additionally to *trans*-fatty acids, stigmasta-3,5-diene is formed from β-sitosterol, a naturally occurring phytosterol. This reaction takes place in the presence of heat as acid-catalyzed reaction with elimination of water (Fig. 23) and also during treatment with strong bleaching earths. The resulting compound, stigmasta-3,5-diene, can be used as indicator for such treatments of oil. While the amount of stigmasta-3,5-diene in crude rapeseed oil should be below 0.15 mg/kg, comparable to the specification in EEC No 2568/91 for native olive oil, in fully refined rapeseed oil remarkably higher amounts can be found indicating a deodorization or bleaching process. Already during

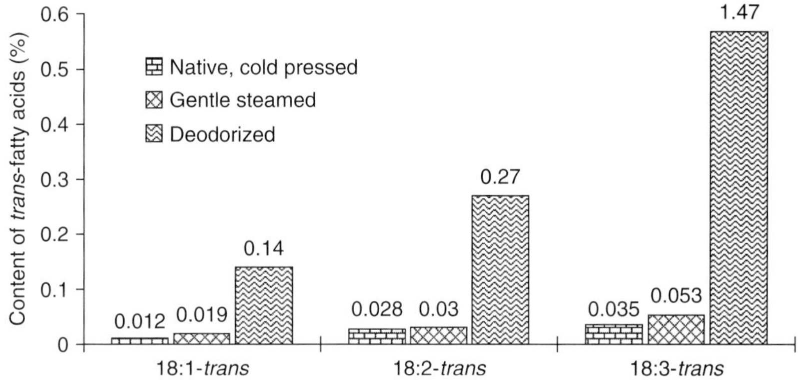

Fig. 22. Formation of *trans*-fatty acids during processing (by courtesy of Dr. L. Brühl, BfEL).

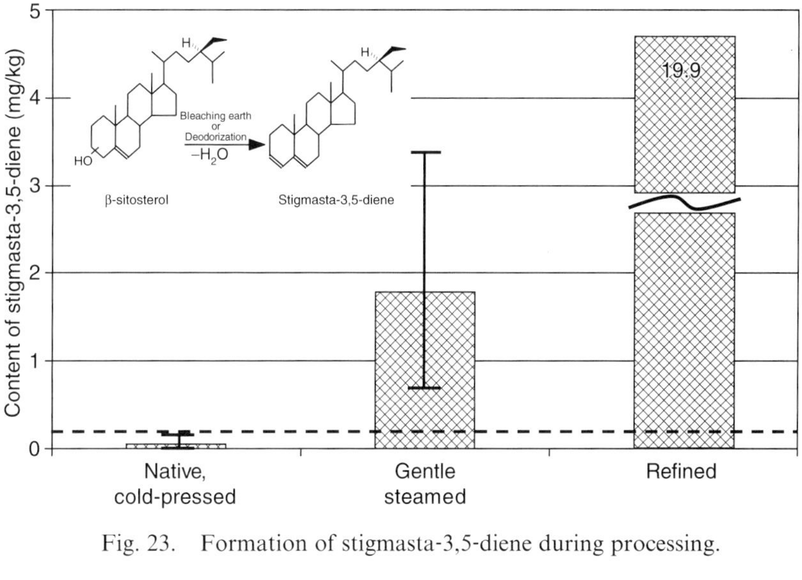

Fig. 23. Formation of stigmasta-3,5-diene during processing.

gentle steaming of the oil to remove undesired aroma compounds, the content of stigmasta-3,5-diene in the oil increases.

Another indicator for the refining process is the amount of oligomer triglycerides that are formed during deodorization. The amount should be below 0.1% in crude or cold pressed rapeseed oil and in fully refined oils amounts up to 0.5% can be found.

Deodorization can be carried out either as batch process or in a semi-continuous or continuous process mode. In modern plants, the latter types are preferred because they have some advantages regarding the efficiency of the process and the consequential better quality of the resulting oil. The batch process needs up to 10% more steam because the oil bed which has to be penetrated by the steam is much higher than in continuously working processes. Additionally, a relatively long processing time (>4 h) makes batch deodorization less interesting to large plants. The advantages of the batch type are lower capital investment and simplicity of operation. Batch deodorization is suitable for small plants with capacities up to 50 tons/day or plants with a wide range of different oils. In principle, the process is carried under different steps such as deaeration, heating, deodorization, and cooling of the oil.

The deodorizer vessel in the batch process is a cylindrical tank. Amounts of 10–20 tons of oil are pumped into the vessel, and after deaeration, it is heated up by heating coils. Deaeration is necessary to ensure a low concentration of oxygen in the vessel to reduce the risk of oxidation and polymerization during the heat treatment. Hot steam necessary for the process is supplied near the bottom of the vessel and a vacuum system provides for low pressure. The process takes about 3–5 h and 3–5% of steam is used. During this time, the temperature ranges between 180 and 240°C and the effect of deodorization starts at about 160°C. Afterward, the vessel is cooled down to 60–80°C under vacuum and the oil is sparged with nitrogen before storing in tanks or bottles.

For continuous working processes, different types of deodorizers are available. One common used type consists of a vertical stainless-steel vessel in which four to seven trays are stacked. At the beginning of the process, the oil is pumped under vacuum into the top tray of the deodorizer, which is heated. The oil is deaerated and heated to about 160°C within 15–20 min and moves down stepwise from one tray to the next countercurrent to a superheated, stripping steam flow. By this way, the oil is further heated to 220–270°C, depending on the conditions and agitated with sparging steam at about 0.5% per hour. In the deodorizing section of the vessel, about 1.5% of sparging steam is used before the oil reaches the final cooling to about 50–60°C. Finally, oil is transferred to a storage tank by passing through a polishing filter (Fig. 24).

Another type is based on a series of horizontal, single-shell, and cylindrical vessels. Hot steam is injected in this apparatus by several pipe distributors running the length of the vessel. The advantage of this type enables the maximal surface for the contact between oil and hot steam.

The material removed from the edible oil is usually recovered in a scrub cooler where it is separated from the steam. Distillates from physical refining mainly consist of free fatty acids and a lower amount of unsaponifiable matter. It can be used for feed industry if the amounts of pesticides are low

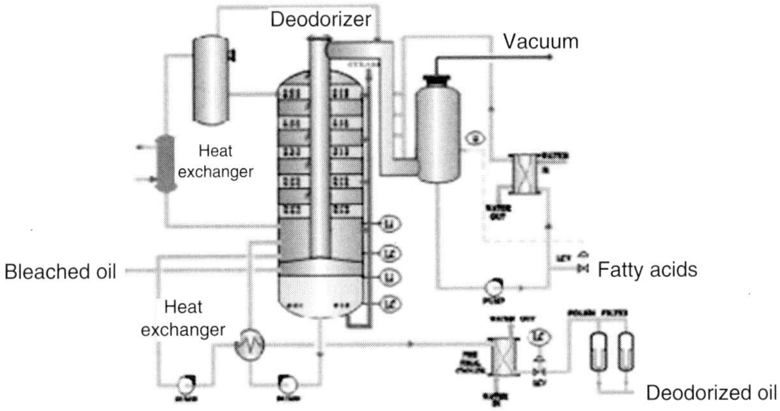

Fig. 24. Deodorization (by courtesy of MUEZ-HEST Process Technologies Pvt. Ltd., Mumbai, India).

enough. Distillates from chemical refining are more valuable because they often contain higher amounts of tocopherols, which can be recovered for the production of vitamin E. On the other hand, higher amounts of tocopherols and sterols make the use of the distillates from rapeseed oil unsuitable for the production of fatty acids for technical purposes, although they are excellent sources for C18 fatty acids.

XXII. EFFECT OF REFINING ON CONTAMINANTS

Like many other products from agriculture oilseeds sometimes contain contaminants such as aflatoxins, polycyclic aromatic hydrocarbons (PAHs), pesticides, or traces of metals. Most of these contaminants are ubiquitous in environment in low concentrations, but sometimes high amounts can be found as a result of improper processing. For example, drying of seeds by open fire results in high amount PAHs in the oil or sometimes farmers use pesticides to improve the result of harvest or to keep the quality of the seeds during storage without following the appropriate waiting periods. In principle, these compounds have no place in food and the aim must be to keep the amount as low as achievable.

Refining provides different possibilities to reduce the amount of contaminants in the resulting oil to a very low level. On the other hand, it must be taken into consideration that these possibilities do not exist during processing of native, cold pressed rapeseed. The producer is particularly responsible for the quality of the raw material; otherwise, everything that is soluble in oil can be found in the product after processing, with no chance to remove it afterward.

Each step of the refining process contributes to the purification of the oil. Removal of aflatoxins is exhaustive during refining. During neutralization, with bases up to 98% of aflatoxins are destroyed because the lactone binding of the molecule is broken. Aflatoxins are removed completely during bleaching, resulting into no aflatoxins in refined oils.

At higher temperatures for a longer time, the removal of PAHs from the oil is possible if the amounts at the beginning are not too high. If the amount of heavy PAHs is high, use of charcoal is the method of choice, while light PAHs are removed during deodorization.

A removal of trace metals as comprehensive as possible is necessary to ensure the oxidative stability of the oil during storage because metals catalyze the degradation of fatty acids. A part of these compounds is removed during degumming and neutralization by carrying away with gums or soaps.

XXIII. MODIFICATION OF RAPESEED OIL

Rapeseed oil is one of the most important vegetable oils, which is widely used as salad oil, in food formulations as prepared salad dressings or mayonnaise and food preparations. The modification of rapeseed oil is a practicable way to change the properties of the oil as desired and required. It is possible to

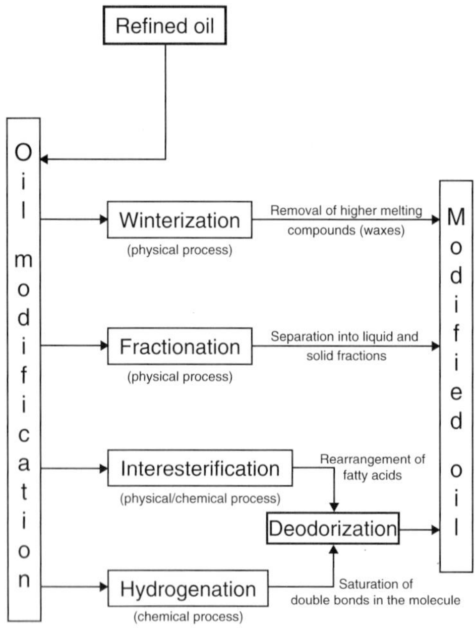

Fig. 25. Different possibilities to modify rapeseed oil to oil with well-defined properties.

modify the properties of rapeseed oil within certain limits, which enables the production of oil with desired quality parameters.

The modification of rapeseed oil requires winterization, fractionation, interesterification, and hydrogenation (Fig. 25). Fractionation and winterization are physical processes that do not change the triglyceride molecule but separate different fractions on basis of different melting points or solubilities. It is possible to build new triglycerides without changing the components of the oil by interesterification. The process of hydrogenation directly interferes into the fatty acid molecule by adding hydrogen into the double bonds. This results into saturated or monounsaturated fatty acids with a higher stability against oxidation reactions and higher melting points.

XXIV. WINTERIZATION

Winterization is a sort of fractionation, which is used to remove traces of higher melting parts such as waxes from the oil. These esters of fatty acids and long-chain alcohols are responsible for cloudiness in the oil during storage at lower temperatures. The name winterization is a leftover from earlier times when oil was stored outside in the winter in barrels to allow higher melting components to solidify.

Oils with a hazy appearance are often disliked by the consumer, although these components have no negative health effect. Nevertheless, industry endeavors to produce clear oils with a good cold stability, which remain clear also during storage in the refrigerator. In that case, the level of higher melting compounds should be below 10 ppm. Another important point is that rapeseed oil is used in the production of salad dressing and mayonnaise. Although the amount of waxes is small in rapeseed oil, fat crystals can destroy the emulsion or the desirable properties since these products are normally stored in the refrigerator.

The principle of this procedure is to cool down the oil to temperatures between 0 and 19°C for at least 6 h. During this time, the crystallization of high melting compounds becomes complete, and after addition of a filter aid, these compounds can be removed by filtration at about 15°C. In most cases, rotary drum filters and horizontal and vertical tank filters are used for the filtration process. Another possibility is centrifugation of the cooled oil. In this process, the oil is introduced into a crystallizer, where waxes have time to agglomerate. Afterward, the oil is gently heated and pumped into the centrifugal separator. This procedure results in excellent oil qualities, but high oil losses have to be taken into consideration and the residual wax content of the product is too high for a good cold stability (50 ppm).

XXV. INTERESTERIFICATION

Interesterification is a method that does not change the composition of rapeseed oil, but changes the distribution of the fatty acids within the triglyceride molecules. That means the fatty acid composition remains the same and the triglyceride composition is changed with changing the melting behavior and physical properties of the oil. The advantage of this procedure in comparison to modification of the oil by hydrogenation is that the equipment is readily available and no explosive gases are necessary. Another important advantage in a nutritional point of view is that the products are very low in *trans*-fatty acids.

The principle of the method is that the oil is separated within a short time into the different components, fatty acids, and glycerin and then these components are rearranged. The result is a new product with new physical specifications. Different types of interesterification are followed, which include (1) interchange between a fat and free fatty acids (acidolysis), (2) interchange between a fat and an alcohol (alcoholysis), and (3) rearrangement of fatty acids within triglycerides (ester interchange) (Going, 1967).

The reaction of the chemical interesterification can take place in a randomized or a directed form. In the random interesterification, the fat or oil is heated in a closed vessel under vacuum or nitrogen to 70–100°C, 0.05–2% catalyst is added, depending on the sort of catalyst and then the mixture is mixed for 1–120 min. Normally the reaction takes place with an amount of catalyst between 0.05% and 0.15% for about 30 min. For the reaction, it is necessary that the oil and the catalyst are completely dry because otherwise the catalyst is inactivated or soaps are formed. Another critical point is free

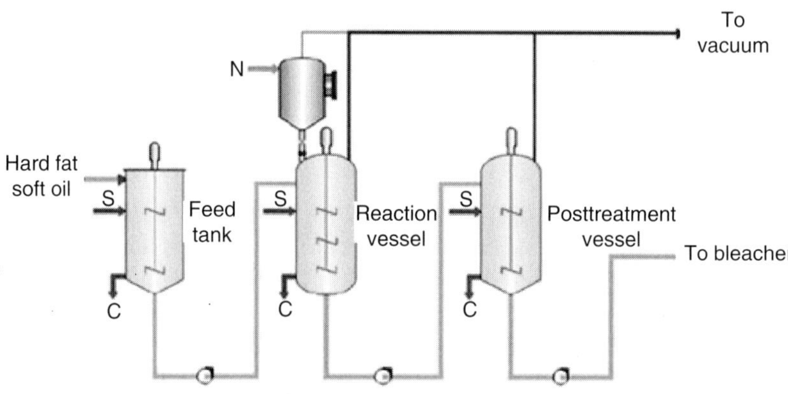

Fig. 26. Interesterification (by courtesy of MUEZ-HEST Process Technologies Pvt. Ltd., Mumbai, India).

fatty acids or peroxides that also will poison the catalyst. The reaction takes place spontaneously and the appearance of a brown coloration, resulting from oxidation products, shows the beginning of the reaction. This brown color has to be removed during the following bleaching process. At the end of the process, the catalyst must be neutralized by addition of diluted mineral acids (Fig. 26).

XXVI. HYDROGENATION

Rapeseed oil contains about 30% of polyunsaturated fatty acids that are 10–100 times more susceptible to oxidative deterioration than monounsaturated or even saturated fatty acids. The melting point of rapeseed oil is also characterized by the high degree of unsaturation. Linolenic, linoleic, oleic, and stearic acids melt at −13, −7, 16, and 70°C, respectively. Thus, for the production of margarines and shortenings or for the use as frying oil, the properties of rapeseed oil have to be changed. Hydrogenation is one method to improve the oxidative stability and to increase the melting point of the oil.

The principle of the procedure is to saturate double bonds within the fatty acid molecule by addition of hydrogen, which eliminates the possibility of reaction with oxygen. The iodine value of rapeseed oil decreases from 110–125 to 75–95 in hydrogenated oil by following the above method. In 1903 W.Normann got a patent for the industrial hydrogenation of animal and vegetable oils. A slight hydrogenation results in higher oxidative stability of rapeseed oil since polyunsaturated fatty acids changes into monounsaturated oleic acid, which is more stable.

The main components of the hydrogenation process are a vessel or converter in which the reaction takes place, equipped with components for heating, cooling, and agitation and means for the supply with hydrogen. Two different methods for the production of hydrogen are in use, either the electrolytic process or the steam hydrocarbon process. For processing, oil is loaded into the vessel and then the oil is heated under agitation and vacuum to remove air and moisture from the oil. The catalyst is added and after the desired hydrogenation temperature is reached, the evacuated vessel is pressurized with hydrogen gas, the oil is cooled down and then filtered to remove the catalyst (Fig. 27). The most important parameters responsible for the degree of hydrogenation are temperature, pressure, concentration of catalyst, properties of catalyst, and agitation.

One disadvantage of hydrogenation, which comes more and more into discussion from a nutritional point of view, is the formation of *trans*-fatty acids during partial hydrogenation. Isomerization occurs at the double bond linkage in the fatty acid molecule comparable to the deodorization process.

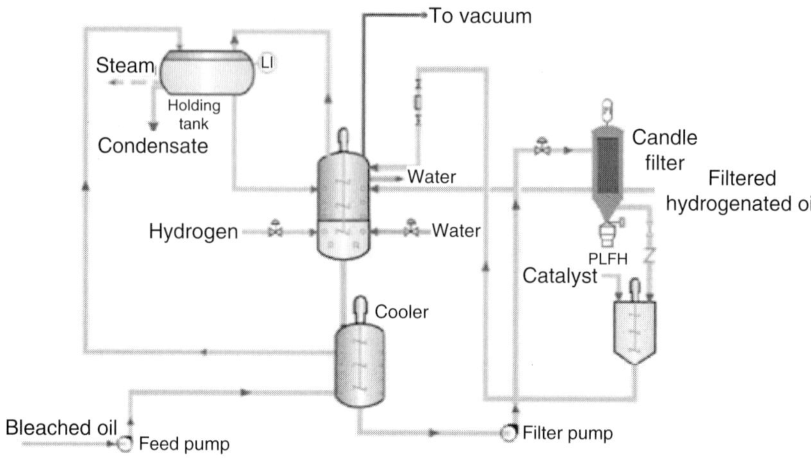

Fig. 27. Hydrogenation (by courtesy of MUEZ-HEST Process Technologies Pvt. Ltd., Mumbai, India).

Another effect that has to be taken into account for the further use of partially hydrogenated rapeseed oil is that the melting point of elaidic acid (44°C), the *trans*-form of oleic acid, is much higher than of the *cis*-form (16° C). This is important for the preparation of products like margarine, shortenings, and confectionery fats, where the melting point is one of the critical parameters.

There are two possibilities to avoid *trans*-fatty acids in rapeseed oil during hydrogenation. The first one is to adjust the operating conditions (temperature and pressure) in a way that the amounts of *trans*-fatty acids remain low and the other way is to harden the oil to a very low iodine value by following interesterification with unhydrogenated oil (List *et al.*, 1995).

The main use of rapeseed oil for nutritional properties is as salad oil and in food formulations and preparations such as margarine, mayonnaise, or salad dressings (Fig. 28). In these cases, the oil is used as refined, bleached, and deodorized oil because it is necessary to ensure no negative impact of the product by the oil with regard to color and taste. Another field of use is the application of rapeseed oil as baking and frying shortenings. Recently, the use of high oleic, low linolenic rapeseed oil with more than 70% of oleic acid and less than 3% of linolenic acid is an interesting alternative to partially hydrogenated rapeseed oil as frying medium. As of the absence of *trans*-fatty acids and a comparable or better performance during frying, this oil can help to overcome the problems of rapeseed oil during frying (Matthäus, 2006).

A relatively new field for rapeseed oil is the use as cold pressed oil similar to native olive oil. The consumers appreciate the *seedlike* and sometimes *nutty*

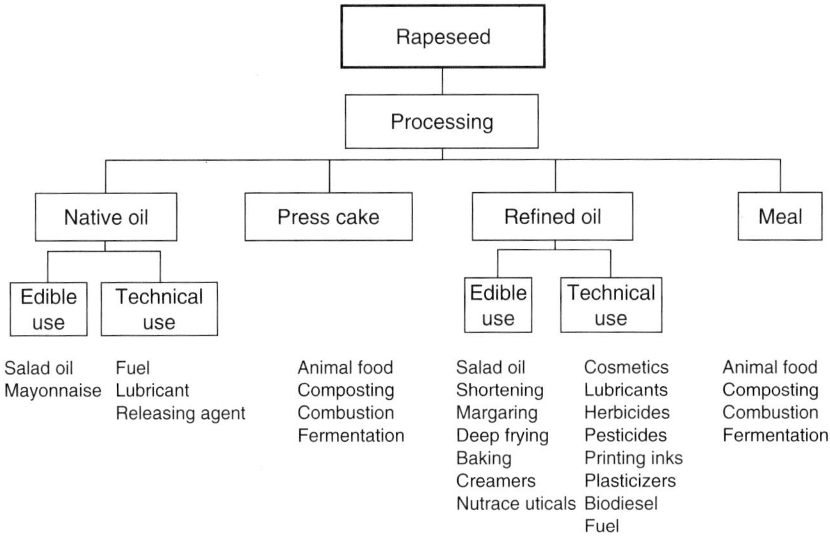

Fig. 28. Use of rapeseed oil.

taste and smell as well as the small processing of this product. The problem is that the quality of the oil in the market is very heterogeneous, and it is not possible for the consumer to recognize which is high-quality cold pressed oil.

The main use of rapeseed oil is in the nonfood sector. Apart from some special varieties with high amounts of erucic acid, which are used as lubricants, conventional rapeseed oil becomes more and more interesting as fuel in diesel engines. There are four different ways by which rapeseed oil can be used as fuel: pure rapeseed oil, mixture between pure rapeseed oil and diesel, rapeseed oil methyl ester (RME, biodiesel), or mixtures of RME and diesel. In most cases, rapeseed oil is used in form of methyl esters or mixtures of RME and diesel because this has been tested for its successfulness for many years. The use of pure rapeseed oil is not widespread, but the basics look promising, although the product has difficulties with quality aspects if native, cold pressed rapeseed oil is used.

XXVII. CONCLUSION

The rapeseed oil is used as refined, bleached, and deodorized oil because it is necessary to ensure no negative impact of the product by the oil with regard to color and taste. The original oil has some disadvantages because of the relative high amount of polyunsaturated fatty acids, but by hydrogenation or interesterification the stability of rapeseed oil for such processes is highly improved with some higher amounts of *trans*-fatty acids in partially hydrogenated rapeseed oil. Recently, the use of high-oleic, low-linolenic rapeseed oil with more than 70% of oleic acid and less than 3% of linolenic acid is an interesting alternative to partially hydrogenated rapeseed oil as frying medium. As of the absence of *trans*-fatty acids and a comparable or better performance during frying, this oil can help to overcome the problems of rapeseed oil during frying (Matthäus, 2006).

A relatively new field for rapeseed oil is the use as cold pressed oil similar to native olive oil. The consumers appreciate the *seedlike* and sometimes *nutty* taste and smell as well as the small processing of this product. The problem is that the quality of the oil in the market is very heterogeneous and it is not possible for the consumer to recognize which is high-quality cold pressed oil.

REFERENCES

Anonymous (2006). "Fuels for vegetable oil compatible combustion engines—Fuel from rapeseed oil—Requirements and test methods—DIN V 51605." Beuth Verlag, Berlin.

Attenberger, A., Matthäus, B., Brühl, L. and Remmele, E. (2005). Research into the influencing factors on the quality of cold pressed rapeseed oil used as edible oil and determination of a quality standard. Technologie- und Förderzentrum. Eigenverlag, Technologie- und Förderzentrum, pp. 84–91.

Beach, C. D. H. (1983). Rapeseed crushing and extraction. *In* "High and Low Erucic Acid Rapeseed Oils" (J. K. G. Kramer, F. D. Sauer and W. J. Pigden, eds.), pp. 181–195. Academic Press, New York.

Bockisch, M. (1993). "Handbuch der Lebensmittel-Technologie Nahrungsfette und –öle," pp. 208–215. Verlag Eugen Ulmer, Stuttgart.

Booth, E. J. (2004). Extraction and refining. *In* "Rapeseed and Canola oil—Production, Processing, Properties and Uses" (F. D. Gunstone, ed.), pp. 17–36. Blackwell Publishing, Oxford.

Carr, R. A. (1995). Processing the seed and oil. *In* "Brassica Oilseeds; Production and Utilization" (D. Kimber and D. I. McGregor, eds.), pp. 267–290. CAB International, Wallingford, Oxon.

Going, L. H. (1967). Interesterification products and processes. *Journal of the American Oil Chemists Society* **44**, 414A–456A.

Kock, M. (1981). Practical experience with a process for enzyme deactivation of soybean flakes before extraction and its influence on the oil quality. American Soybean Association Congress, Antwerpen.

List, G. R., Mounts, T. L., Orthoefer, F. and Neff, W. E. (1995). Margarine and shortening oils by intesterification of liquid and trisaturated triglycerides. *Journal of the American Oil Chemists Society* **72**, 379–382.

Mag, T. (2001). "Canola Seed and Oil Processing." Canola Council of Canada, (http://www.canola-council.org).

Matthäus, B. (1998). Effect of dehulling on the composition of antinutritive compounds in various cultivars of rapeseed. *European Journal of Lipid Science and Technology (formerly Fett/Lipid)* **100**, 295–301.

Matthäus, B. (2006). Utilization of high-oleic rapeseed oil for deep-fat frying of French fries compared to other commonly used edible oils. *European Journal of Lipid Science and Technology* **108**, 200–211.

Niewiadomski, H. (1990). Preliminary technological operations. *In* "Rapeseed-Chemistry and Technology" (H. Niewiadomski, ed.), pp. 123–160. Elsevier, Amsterdam.

Ohlsen, R. and Svensson, C. (1976). Comparison of oxalic acid and phosphoric acid as degumming agents for vegetable oils. *Journal of the American Oil Chemists Society* **53**, 8–11.

Piva, G., Pietri, A., Maccagni, A. and Santi, E. (1985). Fattori antinutrizionali della farina di estrazione di colza. *La rivista Italiana delle sostanze grasse* **62**, 99–103.

Precht, D. and Molkentin, J. (1995). *Trans*-fatty acids: Implications for health, analytical methods, incidence in edible fats and intake. *Die Nahrung* **39**, 343–374.

Schneider, F. H. (1979a). Schälung von Rapssaat durch definierte Verformung. I. Untersuchungen zur Saatanatomie. *Fette Seifen Anstrichmittel* **81**, 11–16.

Schneider, F. H. (1979b). Schälung von Rapssaat durch definierte Verformung. II. Untersuchungen zum Schalverhalten. *Fette Seifen Anstrichmittel* **81**, 53–59.

Schneider, F. H. and Khoo, D. (1986). Trennpressen—Versuch einer Bestandsaufnahme experimenteller Arbeiten. *Fette Seifen Anstrichmittel* **88**, 329–340.

Sosulski, K., Sosulski, F. W. and Coxworth, E. (1988). Carbohydrase hydrolysis of canola to enhance oil extraction with hexane. *Journal of the American Oil Chemists Society* **65**, 357–361.

Stender, S., Dyerberg, J., Holmer, G., Ovesen, L. and Sandström, B. (1995). The influence of trans fatty acids on health: A report from The Danish Nutrition Council. *Clinical Science* **88**, 375–392.

Schumann, W. (2005). "Glucosinolatgehalt von in Deutschland erzeugten und verarbeiteten Rapssaaten und Rapsfuttermitteln," pp. 1–69. UFOP, Berlin.

Unger, E. H. (1990). Commercial processing of canola and rapeseed: Crushing and oil extraction. *In* "Canola and Rapeseed; Production, Chemistry, Nutrition, and Processing Technology" (F. Shahid, ed.), pp. 235–249. van Nostrand Reinhold, New York.

Zschau, W. (1999). Die Bleichung von Speisefetten und Ölen VII. Qualitätskontrolle der Bleichung. *European Journal of Lipid Science and Technology (formerly Fett/Lipid)* **101**, 117–119.

AUTHOR INDEX

Numbers in **bold** refer to pages on which full references are listed.

N

T

U

SUBJECT INDEX

C

H

I

T

Plate 1

Plate 2

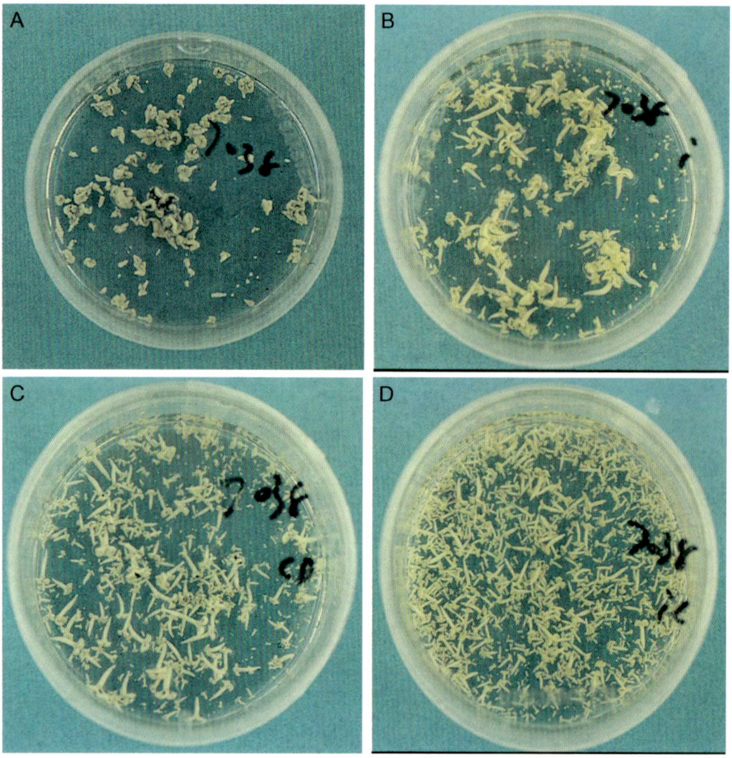

Plate 3

Plate 4

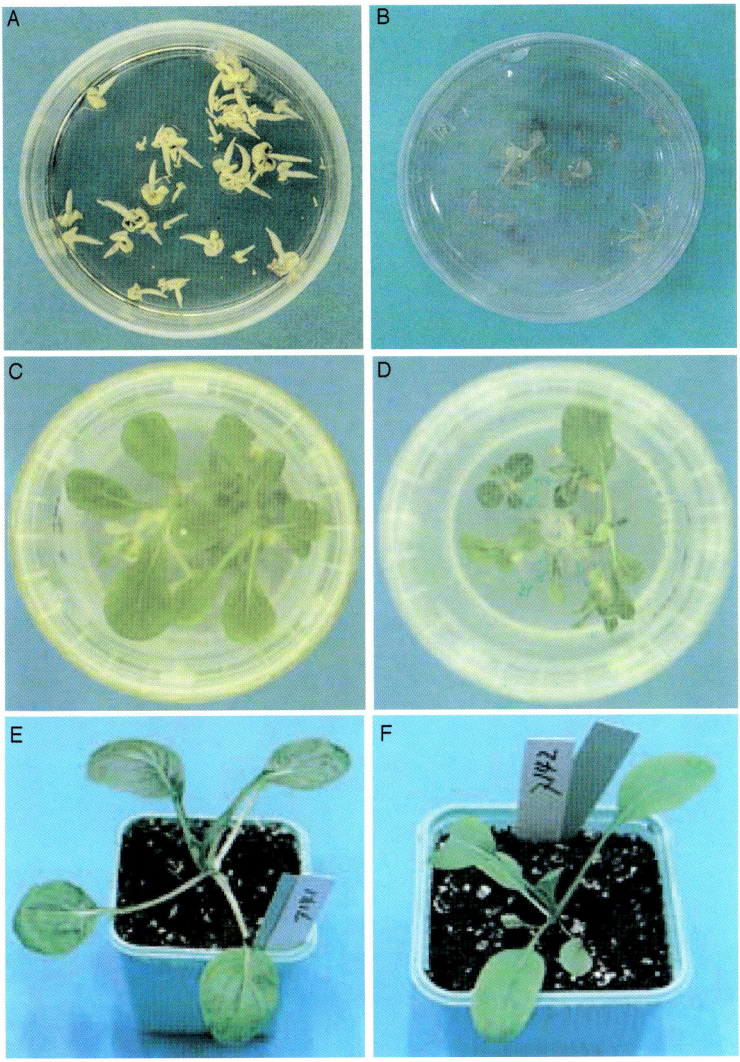

Plate 5

Plate 6

Plate 7

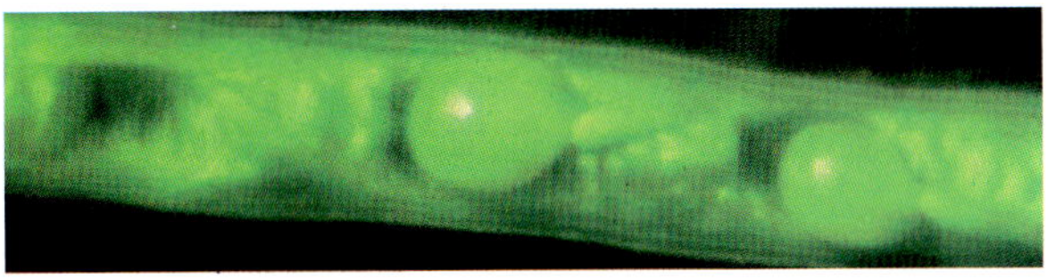

Plate 8

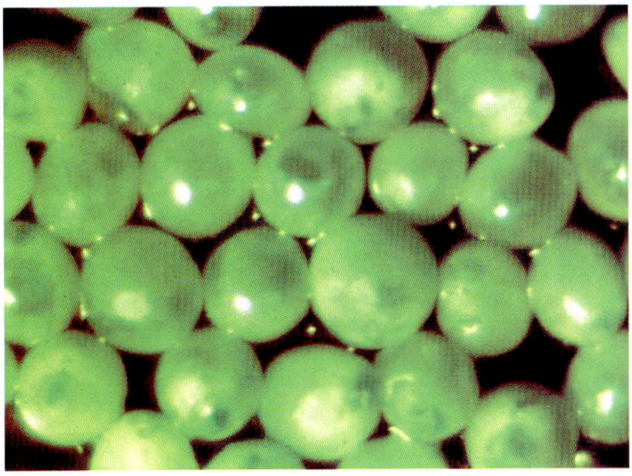

Plate 9

Plate 10

Plate 11

Plate 12

Plate 13

Plate 14

Plate 15

Plate 16

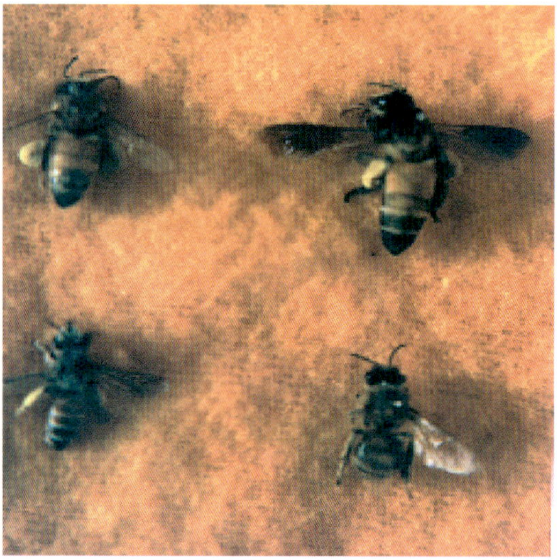

Plate 17

Plate 18

Plate 19

Plate 20

Plate 21

Plate 22